The
Avionics
Handbook

The Electrical Engineering Handbook Series

Series Editor
Richard C. Dorf
University of California, Davis

Titles Included in the Series

The Handbook of Ad Hoc Wireless Networks, Mohammad Ilyas
The Avionics Handbook, Cary R. Spitzer
The Biomedical Engineering Handbook, Second Edition, Joseph D. Bronzino
The Circuits and Filters Handbook, Second Edition, Wai-Kai Chen
The Communications Handbook, Second Edition, Jerry Gibson
The Computer Engineering Handbook, Vojin G. Oklobdzija
The Control Handbook, William S. Levine
The CRC Handbook of Engineering Tables, Richard C. Dorf
The Digital Signal Processing Handbook, Vijay K. Madisetti and Douglas Williams
The Electrical Engineering Handbook, Second Edition, Richard C. Dorf
The Electric Power Engineering Handbook, Leo L. Grigsby
The Electronics Handbook, Jerry C. Whitaker
The Engineering Handbook, Richard C. Dorf
The Handbook of Formulas and Tables for Signal Processing, Alexander D. Poularikas
The Handbook of Nanoscience, Engineering, and Technology, William A. Goddard, III,
    Donald W. Brenner, Sergey E. Lyshevski, and Gerald J. Iafrate
The Handbook of Optical Communication Networks, Mohammad Ilyas and
    Hussein T. Mouftah
The Industrial Electronics Handbook, J. David Irwin
The Measurement, Instrumentation, and Sensors Handbook, John G. Webster
The Mechanical Systems Design Handbook, Osita D.I. Nwokah and Yidirim Hurmuzlu
The Mechatronics Handbook, Robert H. Bishop
The Mobile Communications Handbook, Second Edition, Jerry D. Gibson
The Ocean Engineering Handbook, Ferial El-Hawary
The RF and Microwave Handbook, Mike Golio
The Technology Management Handbook, Richard C. Dorf
The Transforms and Applications Handbook, Second Edition, Alexander D. Poularikas
The VLSI Handbook, Wai-Kai Chen

Forthcoming Titles

The Electrical Engineering Handbook, Third Edition, Richard C. Dorf
The Electronics Handbook, Second Edition, Jerry C. Whitaker
The Engineering Handbook, Second Edition, Richard C. Dorf

The Avionics Handbook

Edited by

CARY R. SPITZER

AvioniCon, Inc.
Williamsburg, Virginia

CRC Press
Boca Raton   London   New York   Washington, D.C.

Library of Congress Cataloging-in-Publication Data

The avionics handbook / edited by Cary R. Spitzer.
        p.   cm. — (Electrical engineering handbook series)
        Includes bibliographical references and index.
        ISBN 0-8493-8348-X (alk. paper)
        1. Avionics, I. Spitzer, Cary R. II. Series.
TL695 .A8163   2000
629.135—dc21
                                                  00-048637

Visit the CRC Press Web site at www.crcpress.com

No claim to original U.S. Government works
International Standard Book Number 0-8493-8348-X
Library of Congress Card Number 00-048637
Printed in the United States of America   3 4 5 6 7 8 9 0
Printed on acid-free paper

Preface

Avionics is the cornerstone of modern aircraft. More and more, vital functions on both military and civil aircraft involve electronic devices. After the cost of the airframe and the engines, avionics is the most expensive item on the aircraft, but well worth every cent of the price.

Many technologies emerged in the last decade that will be utilized in the new millennium. After proof of soundness in design through ground application, advanced microprocessors are finding their way onto aircraft to provide new capabilities that were unheard of a decade ago. The Global Positioning System has enabled satellite-based precise navigation and landing, and communication satellites are now capable of supporting aviation services. Thus, the aviation world is changing to satellite-based communications, navigation, and surveillance for air traffic management. Both the aircraft operator and the air traffic services provider are realizing significant benefits.

Familiar technologies in this book include data buses, one type of which has been in use for over 20 years, head mounted displays, and fly-by-wire flight controls. New bus and display concepts are emerging that may displace these veteran devices. An example is a retinal scanning display.

Other emerging technologies include speech interaction with the aircraft and synthetic vision. Speech interaction may soon enter commercial service on business aircraft as another way to perform some noncritical functions. Synthetic vision offers enormous potential for both military and civil aircraft for operations under reduced visibility conditions or in cases where it is difficult to install sufficient windows in an aircraft.

This book offers a comprehensive view of avionics, from the technology and elements of a system to examples of modern systems flying on the latest military and civil aircraft. The chapters have been written with the reader in mind by working practitioners in the field. This book was prepared for the working engineer and his or her boss and others who need the latest information on some aspect of avionics. It will not make one an expert in avionics, but it will provide the knowledge needed to approach a problem.

Biography

Cary R. Spitzer is a graduate of Virginia Tech and George Washington University. After service in the Air Force, he joined NASA Langley Research Center.

During the last half of his tenure at NASA he focused on avionics. He was the NASA manager of a joint NASA/Honeywell program that made the first satellite-guided automatic landing of a passenger transport aircraft in November 1990. In recognition of this accomplishment, he was nominated jointly by ARINC, ALPA, AOPA, ATA, NBAA, and RTCA for the 1991 Collier Trophy "for his pioneering work in proving the concept of GPS aided precision approaches." He led a project to define the experimental and operational requirements for a transport aircraft suitable for conducting flight experiments and to acquire such an aircraft. Today, that aircraft is the NASA Langley B-757 ARIES flight research platform.

Mr. Spitzer was the NASA representative to the Airlines Electronic Engineering Committee. In 1988 he received the Airlines Avionics Institute Chairman's Special Volare Award. He is only the second federal government employee so honored in over 30 years.

He has been active in the RTCA, including serving as chairman of the Airport Surface Operations Subgroup of Task Force 1 on Global Navigation Satellite System Transition and Implementation Strategy, and as Technical Program Chairman of the 1992 Technical Symposium. He was a member of the Technical Management Committee.

In 1993 Mr. Spitzer founded AvioniCon, an international avionics consulting firm that specializes in strategic planning, business development, technology analysis, and in-house training.

Mr. Spitzer is a Fellow of the Institute of Electrical and Electronics Engineers (IEEE) and an Associate Fellow of the American Institute of Aeronautics and Astronautics (AIAA). He received the AIAA 1994 Digital Avionics Award and an IEEE Centennial Medal and Millennium Medal. He is a Past President of the IEEE Aerospace and Electronic Systems Society. Since 1979, he has played a major role in the highly successful Digital Avionics Systems Conferences, including serving as General Chairman.

Mr. Spitzer presents one-week shortcourses on digital avionics systems and on satellite-based communication, navigation, and surveillance for air traffic management at the UCLA Extension Division. He has also lectured for the International Air Transport Association.

He is the author of *Digital Avionics Systems,* the first book in the field, published by McGraw-Hill and Editor-in-Chief of *The Avionics Handbook,* published by CRC Press.

He and his wife, Laura, have a son, Danny.

His hobbies are working on old Ford products and kite flying.

Contributors

Kathy H. Abbott
Federal Aviation Administration
NASA Langley Research Center
Hampton, VA

Daniel G. Baize
NASA Langley Research Center
Hampton, VA

John G. P. Barnes
Caversham
Reading, U.K.

Gregg F. Bartley
Boeing
Seattle, WA

Douglas Beeks
Rockwell Collins
Cedar Rapids, IA

Barry C. Breen
Honeywell
Monroe, WA

Dominique Briere
Aerospatiale
Toulouse, France

Ronald Brower
United States Air Force
Wright Patterson AFB, OH

Ricky W. Butler
NASA Langley Research Center
Hampton, VA

Christian P. deLong
Honeywell, Defense Avionics
   Systems
Albuquerque, NM

James L. Farrell
VIGIL, Inc.
Severna Park, MD

Christian Favre
Aerospatiale
Toulouse, France

Thomas K. Ferrell
Ferrell and Associates Consulting
Vienna, VA

Uma D. Ferrell
Ferrell and Associates Consulting
Vienna, VA

Lee Harrison
Galaxy Scientific Corp.
Egg Harbor Twp., NJ

Steve Henely
Rockwell Collins
Cedar Rapids, IA

Richard Hess
Honeywell
Phoenix, AZ

Ellis F. Hitt
Battelle
Columbus, OH

Peter Howells
Rockwell Collins Flight Dynamics
Portland, OR

Sally C. Johnson
NASA Langley Research Center
Hampton, VA

Myron Kayton
Kayton Engineering Co.
Santa Monica, CA

Michael S. Lewis
NASA Langley Research Center
Hampton, VA

Thomas M. Lippert
Microvision Inc.
Bothel, WA

Robert P. Lyons, Jr.
United States Air Force
Arlington, VA

James N. Martin
The Aerospace Corporation
Chantilly, VA

Daniel A. Martinec
Aeronautical Radio, Inc. (ARINC)
Annapolis, MD

Frank W. McCormick
Certification Services, Inc.
Eastsound, WA

James Melzer
Kaiser Electro-Optics, Inc.
Carlsbad, CA

Jim Moore
Smiths Industries
Cheltenham, U.K.

Michael J. Morgan
Honeywell
Olathe, KS

Dennis Mulcare
Science Applications
    International Co.
Marietta, GA

Russell V. Parrish
NASA Langley Research Center
Hampton, VA

Michael Pecht
University of Maryland
College Park, MD

J. P. Potocki de Montalk
Airbus Industrie
Blagnac, France

Arun Ramakrishnan
University of Maryland
College Park, MD

Gordon R. A. Sandell
Boeing
Seattle, WA

John Satta
Zycad, Inc.
Dayton, OH

Dennis L. Schmickley
Boeing Helicopter Co.
Mesa, AZ

Grant Stumpf
Zycad, Inc.
Dayton, OH

Cary Spitzer
AvioniCon, Inc.
Williamsburg, VA

Jack Strauss
Zycad, Inc.
Dayton, OH

Toby Syrus
University of Maryland
College Park, MD

Pascal Traverse
Aerospatiale
Toulouse, France

Terry Venema
Zycad, Inc.
Dayton, OH

David G. Vutetakis
Douglas Battery Co.
Winston-Salem, NC

Randy Walter
Smiths Industries
Grand Rapids, MI

Robert B. Wood
Rockwell Collins Flight Dynamics
Portland, OR

Contents

SECTION III   Requirements, Design Analysis, Validation, and Certification

SECTION IV   Software

SECTION V   Implementation

I

Elements

Daniel A. Martinec
ARINC

The basic elements of the avionics suite on aircraft typically relate to the communications, navigation, and surveillance (CNS) functions. The term *CNS* is used widely throughout the aviation industry to address those functions addressed later in this handbook. The elements described in this section constitute the most fundamental "backbones" of the overall avionics suite performing the CNS functions.

Digital data buses provide the necessary onboard digital communications among the avionics elements comprising the overall airborne system. The avionics use digital data buses with (mostly) standardized physical and electrical interfaces to send their internal data to other avionics. The data may comprise sensor information, the results of internal calculations, system commands, information from internal storage, relayed data, or any information that may be generated by a computational device. The overall avionics suite, through the use of these interconnected digital data buses, operates similarly to ground-based networks. A primary difference is the amount of certification required to ensure that the very high level of integrity and safety required for aviation is maintained. Three widely used buses are examined: AS 15531/MIL-STD-1553B Digital Time Division Command/Response Multiplex Data Bus; ARINC 429 Digital Information Transfer System – Mark 33; and the Commercial Standard Digital Bus.

Batteries are an essential element to provide engine starting power and back up, sustaining power for avionics, especially flight critical avionics.

Avionics performing the basic CNS functions are not the only critical elements of aircraft. Crew interfaces play an important role in assuring that the crew can interact with these avionics and that the aircraft can be flown effectively and safely. This section provides a description of some advanced and evolving technologies that can provide the crew situational awareness of the aircraft and the environment in which the aircraft flies. Included are various display technol-ogies and speech recognition along with retinal scanning displays. Guidance is also given on proven techniques for flight deck design, a task often approached in an *ad hoc*, undisciplined manner.

Functions

J. P. Potocki de Montalk
Airbus Industrie

The functions implemented in an avionic installation are there to augment the operational effectiveness
of the aircraft. In many cases operation of the aircraft would not be viable without these functions, either
because the aircraft would become so inefficient that it would become unprofitable or because the aircraft
would be too vulnerable in its intended operational environment.

Perhaps the most important function avionics can perform in aiding the aircraft in its flight is that of
flight controls. Two chapters in this section describe how avionics, rather than moving parts, control

the hydraulic and electric power that moves the aircraft's flying control surfaces. This enables improved handling, reliability, and repairability of the aircraft and helps the crew to better protect the aircraft against out-of-normal flight conditions.

Another important function of avionics is navigation and flight management. Several chapters in this section describe how navigation systems enable the crew to determine the aircraft's altitude, attitude, heading, position, speed, and direction of travel when visual references are no longer available, to navigate toward and along the desired efficient flight path, and to construct this desired path, in a sky that contains many other aircraft and obstacles to safe and efficient flight.

Knowing the environment in which the aircraft is operating is essential. Two chapters describe functions that have their reason for existence in the desire to operate even more safely in the presence of poor visibility and with ever-increasing numbers of other aircraft in proximity. The option to generate a synthetic image of the environment also offers many benefits.

Together, these and other functions help aviators achieve better and better levels of safety and operational effectiveness, enabling improved value for the travel and defense budgets.

Requirements, Design Analysis, Validation, and Certification

Ellis F. Hitt
Battelle

The key to an avionics system or subsystem design, which provides dependable service at a low cost of ownership, is the requirements definition. Requirements engineering [IEEE Software, 1994] is now a recognized field considered by major professional societies such as the IEEE as one of the initial steps in systems engineering as well as subsequent steps in the development process illustratese in Figure III.1.

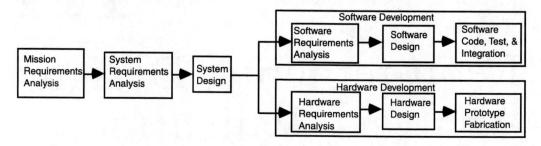

FIGURE III.1   Requirements in system development.

This section presents key processes that result in a reliable and affordable avionics system. These processes are setting requirements, reliability allocation, design analysis using simulation and modeling, validation of the design using formal methods, and processes for certification. The important issue of electromagnetic effects on the design and operation of avionics is also addressed.

IV

Software

Robert P. Lyons, Jr.
United States Air Force

Digital avionics allow much greater integration of functions, unprecedented flexibility, enhanced reliability, and ease of technology upgrade than can be achieved with more classical analog avionics. Indeed, given the cost, volume, weight, prime power, cooling, complexity, and safety constraints of both civil and military aircraft, there may really be no other option than to digitize, and even integrate, avionics functions in ways analog avionics could never achieve. Clearly, advances in electron devices, particularly digital integrated circuits of very large scale and high speed, give nearly unlimited possibilities to the designer of modern avionics. But without concomitant attention to the development and verification of highly complex software and its integration with the underlying hardware, the avionics implementation will simply produce heat, but no useful, much less safe, avionics functions at all.

Although software is an abstract thing dealing with data flows, control flows, algorithms, logical expressions, and the like, it is a building material of today's and tomorrow's aircraft as surely as are silicon, aluminum, titanium, plastics, and composites. And just as all the system and hardware aspects of an aircraft and its avionics require and are amenable to disciplined engineering analysis, synthesis, and verification, so is software. Certainly, the abstract nature of software makes its engineering and implementation less obvious than the fairly concrete methods available to the system and hardware engineering segments of the modern avionics development job. Fortunately, the tools to engineer software do exist.

The two chapters in this section describe two very important aspects of the software engineering applied to avionics. Chapter 26 discusses ISO/IEC 8652-95 Programming Language – Ada. Of all the software languages available to implement very large and very critical avionics functions, Ada-95 and, to a lesser extent, its predecessor Ada-83 have been shown to be the most supportive of software engineering concepts most likely to produce error free, flexible, reliable, complex software programs. One can write bad software in any language; it is just harder to do so with Ada than with other languages such as C and C++. Chapter 26 gives an excellent introduction to the development of the Ada language, its features, and the rudiments of its application.

Chapter 27 covers the important points of one of the most pervasive software standards applicable especially to civil aircraft avionics, but which subsumes other standards usually seen in military avionics developments. DO-178/ED-12 Software Considerations in Airborne Systems and Equipment Certification is a well-coordinated consensus standard combining the best thinking of software experts as well as aircraft certification authorities. Although this standard does not prescribe any particular software development methodology, all elements of the software life cycle are covered under its aegis. DO-178B, the current instantiation of the standard, does for the software part of the avionics sy

V

Implementation

Cary R. Spitzer
AvioniCon

The ultimate goal of any avionics effort is to reduce it to practice, to build it, and to operate it, or, in other words, to implement it. This section presents five examples of current or projected avionics systems along with a discussion of fault tolerance, a key concept in extremely critical (cannot lose its function) avionics.

Fault tolerance is the ability to continue to operate in a satisfactory manner (as defined by the customer) in the presence of multiple temporary or permanent hardware or software faults. Fault tolerance can be achieved in either hardware and/or software. Chapter 28 reviews many techniques for realizing satisfactory operation in the presence of faults.

Two modern commercial transports, the Boeing B-777 and the Airbus A330/340, have some of the most advanced avionics available today. The B-777 Airplane Information Management System (AIMS) is the first significant application of integrated, modular avionics to commercial transport aircraft and is performing admirably in revenue service. The A330/340 has an advanced, second-generation, fly-by-wire flight control system (featuring side stick controllers) and other cutting edge avionics that have proved to be very attractive to the operators in terms of economic payoff.

The Boeing MD-11 is an interesting retrofit architecture that demonstrates what upgraded, modern avionics can do for a vintage aircraft design (the DC-10). The MD-11 avionics not only eliminated the need for the third flight crew member by automating many of the functions of that position, but also offered reduced operating expenses and increased functionality.

The most modern, cutting-edge avionics architecture in military aircraft is the Lockheed F-22 Raptor. Like the B-777 AIMS, the F-22 makes extensive use of integrated, modular avionics for both the vehicle and the mission functions. The Common Integrated Processor (CIP) cabinets have space for up to 66 modules, but have only five different types of modules.

A visionary look at the future is capsulized in the advanced distributed architecture, made possible by the increasing throughput of microprocessors and the ever-growing bandwidth of data busses such as Ethernet. Perhaps the need for centralized processing will be taken over by distributed microprocessors.

1

AS 15531/MIL-STD-1553B Digital Time Division Command/Response Multiplex Data Bus

Chris deLong
Honeywell, Defense Avionics Systems

1.1 Introduction

MIL-STD-1553 is a standard which defines the electrical and protocol characteristics for a data bus. SAE AS-15531 is the commercial equivalent to the military standard. A data bus is similar to what the personal computer and office automation industry have dubbed a "Local Area Network (LAN)." In avionics, a data bus is used to provide a medium for the exchange of data and information between various systems and subsystems.

1.1.1 Background

In the 1950s and 1960s, avionics were simple standalone systems. Navigation, communications, flight controls, and displays consisted of analog systems. Often, these systems were composed of multiple boxes interconnected to form a single system. The interconnections between the various boxes was accomplished with point-to-point wiring. The signals mainly consisted of analog voltages, synchro-resolver signals, and relay/switch contacts. The location of these boxes within the aircraft was a function of operator need, available space, and the aircraft weight and balance constraints. As more and more systems were added,

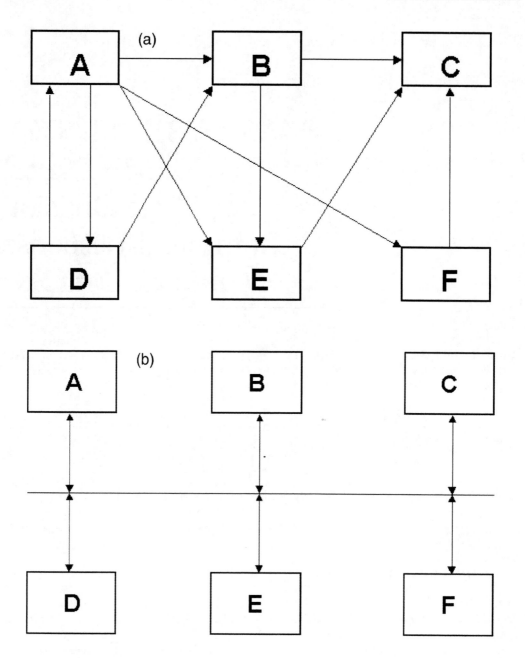

FIGURE 1.1   Systems configurations.

the cockpits became crowded due to the number of controls and displays, and the overall weight of the aircraft increased.

By the late 1960s and early 1970s, it was necessary to share information between various systems to reduce the number of black boxes required by each system. A single sensor providing heading and rate information could provide those data to the navigation system, the weapons system, the flight control system, and pilot's display system (see Figure 1.1a). However, the avionics technology was still basically analog, and while sharing sensors did produce a reduction in the overall number of black boxes, the interconnecting signals became a "rat's nest" of wires and connectors. Moreover, functions or systems that were added later became an integration nightmare as additional connections of a particular signal could have potential system impacts, plus since the system used point-to-point wiring, the system that was the source of the signal typically had to be modified to provide the additional hardware needed to output to the newly added subsystem. As such, intersystem connections were kept to the bare minimums.

By the late 1970s, with the advent of digital technology, digital computers had made their way into avionics systems and subsystems. They offered increased computational capability and easy growth, compared to their analog predecessors. However, the data signals — the inputs and outputs from the sending and receiving systems — were still mainly analog in nature, which led to the configuration of a small number of centralized computers being interfaced to the other systems and subsystems via complex and expensive analog-to-digital and digital-to-analog converters.

As time and technology progressed, the avionics systems became more digital. And with the advent of the microprocessor, things really took off. A benefit of this digital application was the reduction in the number of analog signals, and hence the need for their conversion. Greater sharing of information could be provided by transferring data between users in digital form. An additional side benefit was that digital data could be transferred bidirectionally, whereas analog data were transferred unidirectionally. Serial rather than parallel transmission of the data was used to reduce the number of interconnections within the aircraft and the receiver/driver circuitry required with the black boxes. But this alone was not enough. A data transmission medium which would allow all systems and subsystems to share a single and common set of wires was needed (see Figure 1.1b). By sharing the use of this interconnect, the various subsystems could send data between themselves and to other systems and subsystems, one at a time, and in a defined sequence. Enter the 1553 Data Bus.

1.1.2   History and Applications

MIL-STD-1553(USAF) was released in August of 1973. The first user of the standard was the F-16. Further changes and improvements were made and a tri-service version, MIL-STD-1553A, was released in 1975. The first user of the "A" version of the standard was again the Air Force's F-16 and the Army's new attack helicopter, the AH-64A Apache. With some "real world" experience, it was soon realized that further definitions and additional capabilities were needed. The latest version of the standard, 1553B, was released in 1978.

Today the 1553 standard is still at the "B" level; however, changes have been made. In 1980, the Air Force introduced Notice 1. Intended only for Air Force applications, Notice 1 restricted the use of many of the options within the standard. While the Air Force felt this was needed to obtain a common set of avionics systems, many in industry felt that Notice 1 was too restrictive and limited the capabilities in the application of the standard. Released in 1986, the tri-service Notice 2 (which supersedes Notice 1) places tighter definitions upon the options within the standard. And while not restricting an option's use, it tightly defines how an option will be used if implemented. Notice 2, in an effort to obtain a common set of operational characteristics, also places a minimum set of requirements upon the design of the black box. The military standard was converted to its commercial equivalent as SAE AS 15531, as part of the government's effort to increase the use of commercial products.

Since its inception, MIL-STD-1553 has found numerous applications. Notice 2 even removed all references to "aircraft" or "airborne" so as not to limit its applications. The standard has also been accepted and implemented by NATO and many foreign governments. The U.K. has issued Def Stan 00-18 (Part 2) and NATO has published STANAG 3838 AVS, both of which are versions of MIL-STD-1553B.

1.2   The Standard

MIL-STD-1553B defines the term Time Division Multiplexing (TDM) as "the transmission of information from several signal sources through one communications system with different signal samples staggered in time to form a composite pulse train." For our example in Figure 1.1b, this means that data can be transferred between multiple avionics units over a single transmission media, with the communications between the different avionics boxes taking place at different moments in time, hence time division. Table 1.1 is a summary of the 1553 Data Bus Characteristics. However, before defining how the data are transferred, it is necessary to understand the data bus hardware.

1.2.1 Hardware Elements

The 1553 standard defines certain aspects regarding the design of the data bus system and the black boxes to which the data bus is connected. The standard defines four hardware elements: transmission media, remote terminals, bus controllers, and bus monitors; each of which is detailed as follows.

TABLE 1.1   Summary of the 1553 Data Bus Characteristics

Data Rate	1 MHz
Word Length	20 bits
Data Bits per Word	16 bits
Message Length	Maximum of 32 data words
Transmission Technique	Half-Duplex
Operation	Asynchronous
Encoding	Manchester II Bi-phase
Protocol	Command-Response
Bus Control	Single or Multiple
Message Formats	Controller-to-Terminal (BC-RT)
	Terminal-to-Controller (RT-BC)
	Terminal-to-Terminal (RT-RT)
	Broadcast
	System Control
Number of Remote Terminals	Maximum of 31
Terminal Types	Remote Terminal (RT)
	Bus Controller (BC)
	Bus Monitor (BM)
Transmission Media	Twisted Shielded Pair Cable
Coupling	Transformer or Direct

1.2.1.1 Transmission Media

The transmission media, or data bus, is defined as a twisted shielded pair transmission line consisting of the main bus and a number of stubs. There is one stub for each terminal (system) connected to the bus. The main data bus is terminated at each end with a resistance equal to the cable's characteristic impedance. This termination makes the data bus behave electrically like an infinite transmission line. Stubs, which are added to the main bus in order to connect the terminals, provide "local" loads, and produce an impedance mismatch where added. This mismatch, if not properly controlled, produces electrical reflections and degrades the performance of the main bus. Therefore, the characteristics of both the main bus and the stubs are specified within the standard. Table 1.2 is a summary of the transmission media characteristics.

The standard specifies two stub methods: direct and transformer coupled. This refers to the method in which a terminal is connected to the main bus. Figure 1.2 shows the two methods, the primary difference between the two being that the transformer coupled method utilizes an isolation transformer for connecting the stub cable to the main bus cable. In both methods, two isolation resistors are placed in series with the bus. In the direct coupled method, the resistors are typically located within the terminal, whereas in the transformer coupled method, the resistors are typically located with the coupling transformer in boxes called data bus couplers. A variety of couplers are available, providing single or multiple stub connections.

Another difference between the two coupling methods is the length of the stub. For the direct coupled method, the stub length is limited to a maximum of 1 ft. For the transformer coupled method, the stub can be up to a maximum length of 20 ft. Therefore for direct coupled systems, the data bus must be

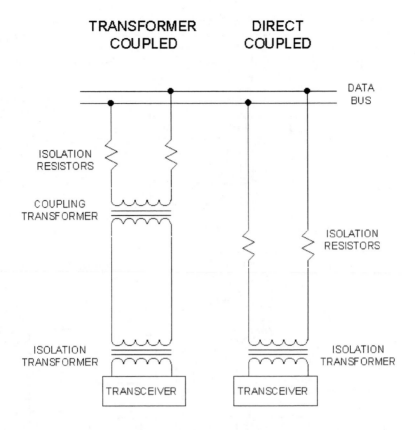

FIGURE 1.2 Terminal connection methods.

routed in close proximity to each of the terminals, whereas for a transformer coupled system, the data bus may be up to 20 ft away from each terminal.

TABLE 1.2 Summary of Transmission Media Characteristics

Cable Type	Twisted Shielded Pair
Capacitance	30.0 pF/ft max — wire to wire
Characteristic Impedance	70.0 to 85.0 ohms at 1 MHz
Cable Attenuation	1.5 dbm/100 ft at 1 MHz
Cable Twists	4 twists per foot maximum
Shield Coverage	90% minimum
Cable Termination	Cable impedance ($\pm 2\%$)
Direct Coupled Stub Length	Maximum of 1 ft
Transformer Coupled Stub Length	Maximum of 20 ft

1.2.1.2 Remote Terminal

A remote terminal is defined within the standard as "All terminals not operating as the bus controller or as a bus monitor." Therefore if it is not a controller, monitor, or the main bus or stub, it must be a remote terminal — sort of a "catch all" clause. Basically, the remote terminal is the electronics necessary to transfer data between the data bus and the subsystem. So what is a subsystem? For 1553 applications, the subsystem is the sender or user of the data being transferred.

In the earlier days of 1553, remote terminals were used mainly to convert analog and discrete data to/from a data format compatible with the data bus. The subsystems were still the sensor which provided the data and computer which used the data. As more and more digital avionics became available, the

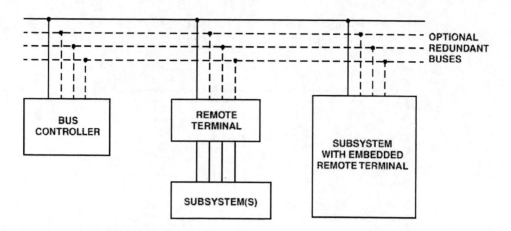

FIGURE 1.3   Simple multiplex architecture.

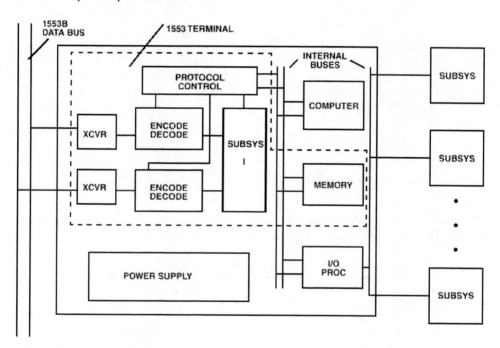

FIGURE 1.4   Terminal definition.

trend has been to embed the remote terminal into the sensor and computer. Today it is common for the subsystem to contain an embedded remote terminal. Figure 1.3 shows the different levels of remote terminals possible.

A remote terminal typically consists of a transceiver, an encoder/decoder, a protocol controller, a buffer or memory, and a subsystem interface. In a modern black box containing a computer or processor, the subsystem interface may consist of the buffers and logic necessary to interface to the computer's address, data, and control buses. For dual redundant systems two transceivers and two encoders/decoders would be required to meet the requirements of the standard.

Figure 1.4 is a block diagram of a remote terminal and its connection to a subsystem. In short, the remote terminal consists of all the electronics necessary to transfer data between the data bus and the user or originator of the data being transferred.

But a remote terminal is more than just a data formatter. It must be capable of receiving and decoding commands from the bus controller, and respond accordingly. It must also be capable of buffering a message-worth of data, be capable of detecting transmission errors and performing validation tests upon the data, and reporting the status of the message transfer. A remote terminal must be capable of performing a few of the bus management commands (referred to as mode commands), and for dual redundant applications it must be capable of listening to and decoding commands on both buses at the same time.

A remote terminal must strictly follow the protocol as defined by the standard. It can only respond to commands received from the bus controller (i.e., it only speaks when spoken to). When it receives a valid command, it must respond within a defined amount of time. If a message does not meet the validity requirements defined, then the remote terminal must invalidate the message and discard the data (not allow the data to be used by the subsystem). In addition to reporting status to the bus controller, most remote terminals today are also capable of providing some level of status information to the subsystem regarding the data received.

1.2.1.3 Bus Controller

The bus controller is responsible for directing the flow of data on the bus. While several terminals may be capable of performing as the bus controller, only one bus controller is allowed to be active at any one time. The bus controller is the only device allowed to issue commands onto the data bus. The commands may be for the transfer of data, or the control and management of the bus (referred to as mode commands).

Typically, the bus controller is a function that is contained within some other computer, such as a mission computer, a display processor, or a fire control computer. The complexity of the electronics associated with the bus controller is a function of the subsystem interface (the interface to the computer), the amount of error management and processing to be performed, and the architecture of the bus controller. There are three types of bus controllers architectures: a word controller, a message controller, and a frame controller.

A *word controller* is the oldest and simplest type. Few word controllers are built today and they are only mentioned herein for completeness. For a word controller, the terminal electronics transfers one word at a time to the subsystem. Message buffering and validation must be performed by the subsystem.

Message controllers output a single message at a time, interfacing with the computer only at the end of the message or perhaps when an error occurrs. Some message controllers are capable of performing minor error processing, such as transmitting once on the alternate data bus, before interrupting the computer. The computer will inform the interface electronics of where the message exists in memory and provide a control word. For each message the control word typically informs the electronics of the message type (e.g., an RT-BC or RT-RT command), which bus to use to transfer the message, where to read or write the data words in memory, and what to do if an error occurs. The control words are a function of the hardware design of the electronics and are not standardized among bus controllers.

A *frame controller* is the latest concept in bus controllers. A frame controller is capable of processing multiple messages in a sequence defined by the computer. The frame controller is typically capable of error processing as defined by the message control word. Frame controllers are used to "off-load" the computer as much as possible, interrupting only at the end of a series of messages or when an error it can not handle is detected.

There is no requirement within the standard as to the internal workings of a bus controller, only that it issue commands onto the bus.

1.2.1.4 Bus Monitor

A bus monitor is just that. A terminal which listens to (monitors) the exchange of information on the data bus. The standard strictly defines what bus monitors may be used for, stating that the information obtained by a bus monitor be used "for off-line applications (e.g., flight test recording, maintenance recording or mission analysis) or to provide a back-up bus controller sufficient information to take over as the bus controller." Monitors may collect all the data from the bus or may collect selected data.

The reason for restricting its use is that while a monitor may collect data, it deviates from the command-response protocol of the standard in that a monitor is a passive device that does not transmit a status word, and therefore can not report on the status of the information transferred. Therefore, bus monitors fall into two categories: a recorder for testing, or as a terminal functioning as a back-up bus controller.

In collecting data, a monitor must perform the same message validation functions as the remote terminal and, if an error is detected, inform the subsystem of the error (the subsystem may still record the data, but the error should be noted). For monitors which function as recorders for testing, the subsystem is typically a recording device or a telemetry transmitter. For monitors which function as back-up bus controllers, the subsystem is the computer.

Today it is common that bus monitors also contain a remote terminal. When the monitor receives a command addressed to its terminal address, it responds as a remote terminal. For all other commands, it functions as a monitor. The remote terminal portion could be used to provide feedback to the bus controller of the monitor's status, such as the amount of memory or time left, or to reprogram a selective monitor as to what messages to capture.

1.2.1.5   Terminal Hardware

The electronic hardware between a remote terminal, bus controller, and bus monitor does not differ much. Both the remote terminal and bus controller (and bus monitor if it is also a remote terminal) must have the transmitters/receivers and encoders/decoders to format and transfer data. The requirements upon the transceivers and the encoders/decoders do not vary between the hardware elements. Table 1.3 lists the electrical characteristics of the terminals.

All three elements have some level of subsystem interface and data buffering. The primary difference lies in the protocol control logic and often this just a different series of micro-coded instructions. For this reason, it is common to find 1553 hardware circuitry that is also capable of functioning as all three devices.

TABLE 1.3   Terminal Electrical Characteristics

Requirement	Transformer Coupled	Direct Coupled	Condition
Input Characteristics			
Input Level	0.86–14.0 V	1.2–20.0 V	p–p, l–l
No Response	0.0–0.2 V	0.0–0.28 V	p–p, l–l
Zero Crossing Stability	±150.0 nsec	±150.0 nsec	
Rise/Fall Times	0 nsec	0 nsec	Sine Wave
Noise Rejection	140.0 mV WGN[a]	200.0 mV WGN	BER 1[b] per 10^7
Common Mode Rejection	±10.0 V peak	±10.0 V peak	line-gnd, DC-2.0 MHz
Input Impedance	1000 ohms	2000 ohms	75 kHz–1 MHz
Output Characteristics			
Output Level	18.0–27.0 V	6.0–9.0 V	p–p, l–l
Zero Crossing Stability	25.0 nsec	25.0 nsec	
Rise/Fall Times	100–300 nsec	100–300 nsec	10%–90%
Maximum Distortion	±900.0 mV	±300.0 mV	peak, l–l
Maximum Output Noise	14.0 mV	5.0 mV	rms, l–l
Maximum Residual Voltage	±250.0 mV	±90.0 mV	peak, l–l

[a] WGN = White Gaussian Noise.

[b] BER = Bit Error Rate.

There is an abundance of "off-the-shelf" components available today from which to design a terminal. These vary from discrete transceivers, encoders/decoders, and protocol logic devices to a single dual redundant hybrid containing everything but the transformers.

1.3 Protocol

The rules under which the transfers occur is referred to as "protocol". The control, data flow, status reporting, and management of the bus is provided by three word types.

1.3.1 Word Types

Three distinct word types are defined by the standard. These are command words, data words, and status words. Each word type has a unique format yet all three maintain a common structure. Each word is 20 bits in length. The first three bits are used as a synchronization field, thereby allowing the decode clock to re-sync at the beginning of each new word. The following 16 bits are the information field and differ among the three word types. The last bit is the parity bit. Parity is based on odd parity for the single word. The three word types are shown in Figure 1.5.

Bit encoding for all words is based on bi-phase Manchester II format. The Manchester II format provides a self-clocking waveform in which the bit sequence is independent. The positive and negative voltage levels of the Manchester waveform is DC balanced (same amount of positive signal as there is negative signal) and as such is well suited for transformer coupling. A transition of the signal occurs at the center of the bit time. A logic "0" is a signal that transitions from a negative level to a positive level. A logic "1" is a signal that transitions from a positive level to a negative level.

The terminal's hardware is responsible for the Manchester encoding and decoding of the word types. The interface that the subsystem sees is the 16-bit information field of all words. The sync and parity fields are not provided directly. However, for received messages, the decoder hardware provides a signal to the protocol logic as to the sync type the word was and as to whether parity was valid or not. For transmitted messages, there is an input to the encoder as to what sync type to place at the beginning of the word, and parity is automatically calculated by the encoder.

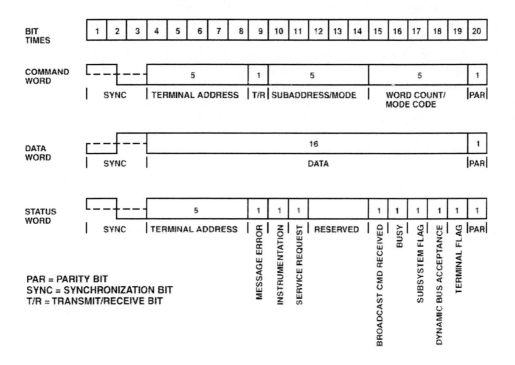

FIGURE 1.5 Word formats.

1.3.1.1   Sync Fields

The first three bit times of all word types is called the sync field. The sync waveform is in itself an invalid Manchester waveform as the transition only occurs at the middle of the second bit time. The use of this distinct pattern allows the decoder to re-sync at the beginning of each word received and maintain the overall stability of the transmissions.

Two distinct sync patterns are used: the command/status sync, and the data sync. The command/status sync has a positive voltage level for the first one and a half bit times, then transitions to a negative voltage level for the second one and a half bit times. The data sync is the opposite — a negative voltage level for the first one and a half bit times, then transitions to a positive voltage level for the second one and a half bit times. The sync patterns are shown in Figure 1.5.

1.3.1.2   Command Word

The Command Word (CW) specifies the function that a remote terminal(s) is to perform. This word is only transmitted by the active bus controller. The word begins with a command sync in the first three bit times. The following 16-bit information field is as defined in Figure 1.5.

The five-bit Terminal Address (TA) field (bit times 4–8) states to which unique remote terminal the command is intended (no two terminals may have the same address). Note that an address of 00000 is a valid address, and that an address of 11111 is reserved for use as the broadcast address. Also note that there is no requirement that the bus controller be assigned an address, therefore the maximum number of terminals the data bus can support is 31. Notice 2 to the standard requires that the terminal address be wire programmable externally to the black box (i.e., an external connector) and that the remote terminal electronics perform a parity test upon the wired terminal address. The Notice basically states that an open circuit on an address line is detected as a logic "1," that connecting an address line to ground is detected as a logic "0," and that odd parity will be used in testing the parity of the wired address field.

The next bit (bit time 9) is the Transmit/Receive (T/R) bit. This defines the direction of information flow and is always from the point of view of the remote terminal. A transmit command (logic 1) indicates that the remote terminal is to transmit data, while a receive command (logic 0) indicates that the remote terminal is going to receive data. The only exceptions to this rule are associated with mode commands.

The following five bits (bit times 10–14) are the Subaddress (SA)/Mode Command (MC) bits. Logic 00000 or 11111 within this field shall be decoded to indicate that the command is a Mode Code Command. All other logic combinations of this field are used to direct the data to different functions within the subsystem. An example might be that 00001 is position and rate data, 00010 is frequency data, 10010 is display information, and 10011 is self-test data. The use of the subaddresses is left to the designer, however, Notice 2 suggests the use of subaddress 30 for data wraparound.

The next five bit positions (bit times 15–19) define the Word Count (WC) or Mode Code to be performed. If the Subaddress/Mode Code field was 00000 or 11111, then this field defines the mode code to be performed. If not a mode code, then this field defines the number of data words either to be received or transmitted depending on the T/R bit. A word count field of 00000 is decoded as 32 data words.

The last bit (bit time 20) is the word parity bit. Only odd parity shall be used.

1.3.1.3   Data Word

The Data Word (DW) contains the actual information that is being transferred within a message. Data words can be transmitted by either a remote terminal (transmit command) or a bus controller (receive command). The first three bit times contain a data sync. This sync pattern is the opposite of that used for command and status words and therefore is unique to the data word type.

The following 16 bits of information are left to the designer to define. The only standard requirement is that the most significant bit (MSB) of the data be transmitted first. While the standard provides no guidance as to their use, Section 80 of MIL-HDBK-1553A and SAE AS-15532 provides guidance and lists the formats (i.e., bit patterns, resolutions, etc.) of the most commonly used data words.

The last bit (bit time 20), is the word parity bit. Only odd parity shall be used.

1.3.1.4 Status Word

The Status Word (SW) is only transmitted by a remote terminal in response to a valid message. The status word is used to convey to the bus controller whether a message was properly received or the state of the remote terminal (i.e., service request, busy, etc.). The status word is defined in Figure 1.5. Since the status word conveys information to the bus controller, there are two views as to the meaning of each bit — what the setting of the bit means to a remote terminal, and what the setting of the bit means to a bus controller. Each field of the status word, and its potential meanings, is examined below.

1.3.1.4.1 *Resetting the Status Word*

The Status Word, with the exception of the remote terminal address, is cleared after receipt of a valid command word. The two exceptions to this rule are if the command word received is a Transmit Status Word Mode Code or a Transmit Last Command Word Mode Code. Conditions which set the individual bits of the word may occur at any time. If after clearing the status word, the conditions for setting the bits still exists, then the bits shall be set again.

Upon detection of a error in the data being received, the Message Error bit is set and the transmission of the status word is suppressed. The transmission of the status word is also suppressed upon receipt of a broadcast message. For an illegal message (i.e., an illegal Command Word), the Message Error bit is set and the status word is transmitted.

1.3.1.4.2 *Status Word Bits*

Terminal Address. The first five bits (bit times 4–8) of the information field are the Terminal Address (TA). These five bits should match the corresponding field within the command word that the terminal received. The remote terminal sets these bit to the address to which it has been programmed. The bus controller should examine these bits to insure that the terminal responding with its status word was indeed the terminal to which the command word was addressed. In the case of a remote terminal to remote terminal message (RT-RT), the receiving terminal should compare the address of the second command word with that of the received status word. While not required by the standard, it is good design practice to insure that the data received are from a valid source.

Message Error. The next bit (bit time 9) is the Message Error (ME) bit. This bit is set to a logic "1" by the remote terminal upon detection of a error in the message or upon detection of an invalid message (i.e., Illegal Command) to the terminal. The error may occur in any of the data words within the message. When the terminal detects an error and sets this bit, none of the data received within the message shall be used. If an error is detected within a message and the ME bit is set, the remote terminal must suppress the transmission of the status word (see Resetting of the Status Word). If the terminal detected an illegal command, the ME bit is set and the status word is transmitted. All remote terminals must implement the ME bit in the status word.

Instrumentation. The Instrumentation bit (bit time 10) is provided so as to differentiate between a command word and a status word (remember, they both have the same sync pattern). The instrumentation bit in the status word is always set to logic "0." If used, the corresponding bit in the command word is set to a logic "1." This bit in the command word is the most significant bit of the subaddress field, and therefore would limit the subaddresses used to 10000–11110, hence reducing the number of subaddresses available from 30 to 15. The instrumentation bit is also the reason why there are two mode code indentifiers (00000 and 11111), the latter required when the instrumentation bit is used.

Service Request. The Service Request bit (bit time 11) is such that the remote terminal can inform the bus controller that it needs to be serviced. This bit is set to a logic "1" by the subsystem to indicate that servicing is needed. This bit is typically used when the bus controller is "polling" terminals to determine if they require processing. The bus controller upon receiving this bit set to a logic "1" typically does one of the following. It can take a predetermined action such as issuing a series of messages, or it can request further data from the remote terminal as to its needs. The later can be accomplished by requesting the terminal to transmit data from a defined subaddress or by using the Transit Vector Word Mode Code.

Reserved. Bit times 12–14 are reserved for future growth of the standard and must be set to a logic "0." The bus controller should declare a message in error if the remote terminal responds with any of these bits set in its status word.

Broadcast Command Received. The Broadcast Command Received bit (bit time 15) indicates that the remote terminal received a valid broadcast command. Upon receipt of a valid broadcast command, the remote terminal sets this bit to logic "1" and suppresses the transmission of its status words. The bus controller may issue a Transmit Status Word or Transmit Last Command Word Mode Code to determine if the terminal received the message properly.

Busy. The Busy bit (bit time 16) is provided as a feedback to the bus controller as to when the remote terminal is unable to move data between the remote terminal electronics and the subsystem in compliance to a command from the bus controller.

In the earlier days of 1553, the Busy bit was required because many of the subsystem interfaces (analogs, synchros, etc.) were much slower compared to the speed of the multiplex data bus. Some terminals were not able to move the data fast enough. So instead of potentially losing data, a terminal was able to set the Busy bit, indicating to the bus controller is could not handle new data at that time, and for the bus controller to try again later. As new systems have been developed, the need for the use of Busy has been reduced. However, there are systems that still need and have a valid use for the Busy bit. Examples of these are radios, where the bus controller issues a command to the radio to tune to a certain frequency. It may take the radio several seconds to accomplish this, and while it is tuning it may set the Busy bit to inform the bus controller that it is doing as it was told.

When a terminal is busy, it does not need to respond to commands in the "normal" way. For receive commands the terminal collects the data, but does not have to pass the data to the subsystem. For transmit commands, the terminal transmits its status word only. Therefore, while a terminal is busy the data it supplies to the rest of the system are not available. This can have an overall effect upon the flow of data within the system and may increase the data latency within time-critical systems (e.g., flight controls).

Some terminals used the Busy bit to overcome design problems, setting the Busy bit whenever needed. Notice 2 to the standard "strongly discourages" the use of the Busy bit. However, as shown in the example above, there are valid needs for its use. Therefore, if used, Notice 2 now requires that the Busy bit may only be set as the result of a particular command received from the bus controller and not due to an internal periodic or processing function. By following this requirement, the bus controller, with prior knowledge of the remote terminal's characteristics, can determine what will cause a terminal to go busy and minimize the effects on data latency throughout the system.

Subsystem Flag. The Subsystem Flag bit (bit time 17) is used to provide "health" data regarding the subsystems to which the remote terminal is connected. Multiple subsystems may logically "OR" their bits together to form a composite health indicator. This single bit is only to serve as an indicator to the bus controller and user of the data that a fault or failure exists. Further information regarding the nature of the failure must be obtained in some other fashion. Typically, a subaddress is reserved for built-in-test (BIT) information, with one or two words devoted to subsystem status data.

Dynamic Bus Control Acceptance. The Dynamic Bus Control Acceptance bit (bit time 18) is used to inform the bus controller that the remote terminal has received the Dynamic Bus Control Mode Code and has accepted control of the bus. For the remote terminal, the setting of this bit is controlled by the subsystem and is based upon passing some level of built-in-test (i.e., a processor passing its power-up and continuous background tests).

The remote terminal upon transmitting its status word becomes the bus controller. The bus controller, upon receipt of the status word from the remote terminal with this bit set, ceases to function as the bus controller and may become a remote terminal or bus monitor.

Terminal Flag. The Terminal Flag bit (bit time 19) is used to inform the bus controller of a fault or failure within the remote terminal circuitry (only the remote terminal). A logic "1" shall indicate a fault condition. This bit is used solely to inform the bus controller of a fault or failure. Further information regarding the nature of the failure must be obtained in some other fashion. Typically, a subaddress is reserved for BIT information, or the bus controller may issue a Transmit BIT Word Mode Code.

Parity. The last bit (bit time 20), is the word parity bit. Only odd parity shall be used.

1.3.2 Message Formats, Validation, and Timing

The primary purpose of the data bus is to provide a common medium for the exchange of data between systems. The exchange of data is based upon message transmissions. The standard defines 10 types of message transmission formats. All of these formats are based upon the three word types just defined. The 10 message formats are shown in Figures 1.6 and 1.7. The message formats have been divided into two groups. These are referred to within the standard as the "information transfer formats" (Figure 1.6) and the "broadcast information transfer formats" (Figure 1.7).

The information transfer formats are based upon the command/response philosophy that all error-free transmissions received by a remote terminal be followed by the transmission of a status word from the terminal to the bus controller. This handshaking principle validates the receipt of the message by the remote terminal.

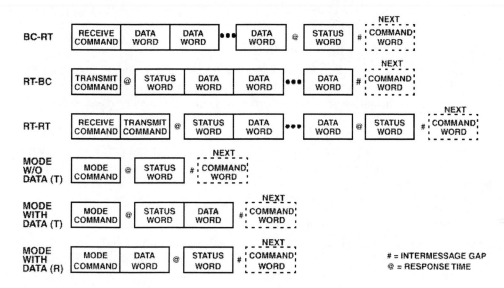

FIGURE 1.6   Information transfer formats.

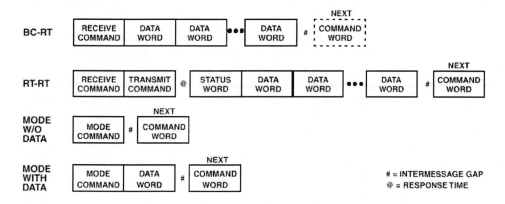

FIGURE 1.7   Broadcast information transfer formats.

Broadcast messages are transmitted to multiple remote terminals at the same time. As such, the terminals suppress the transmission of their status words (not doing so would have multiple boxes trying to talk at the same time and thereby "jam" the bus). In order for the bus controller to determine if a terminal received the message, a polling sequence to each terminal must be initiated to collect the status words.

Each of the message formats is summarized in the following subsections.

1.3.2.1   Bus Controller to Remote Terminal

The bus controller to remote terminal (BC-RT) message is referred to as the receive command since the remote terminal is going to receive data. The bus controller outputs a command word to the terminal defining the subaddress of the data and the number of data words it is sending. Immediately (without any gap in the transmission), the number of data words specified in the command word are sent.

The remote terminal upon validating the command word and all of the data words will issue its status word within the response time requirements (maximum of 12 μsec).

The remote terminal must be capable of processing the next command that the bus controller issues. Therefore the remote terminal has approximately 56 μsec (status word response time 12 μsec, plus status word transmit time 20 μsec, plus intermessage gap minimum 4 μsec, plus command word transmit time 20 μsec, to either pass the data to the subsystem or buffer the data.

1.3.2.2   Remote Terminal to Bus Controller

The remote terminal to bus controller (RT-BC) message is referred to as a transmit command. The bus controller issues only a transmit command word to the remote terminal. The terminal, upon validation of the command word, will first transmit its status word followed by the number of data words requested by the command word.

Since the remote terminal does not know the sequence of commands to be sent and does not normally operate upon a command until the command word has been validated, it must be capable of fetching from the subsystem the data required within approximately 28 μsec (the status word response time 12 μsec, plus the status word transmission time 20 μsec, minus some amount of time for message validation and transmission delays through the encoder and transceiver).

1.3.2.3   Remote Terminal to Remote Terminal

The remote terminal to remote terminal (RT-RT) command is provided to allow a terminal (the data source) to transfer data directly to another terminal (the data sink) without going through the bus controller. The bus controller may, however, collect and use the data.

The bus controller first issues a command word to the receiving terminal immediately followed by a command word to the transmitting terminal. The receiving terminal is expecting data, but instead of data after the command word it sees a command sync (the second command word). The receiving terminal ignores this word and waits for a word with a data sync.

The transmitting terminal ignored the first command word (it did not contain its terminal address). The second word was addressed to it, so it will process the command as an RT-BC command as described above by transmitting its status word and the required data words.

The receiving terminal, having ignored the second command word, again sees a command (status) sync on the next word and waits further. The next word (the first data word sent) now has a data sync and the receiving remote terminal starts collecting data. After receipt of all of the data words (and validating), the terminal transmits its status word.

1.3.2.3.1   RT-RT Validation

There are several things that the receiving remote terminal of an RT-RT message should do. First, Notice 2 requires that the terminal time out in 54 to 60 μsec after receipt of the command word. This is required since if the transmitting remote terminal did not validate its command word (and no transmission occurred) then the receiving terminal will not collect data from some new message. This could occur if the next message is either a transmit or receive message, where the terminal ignores all words with a command/status sync and would start collecting data words beginning with the first data sync. If the same number of data words were being transferred in the follow-on message and the terminal did not test the command/status word contents, then the potential exists for the terminal to collect erroneous data.

The other function that the receiving terminal should do, but is not required by the standard, is to capture the second command word and the first transmitted data word. The terminal could compare the terminal address fields of both words to insure that the terminal doing the transmitting was the one commanded to transmit. This would allow the terminal to provide a level of protection for its data and subsystem.

1.3.2.4 Mode Command Formats

Three mode command formats are provided for. This allows for mode commands with no data words and for the mode commands with one data word (either transmitted or received). The status/data sequencing is as described for the BC-RT or RT-BC messages except that the data word count is either one or zero. Mode codes and their use are described later.

1.3.2.5 Broadcast Information Transfer Formats

The broadcast information transfer formats, as shown in Figure 1.8, are identical to the nonbroadcast formats described above with the following two exceptions. First, the bus controller issues commands to terminal address 31 (11111) which is reserved for this function. And secondly, the remote terminals receiving the messages (those which implement the broadcast option) suppress the transmission of their status word.

The broadcast option can be used with the message formats in which the remote terminal receives data. Obviously, multiple terminals cannot transmit data at the same time, so the RT-BC transfer format and the transmit mode code with data format cannot be used. The broadcast RT-RT allows the bus controller to instruct all remote terminals to receive and then instructs one terminal to transmit, thereby allowing a single subsystem to transfer its data directly to multiple users.

Notice 2 allows the bus controller to only use broadcast commands with mode codes (see Broadcast Mode Codes). Remote terminals are allowed to implement this option for all broadcast message formats. The Notice further states that the terminal must differentiate the subaddresses between broadcast and nonbroadcast messages (see Subaddress Utilization).

1.3.2.6 Command and Message Validation

The remote terminal must validate the command word and all data words received as part of the message. The criteria for a valid command word are that the: word begins with a valid command sync, valid terminal address (matches the assigned address of the terminal or the broadcast address if implemented), all bits are in a valid Manchester code, there are 16 information field bits, and there is a valid parity bit (odd). The criteria for a data word are the same except a valid data sync is required and the terminal address field is not tested. If a command word fails to meet the criteria, the command is ignored. After the command has been validated, and a data word fails to meet the criteria, then the terminal shall set the Message Error bit in the status word and suppress the transmission of the status word. Any single error within a message shall invalidate the entire message and the data shall not be used.

1.3.2.7 Illegal Commands

The standard allows remote terminals the option of monitoring for Illegal Commands. An Illegal Command is one that meets the valid criteria for a command word, but is a command (message) that is not implemented by the terminal. An example is if a terminal only outputs 04 data words to subaddress 01 and a command word was received by the terminal that requested it to transmit 06 data words from subaddress 03, then this command, while still a valid command, could be considered by the terminal as illegal. The standard only states that the bus controller shall not issue illegal or invalid commands.

The standard provides the terminal designer with two options. First, the terminal can respond to all commands as usual (this is referred to as "responding in form"). The data received is typically placed in a series of memory locations which are not accessible by the subsystem or applications programs.

This is typically referred to as the "bit bucket." All invalid commands are placed into the same bit bucket. For invalid transmit commands, the data transmitted is read from the bit bucket. Remember, the bus controller is not supposed to send these invalid commands.

The second option is for the terminal to monitor for Illegal Commands. For most terminal designs, this is as simple as a look-up table with the T/R bit, subaddress, and word count fields supplying the address and the output being a single bit that indicates if the command is valid or not. If a terminal implements Illegal Command detection and an illegal command is received, the terminal sets the Message Error bit in the status word and responds with the status word.

1.3.2.8 Terminal Response Time

The standard states that a remote terminal, upon validation of a transmit command word or a receive message (command word and all data words) shall transmit its status word to the bus controller. The response time is the amount of time the terminal has to transmit its status word. To allow for accurate measurements, the time frame is measured from the mid-crossing of the parity bit of the command word to the mid-crossing of the sync field of the status word. The minimum time is 4.0 μsec, the maximum time is 12.0 μsec. However, the actual amount of "dead time" on the bus is 2 to 10 μsec since half of the parity and sync waveforms are being transmitted during the measured time frame.

The standard also specifies that the bus controller must wait a minimum of 14.0 μsec for a status word response before determining that a terminal has failed to respond. In applications where long data buses are used or where other special conditions exist, it may be necessary to extend this time to 20.0 μsec or greater.

1.3.2.9 Intermessage Gap

The bus controller must provide for a minimum of 4.0 μsec between messages. Again, this time frame is measured from the mid-crossing of the parity bit of the last data word or the status word and the mid-crossing of the sync field of the next command word. The actual amount of "dead time" on the bus is 2 μsec since half of the parity and sync waveforms are being transmitted during the measured time frame.

The amount of time required by the bus controller to issue the next command is a function of the controller type (e.g., word, message, or frame). The gap typically associated with word controllers is between 40 and 100 μsec. Message controllers typically can issue commands with a gap of 10 to 30 μsec. But frame controllers are capable of issuing commands at the 4-μsec rate and often must require a time delay to slow them down.

1.3.2.10 Superseding Commands

A remote terminal must always be capable of receiving a new command. This may occur while operating on a command on bus A and after the minimum intermessage gap, a new command appears, or if operating on bus A and a new command appears on bus B. This is referred to as a Superseding Command. A second valid command (the new command) shall cause the terminal to stop operating on the first command and start on the second. For dual redundant applications, this requirement implies that all terminals must, as a minimum, have two receivers, two decoders, and two sets of command word validation logic.

1.3.3 Mode Codes

Mode codes are defined by the standard to provide the bus controller with data bus management and error handling/recovery capability. The mode codes are divided into two groups: with and without data words. The data words that are associated with the mode codes, and only one word per mode code is allowed, contains information pertinent to the control of the bus and do not generally contain information required by the subsystem (the exception may be the Synchronize with Data Word Mode Code). The mode codes are defined by bit times 15–19 of the command word. The most significant bit (bit 15) can be used to differentiate between the two mode code groups. When a data word is associated with the mode code, the T/R bit determines if the data word is transmitted or received by the remote terminal. The mode codes are listed in Table 1.4.

TABLE 1.4   Mode Code

T/R	Mode Code	Function	Data Word	Broadcast
1	00000	Dynamic Bus Control	No	No
1	00001	Synchronize	No	Yes
1	00010	Transmit Status Word	No	No
1	00011	Initiate Self-Test	No	Yes
1	00100	Transmitter Shutdown	No	Yes
1	00101	Override Transmitter Shutdown	No	Yes
1	00110	Inhibit Terminal Flag Bit	No	Yes
1	00111	Override Inhibit Terminal Flag Bit	No	Yes
1	01000	Reset	No	Yes
1	01001	RESERVED	No	TBD
1	•	•	No	•
1	•	•	No	•
1	01111	RESERVED	No	TBD
1	10000	Transmit Vector Word	Yes	No
0	10001	Synchronize	Yes	Yes
1	10010	Transmit Last Command Word	Yes	No
1	10011	Transmit BIT Word	Yes	No
0	10100	Selected Transmitter Shutdown	Yes	Yes
0	10101	Override Selected Transmitter Shutdown	Yes	Yes
1/0	10110	RESERVED	Yes	TBD
	•	•	Yes	•
	•	•	Yes	•
1/0	11111	RESERVED	Yes	TBD

1.3.3.1   Mode Code Identifier

The mode code identifier is contained in bits 10–14 of the command word. When this field is either 00000 or 11111 then the contents of bits 15–19 of the command word are to be decoded as a mode code. Two mode code identifiers are provided such that the system can utilize the Instrumentation bit if desired. The two mode code identifiers shall not convey different information.

1.3.3.2   Mode Code Functions

The following defines the functionality of each of the mode codes.

Dynamic Bus Control. The Dynamic Bus Control Mode Code is used to provide for the passing of the control of the data bus between terminals, thus providing a "round robin" type of control. Using this methodology, each terminal is responsible for collecting the data it needs from all the other terminals. When it is done collecting, it passes control to the next terminal in line (based on some predefined sequence). This allows the applications program (the end user of the data) to collect the data when it needs it, always insuring that the data collected is from the latest source sample and has not been sitting around in a buffer waiting to be used.

Notices 1 and 2 to the standard forbid the use of Dynamic Bus Control for Air Force applications. This is due to the problems and concerns of what may occur when a terminal, that has passed the control, is unable to perform or does not properly forward the control to the next terminal, thereby forcing the condition of no terminal being in control and having to reestablish control by some terminal. The potential amount of time required to reestablish control could have disastrous effects upon the system (i.e., especially a flight control system).

A remote terminal that is capable of performing as the bus control should be capable of setting the Dynamic Bus Control Acceptance Bit in the terminal's Status Word to logic "1" when it receives the mode code command. Typically, the logic associated with the setting of this bit is based on the subsystem's (computer's) ability to pass some level of confidence test. If the confidence test passes, then the bit is set and the status word is transmitted when the terminal receives the mode command, thereby saying that it will assume the role of bus controller.

The bus controller can only issue the Dynamic Bus Control mode command to one remote terminal at a time. The command obviously is only issued to terminals that are capable of performing as a bus controller. Upon transmitting the command, the bus controller must check the terminal's status word to determine if the Dynamic Bus Control Acceptance Bit is set. If set, the bus controller ceases to function as the controller and becomes either a remote terminal or a bus monitor. If the bit in the status word is not set, the remote terminal which was issued the command is not capable of becoming the bus controller; the current controller must either remain the bus controller or attempt to pass the control to some other terminal.

Synchronize. The synchronize mode code is used to establish some form of timing between two or more terminals. This mode code does not use a data word, therefore the receipt of this command by a terminal must cause some predefined event to occur. Some examples of this event may be the clearing, incrementing, or presetting of a counter; the toggling of an output signal; or the calling of some software routine. Typically, this command is used to time correlate a function such as the sampling of navigation data (i.e., present position, rates, etc.) for flight controls or targeting/fire control systems. Other uses have been for the bus controller to "sync" the back-up controllers (or monitors) to the beginning of a major/minor frame processing.

When a remote terminal receives the Synchronize Mode Command, it should perform its predefined function. For a bus controller, the issuance of the command is all that is needed. The terminals status word only indicates that the message was received, not that the "sync" function was performed.

Transmit Status Word. This is one of the two commands that does not cause the remote terminal to reset or clear its status word. Upon receipt of this command, the remote terminal transmits the status word that was associated with the previous message, not the status word of the mode code message.

The bus controller uses this command for control and error management of the data bus. If the remote terminal had detected an error in the message and suppressed its status word, then the bus controller can issue this command to the remote terminal to determine if indeed the nonresponse was due to an error. As this command does not clear the status word from the previous message, a detected error by the remote terminal in a previous message would be indicated by having the Message Error bit set in the status word.

The bus controller also uses this command when "polling." If a terminal does not have periodic messages, the RT can indicate when it needs communications by setting the Service Request bit in the status word. The bus controller, by requesting the terminal to transmit only its status word, can determine if the terminal is in need of servicing and can subsequently issue the necessary commands. This "polling" methodology has the potential of reducing the amount of bus traffic by eliminating the transmission of unnecessary words.

Another use of this command is when broadcast message formats are used. As all of the remote terminals will suppress their status words, "polling" each terminal for its status word would reveal whether the terminal received the message by having its Broadcast Command Received bit set.

Initiate Self-Test. This command, when received by the remote terminal, shall cause the remote terminal to enter into its self-test. This command is normally used as a ground-based maintenance function, as part of the system power-on tests, or in flight as part of a fault recovery routine. Note that this test is only for the remote terminal, not the subsystem.

In earlier applications, some remote terminals, upon receipt of this command, would enter self-test and go "offline" for long periods of time. Notice 2, in an effort to control the amount of time that a terminal could be "offline," limited the test time to 100.0 μsec following the transmission of the status word by the remote terminal.

While a terminal is performing its self-test, it may respond to a valid command in the following ways: (a) no response on either bus ("off-line"); (b) transmit only the status word with the Busy bit set; or

(c) normal response. The remote terminal may, upon receipt of a valid command received after this mode code, terminate its self-test. As a subsequent command could abort the self-test, the bus controller, after issuing this command, should suspend transmissions to the terminal for the specified amount of time (either a time specified for the remote terminal or the maximum time of 100.0 μsec).

Transmitter Shutdown. This command is used by the bus controller in the management of the bus. In the event that a terminal's transmitter continuously transmits, this command provides for a mechanism to turn the transmitter off. This command is for dual redundant standby applications only.

Upon receipt of this command, the remote terminal shuts down (i.e., turns off) the transmitter associated with the opposite data bus. That is to say if a terminals transmitter is babbling on the A bus, the bus controller would send this command to the terminal on the B bus (a command on the A bus would not be received by the terminal).

Override Transmitter Shutdown. This command is the complement of the previous one in that it provides a mechanism to turn on a transmitter that had previously been turned off. When the remote terminal receives this command, it shall set its control logic such that the transmitter associated with the opposite bus be allowed to transmit when a valid command is received on the opposite bus. The only other command that can enable the transmitter is the Reset Remote Terminal Mode Command.

Inhibit Terminal Flag. This command provides for the control of the Terminal Flag bit in a terminal's status word. The Terminal Flag bit indicates that there is a error within the remote terminal hardware and that the data being transmitted or the data received may be in error. However, the fault within the terminal may not have any effect upon the quality of the data, and the bus controller may elect to continue with the transmissions knowing a fault exists.

The remote terminal receiving this command shall set its Terminal Flag bit to logic "0" regardless of the true state of this signal. The standard does not state that the built-in-test that controls this bit be halted, but only the results be negated to "0."

Override Inhibit Terminal Flag. This command is the complement of the previous one in that it provides a mechanism to turn on the reporting of the Terminal Flag bit. When the remote terminal receives this command, it shall set its control logic such that the Terminal Flag bit is properly reported based upon the results of the terminal's built-in-test functions. The only other command that can enable the response of the Terminal Flag bit is the Reset Remote Terminal Mode Command.

Reset Remote Terminal. This command, when received by the remote terminal, shall cause the terminal electronics to reset to its power-up state. This means that if a transmitter had been disabled or the Terminal Flag bit inhibited, these functions would be reset as if the terminal had just powered up. Again, remember that the reset applies only to the remote terminal electronics and not to the entire box.

Notice 2 restricts the amount of time that a remote terminal can take to reset its electronics. After transmission of its status word, the remote terminal shall reset within 5.0 μsec. While a terminal is resetting, it may respond to a valid command in the following ways: (a) no response on either bus ("offline"); (b) transmit only the status word with the Busy bit set; or (c) normal response. The remote terminal may, upon receipt of a valid command received after this mode code, terminate its reset function. As a subsequent command could abort the reset, the bus controller, after issuing this command, should suspend transmissions to the terminal for the specified amount of time (either a time specified for the remote terminal or the maximum time of 5.0 μsec).

Transmit Vector Word. This command shall cause the remote terminal to transmit a data word referred to as the vector word. The vector word shall identify to the bus controller service request information relating to the message needs of the remote terminal. While not required, this mode code is often tied to the Service Request bit in the Status Word. As indicated, the contents of the data word inform the bus controller of messages that need to be sent.

The bus controller also uses this command when "polling." Though typically used in conjunction with the Service Request bit in the status word, wherein the bus controller requests only the status word (Transmit Status Word Mode Code) and upon seeing the Service Request bit set would then issue the Transmit Vector Word Mode Code, the bus controller can always ask for the Vector Word (always getting the status word anyway) and reduce the amount of time required to respond to the terminal's request.

Synchronize with Data Word. The purpose of this synchronize command is the same as the synchronize without data word, except this mode code provides a data word to provide additional information to the remote terminal. The contents of the data word are left to the imagination of the user. Examples from "real world" applications have used this word to provide the remote terminal with a counter or clock value; to provide a backup controller with a frame identification number (minor frame or cycle number); and to provide a terminal with a new base address pointer used in extending the subaddress capability.

Transmit Last Command Word. This is one of the two commands that does not cause the remote terminal to reset or clear its status word. Upon receipt of this command, the remote terminal transmits the status word that was associated with the previous message and the Last Command Word (valid) that it received.

The bus controller uses this command for control and error management of the data bus. When a remote terminal is not responding properly, then the bus controller can determine the last valid command the terminal received and can re-issue subsequent messages as required.

Transmit BIT Word. This mode command is used to provide detail with regards to the Built-in-Test (BIT) status of the remote terminal. Its contents shall provide information regarding the remote terminal only (remember the definition) and not the subsystem.

While most applications associate this command with the Initiate Self Test Mode Code, the standard requires no such association. Typical use is to issue the Initiate Self Test Mode Code, allow the required amount of time for the terminal to complete its tests, and then issue the Transmit BIT Word Mode Code to collect the results of the test. Other applications have updated the BIT word on a periodic rate based on the results of a continuous background test (e.g., as a data wraparound test performed with every data transmission). This word can then be transmitted to the bus controller, upon request, without having to initiate the test and then wait for the test to be completed. The contents of the data word are left to the terminal designer.

Selected Transmitter Shutdown. Like the Transmitter Shutdown Mode Code, this mode code is used to turn off a babbling transmitter. The difference between the two mode codes is that this mode code has a data word associated with it. The contents of the data word specifies which data bus (transmitter) to shutdown. This command is used in systems which provide more than dual redundancy.

Override Selected Transmitter Shutdown. This command is the complement of the previous one in that it provides a mechanism to turn on a transmitter that had previously been turned off. When the remote terminal receives this command, the data word specifies which data bus (transmitter) shall set its control logic such that the transmitter associated with that bus be allowed to transmit when a valid command is received on that bus. The only other command that can enable the selected transmitter is the Reset Remote Terminal Mode Command.

Reserved Mode Codes. As can be seen from Table 1.4, there are several bit combinations that are set aside as reserved. It was the intent of the standard that these be reserved for future growth. It should also be noticed from the table that certain bit combinations are not listed. The standard allows the remote terminal to respond to these reserved and "undefined" mode codes in the following manner: set the message error bit and respond (see Illegal Commands); or respond in form. The designer of terminal hardware or a multiplex system is forbidden to use the reserved mode codes for any purpose.

1.3.3.3 Required Mode Codes

Notice 2 to the standard requires that all remote terminals implement the following four mode codes: Transmit Status Word, Transmitter Shutdown, Override Transmitter Shutdown, and Reset Remote Terminal. This requirement was levied so as to provide the multiplex system designer and the bus controller with a minimum set of commands for managing the multiplex system. Note that the above requirement was placed on the remote terminal. Notice 2 also requires that a bus controller be capable of implementing all of the mode codes, however, for Air Force applications, the Dynamic Bus Control Mode Code shall never be used.

1.3.3.4 Broadcast Mode Codes

Notice 2 to the standard allows the broadcast of mode codes (see Table 1.4). The use of the broadcast option can be of great assistance in the areas of terminal synchronization. Ground maintenance and troubleshooting can take advantage of broadcast Reset Remote Terminal or Initiate Self, but these two commands can have disastrous effects if used while in flight. The designer must provide checks to insure that commands such as these are not issued by the bus controller or operated upon by a remote terminal when certain conditions exists (e.g., in flight).

1.4 Systems-Level Issues

The standard provides very little guidance in how it is applied. Lessons learned from "real world" applications have led to design guides, application notes, and handbooks that provide guidance. This section will attempt to answer some of the systems-level questions and identify implied requirements that, while not specifically called out in the standard, are required nonetheless.

1.4.1 Subaddress Utilization

The standard provides no guidance on how to use the subaddresses. The assignment of subaddresses and their functions (the data content) is left to the user. Most designers automatically start assigning subaddresses with 01 and count upwards. If the Instrumentation bit is going to be used, then the subaddresses must start at 16.

The standard also requires that normal subaddresses be separated from broadcast subaddresses. If the broadcast option is implemented, then an additional memory block is required to receive broadcast commands.

1.4.1.1 Extended Subaddressing

The number of subaddresses that a terminal has is limited to 60 (30 transmit and 30 receive). Therefore, the number of unique data words available to a terminal is 1920 (60 $\times$ 32). For earlier applications, where data being transferred were analog sensor data and switch settings, this was more than sufficient. However, in some of today's applications, in which digital computers exchanging data, or for a video sensor passing digitized video data, the number of words is too limited.

Most terminal designs establish a block of memory for use by the 1553 interface circuitry. This block contains a address start pointer and then the memory is offset by the subaddress number and the word count number to arrive at a particular memory address.

A methodology of extending the range of the subaddresses has been successfully utilized. This method uses either a dedicated subaddress and data word, or makes use of the synchronize with data word mode code. The data word associated with either of these contains an address pointer which is used to reestablish the starting address of the memory block. The changing of the blocks is controlled by the bus controller and can be done based on numerous functions. Examples are operational modes, wherein one block is used for startup messages, a different block for take-off and landing, a different block for navigation and cruise, a different block for mission functions (i.e., attack or evade modes), and a different block for maintenance functions.

Another example is that the changing of the start address could also be associated with minor frame cycles. Eight minor frames could have a separate memory block for each frame. The bus controller could synchronize frames and change memory pointers at the beginning of each new minor frame.

For computers exchanging large amounts of data (e.g., GPS Almanac Tables) or for computers that receive program loads via the data bus at power-up, the bus controller could set the pointers at the beginning of a message block, send 30, 32-word messages, move the memory pointer to the last location in the remote terminals memory that received data, then send the next block of 30, 32-word messages, continuing this cycle until the memory is loaded. The use is left to the designer.

1.4.2   Data Wraparound

Notice 2 to the standard does require that the terminal is able to perform a data wraparound and subaddress 30 is suggested for this function. Data wraparound provides the bus controller with a methodology of testing the data bus from its internal circuitry, through the bus media, to the terminal's internal circuitry. This is done by the bus controller sending the remote terminal a message block and then commanding the terminal to send it back. The bus controller can then compare the sent data with that received to determine the state of the data link. There are no special requirements upon the bit patterns of the data being transferred.

The only design requirements are placed upon the remote terminal. These are that the terminal, for the data wraparound function, be capable of sending the number of data words equal to the largest number of data words sent for any transmit command. This means that if a terminal maximum data transmission is only four data words, it need only provide for four data words in its data wraparound function.

The other requirement is that the remote terminal need only hold the data until the next message. The normal sequence is for the bus controller to send the data, then in the next message it asks for it back. If another message is received by the remote terminal before the bus controller requests the data, the terminal can discard the data from the wraparound message and operate on the new command.

1.4.3   Data Buffering

The standard specifies that the any error within a message shall invalidate the entire message. This implies that the remote terminal must store the data within a message buffer until the last data word has been received and validated before allowing the subsystem access to the data. To insure that the subsystem always has the last message of valid data received to work with would require the remote terminal to, as a minimum, double buffer the received data.

There are several methods to accomplish this in hardware. One method is for the terminal electronics to contain a First-In First-Out (FIFO) memory that stores the data as it is received. Upon validation of the last data word, the terminal's subsystem interface logic will move the contents of the FIFO into memory accessible by the subsystem. If an error occurred during the message, the FIFO is reset.

A second method establishes two memory blocks for each message in common memory. The subsystem is directed to read from one block (block A) while the terminal electronics writes to the other (Block B). Upon receipt of a valid message, the terminal will switch pointers, indicating that the subsystem is to read from the new memory block (block B) while the terminal will now write to block B. If an error occurs within the message, the memory blocks are not switched.

Some of the "off-the-shelf" components available provide for data buffering. Most provide for double buffering, while some provided for multilevels of buffering.

1.4.4   Variable Message Blocks

Remote terminals should be able to transmit any subset of any message. This means that if a terminal has a transmit message at subaddress 04 of 30 data words, it should be capable of transmitting any number of those data words (01–30) if so commanded by the bus controller. The order in which the subset is transmitted should be the same as if the entire message is being transmitted, that being the contents of data word 01 is the same regardless of the word count.

Terminals which implement Illegal Command detection should not consider subsets of a message as illegal. That is to say, if in our example above a command is received for 10 data words, this should not be illegal. But, if a command is received for 32 data words, this would be considered as an illegal command.

1.4.5   Sample Consistency

When transmitting data, the remote terminal needs to ensure that each message transmitted is of the same sample set and contains mutually consistent data. Multiple words used to transfer multiple precision parameters or functionally related data must of the same sampling.

If a terminal is transmitting pitch, roll, and yaw rates, and while transmitting the subsystem updates these data in memory, but this occurs after pitch and roll had been read by the terminal's electronics, then the yaw rate transmitted would be of a different sample set. Having data from different sample rates could have undesirable effects on the user of the data.

This implies that the terminal must provide some level of buffering (the reverse of what was described above) or some level of control logic to block the subsystem from updating data while being read by the remote terminal.

1.4.6   Data Validation

The standard tightly defines the criteria for the validation of a message. All words must meet certain checks (i.e., valid sync, Manchester encoding, number of bits, odd parity, etc.) in order for each word and each message to be valid. But what about the contents of the data word? MIL-STD-1553 provides the checks to insure the quality of the data transmission from terminal to terminal, sort of a "data in equals data out," but is not responsible for the validation tests of the data itself. This is not the responsibility of the 1553 terminal electronics, but of the subsystem. If bad data are sent, then "garbage in equals garbage out." But the standard does not prevent the user from providing additional levels of protection. The same techniques used in digital computer interfaces (i.e., disk drives, serial interfaces, etc.) can be applied to 1553. These techniques include checksums, CRC words, and error detection/correction codes. Section 80 of MIL-HDBK-1553A which covers data word formats even offers some examples of these techniques.

But what about using the simple indicators embedded within the standard. Each remote terminal provides a status word — indicating not only the health of the remote terminal's electronics, but also that of the subsystem. However, in most designs, the status word is kept within the terminal electronics and not passed to the subsystems. In some "off-the-shelf" components, the status word is not even available to be sent to the subsystem. But two bits from the status word should be made available to the subsystem and the user of the data for further determination as to the validity of the data. These are the Subsystem Flag and the Terminal Flag bits.

1.4.7   Major and Minor Frame Timing

The standard specifies the composition of the words (command, data, and status) and the messages (information formats and broadcast formats). It provides a series of management messages (mode codes), but it does not provide any guidance on how to apply these within a system. This is left to the imagination of the user.

Remote terminals, based upon the contents of their data, will typically state how often data are collected and the fastest rate they should be outputted. For input data, the terminal will often state how often it needs certain data to either perform its job or maintain a certain level of accuracy. The rates are referred to as the transmission and update rates. It is the system designer's job to examine the data needs of all of the systems and determine when data are transferred from whom to whom. These data are subdivided into periodic messages — those which must be transferred at some fixed rate, and aperiodic messages, those which are typically either event driven (i.e., the operator pushes a button) or data driven (i.e., a value is now within range).

A major frame is defined such that all periodic messages are transferred at least once. This is therefore defined by the message with the slowest transmission rate. Typical major frame rates used in today's applications vary from 40 to 640 μsec. There are some systems that have major frame rates in the 1- to 5-sec range, but these are the exceptions, not the norm. Minor frames are then established to meet the requirements of the higher update rate messages.

The sequence of messages within a minor frame is again left undefined. There are two methodologies that are predominately used. In the first method, the bus controller starts the frame with the transmission of all of the periodic messages (transmit and receive) to be transferred in that minor frame. At the end of the periodic messages, the bus controller is either finished (resulting in dead bus time — no transmissions) until the beginning of the next frame, or the bus controller can use this time to transfer aperiodic messages, error handling messages, or transfer data to the back-up bus controller(s).

In the second method (typically used in a centralized processing architecture), the bus controller issues all periodic and aperiodic transmit messages (collects the data), then processes the data (possibly using dead time during this processing), and then issues all the receive messages (outputting the results of the processing). Both methods have been used successfully.

1.4.8   Error Processing

The amount and level of error processing is typically left to the systems designer but may be driven by the performance requirements of the system. Error processing is typically only afforded to critical messages, wherein the noncritical messages just await the next normal transmission cycle. If a data bus is 60% loaded and each message received an error, the error processing would exceed 100% of available time and thereby cause problems within the system.

Error processing is again a function of the level of sophistication of the bus controller. Some controllers (typically message or frame controllers) can automatically perform some degree of error processing. This usually is limited to a retransmission of the message either once on the same bus or once on the opposite bus. Should the retried message also fail, the bus controller software is informed of the problem. The message may then be retried at the end of the normal message list for the minor frame.

If the error still persists, then it may be necessary to stop communicating with the terminal, especially if the bus controller is spending a large amount of time performing error processing. Some systems will try to communicate with a terminal for a predefined number of times on each bus. After this, all messages to the terminal are removed from the minor frame lists, and substituted with a single transmit status word mode code.

An analysis should be performed on the critical messages to determine the effects upon the system if they are not transmitted or the effects of data latency if they are delayed to the end of the frame.

1.5   Testing

The testing of a MIL-STD-1553 terminal or system is not a trivial task. There are a large number of options available to the designer including message formats, mode commands, status word bits, and coupling methodology. In addition, history has shown that different component manufacturers and designers have made different interpretations regarding the standard, thereby introducing products that implement the same function quite differently.

For years, the Air Force provided for the testing of MIL-STD-1553 terminals and components. Today this testing is the responsibility of industry. The Society of Automotive Engineers (SAE), in conjunction with the government, has developed a series of Test Plans for all 1553 elements. These Test Plans are listed in Table 1.6.

TABLE 1.6   SAE 1553 Test Plans

AS-4111	Remote Terminal Validation Test Plan
AS-4112	Remote Terminal Production Test Plan
AS-4113	Bus Controller Validation Test Plan
AS-4114	Bus Controller Production Test Plan
AS-4115	Data Bus System Test Plan
AS-4116	Bus Monitor Test Plan
AS-4117	Bus Components Test Plan

Further Information

In addition to the SAE Test Plans listed in Table 1.6, there are other documents that can provide a great deal of insight and assistance in designing with MIL-STD-1553:

MIL-STD-1553B Digital Time Division Command/Response Multiplex Data Bus
MIL-HDBK-1553A Multiplex Applications Handbook
SAE AS-15531 Digital Time Division Command/Response Multiplex Data Bus
SAE AS-15532 Standard Data Word Formats
SAE AS-12 Multiplex Systems Integration Handbook
SAE AS-19 MIL-STD-1553 Protocol Reorganized
DDC 1553 Designers Guide
UTMC 1553 Handbook

And lastly, there is the SAE 1553 Users Group. This is a collection of industry and military experts in 1553 who provide an open forum for information exchange, and provide guidance and interpretations/clarifications with regard to the standard. This group meets twice a year as part of the SAE Avionics Systems Division conferences.

2

ARINC 429

Daniel A. Martinec
ARINC

2.1   Introduction

ARINC Specifications 419, 429, and 629 and Project Paper 453 are documents prepared by the Airlines Electronic Engineering Committee (AEEC) and published by Aeronautical Radio, Inc. These are among over 300 air transport industry avionics standards published since 1949. These documents, commonly referred to as ARINC 419, ARINC 429, ARINC 453, and ARINC 629, describe data communication systems used primarily on commercial transport airplanes. A limited number of general aviation and military airplanes also use these data systems. The differences between the systems are described in detail in the subsequent sections.

2.2   ARINC 419

ARINC Specification 419, "Digital Data Compendium," provides detailed descriptions of the various interfaces used in the ARINC 500 series of avionics standards prior to 1980. ARINC Specification 419 is often incorrectly assumed to be a standalone bus standard. ARINC Specification 419 provides a summary of electrical interfaces, protocols, and data standards for avionics built prior to the airlines' selection of a single standard, i.e., ARINC 429, for the distribution of digital information aboard aircraft.

2.3   ARINC 429

2.3.1   General

ARINC Specification 429, "Digital Information Transfer System (DITS)," was first published in 1977 and has since become the ARINC standard most widely used by the airlines. The title of this airline standard was chosen so as not to describe it as a "data bus." Although ARINC 429 is a vehicle for data transfer, it does not fit the normal definition of a data bus. A typical data bus provides multidirectional transfer of data between multiple points over a single set of wires. ARINC 429's simplistic one-way flow of data significantly limits this capability, but the associated low cost and the integrity of the installations have provided the airlines with a system exhibiting excellent service for more than two decades. Additional information regarding avionics standards may be found at URL http://www.arinc.com/aeec.

2.3.2   History

In the early 1970s the airlines recognized the potential advantage of implementation of digital equipment. Some digital equipment had already been implemented to a certain degree on airplanes existing at that time. However, there were three new transport airplanes on the horizon. These were the Airbus A-310 and the Boeing B-757 and B-767. The airlines, along with the airframe and equipment manufacturers, established a goal to create an all-new suite of avionics using digital technology.

Obviously, with digital avionics came the need for an effective means of data communications among the avionics units. The airlines recognized that the military was also in the early stages of development of a data bus that could perform the data transfer functions among military avionics. The potential for a joint program to produce a data bus common to the air transport industry and the military exhibited a potential for significant economical benefits.

The early work to develop the military's data bus was taken on by the Society of Automotive Engineers (SAE). Participants in the SAE program emanated from many parts of the military and private sectors of aviation. A considerable effort went into defining all aspects of the data bus with the goal of meeting the needs of both the military and air transport users. That work culminated in the development of the early version of the data bus identified by Mil-Std 1553 (see Chapter 1).

Early in the process of the Mil-Std 1553 development, representatives from the air transport industry realized that the stringent and wide range of military requirements would cause the Mil-Std 1553 to be overly complex for the commercial user and would not exhibit the flexibility to accommodate the varying applications of transport airplanes. Difficulty in certification also was considered a potential problem. The decision was made to abandon a cooperative data bus development program with the military and pursue work on a data bus to more closely reflect commercial airplane requirements.

Numerous single transmitter/multiple receiver data transfer systems were being used on airplanes built in the early 1970s. These proved to be reliable and efficient compared to the more complex data buses of the time. These transfer systems, described in ARINC Specification 419, were considered as candidates for the new digital aircraft.

While none of the systems addressed in the ARINC Specification could adequately perform the task, each exhibited desirable characteristics that could be applied to a new design. The result was the release of a new data transfer system exhibiting a high level of efficiency, extremely good reliability, and ease of certification. ARINC 429 became the industry standard. Subsequent to release of the standard, numerous low-cost integrated circuits were produced by solid-state component manufacturers. ARINC 429 was used widely by the air transport industry and even found applications in non-aviation commercial and military applications. ARINC 429 has been used as the standard for virtually all ARINC 700-series standards for "digital avionics" used by the air transport industry.

Aeronautical Radio Inc. has maintained and provided the necessary routine updates for new data word assignments and formats. There were no significant changes in the basic design until 1980 when operational experience showed that certain shorted wire conditions would allow the bus to operate in a faulty condition. The bus would operate in this condition with much reduced noise immunity. This condition also proved to be very difficult to locate during routine maintenance. In response, the airlines suggested that the design be changed in order to ensure that the bus would not continue to operate when this condition occurred. A change to the receiver voltage thresholds and impedances solved this problem.

No basic changes to the design have been made since that time. ARINC 429 has remained a reliable system and even today is used extensively in the most modern commercial airplanes.

2.3.3   Design Fundamentals

2.3.3.1   Equipment Interconnection

A single transmitter is connected with up to 20 data receivers via a single twisted and shielded pair of wires. The shields of the wires are grounded at both ends and at any breaks along the length of the cable. The shields are kept as short as possible.

2.3.3.2   Modulation

Return-To-Zero (RZ) modulation is used. The voltage levels are used for this modulation scheme.

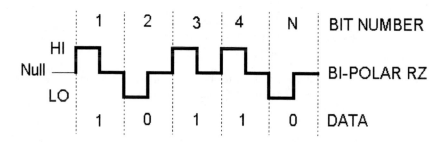

2.3.3.3   Voltage Levels

The differential output voltages across the transmitter output terminal with no load is described in the following table:

	HI(V)	NULL(V)	LO(V)
Line A to Line B	$+10 \pm 1.0$	0 ± 0.5	-10 ± 1.0
Line A to Ground	5 ± 0.5	0 ± 0.25	-5 ± 0.5
Line B to Ground	-5 ± 0.5	0 ± 0.25	$+5 \pm 0.5$

The differential voltage seen by the receiver will depend on wire length, loads, stubs, etc. With no noise present on the signal lines the nominal voltages at the receiver terminals (A and B) would be

$$\text{HI} \qquad +7.25\text{V to } +11\text{V}$$

$$\text{NULL} \qquad +0.5\text{V to } -0.5\text{V}$$

$$\text{LO} \qquad -7.25\text{V to } -11\text{V}$$

In practical installations impacted by noise, etc. The following voltages ranges will be typical across the receiver input (A and B):

HI	+6.5V to +13V
NULL	+2.5V to −2.5V
LO	−6.5V to −13V

Line (A or B) to ground voltages are not defined.

Receivers are expected to withstand without damage steady-state voltages of 30 VAC RMS applied across terminals A and B, or ±29 VDC applied between terminal A or B and the ground.

2.3.3.4 Impedance Levels

2.3.3.4.1 Transmitter Output Impedance

The transmitter output impedance is 70 to 80 (nominal 75) ohms and is divided equally between lines A and B for all logic states and transitions between those states.

2.3.3.4.2 Receiver Input Impedance

The typical receiver input characteristics are as follows:

Differential Input Resistance R_I = 12,000 ohms minimum
Differential Input Capacitance C_I = 50 pF maximum
Resistance to Ground R_H and R_G ≥ 12,000 ohms
Capacitance to Ground C_H and C_G ≤ 50 pF

The total receiver input resistance including the effects of R_I, R_H and R_G in parallel is 8000 ohms minimum (400 ohms minimum for 20 receivers). A maximum of 20 receivers is specified for any one transmitter. See below for the circuit standards.

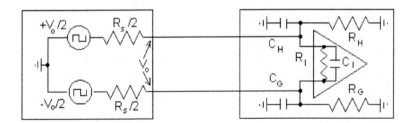

2.3.3.4.3 Cable Impedance

The wire gauges used in the interconnecting cable will typically vary between 20 and 26 depending on desired physical integrity of the cable and weight limitations. Typical characteristic impedances will be in the range of 60 to 80 ohms. The transmitter output impedance was chosen at 75 ohms nominal to match this range.

2.3.3.5 Fault Tolerance

The electrical power on an airplane is provided by a generator on each engine. The airplane electrical system is designed to take into account any variation in engine speeds, phase differentials, power bus switching, etc. However, it is virtually impossible to ensure that the power source will be perfect at all times. Failures within a system can also cause erratic power levels. The design of the ARINC 429 components take power variation into account and are not generally susceptible to either damage or erratic operation when those variations occur. The ranges of those variations are provided in the following sections.

2.3.3.5.1 Transmitter External Fault Voltage

Transmitter failures caused by external fault voltages will not typically cause other transmitters or other circuitry in the unit to function outside of their specification limits or to fail.

2.3.3.5.2 Transmitter External Fault Load Tolerance

Transmitters should indefinitely withstand without sustaining damage a short circuit applied:

- a. across terminals A and B, or
- b. from terminal A to ground, or
- c. from terminal B to ground, or
- d. b and c above, simultaneously.

2.3.3.6 Fault Isolation

2.3.3.6.1 Receiver Fault Isolation

Each receiver incorporates isolation provisions to ensure that the occurrence of any reasonably probable internal LRU or bus receiver failure does not cause any input bus to operate outside its specification limits (both undervoltage or overvoltage).

2.3.3.6.2 Transmitter Fault Isolation

Each transmitter incorporates isolation provisions to ensure that it does not under any reasonably probable equipment fault condition provide an output voltage in excess of:

- a. a voltage greater than 30 VAC RMS between terminal A and B, or
- b. greater than ±29VDC between A and ground, or
- c. greater than ±29 VDC between B and ground.

2.3.3.7 Logic-Related Elements

This section describes the digital transfer system elements considered to be principally related to the logic aspects of the signal circuit.

2.3.3.7.1 Digital Language

Numeric Data — The ARINC 429 accommodates numeric data encoded in two digital languages, (a) BNR expressed in two's complement fractional notation, and (b) BCD per the numerical subset of ISO Alphabet No. 5. An information item encoded in both languages is assigned a unique address for each (see Section 2.4.3).

Discrete Data — In addition to handling numeric data as specified above, ARINC 429 is also capable of accommodating discrete items of information either in the unused (pad) bits of data words or, when necessary, in dedicated words.

The rule in the assignment of bits in discrete numeric data words is to start with the least significant bit of the word and to continue towards the most significant bit available in the word. There are two types of discrete words. These are general purpose discrete words, and dedicated discrete words. Seven labels (270 XX–276 XX) are assigned to the general purpose discrete words. These words are assigned in ascending label order (starting with 270 XX), where XX is the equipment identifier.

32	31 30	29 28 27 26 25 24 23 22 21 20 19 18 17 16 15 14 13 12 11	10 9	8 7 6 5 4 3 2 1
P	SSM	DATA→ ←PAD ←DISCRETES	SDI	LABEL
		MSB LSB		

Generalized BCD Word Format

P	SSM	BCD CH #1	BCD CH #2	BCD CH #3	BCD CH #4	BCD CH #5	SDI	8 7 6 5 4 3 2 1
		4 2 1	8 4 2 1	8 4 2 1	8 4 2 1	8 4 2 1		
0	0 0 0	0 1 0	0 1 0 1	0 1 1 1	1 0 0 0	0 1 1 0	0 0	1 0 0 0 0 0 0 1
Example		2	5	7	8	6	DME DISTANCE	

BCD Word Format Example (No Discretes)

32	31 30 29	28 27 26 25 24 23 22 21 20 19 18 17 16 15 14 13 12 11	10 9	8 7 6 5 4 3 2 1
P	SSM	DATA→ ←PAD ←DISCRETES	SDI	LABEL
		MSB LSB		

Generalized BCD Word Format

Maintenance Data (General Purpose)—The general purpose Maintenance words are assigned labels in sequential order as are the labels for the general purpose Discrete words. The lowest octal value label assigned to the Maintenance words is used when only one Maintenance word is transmitted. When more than one word is transmitted the lowest octal value label is used first and the other labels used sequentially until the message has been completed. The General Purpose Maintenance words may contain Discrete, BCD, or BNR Numeric data. They do not contain ISO Alphabet No. 5 messages. The General Purpose Maintenance words are formatted according to the layouts of the corresponding BCD/BNR/Discrete data words shown above.

2.4 Message and Word Formatting

2.4.1 Direction of Information Flow

The information output of a system element is transmitted from a designated port (or ports) to which the receiving ports of other system elements in need of that information are connected. In no case does information flow into a port designated for transmission. A separate data bus (twisted and shielded pair of wires) is used for each direction when data are required to flow both ways between two system elements.

2.4.2 Information Element

The basic information element is a digital word containing 32 bits. There are five application groups for such words, BNR data, BCD data, Discrete data, Maintenance data (general) and Acknowledgment, ISO Alphabet No. 5 and Maintenance (ISO Alphabet No. 5) data (AIM). The relevant data handling rules are set forth in Section 2.4.6. When less than the full data field is needed to accommodate the information conveyed in a word in the desired manner, the unused bit positions are filled with binary zeros or, in the case of BNR/BCD numeric data, valid data bits. If valid data bits are used, the resolution may exceed the accepted standard for an application.

2.4.3 Information Identifier

The type of information contained in a word is identified by a six-character label. The first three characters are octal characters coded in binary in the first eight bits of the word. The eight bits will identify the information contained within BNR and BCD numeric data words (e.g., DME distance, static air temperature, etc.) and identify the word application for Discrete, Maintenance, and AIM data.

The last three characters of the six-character label are hexadecimal characters used to provide for identification of ARINC 429 bus sources. Each triplet of hexadecimal characters identifies a system element with one or more DITS ports. Each three character code (and black box) may have up to 255 eight-bit labels assigned to it. The code is used administratively to retain distinction between unlike parameters having like labels assignments.

Octal label 377 has been assigned for the purpose of electrically identifying the system element. The code appears in the three least significant digits of the 377 word in a BCD Word format. The transmission of the equipment identifier word on a bus will permit receivers attached to the bus to recognize the source of the DITS information. Since the transmission of the equipment identifier word is optional, receivers should not depend on that word for correct operation.

2.4.4 Source/Destination Identifier

Bit numbers 9 and 10 of numeric data words are used for a data source/destination identification function. They are not available for this function in alpha/numeric (ISO Alphabet No. 5) data words of this document or when the resolution needed for numeric (BNR/BCD) data necessitates their use for valid data. The source/destination identifier function may find application when specific words need to be directed to a specific system of a multisystem installation or when the source system of a multisystem installation needs to be recognizable from the word content. When it is used, a source equipment encodes its aircraft installation number in bits 9 and 10 as shown in the following table. A sink equipment will recognize words containing its own installation number code *and* words containing code "00," the "all-call" code.

Equipment will fall into the categories of source only, sink only, or both source and sink. Use of the SDI bits by equipment functioning only as a source or only as a sink is described above. *Both* the source and sink texts above are applicable to equipment functioning as both a source and a sink. Such equipment will recognize the SDI bits on the inputs and also encode the SDI bits, as applicable, on the outputs. DME, VOR, ILS, and other sensors, are examples of source and sink equipment generally considered to be only source equipment. These are actually sinks for their own control panels. Many other types of equipment are also misconstrued as source only or sink only. If a unit has a 429 input port and a 429 output port, it is a source and sink! With the increase of equipment consolidation, e.g., centralized control panels, the correct use of the SDI bits cannot be overstressed.

Bit No.		Installation No.
10	9	See text
0	0	
0	1	1
1	0	2
1	1	3

Note: In certain specialized applications of the SDI function the all-call capability may be forfeited so that code "00" is available as an "installation no. 4" identifier.

When the SDI function is not used, binary zeros or valid data should be transmitted in bits 9 and 10.

2.4.5   Sign/Status Matrix

This section describes the coding of the Sign/Status Matrix (SSM) field. In all cases the SSM field uses bits 30 and 31. For BNR data words, the SSM field also includes bit 29.

The SSM field is used to report hardware equipment condition (fault/normal), operational mode (functional test), or validity of data word content (verified/no computed data). The following definitions apply:

Invalid Data—Is defined as any data generated by a source system whose fundamental characteristic is the inability to convey reliable information for the proper performance of a user system. There are two categories of invalid data, namely, "No Computed Data" and "Failure Warning."

No Computed Data—Is a particular case of data invalidity where the source system is unable to compute reliable data for reasons other than system failure. This inability to compute reliable data is caused exclusively by a definite set of events or conditions whose boundaries are uniquely defined in the system characteristic.

Failure Warning—Is a particular case of data invalidity where the system monitors have detected one or more failures. These failures are uniquely characterized by boundaries defined in the system characteristic.

Displays are normally "flagged invalid" during a "Failure Warning" condition. When a "No Computed Data" condition exists, the source system indicates that its outputs are invalid by setting the sign/status matrix of the affected words to the "No Computed Data" code, as defined in the subsections which follow. The system indicators may or may not be flagged depending on system requirements.

While the unit is in the functional test mode, all output data words generated within the unit (i.e., pass-through words are excluded) are coded with "Functional Test." Passthrough data words are those words received by the unit and retransmitted without alteration.

When the SSM code is used to transmit status and more than one reportable condition exists, the condition with the highest priority is encoded in bits number 30 and 31. The order of condition priorities is shown in the table below.

Failure Warning	Priority 1
No Computed Data	Priority 2
Functional Test	Priority 3
Normal Operation	Priority 4

Each data word type has its own unique utilization of the SSM field. These various formats are described in the following sections.

2.4.5.1   BCD Numeric

When a failure is detected within a system which would cause one or more of the words normally output by that system to be unreliable, the system stops transmitting the affected word or words on the data bus.

Some avionic systems are capable of detecting a fault condition which results in less than normal accuracy. In these systems, when a fault of this nature (for instance, partial sensor loss which results in degraded accuracy) is detected, each unreliable BCD digit is encoded "1111" when transmitted on the data bus. For equipment having a display, the "1111" code should, when received, be recognized as representing an inaccurate digit and a "dash" or equivalent symbol is normally displayed in place of the inaccurate digit.

The sign (plus/minus, north/south, etc.) of BCD Numeric Data is encoded in bits 30 and 31 of the word as shown in the table below. Bits 30 and 31 of BCD Numeric Data words are "zero" where no sign is needed.

The "No Computed Data" code is annunciated in the affected BCD Numeric Data word(s) when a source system is unable to compute reliable data for reasons other than system failure. When the "Functional Test" code appears in bits 30 and 31 of an instruction input data word, it is interpreted as a command to perform a functional test.

BCD Numeric Sign/Status Matrix

Bit No		Function
31	30	
0	0	Plus, North, East, Right, To, Above
0	1	No Computed Data
1	0	Functional Test
1	1	Minus, South, West, Left, From, Below

2.4.5.2   BNR Numeric Data Words

The status of the transmitter hardware is encoded in the Status Matrix field (bit numbers 30 and 31) of BNR Numeric Data words as shown in the table below.

A source system annunciates any detected failure that causes one or more of the words normally output by that system to be unreliable by setting bit numbers 30 and 31 in the affected word(s) to the "Failure Warning" code defined in the table below. Words containing this code continue to be supplied to the data bus during the failure condition.

The "No Computed Data" code is annunciated in the affected BNR Numeric Data word(s) when a source system is unable to compute reliable data for reasons other than system failure.

When the "Functional Test" code appears as a system output, it is interpreted as advice that the data in the word result from the execution of a functional test. A functional test produces indications of 1/8 of positive full-scale values unless indicated otherwise in an ARINC equipment Characteristic.

If, during the execution of a functional test, a source system detects a failure which causes one or more of the words normally output by that system to be unreliable, it changes the states of bits 30 and 31 in the affected words such that the "Functional Test" annunciation is replaced with a "Failure Warning" annunciation.

Status Matrix

Bit No		Function
31	30	
0	0	Failure Warning
0	1	No Computed Data
1	0	Functional Test
1	1	Normal Operation

The sign (plus, minus, north, south, etc.) of BNR Numeric Data words are encoded in the Sign Matrix field (bit 29) as shown in the table below. Bit 29 is "zero" when no sign is needed.

Status Matrix

Bit No. 29	Function
0	Plus, North, East, Right, To, Above
1	Minus, South, West, Left, From, Below

Some avionic systems are capable of detecting a fault condition which results in less than normal accuracy. In these systems, when a fault of this nature (for instance, partial sensor loss which results in degraded accuracy) is detected, the equipment will continue to report "Normal" for the sign status matrix while indicating the degraded performance by coding bit 11 as follows:

Accuracy Status

Bit No. 11	Function
0	Nominal Accuracy
1	Degraded Accuracy

This implies that degraded accuracy can be coded only in BNR words not exceeding 17 bits of data.

2.4.5.3 Discrete Data Words

A source system annunciates any detected failure that could cause one or more of the words normally output by that system to be unreliable. Three methods are defined. The first method is to set bits 30 and 31 in the affected word(s) to the "Failure Warning" code defined in the table below. Words containing the "Failure Warning" code continue to be supplied to the data bus during the failure condition. When using the second method, the equipment may stop transmitting the affected word or words on the data bus. This method is used when the display or utilization of the discrete data by a system is undesirable. The third method applies to data words which are defined such that they contain failure information within the data field. For these applications, the associated ARINC equipment Characteristic specifies the proper SSM reporting. Designers are urged not to mix operational and BITE data in the same word.

The "No Computed Data" code is annunciated in the affected Discrete Data word(s) when a source system is unable to compute reliable data for reasons other than system failure. When the "Functional

Test" code appears as a system output, it is interpreted as advice that the data in the Discrete Data word contents are the result of the execution of a functional test.

Discrete Data Words

Bit No.		
31	30	Function
0	0	Verified Data, Normal Operation
0	1	No Computed Data
1	0	Functional Test
1	1	Failure Warning

2.4.6 Data Standards

The units, ranges, resolutions, refresh rates, number of significant bits, pad bits, etc. for the items of information to be transferred by the Mark 33 DITS are administered by the AEEC and tabulated in ARINC Characteristic 429.

ARINC Characteristic 429 calls for numeric data to be encoded in BCD and binary, the latter using two's complement fractional notation. In this notation, the most significant bit of the data field represents one half of the maximum value chosen for the parameter being defined. Successive bits represent the increments of a binary fraction series. Negative numbers are encoded as the two's complements of positive value and the negative sign is annunciated in the sign/status matrix.

In establishing a given parameter's binary data standards, the unit's maximum value and resolution are first determined in that order. The least significant bit of the word is then given a value equal to the resolution increment, and the number of significant bits is chosen such that the maximum value of the fractional binary series just exceeds the maximum value of the parameter, i.e., equals the next whole binary number greater than the maximum parameter value less one least significant bit value. For example, to transfer altitude in units of feet over a range of zero to 100,000 ft with a resolution of 1 ft, the number of significant bits is 17 and the maximum value of the fractional binary series is 131,071 (i.e., 131,072 − 1).

Note that because accuracy is a quality of the measurement process and not the data transfer process, it plays no part in the selection of word characteristics. Obviously, the resolution provided in the data word should equal or exceed the accuracy in order not to degrade it.

For the binary representation of angular data, the ARINC 429 employs "degrees divided by 180°" as the unit of data transfer and ±1 (semicircle) as the range for two's complement fractional notation encoding (ignoring, for the moment, the subtraction of the least significant bit value). Thus the angular range 0 through 359.XXX degrees is encoded as 0 through ±179.XXX degrees, the value of the most significant bit is one half semicircle and there are no discontinuities in the code.

This may be illustrated as follows. Consider encoding the angular range 0° to 360° in 1° increments. Per the general encoding rules above, the positive semicircle will cover the range 0° to 179° (one least significant bit less than full range). All the bits of the code will be "zeros" for 0° and "ones" for 179°, and the sign/status matrix will indicate the positive sign. The negative semicircle will cover the range 180° to 359°. All the bits will be "zeros" for 180°. The codes for angles between 181° to 359° will be determined by taking the two's complements of the fractional binary series for the result of subtracting each value from 360. Thus, the code for 181° is the two's complement of the code for 179°. Throughout the negative semicircle, which includes 180°, the sign/status matrix contains the negative sign.

2.5 Timing-Related Elements

This section describes the digital data transfer system elements considered to be principally related to the timing aspects of the signal circuit.

2.5.1 Bit Rate

2.5.1.1 High-Speed Operation

The bit rate for high-speed operation of the system is 100 kilobits per second (kbps) ±1%.

2.5.1.2 Low-Speed Operation

The bit rate for low-speed operation of the system is within the range 12.0 to 14.5 kbps. The selected rate is maintained within 1%.

2.5.2 Information Rates

The minimum and maximum transmit intervals for each item of information are specific by ARINC Specification 429. Words with like labels but with different SDI codes are treated as unique items of information. Each and every unique item of information is transmitted once during an interval bounded in length by the specified minimum and maximum values. Stated another way, a data word having the same label and four different SDI codes will appear on the bus four times (once for each SDI code) during that time interval.

Discrete bits contained within data words are transferred at the bit rate and repeated at the update rate of the primary data. Words dedicated to discretes should be repeated continuously at specified rates.

2.5.3 Clocking Method

Clocking is inherent in the data transmission. The identification of the bit interval is related to the initiation of either a HI or LO state from a previous NULL state in a bipolar RZ code.

2.5.4 Word Synchronization

The digital word should be synchronized by reference to a gap of four bit times (minimum) between the periods of word transmissions. The beginning of the first transmitted bit following this gap signifies the beginning of the new word.

2.5.5 Timing Tolerances

The waveform timing tolerances are shown below:

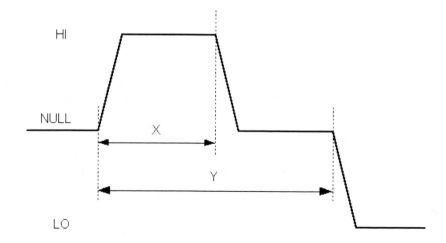

Parameter	High-Speed Operation	Low-Speed Operation
Bit Rate	100 kbps ± 1%	12–14.5 kbps
Time Y	10 μsec ± 2.5%	Z^a μsec ± 2.5%
Time X	5 μsec ± 5%	Y/2 ± 5%
Pulse Rise Time	1.5 ± 0.5 μsec	10 ± 5 μsec
Pulse Fall Time	1.5 ± 0.5 μsec	10 ± 5 μsec

Note: Pulse rise and fall times are measured between the 10% and 90% voltage amplitude points on the leading and trailing edges of the pulse and include time skew between the transmitter output voltages A-to-ground and B-to-ground.
^{a}Z = 1/R where R = bit rate selected from 12–14.5 kbps range.

2.6 Communications Protocols

2.6.1 Development of File Data Transfer

ARINC Specification 429 was adopted by AEEC in July 1977. Specification 429 defined a broadcast data bus. General provisions were made for file data transfer. In October 1989, AEEC updated a file data transfer procedure with a more comprehensive process that will support the transfer of both bit- and character-oriented data. The new protocol became known popularly as the "Williamsburg Protocol."

2.6.1.1 File Data Transfer Techniques

This "File Data Transfer Techniques" specification describes a system in which an LRU may generate binary extended length messages "on demand." Data is sent in the form of Link Data Units (LDU) organized in 8-bit octets. System Address Labels (SAL) are used to identify the recipient. Two data bus speeds are supported.

2.6.1.2 Data Transfer

The same principles of the physical layer implementation apply to file data transfer. Any avionics system element having information to transmit does so from a designated output port over a single twisted and shielded pair of wires to all other system elements having need of that information. Unlike the simple broadcast protocol that can deliver data to multiple recipients in a single transmission, the file transfer technique can be used only for point-to-point message delivery.

2.6.1.3 Broadcast Data

The broadcast transmission technique described above can be supported concurrently with file data transfer.

2.6.1.4 Transmission Order

The most significant octet of the file and least significant bit (LSB) of each octet should be transmitted first. The label is transmitted ahead of the data in each case. It may be noted that the Label field is encoded in reverse order, i.e., the least significant bit of the word is the most significant bit of the label. This "reversed label" characteristic is a legacy from past systems in which the octal coding of the label field was, apparently, of no significance.

2.6.1.5 Data Bit Encoding Logic

A HI state after the beginning of the bit interval returning to a NULL state before the end of the same bit interval signifies a logic "one." A LO state after the beginning of the bit interval returning to a NULL state before the end of the same bit interval signifies a logic "zero."

2.6.1.6 Bit-Oriented Protocol Determination

An LRU will require logic to determine which protocol (character- or bit-oriented) and which version to use when prior knowledge is not available.

2.6.2 Bit-Oriented Communications Protocol

This subsection describes Version 1 of the bit-oriented (Williamsburg) protocol and message exchange procedures for file data transfer between units desiring to exchange bit-oriented data assembled in data files. The bit-oriented protocol is designed to accommodate data transfer between sending and receiving units in a form compatible with the Open Systems Interconnect (OSI) model developed by the International Standards Organization (ISO). This document directs itself to an implementation of the Link layer, however, an overview of the first four layers (Physical, Link, Network, and Transport) is provided.

Communications will permit the intermixing of bit-oriented file transfer data words (which contain System Address Labels [SALs]) with conventional data words (which contain label codes). If the sink should receive a conventional data word during the process of accepting a bit-oriented file transfer message, the sink should accept the conventional data word and resume processing of the incoming file transfer message.

The data file and associated protocol control information are encoded into 32-bit words and transmitted over the physical interface. At the Link layer, data are transferred using a transparent bit-oriented data file transfer protocol designed to permit the units involved to send and receive information in multiple word frames. It is structured to allow the transmission of any binary data organized into a data file composed of octets.

1. *Physical Medium.* The physical interface is described above.
2. *Physical Layer.* The Physical layer provides the functions necessary to activate, maintain, and release the physical link which will carry the bit stream of the communication. The electrical interface, voltage, and timing, described above, is used by the interfacing units. Data words will contain 32 bits; bits 1 through 8 will contain the System Address Label (SAL) and bit 32 will be the parity (odd) bit.
3. *Link Layer.* The Link layer is responsible for transferring information from one logical network entity to another and for enunciating any errors encountered during transmission. The Link layer provides a highly reliable virtual channel and some flow control mechanisms.
4. *Network Layer.* It is the responsibility of the Network layer to ensure that data packets are properly routed between any two terminals. The Network layer performs a number of functions. The Network layer expects the Link layer to supply data from correctly received frames. The Network layer provides for the decoding of information up to the packet level to determine which node (unit) the message should be transferred to. To obtain interoperability, this process, though simple in this application, must be reproduced using the same set of rules throughout all the communications networks (and their subnetworks) on-board the aircraft and on the ground. The bit-oriented data link protocol was designed to operate in a bit-oriented Network layer environment. Specifically, ISO 8208 would typically be selected for the Subnetwork layer protocol for air/ground subnetworks. There are, however, some applications where the bit-oriented file transfer protocol will be used under other Network layer protocols.
5. *Transport Layer.* The Transport layer controls the transportation of data between a source end-system to a destination end-system. It provides "network independent" data delivery between these processing end-systems. It is the highest order of function involved in moving data between systems. It relieves higher layers from any concern with the pure transportation of information between them.

2.6.2.1 Link Data Units (LDU)

A Link Data Unit (LDU) contains binary encoded octets. The octets may be set to any possible binary value. The LDU may represent raw data, character data, bit-oriented messages, character-oriented messages, or any string of bits desired. The only restriction is that the bits be organized into full 8-bit octets. The interpretation of those bits is not a part of this Link layer protocol. The LDUs are assembled to make up a data file.

LDUs consist of a set of contiguous ARINC 429 32-bit data words, each containing the System Address Label (see Section 2.6.2.3) of the sink. The initial data word of each LDU is a Start of Transmission

(SOT). The data described above are contained within the data words which follow. The LDU is concluded with an End of Transmission (EOT) data word. No data file should exceed 255 LDUs.

Within the context of this document, LDUs correspond to frames and files correspond to packets.

2.6.2.2 Link Data Unit (LDU) Size and Word Count

The Link Data Unit (LDU) may vary in size from 3 to 255 ARINC 429 words including the SOT and EOT words. When a LDU is organized for transmission, the total number of ARINC 429 words to be sent (word count) is calculated. The word count is the sum of the SOT word, the data words in the LDU, and the EOT word.

In order to obtain maximum system efficiency, the data is typically encoded into the minimum number of LDUs.

The word count field is 8 bits in length. Thus the maximum number of ARINC 429 words that can be counted in this field is 255. The word count field appears in the RTS and CTS data words. The number of LDUs needed to transfer a specific data file will depend upon the method used to encode the data words.

2.6.2.3 System Address Labels (SALs)

LDUs are sent point-to-point, even though other systems may be connected and listening to the output of a transmitting system. In order to identify the intended recipient of a transmission, the Label field (bits 1–8) is used to carry a System Address Label (SAL). Each on-board system is assigned a SAL. When a system sends an LDU to another system, the sending system (the "source") addresses each ARINC 429 word to the receiving system (the "sink") by setting the Label field to the SAL of the sink. When a system receives any data containing its SAL that is not sent through the established conventions of this protocol, the data received are ignored.

In the data transparent protocol, data files are identified by content rather than by ARINC 429 label. Thus, the label field loses the function of parameter identification available in broadcast communications.

2.6.2.4 Bit Rate and Word Timing

Data transfer may operate at either high speed or low speed. The source introduces a gap between the end of each ARINC 429 word transmitted and the beginning of the next. The gap should be 4 bit times (minimum). The sink should be capable of receiving the LDU with the minimum word gap of 4 bit times between words. The source should not exceed a maximum average of 64 bit times between data words of an LDU.

The maximum average word gap is intended to compel the source to transmit successive data words of an LDU without excessive delay. This provision prevents a source that is transmitting a short message from using the full available LDU transfer time. The primary value of this provision is realized when assessing a maximum LDU transfer time for short fixed-length LDUs, such as for Automatic Dependence Surveillance (ADS).

If a Williamsburg source device were to synchronously transmit long length or full LDUs over a single ARINC 429 data bus to several sink devices, the source may not be able to transmit the data words for a given LDU at a rate fast enough to satisfy this requirement because of other bus activity. In aircraft operation, given the asynchronous burst mode nature of Williamsburg LDU transmissions, it is extremely unlikely that a Williamsburg source would synchronously begin sending a long length or full LDU to more than two Williamsburg sink devices. A failure to meet this requirement will either result in a successful (but slower) LDU transfer, or an LDU retransmission due to an LDU transfer time-out.

2.6.2.5 Word Type

The Word Type field occupies bit 31–29 in all bit-oriented LDU words. The Word Type field is used to identify the function of each ARINC 429 data word used by the bit-oriented communication protocol.

2.6.2.6 Protocol Words

The protocol words are identified with a Word Type field of "100" and are used to control the file transfer process.

2.6.2.6.1 *Protocol Identifier*

The protocol identifier field occupies bits 28–25 of the protocol word and identifies the type of protocol word being transmitted. Protocol words with an invalid protocol identifier field are ignored.

2.6.2.6.2 *Destination Code*

Some protocol words contain a Destination Code. The Destination Code field (bits 24–17) indicates the final destination of the LDU. If the LDU is intended for the use of the system receiving the message, the destination code may be set to null (hex 00). However, if the LDU is a message intended to be passed on to another on-board system, the Destination Code will indicate the system to which the message is to be passed. The Destination Codes are assigned according to the applications involved. The codes are used in the Destination Code field to indicate the address of the final destination of the LDU.

In an OSI environment, the Link layer protocol is not responsible for validating the destination code. It is the responsibility of the higher-level entities to detect invalid destination codes and to initiate error logging and recovery.

Within the pre-OSI environment, the Destination Code provides Network layer information. In the OSI environment, this field may contain the same information for routing purposes between OSI and non-OSI systems.

2.6.2.6.3 *Word Count*

Some protocol words contain a Word Count field. The Word Count field (bits 16–9) reflects the number of ARINC 429 words to be transmitted in the subsequent LDU. The maximum word count value is 255 ARINC 429 words and the minimum word count value is 3 ARINC 429 words. A LDU with the minimum word count value of 3 ARINC 429 words would contain a SOT word, one data word, and an EOT word. A LDU with the maximum word count value of 255 ARINC 429 words would contain a SOT word, 253 data words, and an EOT word.

2.7 Applications

2.7.1 Initial Implementation

ARINC 429 was first used in the early 1980s on the Airbus A-310 and Boeing B-757 and B-767 airplanes. Virtually all data transfer on these airplanes was accommodated by approximately 150 separate buses interconnecting computers, radios, displays, controls, and sensors. Most of these buses operate at the lower speed. The few that operate at the higher speed of 100 kbps are typically connected to critical navigation computers.

2.7.2 Evolution of Controls

The first applications of ARINC 429 for controlling devices were based on the federated avionics approach used on airplanes which comprised mostly analog interfaces. Controllers for tuning communications equipment used an approach defined as two-out-of-five tuning. Each digit of the desired radio frequency was encoded on each set of five wires. Multiple digits dictated the need for multiple sets of wires for each radio receiver.

The introduction of ARINC 429 proved to be a major step toward reduction of wires. A tuning unit needed only one ARINC 429 bus to tune multiple radios of the same type. An entire set of radios and navigation receivers could be tuned with a few control panels, using approximately the same number of wires previously required to tune a single radio.

As cockpit space became more critical, the need to reduce the number of control panels became critical. The industry recognized that a single control panel, properly configured, could replace most of the existing control panels. The Multi-Purpose Control/Display Unit (MCDU) emanated from the industry effort. The MCDU was derived essentially from the control and display approach used by the rather sophisticated controller for the Flight Management System. For all intents and purposes, the MCDU became the cockpit controller.

A special protocol had to be developed for ARINC 429 to accommodate the capability of addressing different units connected to a single ARINC 429 bus from the MCDU. The protocol employed two-way communications using two pairs of wires between the controlling unit and the controlled device. An addressing scheme provided for selective communications between the controlling unit and any one of the controlled units. Only one output bus from the controller is required to communicate addresses and commands to the receiving units. With the basic ARINC 429 design, up to 20 controlled units could be connected to the output of the controller. Each of the controlled units is addressed by an assigned SAL.

2.7.3   Longevity of ARINC 429

New airplane designs in the 21st century continue to employ the ARINC 429 bus for data transmission. The relative simplicity and integrity of the bus, as well as the ease of certification are characteristics that contribute to the continued selection of the ARINC 429 bus when the required data bandwidth is not critical. The ARINC 629 data bus developed as the replacement for ARINC 429 is used in applications where a large amount of data must be transferred or where many sources and sinks are required on a single bus.

2.8   ARINC 453

ARINC Project Paper 453 was developed by the Airlines Electronic Engineering Committee (AEEC) in response to an anticipated requirement for data transfer rates higher than achievable with ARINC 429. The original drafts of Project Paper 453 were based on techniques already employed at that time. The electrical characteristics, including the physical medium, voltage thresholds, and modulation techniques were based on Mil-Std 1553. The data protocols and formats were based on those used in ARINC Specification 429.

During the preparation of the drafts of Project Paper 453, the Boeing Company petitioned AEEC to consider the use of the Digital Autonomous Terminal Access Communications (DATAC) Bus developed by the Boeing Company to accommodate higher data throughput. AEEC accepted Boeing's recommendation for the alternative. ARINC 629 was based on the original version of the Boeing DATAC Bus. The work on Project 453 was then curtailed. The latest draft of Project Paper 453 is maintained by ARINC for reference purposes only.

3

Commercial Standard Digital Bus

Lee H. Harrison
Galaxy Scientific Corp.

3.1 Introduction

The Commercial Standard Digital Bus (CSDB) is one of three digital serial integration data buses that currently predominate in civilian aircraft. The CSBD finds its primary implementations in the smaller business and private General Aviation (GA) aircraft, but has also been used in retrofits of some commercial transport aircraft.

CSDB, a unidirectional data bus, was developed by the Collins General Aviation Division of Rockwell International. The bus used in a particular aircraft is determined by which company the airframe manufacturer chooses to supply the avionics. Collins is one of only a handful of major contributors to avionics today.

CSDB is an **asynchronous linear** broadcast bus, specifying the use of a twisted, shielded pair cable for device interconnection. Two bus speeds are defined in the CSDB specification. A low-speed bus operates at 12,500 bits per second (bps) and a high-speed bus operates at 50,000 bps. The bus uses twisted, unterminated, shielded pair cable and has been tested to lengths of 50 m.

The CSDB standard also defines other physical characteristics such as modulation technique, voltage levels, load capacitance, and signal rise and fall times. Fault protection for short-circuits of the bus conductors to both 28 VDC and 115 VAC is defined by the standard.

3.2 Bus Architecture

Only one transmitter can be attached to the bus, while it can accommodate up to ten receivers. Figure 3.1 illustrates the unidirectional bus architecture.

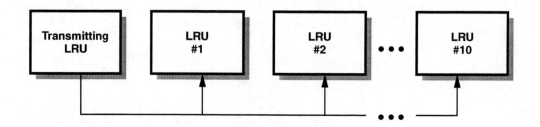

FIGURE 3.1   Unidirectional CSDB communication.

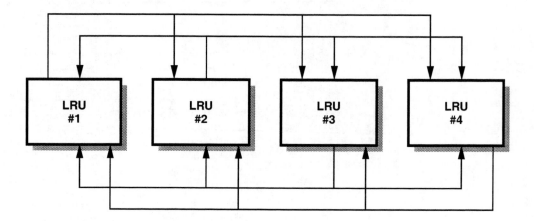

FIGURE 3.2   Bidirectional CSDB communication.

Bidirectional transmission can take place between two bus users. If a receiving bus user is required to send data to any other bus user, a separate bus must be used. Figure 3.2 shows how CSDB may implement bidirectional transmissions between bus users. It can be seen that if each bus user is required to communicate with every other bus user, a significantly greater amount of cabling would be required. In general, total interconnectivity has not been a requirement for CSDB-linked bus users.

It is possible to interface CSDB to other data buses. When this is done, a device known as a **gateway** interfaces to CSDB and the other bus. If the other bus is ARINC 429 compliant, then messages directed through the gateway from CSDB are converted to the ARINC 429 protocol (see Chapter 2), and vice versa. The gateway would handle bus timing, error checking, testing, and other necessary functions. The system designers would ensure that data latency introduced by the gateway would not cause a "stale data" problem, resulting in a degradation of system performance. Data are stale when they do not arrive at the destination line replaceable unit (LRU) when required, as specified in the design.

3.3   Basic Bus Operation

In Section 2.1.4 of the CSDB standard, three types of transmission are defined:

- Continuous repetition,
- Noncontinuous repetition, and
- "Burst" transmissions

Continuous repetition transmission refers to the periodic updates of certain bus messages. Some messages on CSDB are transmitted at a greater repetition rate than others. The CSDB standard lists these

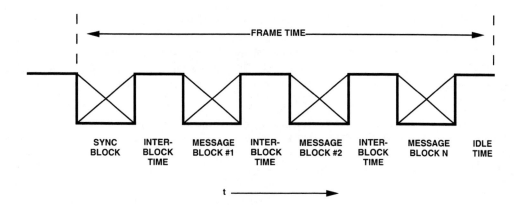

FIGURE 3.3 CSDB data frame structure.

update rates, along with the message address and message block description. Noncontinuous repetition is used for parameters that are not always valid, or available, such as mode or test data. When noncontinuous repetition transmission is in use, it operates the same as continuous repetition. Burst transmission initiates an action (such as radio tuning), or may be used to announce a specific event. Operation in this mode initiates 16 repetitions of the action in each of 16 successive **frames**, using an update rate of 20 per second.

For CSDB, bytes consist of 11 bits: a start bit, 8 data bits, a **parity** bit, and a stop bit. The least significant bit (bit 0) follows the start bit. The CSDB standard defines the message block as "a single serial message consisting of a fixed number of bytes transmitted in a fixed sequence." Essentially, a message block consists of a number of bytes concatenated together, with the first byte always being an address byte. A status byte may or may not be included in the message block. When it is, it immediately follows the address byte. The number of data bytes in a message block vary.

Data are sent as frames consisting of a **synchronization block** followed by a number of message blocks. A particular frame is defined from the start of one synchronization block to the start of the next synchronization block. A "sync" block consists of N bytes of the sync character, which is defined as the hexadecimal character "A5." The sync character is never used as an address. While the data may contain a sync character, it may occur in the data a maximum of N − 1 times. Frames consist of message blocks, preceded by a sync block. The start of one sync block to the start of the next sync block is one frame time. Figure 3.3 shows what transpires during a typical frame time.

3.4 CSDB Bus Capacity

The CSDB is similar to the ARINC 429 data bus in that it is an asynchronous broadcast bus and operates using character-oriented protocol. Data are sent as frames consisting of a synchronization block followed by a number of message blocks. A particular frame is defined from the start of one synchronization block to the start of the next synchronization block. A message block contains an address byte, a status byte, and a variable number of data bytes. The typical byte consists of one start bit, eight data bits, a parity bit, and a stop bit.

The theoretical bus data rate for a data bus operating at 50,000 bps with an 11-bit data byte, is 4545 bytes per second. For CSDB, the update rate is reduced by the address byte and synchronization block overhead required by the standard.

The CSDB Interblock and Interbyte times also reduce bus throughput. According to the specification, there are no restrictions on these idle times for the data bus. These values, however, are restrained by the defined update rate chosen by the designer. If the update rate needs to be faster, the Interblock time and the Interbyte time can be reduced as required, until bus utilization reaches a maximum.

3.5   CSDB Error Detection and Correction

Two methods of error detection are referenced in the standard. They are the use of parity and **checksums**. A parity bit is appended after each byte of data in a CSDB transmission. The "burst" transmission makes use of the checksum for error detection. As the General Aviation Manufacturers Association (GAMA) specification states:

> It is expected that the receiving unit will accept as a valid message the first message block which contains a verifiable checksum. (GAMA CSDB 1986.)

3.6   Bus User Monitoring

Although many parameters are defined in the CSDB specification, there is no suggestion that they be monitored by receivers. The bus frame, consisting of the synchronization block and message block, may be checked for proper format and content. A typical byte, consisting of start, stop, data, and parity bits, may be checked for proper format.

The bus hardware should include the functional capability to monitor these parameters. Parity, frame errors, and buffer overrun errors are typically monitored in the byte format of character-oriented protocols. The message format can be checked and verified by the processor if the hardware does not perform these checks.

3.7   Integration Considerations

The obvious use of a data bus is for integrating various LRUs that need to share data or other resources. In the following sections, integration considerations for CSDB are examined at various levels. These include physical, logical, software, and functional considerations.

3.7.1   Physical Integration

The physical integration of LRUs connected to the CSDB is addressed by the standardization of the bus medium and connectors. These must conform to the Electronic Industries Association (EIA) Recommended Standard (RS)-422-A (1978), "Electrical Characteristics of Balanced Voltage Digital Interface Circuits." The CSDB standard provides for the integration of up to 10 receivers on a single bus, which can be up to 50 m in length. No further constraints or guidelines on the physical layout of the bus are given.

Each LRU connected to a CSDB must satisfy the electrical signals and bit timing that are specified in the EIA RS-422-A. Physical characteristics of the CSDB are given in Table 3.1. The non-return to zero (NRZ) data format used by CSDB LRUs is shown in Figure 3.4. NRZ codes remain constant throughout a bit interval and either use absolute values of the signal elements or differential encoding where the polarity of adjacent elements is compared to determine the bit value.

TABLE 3.1   CSDB Physical Characteristics

Modulation Technique	Non-Return to Zero (NRZ)
Logic Sense for Logic "0"	Line B Positive with Respect to Line A
Logic Sense for Logic "1"	Line A Positive with Respect to Line B
Bus Receiver	High Impedance, Differential Input
Bus Transmitter	Differential Line Driver
Bus Signal Rates	Low Speed: 12,500 bps
	High Speed: 50,000 bps
Signal Rise-Time and Fall-Time	Low Speed: 8 μs
	High-Speed: 0.8–1.0 μs
Receiver Capacitance Loading	Typical: 600 pF
	Maximum: 1,200 pF
Transmitter Driver Capability	Maximum: 12,000 pF

Source: Commercial Standard Digital Bus, 8th ed., Collins General Aviation Division, Rockwell International Corporation, Cedar Rapids, IA, January 30, 1991.

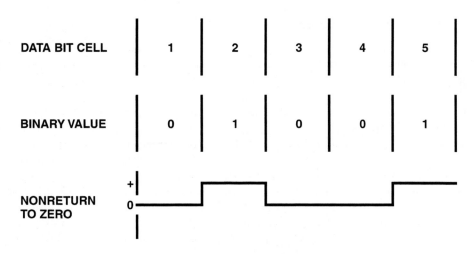

FIGURE 3.4 Non-Return to Zero data example.

Typical circuit designs for transmitter and receiver interfaces are given in the CSDB standard. Protection against short-circuits is also specified for receivers and transmitters. Receiver designs should include protection against bus conductor shorts to 28 VDC and to 115 VAC. Transmitter designs should afford protection against faults propagating to other circuits of the LRU in which the transmitter is located.

To ensure successful integration of CSDB LRUs, and avoid potential future integration problems, the electrical load specification must be applied to a fully integrated system, even if the initial design does not include a full complement of receivers. As a result, additional receivers can be integrated at a later time without violating the electrical characteristics of the bus.

3.7.2 Logical Integration

The logical integration of the hardware is controlled by the CSDB standard, which establishes the bit patterns that initiate a message block, and the start bit, data bits, parity bit, and stop bit pattern that comprises each byte of the message. The system designer, however, must control the number of bytes in each message and ensure that all the messages on a particular bus are of the same length.

3.7.3 Software Integration

Many software integration tasks are left to the system designer for implementation. Hence, CSDB does not fully specify software integration. The standard is very thorough in defining the authorized messages and in constraining their signaling rate and update rate. The synchronization message that begins a new frame of messages is also specified. However, the determination of which messages are sent within a frame for a particular bus is unspecified. Also, there are no guidelines given for choosing the message sequence or frame loading. The frame design is left to the system designer.

In general, the sequencing of the messages does not present an integration problem since receivers recognize messages by the message address, not by the sequence. However, this standard does not disallow an LRU from depending on the message sequence for some other purpose. The system designer must be aware of whether any LRU is depending on the sequence for something other than message recognition since once the sequence is chosen, it is fixed for every frame.

The bus frame loading is more crucial. There are three types of messages that can occur within a frame: continuous repetition, noncontinuous repetition, and burst transmissions. The system designer must specify which type of transmission to use for each message and ensure that the worst maximum coincidence of the three types within one frame does not exhaust the frame time. The tables of data needed to support this system design are provided, but the system designer must ensure that no parts of the CSDB standard are violated.

3.7.4 Functional Integration

The CSDB standard provides much of the data needed for functional integration. The detailed message block definitions give the interpretation of the address, status byte, and data words for each available message. Given that a particular message is broadcast, the standard completely defines the proper interpretation of the message. The standard even provides a system definition, consisting of a suite of predefined buses which satisfy the integration needs of a typical avionics system.

If this predefined system is applicable, most of the system integration questions are already answered. But if there is any variation from the standard, the designer of a subsystem in a CSDB integrated system must inquire to find out which LRUs are generating the messages that the subsystem needs, on which bus each message is transmitted, at what bus speed the messages are transmitted, and the type of transmission. The designer must also ensure that the subsystem provides the messages required by other LRUs. The system designer needs to coordinate this information accurately and comprehensively. The system design must ensure that all the messages on a particular bus are of the same length. It must also control the data latencies that may result as data are passed from bus to bus by various LRUs. All testing is left to the system designer.

There are no additional guidelines published for the CSDB. Whatever problems are unaddressed by the standard are addressed by Collins during system integration. Furthermore, Collins has not found the need to formalize their integration and testing in internal documents since this work is done by CSDB-experienced engineers.

3.8 Bus Integration Guidelines

The CSDB, like the ARINC 429 bus, has only one LRU that is capable of transmitting with (usually) multiple LRUs receiving the transmission. Thus, the CSDB has few inherent subsystem integration problems. However, the standard does not address them. The preface to the CSDB standard clearly states its position concerning systems integration:

> This specification pertains only to the implementation of CSDB as used in an integrated system. Overall systems design, integration, and certification remain the responsibility of the systems integrator. (GAMA CSDB 1986.)

Although this appears to be a problem for the reliability of CSDB-integrated systems, the GA scenario is quite different from the air transport market. The ARINC standards are written to allow any manufacturer to independently produce a compatible LRU. In contrast, the General Aviation Manufacturers Association standard states the following in the preface:

> This specification ... is intended to provide the reader with a basic understanding of the data bus and its usage. (CSDB 1986.)

The systems integrator for all CSDB installations is the Collins General Aviation Division of Rockwell International. That which is not published in the standard is still standardized and controlled because the CSDB is a sole source item.

Deviations from the standard are allowed, however, for cases where there will be no further interfaces to other subsystem elements. When variations are made, the change must first be approved in a formal design review and the product specification is then updated accordingly. Integration standards and guidelines for CSDB include the CSDB standard and EIA RS-422-A by the Electronic Industries Association.

3.9 Bus Testing

The CSDB connects avionic LRUs point-to-point to provide an asynchronous broadcast method of transmission. Before the bus was used in the avionic environment, it was put through validation tests similar to those used on other commercial data buses. These included the environmental tests presented in RTCA DO-160 and failure analyses. Most environmental tests were done transparently on the bus after it was installed in an aircraft.

As with other avionic data buses, Rockwell's Collins Division had to develop external tests to show that the bus satisfied specifications in the standard. Test procedures of this nature are not included in the CSDB standard.

Internal bus tests that the CSDB standard describes include a checksum test and a parity check. Both of these are used to ensure the integrity of the bus's data. Care should be taken when using these tests because their characteristics do not allow them to be used in systems of all criticality levels.

Simulation is used for development and testing of LRUs with a CSDB interface. Manufacturers make black box testers that are used to simulate an LRU connection to the bus. They are made to generate and evaluate messages according to the electrical and logical standards for the bus. They consist of a general purpose computer connected to bus interface cards. The simplest ones may simulate a single LRU transmitting or receiving. The more complex ones may be able to simulate multiple LRUs simultaneously.

These are not the only external and internal tests that the CSDB manufacturer can perform. Many more characteristics which may require testing are presented in the CSDB specification. Again, it remains the manufacturer's responsibility to prove that exhaustive validation testing of the bus and its related equipment has met all the requirements of the Federal Aviation Regulations.

3.10 Aircraft Implementations

This section gives a sampling of the aircraft in which the CSDB is installed. Table 3.2 lists some of the commercial transport aircraft and regional airliners using CSDB. CSDB is used both in retrofit installations and as the main integration bus. Additionally, a number of rotorcraft use the CSDB to communicate between the Collins-supplied LRUs.

TABLE 3.2   Aircraft and Their Use of the CSDB

Boeing 727	Retrofit
Boeing 737	Retrofit
McDonnell-Douglas DC-8	Retrofit
Saab 340, Saab 2000	Primary Integration Bus
Embraer	Primary Integration Bus
Short Brothers SD330 and SD360	Primary Integration Bus
ATR42 and ATR72	Primary Integration Bus
De Haviland Dash 8	Primary Integration Bus
Canadair Regional	Primary Integration Bus

Source: Collins Division of Rockwell International, Cedar Rapids, IA.

Defining Terms

Asynchronous: Operating at a speed determined by the circuit functions rather than by timing signals.

Checksum: An error detection code produced by performing a binary addition, without carry, of all the words in a message.

Frame: A formatted block of data words or bits used to construct messages.

Gateway: A bus user that is connected to more than one bus for the purpose of transferring bus messages from one bus to another, where the buses do not follow the same protocol.

Linear Bus: A bus where users are connected to the bus medium, one on each end, with the rest connected in-between.

Parity: An error detection method that adds a bit to a data word based on whether the number of "one" bits is even or odd.

Synchronization Block: A special bus pattern, consisting of a certain number of concatenated "sync byte" data words, used to signal the start of a new frame.

References

1. GAMA, *Commercial Standard Digital Bus (CSDB)*, General Aviation Manufacturers Association, Washington, D.C., June 10, 1986.
2. Eldredge, D. and E. F. Hitt, "Digital System Bus Integrity," DOT/FAA/CT-86/44, Federal Aviation Administration Technical Center, Atlantic City International Airport, NJ, March 1987.
3. Elwell, D., L. Harrison, J. Hensyl, and N. VanSuetendael, "Avionic Data Bus Integration Technology," DOT/FAA/CT-91-19, Federal Aviation Administration Technical Center, Atlantic City International Airport, NJ, December 1991.
4. Collins, "Serial Digital Bus Specification," Part No 523-0772774, Collins General Aviation Division/ Publications Dept, 1100 West Hibiscus Blvd., Melbourne, FL 32901.

Further Information

The most detailed information available for CSDB is the GAMA CSDB Standard, Part Number 523-0772774. It is available from the Collins Division of Rockwell International, Cedar Rapids, IA.

Bibliography

ARINC Specification 600-7, "Air Transport Avionics Equipment Interfaces," Aeronautical Radio, Inc., Annapolis, MD, January 1987.

ARINC Specification 607, "Design Guidance for Avionic Equipment," Aeronautical Radio, Inc., Annapolis, MD, February 17, 1986.

ARINC Specification 607, "Design Guidance for Avionic Equipment," Supplement 1, Aeronautical Radio, Inc., Annapolis, MD, July 22, 1987.

ARINC Specification 617, "Guidance for Avionic Certification and Configuration Control," Draft 4, Aeronautical Radio, Inc., Annapolis, MD, December 12, 1990.

Card, M. Ace, "Evolution of the Digital Avionic Bus," *Proceedings of the IEEE/AIAA 5th Digital Avionics Systems Conference*, Institute of Electrical and Electronics Engineers, New York, NY, 1983.

Eldredge, Donald and Ellis F. Hitt, "Digital System Bus Integrity," DOT/FAA/CT-86/44, U.S. Department of Transportation, Federal Aviation Administration, March 1987.

Eldredge, Donald and Susan Mangold, "Digital Data Buses for Aviation Applications," *Digital Systems Validation Handbook, Volume II*, Chapter 6, DOT/FAA/CT-88/10, U.S. Department of Transportation, Federal Aviation Administration, February 1989.

GAMA, "Commercial Standard Digital Bus (CSDB)," General Aviation Manufacturers Association, Washington, DC, June 10, 1986.

Hubacek, Phil, "The Advanced Avionics Standard Communications Bus," Business and Commuter Aviation Systems Division, Honeywell, Inc., Phoenix, Arizona, July 10, 1990.

Jennings, Randle G., "Avionics Standard Communications Bus - Its Implementation And Usage," *Proceedings of the IEEE/AIAA 7th Digital Avionics Systems Conference*, Institute of Electrical and Electronics Engineers, New York, NY, 1986.

RS-422-A, "Electrical Characteristics of Balanced Voltage Digital Interface Circuits," Electronic Industries Association, Washington, DC, December 1978.

RTCA DO-160C, "Environmental Conditions and Test Procedures for Airborne Equipment," Radio RTCA DO Technical Commission for Aeronautics, Washington, DC, December 1989.

RTCA DO-178A, "Software Considerations in Airborne Systems and Equipment Certification," Radio RTCA DO Technical Commission for Aeronautics, Washington, DC, March 1985.

Runo, Steven C., "Gulfstream IV Flight Management System," *Proceedings of the 1990 AIAA/FAA Joint Symposium on General Aviation Systems*, DOT/FAA/CT-90/11, U.S. Department of Transportation, Federal Aviation Administration, May 1990.

Spitzer, Cary R., "Digital Avionics Architectures — Design and Assessment," *Tutorial of the IEEE/AIAA 7th Digital Avionics Systems Conference,* Institute of Electrical and Electronics Engineers, New York, NY, 1986.

Spitzer, Cary R., *Digital Avionics Systems,* Prentice Hall, Englewood Cliffs, NJ, 1987.

Thomas, Ronald E., "A Standardized Digital Bus For Business Aviation," *Proceedings of the IEEE/AIAA 5th Digital Avionics Systems Conference,* Institute of Electrical and Electronics Engineers, New York, NY, 1983.

"WD193X Synchronous Data Link Controller," Western Digital Corporation, Irvine, CA, 1983.

"WD1931/WD1933 Compatibility Application Notes," Western Digital Corporation, Irvine, CA, 1983.

"WD1993 ARINC 429 Receiver/Transmitter and Multi-Character Receiver/Transmiter," Western Digital Corporation, Irvine, CA, 1983.

4

Head-Up Displays

Robert B. Wood
Rockwell Collins Flight Dynamics

Peter J. Howells
Rockwell Collins Flight Dynamics

4.1 Introduction

During early military Head-Up Display (HUD) development, it was found that pilots using HUDs could operate their aircraft with greater precision and accuracy than they could with conventional flight instrument systems.[1,2] This realization eventually led to the development of the first HUD systems intended specifically to aid the pilot during commercial landing operations. This was first accomplished by Sextant Avionique for the Dassault Mercure aircraft in 1975, and then by Sundstrand and Douglas Aircraft Company for the MD80 series aircraft in the late 1970s (see Figure 4.1).

In the early 1980s, Flight Dynamics developed a holographic optical system to display an inertially derived aircraft flight path along with precision guidance, thus providing the first wide field-of-view (FOV) head-up guidance system. Subsequently, Alaska Airlines became the first airline to adopt this technology and perform routine fleet-wide manually flown CAT IIIa operations on B-727-100/200 aircraft using the Flight Dynamics system (see Figure 4.2). Once low-visibility operations were successfully demonstrated using a HUD in lieu of a fail passive autoland system, regional airlines opted for this technology to help maintain their schedules when the weather fell below CAT II minimums, and to help improve situational awareness.

By the end of the century, many airlines had installed head-up guidance systems, and thousands of pilots were fully trained in their use. HUD-equipped aircraft had logged more than 6,000,000 flight hours and completed over 30,000 low-visibility operations. HUDs are now well-established additions to aircraft cockpits, providing both additional operational capabilities and enhanced situational awareness, resulting in improved aircraft safety.

4.2 HUD Fundamentals

All head-up displays require an image source, generally a high-brightness cathode-ray tube, and an optical system to project the image source information at optical infinity. The HUD image is viewed by the pilot after reflecting from a semitransparent element referred to as the HUD combiner. The combiner is

FIGURE 4.1   Early commercial HUD.[3]

FIGURE 4.2   Commercial manually flown CAT IIIa HUD installed in a B-737-800.

located between the pilot's eyes and the aircraft windshield and is angled to reflect image-source light rays to the pilot for viewing. Special coatings on the combiner simultaneously reflect the HUD information, and transmit the real-world scene, enabling the pilot to view both the outside world and the collimated display.

Head-up display systems are comprised of two major subsystems: the pilot display unit (PDU), and the HUD processor or HUD computer. The PDU interfaces electrically and mechanically with the aircraft structure and provides the optical interface to the pilot. The HUD processor interfaces electronically with aircraft sensors and systems, runs a variety of algorithms related to data verification and formatting, and generates the characters and symbols making up the display. Modern HUD processors are capable of generating high-integrity guidance commands and cues for precision low-visibility take-off, approach, landing (flare), and rollout. The interface between the HUD processor and the PDU can be either a serial digital display list or analog X and Y deflection and Z-axis video bright-up signals for controlling the display luminance.

The PDU is located within the cockpit to allow a pilot positioned at the cockpit Design Eye Position (DEP) to view HUD information which is precisely positioned with respect to the outside world. This allows, for example, the computer-generated and displayed horizon line to overlay the real-world horizon in all phases of flight.

The cockpit DEP is defined as the optimum cockpit location that meets the requirements of FAR 25.773[4] and 25.777.[5] From this location the pilot can easily view all relevant head-down instruments and the outside world scene through the aircraft windshield, while being able to access all required cockpit controls. The HUD "eyebox," is always positioned with respect to the cockpit DEP, allowing pilots to fly the aircraft using the HUD from the same physical location as a non-HUD-equipped aircraft would be flown.

4.2.1 Optical Configurations

The optics in head-up display systems are used to "collimate" the HUD image so that essential flight parameters, navigational information, and guidance are superimposed on the outside world scene.

The four distinct FOV characteristics used to fully describe the characteristics of the angular region over which the HUD image is visible to the pilot are illustrated in Figure 4.3, and summarized as follows:

Total FOV (TFOV) — The maximum angular extent over which symbology from the image source can be viewed by the pilot with either eye allowing vertical and horizontal head movement within the HUD eyebox.

Instantaneous FOV (IFOV) — The union of the two solid angles subtended at each eye by the clear apertures of the HUD optics from a fixed head position within the HUD eyebox. Thus, the instantaneous FOV is comprised of what the left eye sees plus what the right eye sees from a fixed head position within the HUD eyebox.

Binocular overlapping FOV — The binocular overlapping FOV is the intersection of the two solid angles subtended at each eye by the clear apertures of the HUD optics from a fixed head position within the HUD eyebox. The binocular overlapping FOV thus defines the maximum angular extent of the HUD display that is visible to both eyes simultaneously.

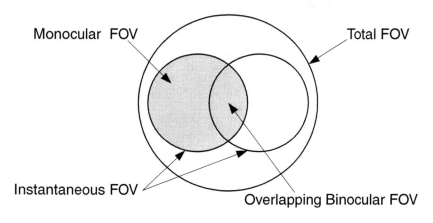

FIGURE 4.3 HUD fields-of-view defined.

Monocular FOV — The solid angle subtended at the eye by the clear apertures of the HUD optics from a fixed eye position. Note that the monocular FOV size and shape may change as a function of eye position within the HUD eyebox.

The FOV characteristics are designed and optimized for a specific cockpit geometric configuration based on the intended function of the HUD. In some cases, the cockpit geometry may impact the maximum available FOV.

One of the most significant advances in HUD optical design in the last 20 years is the change from optical systems that collimate by refraction to systems that collimate by reflection or, in some cases, by diffraction. The move towards more complex (and expensive) reflective collimation systems has resulted in larger display fields-of-view which expand the usefulness of HUDs as full-time primary flight references.

4.2.1.1 Refractive Optical Systems

Figure 4.4 illustrates the optical configuration of a refractive HUD system. This configuration is similar to the basic HUD optical systems in use since the 1950s.[6] In this optical configuration, the CRT image is collimated by a combination of refractive lens elements designed to provide a highly accurate display over a moderate display field of view. Note that an internal mirror is used to fold the optical system to reduce the physical size of the packaging envelope of the HUD. Also shown in Figure 4.4 is the HUD combiner glass, a flat semitransparent plate designed to reflect approximately 25% of the collimated light from the CRT, and transmit approximately 70% of the real-world luminance.

Note that the vertical instantaneous FOV can be increased by adding a second flat combiner glass, displaced vertically above and parallel with the first.

4.2.1.2 Reflective Optical Systems

In the late 1970s, HUD optical designers looked at ways to significantly increase the display total and instantaneous FOVs.[7,8] Figure 4.5 illustrates the first overhead-mounted reflective HUD optical system (using a holographically manufactured combiner) designed specifically for a commercial transport cockpit.[9] As in the classical refractive optical system, the displayed image is generated on a small CRT, about 3 in. in diameter. The reflective optics can be thought of as two distinct optical subsystems. The first is a relay lens assembly designed to re-image and pre-aberrate the CRT image source to an intermediate aerial image, located at one focal length from the optically powered combiner/collimator element.

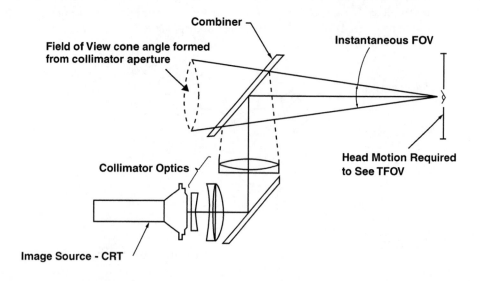

FIGURE 4.4   Refractive optical systems.

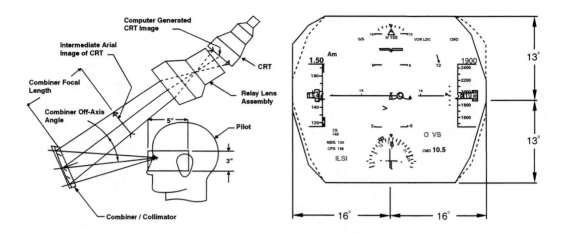

Reflective HUD FOV (Typical)

FIGURE 4.5 Reflective optical systems (overhead mounted).

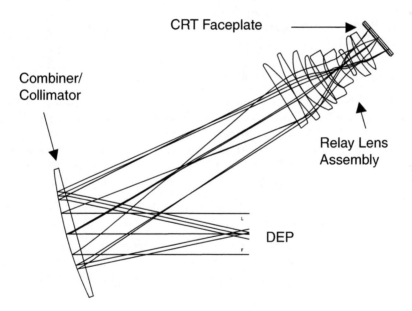

FIGURE 4.6 Reflective optical system raytrace.

The second optical subsystem is the combiner/collimator element that re-images and collimates the intermediate aerial image for viewing by the pilot. As in the refractive systems, the pilot's eyes focus at optical infinity, looking through the combiner to see the virtual image. To prevent the pilot's head from blocking rays from the relay lens to the combiner, the combiner is tilted off-axis with respect to the axial chief ray from the relay lens assembly. The combiner off-axis angle, although required for image viewing reasons, significantly increases the optical aberrations within the system, which must be compensated in the relay lens to have a well-correlated, accurate virtual display.

Figure 4.6 illustrates the optical raytrace of a typical reflective HUD system showing the complexity of the relay lens assembly. (This is the optical system used on the first manually flown CAT IIIa HUD system ever certified.)

The complexity of the relay lens, shown in Figure 4.6, provides a large instantaneous FOV over a fairly large eyebox, while simultaneously providing low display parallax and high display accuracy.

The reflective optical system can provide an instantaneous and binocular overlapping FOV that is equal to the total FOV, allowing the pilot to view all of the information displayed on the CRT with each eye with no head movement. Table 4.1 summarizes typical field-of-view performance characteristics for HUD systems.

TABLE 4.1 Typical HUD Fields-of-View

	Refractive HUD FOV Characteristics[a]		Reflective HUD Optics FOV Characteristics
	Single Combiner	Dual Combiners	
Total Field Of View	20°–25° Diameter	25°–30° Diameter	22–28° V × 28–34° H
Instantaneous FOV	12° V × 17.8° H	16° V × 17.8° H	22–28° V × 28–34° H
Overlapping	11° V × 6° H	16° V × 6° H	22–26° V × 25–30° H
Monocular FOV	12° Diameter	16° V × 12° H	22–28° V × 30° H

[a] Calculations assume a collimator exit aperture diameter of 5.0″, and a distance of 24″ between the pilot and the HUD collimator exit aperture.

All commercially certified HUD systems in airline operation today use reflective optical systems because of the improved display FOV characteristics compared with refractive systems.

4.2.2 Significant Optical Performance Characteristics

This section summarizes other important optical characteristics associated with conformal HUD systems. It is clear that the HUD FOV, luminance, and display line width characteristics must meet basic performance requirements.[10] However, optical system complexity and cost are driven by HUD eyebox size, combiner off-axis angle, display accuracy requirements, and optical parallax errors. Without a well-corrected optical system, conformal symbology will not properly overlay the outside world view and symbology will not remain fixed with respect to the real-world view as the head is moved around within the HUD eyebox.

4.2.2.1 Display Luminance and Contrast Ratio

The HUD should be capable of providing a usable display under all foreseeable ambient lighting conditions, including a sun-lit cloud with a luminance of 10,000 foot-Lamberts (ft-L)(or 34,000 cd/m²), and a night approach to a sparsely lit runway. HUD contrast ratio is a measure of the relative luminance of the display with respect to the real-world background and is defined as follows:

$$\text{HUD Contrast Ratio} = \frac{\text{Display Luminance} + \text{Real World Luminance}}{\text{Real World Luminance}}$$

The display luminance is the photopically weighted CRT light output that reaches the pilot's eyes. Real-world luminance is the luminance of the real world as seen through the HUD combiner. (By convention, the transmission of the aircraft windshield is left out of the real-world luminance calculation.)

It is generally agreed that a contrast ratio (CR) of 1.2 is adequate for display viewing, but that a CR of 1.3 is preferable. A HUD contrast ratio of 1.3 against a 10,000-ft-L cloud seen through a combiner with an 80% photopic transmission requires a display luminance at the pilot's eye of 2400 ft-L, a luminance about 10 times higher than most head-down displays. (This luminance translates to a CRT faceplate brightness of about 9000 ft-L, a luminance easily met with high-brightness monochrome CRTs.)

4.2.2.2 Head Motion Box

The HUD head motion box, or "eyebox," is a three-dimensional region in space surrounding the cockpit DEP in which the HUD can be viewed with at least one eye. The center of the eyebox can be displayed forward or aft, or upward or downward, with respect to the cockpit DEP to better accommodate the actual sitting position of the pilot. The positioning of the cockpit eye reference point[11] or DEP is dependent on a number of ergonomically related cockpit issues such as head-down display visibility, the over-the-nose down-look angle, and the physical location of various controls such as the control yoke and the landing gear handle.

The HUD eyebox should be as large as possible to allow maximum head motion without losing display information. The relay lens exit aperture, the spacing between the relay lens and combiner and the combiner to DEP, and the combiner focal length all impact the eyebox size. Modern HUD eyebox dimensions are typically 5.2 in lateral, 3.0 in vertical, and 6.0 in longitudinal.

In all HUDs, the monocular instantaneous FOV is reduced (or vignettes) with lateral or vertical eye displacement, particularly near the edge of the eyebox. Establishing a minimum monocular FOV from the edge of the eyebox thus ensures that even when the pilot's head is de-centered so that one eye is at the edge of the eyebox, useful display FOV is still available. A 10° horizontal by 10° vertical monocular FOV generally can be used to define the eyebox limits. In reflective HUDs, relatively small head movements (>1.5 in laterally) will cause one eye to be outside of the eyebox and see no display. Under these conditions, the other eye will see the total FOV, so no information is lost to the pilot.

4.2.2.3 HUD Display Accuracy

Display accuracy is a measure of how precisely the projected HUD image overlays the real-world view seen through the combiner and windshield from any eye position within the eyebox. Display accuracy is a monocular measurement and, for a fixed display location, is numerically equal to the angular difference between a HUD-projected symbol element and the corresponding real-world feature as seen through the combiner and windshield. The total HUD system display accuracy error budget includes optical errors, electronic gain and offset errors, errors associated with the CRT and yoke, Overhead to Combiner misalignment errors, windshield variations, environmental conditions (including temperature), assembly tolerances, and installation errors. Optical errors are both head-position and field-angle dependent.

The following display accuracy values are achievable in commercial HUDs when all the error sources are accounted for:

Boresight	+/− 3.0 milliradians (mrad)
Total Display Accuracy	+/− 7.0 milliradians (mrad)

The boresight direction is used as the calibration direction for zeroing all electronic errors. Boresight errors include the mechanical installation of the HUD hardpoints to the airframe, electronic drift due to thermal variations, and manufacturing tolerances for positioning the combiner element. Refractive HUDs with integrated combiners (i.e., F-16) are capable of achieving display accuracies of about half of the errors above.

4.2.2.4 HUD Parallax Errors

Within the binocular overlapping portion of the FOV, the left and right eyes view the same location on the CRT faceplate. These slight angular errors between what the two eyes see are binocular parallax errors or collimation errors. The binocular parallax error for a fixed field point within the total FOV is the angular difference in rays entering two eyes separated horizontally by the interpupillary distance, assumed to be 2.5 in. If the projected virtual display image were perfectly collimated at infinity from all eyebox positions, the two ray directions would be identical, and the parallax errors would be zero. Parallax errors consist of both horizontal and vertical components.

TABLE 4.2   HUD Optical System Summary (Typical Reflective HUD)

1. Combiner Design	Wide Field-of-view Wavelength Selective
	Stowable
	Inertial Break-away (HIC*-Complaint)
2. DEP to Combiner Distance	9.5 to 13.5 in. (Cockpit geometry dependent)
3. Display Fields-of-View	
Total Display FOV	24–28° Vertical × 30–34° Horizontal
Instantaneous FOV	24–28° Vertical × 30–34° Horizontal
Overlapping Binocular FOV	22–24° Vertical × 24–32° Horizontal
4. Head Motion Box or Eyebox	Typical Dimensions (Configuration
	dependent)
Horizontal	4.7 to 5.4 in.
Vertical	2.5 to 3.0 in.
Depth (fore/aft)	4.0 to 7.0 in.
5. Head Motion Needed to View TFOV	None
6. Display Parallax Errors (Typical)	
Convergence	95% of data points <2.5 mrad
Divergence	95% of data points <1.0 mrad
Dipvergence	93% of data points <1.5 mrad
7. Display Accuracy (2 sigma)	
Boresight	<2.5–4.0 mrad
Total Field-of-view	<5.0–9.0 mrad
8. Combiner Transmission and	78–82% photopic (day-adapted eye)
Coloration	84% Scotopic (night-adapted eye)
	<0.03 Color shift u'v' coordinates
9. Display Luminance and Contrast	
Ratio	
Stroke Only	1,600–2,400 foot-Lambert (ft-L)
Raster	600–1,000 ft-L
Display Contrast Ratio	1.2 to 1.3:1 (10,000 ft-L ambient background)
10. Display Line Width	0.7–1.2 mrads
11. Secondary Display Image Intensity	<0.5% of the primary image from eyebox

*Head Injury Criteria

Parallax errors in refractive HUDs are generally less than about 1.0 mrad due to the rotational symmetry of the optics, and because of the relatively small overlapping binocular FOV.

4.2.2.5   Display Line Width

The HUD line width is the angular dimension of displayed symbology elements. Acceptable HUD line widths are between 0.7 and 1.2 mrad when measured at the 50% intensity points. The displayed line width is dependent on the effective focal length of the optical system and the physical line width on the CRT faceplate. A typical wide FOV reflective HUD optical system with a focal length of 5 in. will provide a display line width of about 1 mrad given a CRT line width of 0.005 in. The display line width should be met over the full luminance range of the HUD, often requiring a high-voltage power supply with dynamic focus over the total useful screen area of the CRT.

HUD optical system aberrations will adversely affect apparent display line width. These aberrations include uncorrected chromatic aberrations (lateral color) and residual uncompensated coma and astigmatism. Minimizing these optical errors during the optimization of the HUD relay lens design will also help meet the parallax error requirements. Table 4.2 summarizes the optical performance characteristics of a commercial wide-angle reflective HUD optical system.

4.2.3   HUD Mechanical Installation

The intent of the HUD is to display symbolic information which overlays the real world as seen by the pilot. To accomplish this, the HUD pilot display unit must be very accurately aligned with respect to the pitch, roll, and heading axis of the aircraft. For this reason, the angular relationship of the HUD PDU with respect to the cockpit coordinates is crucial. The process of installing and aligning the HUD

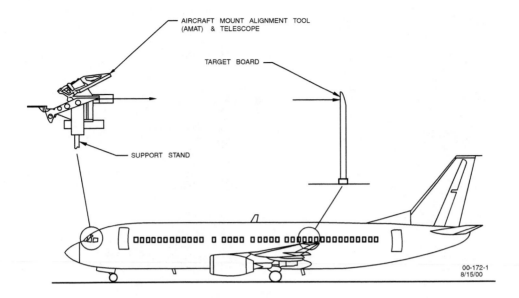

FIGURE 4.7 Boresighting the HUD hardpoints.

attachment bushings or hardpoints into the aircraft is referred to as "boresighting" and occurs when the aircraft is first built. (Although the alignment of the HUD hardpoints may be checked occasionally, once installed, the hardpoints are permanent and rarely need adjustment.)

Some reflective HUDs utilize mating bushings for the PDU hardware which are installed directly to the aircraft structure. Once the bushings are aligned and boresighted to the aircraft axis, they are permanently fixed in place using a structural epoxy. Figure 4.7 illustrates this installation method for HUD boresighting. In this case, the longitudinal axis of the aircraft is used as the boresight reference direction. Using special tooling, the Overhead Unit and Combiner bushings are aligned with a precisely positioned target board located near the aft end of the fuselage. This boresighting method does not require the aircraft to be jacked and leveled.

Other HUD designs utilize a tray that attaches to the aircraft structure and provides an interface to the HUD LRUs. The PDU tray must still be installed and boresighted to the aircraft axis.

4.2.4 HUD System Hardware Components

A typical commercial HUD system includes four principal line replaceable units (LRUs). (HUD LRUs can be interchanged on the flight deck without requiring any alignment or recalibration.) The cockpit-mounted LRUs include the Overhead Unit and Combiner (the Pilot Display Unit) and the HUD Control Panel. The HUD computer is located in the electronics bay or other convenient location. A HUD interconnect diagram is shown in Figure 4.8.

4.2.4.1 HUD Overhead Unit

The Overhead Unit (OHU), positioned directly above the pilot's head, interfaces with the HUD computer receiving either analog X and Y deflection and Z-video data or a serial digital display list, as well as control data via a serial interface. The OHU electronics converts the deflection and video data to an image on a high-brightness cathode ray tube (CRT). The CRT is optically coupled to the relay lens assembly which re-images the CRT object to an intermediate aerial image one focal length away from the combiner LRU, as illustrated in the optical schematic in Figure 4.5. The combiner re-images the intermediate image at optical infinity for viewing by the pilot. The OHU includes all of the electronics necessary to drive the CRT and monitor the built-in-test (BIT) status of the LRU. The OHU also provides the electronic interfaces to the Combiner LRU.

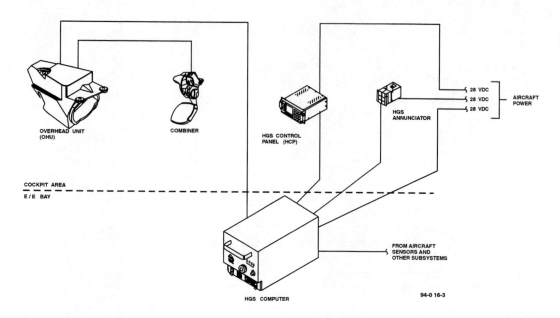

FIGURE 4.8   HUD interconnect diagram (RCFD HGS-4000).

A typical Overhead Unit is illustrated in Figure 4.9. This LRU contains all electronic circuitry required to drive the high-brightness CRT, and all BIT-related functions. The following are the major OHU subsystems:

- Relay Lens Assembly
- Desiccant Assembly (prevents condensation within the relay lens)
- Cathode Ray Tube Assembly
- High-voltage power supplies
- Low-voltage power supplies and energy storage
- Deflection amplifiers (X and Y)
- Video amplifier
- Built-In-Test (BIT) and monitoring circuits
- Motherboard Assembly
- OHU Chassis

In some HUD systems, the PDU may provide deflection data back to the HUD computer as part of the "wraparound" critical symbol monitor feature.[12] Real-time monitoring of certain critical symbol elements (i.e., horizon line) provides the high integrity levels required for certifying a HUD as a primary flight display. Other monitored critical data on the HUD may include ILS data, airspeed, flight path vector, and low-visibility guidance symbology.

4.2.4.2   HUD Combiner

The combiner is an optical-mechanical LRU consisting of a precision support structure for the wavelength-selective combiner element, and a mechanism allowing the combiner to be stowed and to breakaway. The combiner LRU interfaces with a precision pre-aligned mating interface permanently mounted to the aircraft structure. The combiner glass support structure positions the combiner with respect to the cockpit DEP and the Overhead Unit. The combiner mechanism allows the glass to be stowed upward when not in use, and to break away during a rapid aircraft deceleration, thus meeting the newly defined cockpit "head injury criteria" or HIC.[13] The combiner locks into both the stowed and breakaway positions and requires positive actions by the pilot to return it to the deployed position. Many HUD combiner

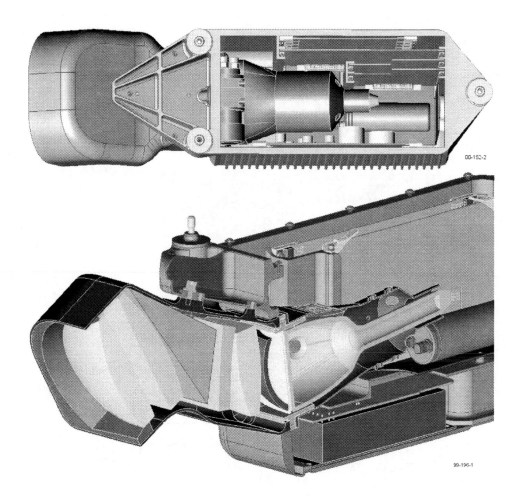

FIGURE 4.9 HUD overhead unit chassis (WFOV reflective optics).

assemblies include a built-in alignment detector that monitors the glass position in real time. Figure 4.10 shows a commercial HUD PDU and a wavelength-selective combiner. The combiner usually includes the HUD optical controls (brightness and contrast).

4.2.4.3 HUD Computer

The HUD computer interfaces with the aircraft sensors and systems, performs data conversions, validates data, computes command guidance (if applicable), positions and formats symbols, generates the display list, and converts the display list into X, Y, and Z waveforms for display by the PDU. In some commercial HUD systems, the HUD computer performs all computations associated with low-visibility take-off, approach, landing, and rollout guidance, and all safety-related performance and failure monitoring. Because of the critical functions performed by these systems, the displayed data must meet the highest integrity requirements. The HUD computer architecture is designed specifically to meet these requirements.

One of the key safety requirements for a full flight regime HUD is that the display of unannunciated, hazardously misleading attitude on the HUD must be improbable, and that the display of unannunciated hazardously misleading low-visibility guidance must be extremely improbable. An analysis of these requirements leads to the system architecture shown in Figure 4.11.

In this architecture, primary data are brought into the HUD computer via dual independent input/output (I/O) subsystems from the primary sensors and systems on the aircraft. (The avionics interface

FIGURE 4.10   HUD PDU.

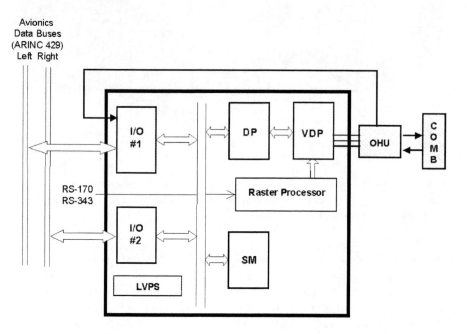

FIGURE 4.11   High integrity HUD computer architecture.

for a specific HUD computer depends on the avionics suite, and can include a combination of any of the following interfaces: ARINC 429, ARINC 629, ASCB-A, B, C, or D, or MIL STD 1553B.) Older aircraft will often include analog inputs as well as some synchro data. The I/O subsystem also includes the interfaces required for the Overhead Unit and Combiner and will often include outputs to the flight data recorder and central maintenance computer.

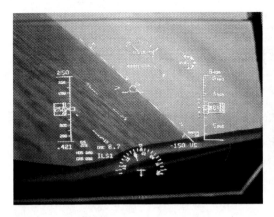 

FIGURE 4.12 Commercial HUD symbology.

Figure 4.12 is a photograph of a typical commercial HUD symbology set. The aircraft sensor data needed to generate this display are in Table 4.3. In general two sources of the critical data are required to meet the safety and integrity requirements.

The Display Processor converts all input into engineering units, verifies the validity of the data, compares like data from the dual sensors, runs the control law algorithms, computes the display element locations, and generates a display list. The Video Display Processor (VDP) converts the display list into X, Y, and Z signals that are output to the OHU.

The System Monitor processor (SM) verifies the display path by monitoring the displayed position of critical symbols using an inverse function algorithm,[12] independently computes the guidance algorithms using off-side data for comparison to the guidance solution from the Display Processor, and monitors the approach parameters to ensure a safe touchdown. The critical symbol monitor is a wraparound monitor that computes the state of the aircraft based on the actual display information on the CRT. The displayed state is compared to the actual aircraft state based on the latest I/O data. A difference between the actual state and the computed state causes the System Monitor to blank the display through two independent channels, since any difference in states could indicate a display processor fault. All software in the HUD computer is generally developed to DO-178B Level A requirements due to the critical functions performed.

Also shown in Figure 4.11 is the Raster Processor subassembly, used in HUD Systems that interface with Enhanced Vision sensors. This subsystem converts standard raster sensor video formats (RS-170 or RS-343) into a display format that is optimized for display on the HUD. In most raster-capable HUDs there is a trade-off between how much time is available for writing stroke information, and how much time is available for writing the raster image (the frame rate is fixed at 60 Hz, corresponding to 16.67 msec per frame). Some HUD systems "borrow" video lines from the raster image to provide adequate time to draw the stroke display (a technique called "line stealing"). The alternative is to limit the amount of stroke information that can be written on top of the raster image. Neither approach is optimal for a primary flight reference HUD required to display both stroke and raster images.

One solution is to convert the standard raster image format to a display format that is more optimized for HUD display. Specifically, the video input is digitized and scan-converted into a bi-directional display format, thus saving time from each horizontal line (line overscan, and flyback). This technique increases the time available for writing stroke information in the stroke-raster mode from about 1.6 msec to about 4.5 msec, adequate enough to write the entire worst-case stroke display. The bi-directional raster scan technique is illustrated in Figure 4.13, along with a photograph of a full-field raster image.

Figure 4.14 is a photograph of a HUD computer capable of computing take-off guidance, manual CAT IIIa landing guidance, rollout guidance, and raster image processing.

TABLE 4.3 Sensor Data Required for Full Flight Regime Operation

Input Data	Data Source
Attitude	Pitch and Roll Angles — 2 independent sources
Airspeed	Calibrated Airspeed
	Low Speed Awareness Speed(s) (e.g., Vstall)
	High Speed Awareness Speed(s) (e.g., Vmo)
Altitude	Barometric Altitude (pressure altitude corrected with altimeter setting)
	Radio Altitude
Vertical Speed	Vertical Speed (inertial if available, otherwise raw air data)
Slip/Skid	Lateral Acceleration
Heading	Magnetic Heading
	True Heading or other heading (if selectable)
	Heading Source Selection (if other than Magnetic selectable)
Navigation	Selected Course
	VOR Bearing/Deviation
	DME Distance
	Localizer Deviation
	Glideslope Deviation
	Marker Beacons
	Bearings/Deviations/Distances for any other desired nav signals (e.g., ADF, TACAN, RNAV/FMS)
Reference Information	Selected Airspeed
	Selected Altitude
	Selected Heading
	Other Reference Speed Information (e.g., V_1, V_R, Vapch)
	Other Reference Altitude Information (e.g., landing minimums [DH/MDA], altimeter setting)
Flight Path	Pitch Angle
	Roll Angle
	Heading (Magnetic or True, same as Track)
	Ground Speed (inertial or equivalent)
	Track Angle (Magnetic or True, same as Heading)
	Vertical Speed (inertial or equivalent)
	Pitch Rate, Yaw Rate
Flight Path Acceleration	Longitudinal Acceleration
	Lateral Acceleration
	Normal Acceleration
	Pitch Angle
	Roll Angle
	Heading (Magnetic or True, same as Track)
	Ground Speed (inertial or equivalent)
	Track Angle (Magnetic or True, same as Heading)
	Vertical Speed (inertial or equivalent)
Automatic Flight Control System	Flight Director Guidance Commands
	Autopilot/Flight Director Modes
	Autothrottle Modes
Miscellaneous	Wind Speed
	Wind Direction (and appropriate heading reference)
	Mach
	Windshear Warning(s)
	Ground Proximity Warning(s)
	TCAS Resolution Advisory Information

4.2.4.4   HUD Control Panel

Commercial HUD systems used for low-visibility operations often require some pilot-selectable data not available on any aircraft system bus as well as a means for the pilot to control the display mode. Some HUD operators prefer to use an existing flight deck control panel, e.g., an MCDU, for HUD data entry

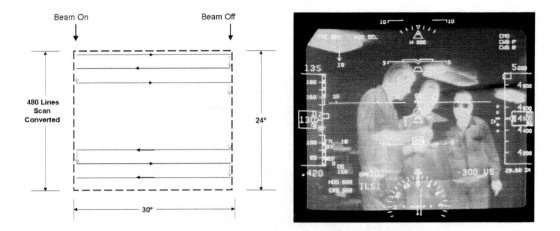

FIGURE 4.13 Bi-directional scan-converted raster image.

FIGURE 4.14 High integrity HUD computer.

and control. Other operators prefer a standalone control panel, dedicated to the HUD function. Figure 4.15 illustrates a standalone HUD control panel certified for use in CAT IIIa HUD systems.

4.2.5   Aspects of HUD Certification

Certification requirements for a HUD system depend on the functions performed. As the role of HUDs have expanded from CAT IIIa landing devices to full flight regime primary flight references including take-off and rollout guidance, the certification requirements have become more complex. It is beyond the scope of this chapter to describe all the certification issues and requirements for a primary flight display HUD, however, the basic requirements are not significantly different from PFD head-down display certification requirements.

The FAA has documented the requirements for systems providing guidance in low-visibility conditions in Advisory Circular AC 120-28, "Criteria for Approval of Category III Weather Minima for Takeoff, Landing, and Rollout." The certification of the landing guidance aspects of the HUD are fundamentally different from automatic landing systems because the human pilot is in the active control loop during

FIGURE 4.15   HUD control and data entry panel.

the beam tracking and flare. The following summarizes the unique aspects of the certification process for a manual Category III system.

1. *Control Law Development* — The guidance control laws are developed and optimized based on the pilot's ability to react and respond. The control laws must be "pilot centered" and tailored for a pilot of average ability. The monitors must be designed and tuned to detect approaches that will be outside the footprint requirement, yet they cannot cause a go-around rate greater than about 4%.

2. *Motion-Based Simulator Campaign* — Historically, approximately 1400 manned approaches in an approved motion-based simulator, with at least 12 certification authority pilots, are required for performance verification for a FAA/JAA certification. The Monte Carlo test case ensemble is designed to verify the system performance throughout the envelope expected in field operation. Specifically, the full environment must be sampled (head winds, cross winds, tail winds, turbulence, etc.) along with variations in the airfield conditions (sloping runways, ILS beam offsets, beam bends, etc.). Finally, the sensor data used by the HUD must be varied according to the manufacturer's specified performance tolerances. Failure cases must also be simulated. Time history data for each approach, landing, and rollout is required to perform the required data reduction analysis. A detailed statistical analysis is required to demonstrate, among other characteristics, the longitudinal, lateral, and vertical touchdown footprint. Finally, the analysis must project out the landing footprint to a one-in-a-million (10^{-6}) probability.

3. *Aircraft Flight Test* — Following a successful simulator performance verification campaign, the HUD must be demonstrated in actual flight trials on a fully equipped aircraft. As in the simulator case, representative head winds, cross winds, and tail winds must be sampled for certification. Failure conditions are also run to demonstrate system performance and functionality.

This methodology has been used to certify head up display systems providing manual guidance for take-off, landing, and rollout on a variety of different aircraft types.

4.3   Applications and Examples

This section describes how the HUD is used on a typical aircraft. This includes the typical symbology sets that are displayed to a pilot in specific phases of flight. The symbology examples used in this section are taken from a Rockwell Collins Flight Dynamics Head-Up Guidance System (HGS®) installed on an in-service aircraft.

In addition to symbology, this section also discusses the pilot-in-the-loop optimized guidance algorithms that are provided as part of a HGS. Another feature of some HUDs is the display of video images on the HUD and the uses of this feature—where the HUD is only a display device—are discussed.

4.3.1 Symbology Sets and Modes

To optimize the presentation of information, the HUD has different symbology sets that present only the information needed by the pilot in that phase of flight. For example, the aircraft pitch information is not important when the aircraft is on the ground. These symbology sets are either selected as modes by the pilot or are displayed automatically when a certain condition is detected.

4.3.1.1 Primary Mode

The HGS Primary (PRI) mode can be used during all phases of flight from take-off to landing. This mode supports low-visibility take-off operations, all en route operations, and approaches to CAT I or II minimums using FGS Flight Director guidance.

The HGS Primary mode display is very similar to the Primary Flight Display (PFD) to enhance the pilot's transition from head down instruments to headup symbology. Figure 4.16 shows a typical in-flight Primary mode display that includes the following symbolic information:

- Aircraft Reference (boresight) symbol
- Pitch — scale and horizon relative to boresight
- Roll — scale and horizon relative to boresight
- Heading — horizon, HIS, and digital readouts

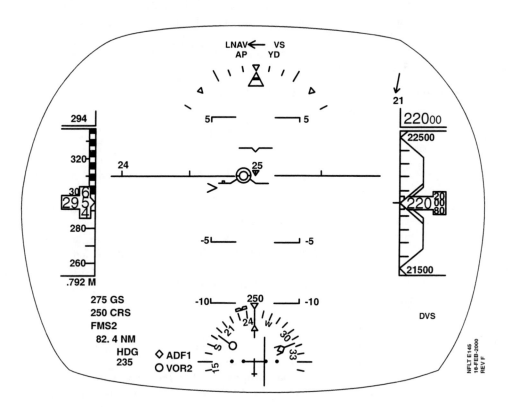

FIGURE 4.16 HGS primary mode symbology: in-flight.

- Speeds — CAS (tape), vertical speed, ground speed, speed error tape
- Altitudes — barometric altitude (tape), digital radio altitude
- Flight Path (inertial)
- Flight Path acceleration
- Slip/Skid Indicators
- FGS Flight Director (F/D) guidance cue and modes
- Flight Director armed and capture modes
- Navigation data — ILS, VOR, DME, FMS, marker beacons
- Wind — speed and direction
- Selected parameters — course, heading, airspeed, and altitude
- Attitude
- Altitude
- Airspeed
- Navigation Data
- Warning and Advisory

When the aircraft is on the ground, several symbols are removed or replaced as described in the following sections. After take-off rotation, the full, in-flight set of symbols is restored.

The Primary mode is selectable at the HCP or by pressing the throttle go-around switch during any mode of operation.

4.3.1.1.1 Primary Mode: Low-Visibility Take-off (HGS Guidance)

The Primary mode includes special symbology used for a low-visibility take-off as shown in Figure 4.17. The HGS guidance information supplements visual runway centerline tracking and enhances situational awareness.

For take-off operation, the HSI scale is removed from the Primary display until the aircraft is airborne. Additional symbols presented during low-visibility take-off operation are

- Ground Roll Reference Symbol (fixed position)
- Ground Localizer Scale and Index
- Ground Roll Guidance Cue (HGS-derived steering command)
- TOGA Reference Line

The Ground Localizer Scale and Index provide raw localizer information any time the aircraft is on the ground. For a low-visibility take-off, the general operating procedure is to taxi the aircraft into take-off position over the runway centerline. The selected course is adjusted as necessary to overlay the Selected Course symbol on the actual runway centerline at the furthest point of visibility. Take-off roll is started and the captain uses rudder control to center the Ground Roll Guidance Cue in the Ground Roll Reference symbol (concentric circles). If the cue is to the right of the Ground Roll Reference symbol then the pilot would need to apply right rudder to again center the two symbols. (At rotation, the Ground Roll Reference and Guidance Cue symbols are replaced by the Flight Path symbol and the Flight Director Guidance Cue.)

4.3.1.1.2 Primary Mode: Climb

At rotation, a number of changes take place on the display (see Figure 4.16). Flight Path Acceleration, now positioned relative to Flight Path controlling the aircraft, is particularly useful in determining a positive climb gradient and in optimizing climb performance. With the appropriate airspeed achieved, to null Flight Path Acceleration will maintain airspeed. Alternately, the Flight Director commands can be followed.

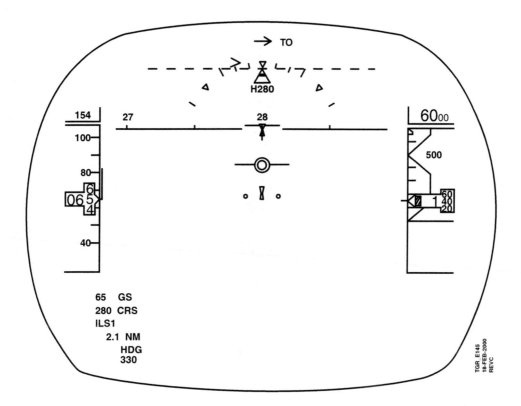

FIGURE 4.17 HGS primary mode: low-visibility take-off.

4.3.1.1.3 *Primary Mode: Cruise*

Figure 4.16 shows a typical HGS display for an aircraft in straight and level flight at 22,000 ft, 295 kn, and Mach .792. Ground Speed is reduced to 275 kn as a result of a 21-kn, right-quartering headwind indicated by the wind arrow.

The aircraft is being flown by the autopilot with LNAV and VS modes selected. Holding the center of the Flight Path symbol level on the horizon, and the Flight Path Acceleration symbol ($>$) on the Flight Path wing will maintain level flight.

4.3.2 AIII Approach Mode

The HGS AIII mode is designed for precision, manual ILS approach, and landing operations to CAT III minimums. Additionally, the AIII mode can be used for CAT II approaches at Type I airfields if operational authorization has been obtained (see Figure 4.18). The display has been de-cluttered to maximize visibility by removing the altitude and airspeed tape displays and replacing them with digital values. The HSI is also removed, with ILS raw data (localizer and glideslope deviation) now being displayed near the center of the display. (In the AIII mode, guidance information is shown as a circular cue whose position is calculated by the HGS.)

Tracking the HGS Guidance Cue, and ultimately the ILS, is achieved by centering and maintaining the Flight Path symbol over the cue. Monitoring localizer and glideslope lines relative to their null positions helps to minimize deviations and to anticipate corrections. Airspeed control is accomplished by nulling the Speed Error Tape (left wing of Flight Path symbol) using the Flight Path Acceleration caret to lead the airspeed correction. Any deviations in ILS tracking or airspeed error are easily identified by these symbolic relationships.

Following touchdown, the display changes to remove unnecessary symbology to assist with the landing rollout. The centerline is tracked while the aircraft is decelerated to exit the runway.

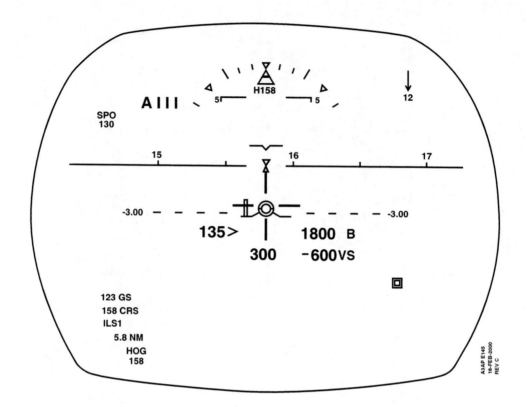

FIGURE 4.18 HGS AIII approach display.

4.3.2.1 AIII Mode System Monitoring

The HGS computer contains an independent processor, the system monitor, which verifies that HGS symbology is positioned accurately and that the approach is flown within defined limits. If the System Monitor detects a failure within the HGS or in any required input, it disables the AIII status, and an Approach Warning is annunciated to both crew members.

4.3.2.2 Unusual Attitude

The HGS Unusual Attitude (UA) display is designed to aid the pilot in recognition of and recovery from unusual attitude situations. When activated, the UA display replaces the currently selected operational mode symbology, and the HCP continues to display the currently selected operational mode that will be reactivated once the aircraft achieves a normal attitude.

The UA symbology is automatically activated whenever the aircraft exceeds operational roll or pitch limits, and deactivated once the aircraft is restored to controlled flight, or if either pitch or roll data becomes invalid. When the UA symbology is deactivated, the HGS returns to displaying the symbology for the currently selected operational mode.

The UA symbology includes a large circle (UA Attitude Display Outline) centered on the combiner (see Figure 4.19). The circle is intended to display the UA attitude symbology in a manner similar to an Attitude Direction Indicator (ADI). The UA Horizon Line represents zero degrees pitch attitude and is parallel to the actual horizon. The UA Horizon Line always remains within the outline to provide a sufficient sky/ground indication, and always shows the closest direction to and the roll orientation of the actual horizon. The Aircraft Reference symbol is displayed on top of a portion of the UA Horizon Line and UA Ground Lines whenever the symbols coincide.

The three UA Ground Lines show the ground side of the UA Horizon Line corresponding to the brown side on an ADI ball or EFIS attitude display. The Ground Lines move with the Horizon Line and are angled to simulate a perspective view.

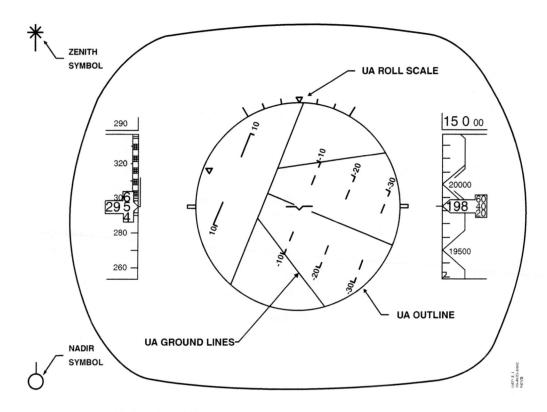

FIGURE 4.19 Unusual Attitude display.

The UA Pitch Scale range is from $-90°$ through $+90°$ with a zenith symbol displayed at the $+90°$ point, and a nadir symbol displayed at the $-90°$ point.

The UA Roll Scale is positioned along the UA Attitude Display Outline, with enhanced tic marks at $\pm90°$. The UA Roll Scale Pointer rotates about the UA Aircraft Reference symbol to always point straight up in the Earth frame.

4.3.3 Mode Selection and Data Entry

The data entry needs of the HUD are limited to mode selection and runway information for the guidance algorithms to work effectively. Data entry can be via a dedicated control panel or a Multipurpose Control Display Unit (MCDU), such as that defined by ARINC 739.

4.3.3.1 Mode Selection

On most aircraft the pilot has a number of ways to configure the HUD for an approach and landing based on the visibility conditions expected at the airport. In good weather, where the cloud "ceiling" is high and the runway visual range (RVR) is long, the pilot may leave the HUD in the Primary mode or select a landing mode such as VMC, which removes some symbol groups, but has no guidance information. As the ceiling and/or RVR decreases the pilot may select the IMC mode to display FGS guidance (usually from the Flight Management System). If the visibility is at or near the Category III limit the pilot will select AIII mode, which requires an Instrument Landing System and special approach guidance. To reduce workload, the HUD can be configured to automatically select the appropriate landing mode when certain conditions are met, such as the landing system deviations become active.

Another mode that is available for selection, but only on the ground, is the test mode where the pilot, or more usually a maintenance person, can verify the health of the HUD and the sensors that are connected to the system.

4.3.3.2 Data Entry

To make use of the HUD-based guidance the pilot must enter the following information:

- Runway Elevation — the altitude of the runway threshold
- Runway Length — official length of the runway in feet or meters
- Reference Glideslope — the published descent angle to the runway, e.g., 3°

On some aircraft, these data may be sent from the FMS and confirmed by the pilot.

4.3.4 HUD Guidance

On some aircraft the HUD can provide a pilot-in-the-loop low-visibility landing capability that is more cost-effective than that provided by an autoland system.

Huds that compute guidance to touchdown use deviations from the ILS to direct the pilot back to the center of the optimum landing path. The method for guiding the pilot is the display of a guidance cue that is driven horizontally and vertically by the guidance algorithms. The goal of the pilot is to control the aircraft so that the Flight Path symbol overlays the guidance cue. The movement of the guidance cue is optimized for pilot-in-the-loop flying. This optimization includes:

- Limiting the movement of the cue to rates that are achievable by a normal pilot
- Anticipating the natural delay between the movement of the cue and reaction of the pilot/aircraft
- Filtering out short-term cue movements that may be seen in turbulent air

In addition to approach guidance where the goal is to keep the aircraft in the center of the ILS beam, guidance is also provided for other phases of the approach. During the flare phase — a pitch-up maneuver prior to touchdown — the guidance cue must emulate the normal rate and magnitude of pull-back that the pilot would use during a visual approach. During the rollout phase — where the goal is to guide the aircraft down the centerline of the runway — the pilot is given smooth horizontal commands that are easy to follow.

All these algorithms have to work for all normal wind and turbulence conditions. As following the guidance is critical to the safety of the aircraft, the algorithms include monitors to ensure that the information is not misleading and monitors to ensure that the pilot is following the commands. If the system detects the pilot is significantly deviating from the path or speed target the system will display an Approach Warning message that requires the pilot to abort the landing.

4.3.4.1 Annunciations

An important element of any system is the annunciations that inform or alert the pilots to problems that require their action. In a well-managed flight deck the role of each of the pilots is designed to be complementary. The pilot flying (PF) is responsible for control of the aircraft. The pilot not flying (PNF) is responsible for navigation and communication as well as monitoring the performance of the PF.

All the status information needed to safely fly the aircraft is displayed on the HUD for the pilot including:

- Mode Status — modes of the HGS guidance or the guidance source.
- Cautions — approaching operating limitations or loss of a sensor.
- Warnings — loss of a critical sensor requiring immediate action.
- System Failure — HUD has failed and the pilot should not use the system.

Because of the technology used in the HUD, the PNF can not directly monitor these annunciations. To support PNF monitoring the HUD outputs some or all of these annunciations to either a flight deck central warning system or to a dedicated annunciator panel in front of the other pilot.

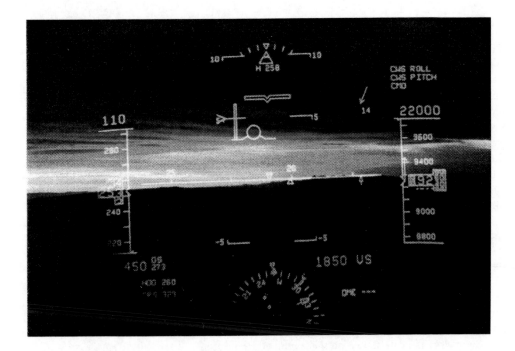

FIGURE 4.20 Effects of background color on perceived display color.

4.3.5 Recent Developments

4.3.5.1 Color HUD

Due to the complexity of wide field-of-view reflective HUD optical systems, the optical designer must use all means available to meet display accuracy and parallax error requirements. All certified reflective HUDs today are monochromatic, generally using a narrow-band green emitting phosphor. The addition of a second color to the HUD is a desirable natural progression in HUD technology, however, one of the technical challenges associated with adding a second (or third) display color is maintaining the performance standards available in monochrome displays. One method for solving this problem uses a collimator with two independent embedded curvatures, one optimized for green symbology, the other optimized for red symbology, each with a wavelength-selective coating.[14]

One fundamental issue associated with color symbology on HUDs is the effects of the real-world background color "adding" to the display color (green), resulting in an unintended perceived display color. Figure 4.20 illustrates the effects of additive color. Clearly, warnings and annunciations on the color HUD must be carefully designed to preclude a misinterpretation due to ambient background color.

4.3.5.2 Display of Enhanced Vision Sensor Images

Many modern HUD systems are capable of simultaneously displaying a real-time external video image and stroke symbology and guidance overlay. Given a sensor technology capable of imaging the real world through darkness, haze, or fog, the Enhanced Vision System (EVS) provides an image of the real world to the pilot while continuing to provide standard HUD symbology. This capability could provide benefit to the operator during taxi operations, low-visibility take-off, rollout, and perhaps during low-visibility approaches.

The interface between the EVS sensor and the HUD can be a standard video format (i.e., RS-170 or RS-343) or can be customized (i.e., serial digital). Sensor technologies that are candidates for EVS include:

- Forward-looking infrared, either cooled (InSb) or uncooled (InGaAs or microbolometer)
- MMW radar (mechanical or electronic scan)

- MMW radiometers (passive camera)
- UV sensors

Although the concept of interfacing a sensor with a HUD to achieve additional operational credit is straightforward, there are a number of technical and certification issues which must be overcome including pilot workload, combiner see-through with a raster image, sensor boresighting, integrity of the sensor, and potential failure modes. In addition, the location of the sensor on the aircraft can affect both parallax between the sensor image and the real world, and the aircraft aerodynamic characteristics.

Synthetic vision is an alternative approach to improving the pilot's situational awareness. In this concept, an onboard system generates a "real-world-like view" of the outside scene based on a terrain database using GPS position, track, and altitude. Some HUD systems today generate "artificial runway outlines" to improve the pilot's awareness of ground closure during low-visibility approach modes, a simple application of synthetic vision.

Defining Terms

Boresight: The aircraft longitudinal axis, used to position the HGS during installation and as a reference for symbol positioning. The process of aligning the HUD precisely with respect to the aircraft reference frame.

Collimation: The optical process of producing parallel rays of light, providing an image at infinity.

Eyebox: The HUD eyebox is a three-dimensional area around the flight deck eye reference point (ERP) where all of the data shown on the combiner can be seen.

References

1. Naish, J. Michael, Applications of the Head-Up Display (HUD) to a Commercial Jet Transport, *J. Aircraft*, August 1972, Vol. 9, No. 8, pp 530–536.
2. Naish, J. Michael, Combination of Information in Superimposed Visual Fields, *Nature*, May 16, 1964, Vol. 202, No. 4933, pp 641–46.
3. Sundstrand Data Control, Inc. (1979), *Head Up Display System*.
4. Part 25 — Airworthiness Standards: Transport Category Airplanes, Special Federal Aviation Regulation No. 13, Subpart A — General, Sec. 25.773 Pilot Compartment View.
5. Part 25 — Airworthiness Standards: Transport Category Airplanes, Special Federal Aviation Regulation No. 13, Subpart A — General, Sec. 25.775 Windshield and Windows.
6. Vallance, C.H. (1983). The approach to optical system design for aircraft head up display, *Proc. SPIE*, 399:15–25.
7. Hughes, U.S. Patent 3,940,204 (1976), Optical Display Systems Utilizing Holographic Lenses.
8. Marconi, U.S. Patent 4,261,647 (1981), Head Up Displays.
9. Wood, R. B. (1988), Holographic and classical head up display technology for commercial and fighter aircraft, *Proc. SPIE*, 883:36–52.
10. SAE (1998), AS8055 Minimum Performance Standard for Airborne Head Up Display (HUD).
11. Stone, G. (1987), The design eye reference point, SAE 6th Aerospace Behavioral Eng. Technol. Conf. Proc., *Human/Computer Technology: Who's in Charge?*, pp. 51–57.
12. Desmond, J., U.S. Patent 4,698,785 (1997), Method And Apparatus For Detecting Control System Data Processing Errors.
13. Part 25 — Airworthiness Standards: Transport Category Airplanes, Special Federal Aviation Regulation No. 13, Subpart A — General, Sec. 25.562 Emergency Landing Dynamic Conditions.
14. Gohman et al., U. S. Patent 5,710,668 (1988), Multi-Color Head-Up Display System.

5

Head-Mounted Displays

James E. Melzer
Kaiser Electro-Optics Inc.

5.1 Introduction

Head-Mounted Displays (HMD)* are personal information-viewing devices that can provide information in a way that no other display can. While they can be used as hands-off information sources, the displayed video can also be made reactive to head and body movements, replicating the way we view, navigate through, and explore the world. This unique capability lends itself to applications such as Virtual Reality for creating artificial environments,[1] to medical visualization as an aid in surgical procedures, [2,3] to military vehicles for viewing sensor imagery,[4] to airborne workstation applications reducing size, weight, and power over conventional displays,[5] to aircraft simulation and training,[6–8] and (central to this chapter) for fixed and rotary wing avionics display applications.[9,10]

In some applications, such as the medical and soldier's displays in Figure 5.1, the HMD is used solely as a hands-off information source. To truly reap the benefits of the HMD as part of an avionics application, however, it must be part of a Visually Coupled System (or VCS) that includes the HMD, a head position tracker, and a graphics engine or video source.[11,12] As the pilot turns his/her head, the tracker relays the orientation data to the mission computer, which updates the displayed information accordingly. This gives the pilot a myriad of real-time data that is *linked to head orientation*. In a fixed-wing fighter, a missile's sensor can be slaved to the pilot's head line-of-sight, allowing the pilot to designate targets away from the forward line-of-sight of the aircraft. In a helicopter, the pilot can point sensors such as forward-looking infrared (FLIR)** and fly at night.

The U.S. military introduced HMDs into fixed-wing aircraft in the early 1970s for targeting air-to-air missiles. Several hundred of the Visual Targeting Acquisition Systems (VTAS) were fielded on F-4 Phantom fighter jets between 1973 and 1979.[10,13] This program was eventually abandoned because the HMD

---

*The term Head-Mounted Display is used in this chapter as a more generic term than Helmet-Mounted Display which more often refers to military-oriented hardware. Helmet-Mounted Sight (HMS) is another term often used referring to an HMD that provides only a simple targeting reticle.

**Forward-Looking Infrared (FLIR) is a sensor technology that creates shades-of-grey imagery of objects from slight differences in black-body thermal emissions.

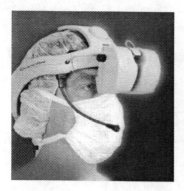  

FIGURE 5.1   Three different applications for HMDs: the CardioView® for minimally invasive cardiac surgery (photo courtesy of Vista Medical Technologies, Inc.), a prototype of the U.S. Army Land Warrior HMD (photo courtesy of Program Manager, Soldier, U.S. Army), and the SIM EYE XL100 for aviation training (photo courtesy of Kaiser Electro-Optics).

capabilities were not matched by missile technology of the day.* HMDs were given new life when a Soviet MiG-29 was photographed in 1985 showing a simple helmet-mounted sight for off-axis targeting of the Vympel R-73 missile — also called the AA-11 Archer. With this revelation, the Israelis initiated a fast-paced program that deployed the Elbit DASH HMD for off-axis targeting of the Rafael Python 4 missile in 1993-94.[14]

Two doestic studies — Vista Sabre[15] and Vista Sabre II[16] — demonstrated the clear advantages for a pilot equipped with an HMD for missile targeting over one using only his HUD. Encouraged by these and by a post-Berlin Wall examination of the close-combat capabilities of the HMD-equipped MiG-29,[17] the U.S. military initiated their own off-boresight missile targeting program. The result is the Joint Helmet Mounted Cueing System (JHMCS, built by Vision Systems International) scheduled to deploy on the U.S. Navy F/A-18, the U.S. Air Force F-15 and F-22, and on both domestic and international versions of the F-16 early in the 21st century. The JHMCS will give pilots off-axis targeting symbology for the AIM-9X missile, aircraft status,[18] and provide them with improved situational awareness of the airspace surrounding the aircraft.

The U.S. Army has taken a more aggressive approach with HMD technology, putting it on rotary wing aircraft starting with the AH-1S Cobra helicopter gunship in the 1970s. A turreted machine gun is slaved to the pilot's head orientation via a mechanical linkage attached to his helmet. The pilot aims the weapon by superimposing a small helmet-mounted reticle on the target.[19]

In the 1980s, the Army adopted the Integrated Helmet and Sighting System (IHADSS) for the AH-64 Apache helicopter. This monocular helmet-mounted display gives the pilot the ability — similar to the Cobra gunship — to target head-slaved weapons. The IHADSS has the added ability to display head-tracked FLIR imagery for nighttime flying. Over 5000 of these CRT-based, monochrome systems have been delivered by Honeywell on this very successful program for the Army.[10]

Using an HMD as a key interface to the aircraft has proven so effective that the Army's newest helicopter, the RAH-66 Comanche will field the binocular Helmet Integrated Display Sighting System (HIDSS) when it is deployed early in the 21st century.

In addition to these domestic applications, HMD-based pilotage systems are being adopted throughout the international aviation community on platforms such as Eurocopter's Tiger helicopter scheduled for deployment early in the 21st century. The U.S. Army also has extensive experience using helmet-mounted Night Vision Goggles (NVGs) in aviation environments. These devices have their own unique set of performance, interface, and visual issues [20–23] and are discussed in more detail elsewhere in this book.

---

*There was also a Memorandum of Understanding signed in 1980 that relegated the development of short-range missile technology (and therefore HMDs) to the Europeans.

FIGURE 5.2 The U.S. Air Force and Navy's Joint Helmet Mounted Cueing System helmet-mounted display that will go into service early in the 21st century. (Photo courtesy of Vision Systems International, used with permission.)

Head tracker sensors

See-through combiner

Collimating optics

CRT

Helmet mount

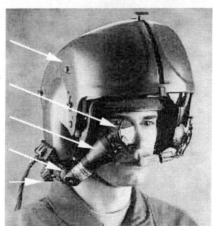

FIGURE 5.3 The Honeywell IHADSS is a monocular, monochrome, CRT-based, head-tracked, see-through helmet-mounted display used on the U.S. Army AH-64 Apache helicopter. (Photo courtesy of Honeywell Electronics, used with permission.)

5.2   What Is an HMD?

In its simplest incarnation, an HMD consists of one or more image sources, collimating optics, and a means to mount the assembly on the head. In the IHADSS HMD shown in Figure 5.3, the image source is a single, high-brightness cathode ray tube (CRT). The monocular optics create and relay a virtual image of the CRT surface, projecting the imagery onto the see-through combiner to the pilot's eye. This display module is attached to the right side of the aviator's protective helmet with adjustments that let the pilot position the display to see the entire image.

The early VTAS and Cobra helicopter HMDs used a simple targeting reticle to point weapons similar to the one shown on the left in Figure 5.5. The JHMCS HMD has a more sophisticated targeting capability including "look-to" and shoot cues (similar to the one shown on the right side of the same figure), as well as altitude, airspeed, compass heading, and artificial horizon data. With the IHADSS, the AH-64 Apache helicopter pilot sees a similar symbology set augmented with head-tracked FLIR data.

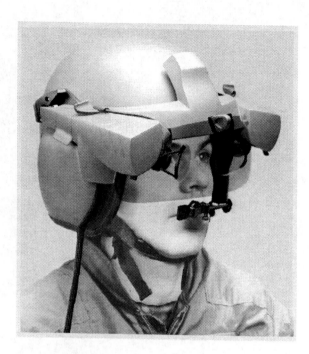

FIGURE 5.4 A prototype of the Kaiser Electronics' HIDSS for the RAH-66 Comanche helicopter. (Photo courtesy of Kaiser Electronics, used with permission.)

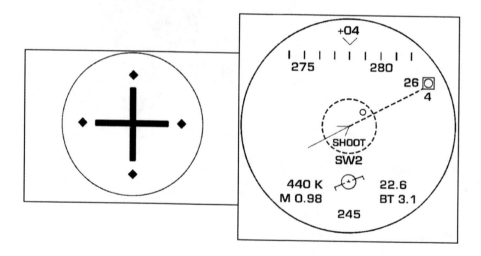

FIGURE 5.5 Comparison of early HMD reticle imagery (left) with a more capable symbology set (right) to be used with the HMDs such as the JHMCS.

This collection of components, though deceptively simple, has at its core a complex interaction of system and hardware issues as well as visual, anthropometric, physical, and display issues. These in turn are viewed by an equally complex *human perceptual system*.[24] The design task is complicated further in the aircraft environment, because the HMD — now a *helmet*-mounted display — provides both display and life support for the pilot. Issues of luminance, contrast, alignment, and focus must be addressed while not impacting pilotage or crash safety. For all these reasons, HMD design requires a careful balancing — a *suboptimization* — of both display and physical requirements.

The next sections will examine the important components or features in an HMD.

5.2.1 Image Sources for HMDs

As of the year 2000, almost all of the HMDs deployed use CRTs as image sources, primarily because the technology is the most mature. It can provide the required high luminance and the HMDs can be ruggedized to withstand the harsh military environment.[25] Over the last decade, however, small, flat-panel image sources have improved to where they are being considered as alternatives to CRTs because of their reduced size, weight, and power requirements. [26,27]

There are two major categories of image sources, *emissive* and *nonemissive* (see Table 5.1). The *nonemissive image sources* modulate a separate illumination on a pixel-by-pixel basis to create the desired imagery. Examples are

- *Transmissive Liquid Crystal Displays (LCD)* — The pixel matrix is illuminated from the rear. A modulated electric field controls the transmission of the backlight through the individual liquid crystal-filled cells. Quality transmissive LCDs are manufactured in large quantity in Japan, though in limited quantity domestically.

- *Reflective Liquid Crystal on Silicon Displays (LCOS)* — This is the same as the transmissive device except that the image source is illuminated from the front. The light transmits through the cell and reflects off a mirror-like surface when the pixel is transmitting and is scattered when the pixel is turned off. This is a fast-growing area of development in the U.S. because the manufacturing technology is similar to silicon wafer fabrication.

- *Scanning Display* — A point source (such as a laser) or line of point sources (such as LEDs) is modulated in one or more directions using resonance scanners or opto-acoustic modulators to produce imagery. One example is the Retinal Scanning Display (RSD).[28,29]

Emissive devices represent a large category of image sources in which the image plane of the device emits light without the need for supplemental illumination. Such devices include:

- *Active Matrix Electroluminescent (AMEL)* — A thin-film layer of luminescent phosphor is sandwiched between two electrodes, one transparent, in a pixilated array. The pixels are digitally addressed using high-frequency pulses to achieve grayscale. Recent improvements use a quasi-analog addressing to achieve greater grayscale range and improved luminance. These are compact and very rugged devices.[30]

- *Cathode Ray Tube (CRT)* — This is a vacuum tube with an electron gun at one end and a phosphor screen at the other. A beam from the electron gun is modulated by deflection grids and directed onto the screen. The incident electrons excite the phosphor, emitting visible light.[25] CRTs can be very bright and very rugged for the aviation environment, though they are larger than flat-panel displays and require high voltage.

- *Vacuum Fluorescent Display (VFD)* — Most commonly seen in alphanumeric displays, the VFD uses a vacuum package containing phosphors that are excited by a series of filaments. These capabilities are being expanded as imaging devices. Though currently available only in low-resolution devices, VFDs may in time become more prevalent.[31]

- *Organic Light Emitting Diodes (OLED)* — A low-voltage drive across a thin layer of organic material causes it to emit visible light when the charge carriers recombine within the material. A very promising technology, though as of this writing it is still in the developmental stages.

The choice of an image source for an HMD is not easy. Depending on the application, it may be preferable to have a backlit (i.e., transmissive) LCD over a reflective one for size, power, or packaging reasons. Or, it may be preferable to have a self-emissive device such as an AMEL with its minimum package size. Another consideration is that liquid crystal-based image sources have a finite area over which the image is observable. Collimating optics with a very short focal length may lose part of the image. When considering which image source to use, designers must be concerned with numerous

TABLE 5.1 Categories of Miniature Image Sources Suitable for HMDs

Technology	Transmissive	Reflective	Self-emissive	Scanning
Description	Light source illuminates the display from the rear. Pixels are turned on/off or partially on for gray scale. Transistors along the sides of the pixels.	Light source illuminates the front of the display with a reflective surface under each pixel. Pixels are turned on off or partially on for gray scale, blanking out the incident light. Transistors underneath the pixels.	Individual pixels are turned on/off or partially on for gray scale. Transistors underneath the pixels (AMELs, OLEDs). Drive electronics are remote from the image source (CRTs).	Image source (LED or laser) scans across the image plane. Drive electronics are remote from image source surface.
Examples	Active Matrix Liquid Crystal Display (AMLCD)	Reflective Liquid Crystal on Silicon (LCOS) Digital Micromirror Display (DMD)	Active Matrix Electroluminescent (AMEL) Cathode Ray Tube (CRT) Vacuum Fluorescent Display (VFD) Organic Light Emitting Diode (OLED)	Retinal Scanning Display (RSD) Scanning Light-Emitting Diode
Advantages	Very simple illumination design High quality imagery Available commercially in quantity	High luminous efficiency High fill factor (transistors under the pixel)	Smallest package Lightest weight High fill factor (transistors under the pixel) Wide temperature range (AMEL)	High luminance Saturated colors Potential for image plane distortion (RSD)
Disadvantages	Less efficient fill factor Transmission loss through LCD Requires spatial or temporal integration for color (Post) Limited temperature range (LCD) Slower response time (LCD)	Front illumination is more difficult to package Scattered light management is very important Temporal integration for color	Limited luminance Color by temporal integration	Limited availability (RSD) Limited resolution (LED) Packaging limitations

issues such as:

- *Size* — What is the size of the image source itself? If a supplemental illumination source is required, how large is it? How large is the active area of the display? What is the size of the required drive electronics?

- *Weight* — What is the weight of the image source and any required supplemental illumination? If electronic components must be within close proximity to the image source (i.e., head-mounted), how much do they weigh? Can they be taken off the head or moved to a more favorable location on the head? (See Section 5.2.3).

- *Power* — Some image source technologies such as CRTs and AMELs require a high voltage drive. Image sources such as some LCDs have low transmission, requiring a brighter backlight to meet the display luminance requirements.

- *Resolution* — How many pixels can be displayed? Is the image generator or sensor video compatible with this resolution? Is the response time of the image source fast enough to meet pilotage performance requirements?[32] If not, can measures be taken to improve the response time? [33,34]

- *Addressability* — CRTs are considered infinitely addressable because the imagery is drawn in calligraphic fashion. Pixilated devices such as LCDs, AMELs, and OLEDs are considered finite addressable displays because the pixel location is fixed. This limits their ability to compensate for image plane distortion.

- *Aspect ratio* — Most miniature CRTs have a circular format, while most of the solid-state pixilated devices such as LCDs and AMELs have a rectangular format. Flat-panel devices with VGA, SVGA, or XGA resolution have a 4:3 horizontal-to-vertical aspect ratio. SXGA resolution devices have a 5:4 aspect ratio.* This is an important consideration when choosing an image source because it determines the field of view of the display.

- *Luminance and contrast* — It is important that the image source be capable of providing a display luminance that is compatible with viewing against bright ambient backgrounds typically found in aviation environment. (See Section 5.4.3)

- *Color* — Is the image source capable of producing color imagery?[35] Because of the advantage that data color-coding provides to the pilot,[36] color is becoming more prevalent in head-down displays. Though not in widespread use in head-up and head-mounted displays, it may become more important because there are some preliminary indications that high *g*-forces can alter color perception in the cockpit.[37]

As of this writing, most cockpit displays — head-up, head-down, and helmet-mounted — are still CRT-based, though there is movement towards backlit LCDs and some new projection approaches. The Microvision RSD is showing promise because of its potential for a very high luminance output, though still a bit bulky and with limited availability for some HMD applications. It is likely that the first deployed use of a flat-panel image source in an HMD will be for the Comanche HIDSS, using a small, high resolution AMLCD from Kopin.[34]

5.2.2 Optical Design

The purpose of the optics in an HMD is threefold:

- Collimate the image source — creating a *virtual image*, which appears to be farther away than just a few inches from the face.

- Magnify the image source — making the imagery appear larger than the actual size of the image source.

- Relay the image source — creating the virtual image away from the image source, away from the front of the face.

---

* VGA is 640 horizontal pixels by 480 vertical rows. SVGA is 800 horizontal pixels by 600 vertical rows. XGA is 1024 horizontal pixels by 768 vertical rows. SXGA is 1280 horizontal pixels by 1024 vertical rows.

There are two optical design approaches common in HMDs. The first is the *non-pupil-forming design* — a simple magnifying lens — hence the term *simple magnifier.*[38,39] It is the easiest to design, the least expensive to fabricate, the lightest and the smallest, though it does suffer from a short throw distance between the image source and the virtual image, putting the whole assembly on the front of the head, close to the eyes. This approach is typically used for simple viewing applications such as the medical HMD (Figure 5.1a) and the Land Warrior display (Figure 5.1b).

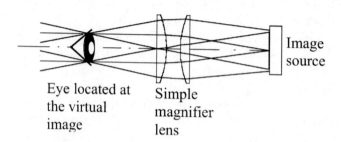

FIGURE 5.6   Diagram of a simple magnifier, or non-pupil-forming lens.

The second optical approach is a bit more complex, the *pupil-forming design.* This is more like the *compound microscope,* or a submarine periscope in which a first set of lenses creates an intermediate image of the image source. This intermediate image is *relayed* by another set of lenses to where it creates a pupil, or a hard image of the intermediate image.

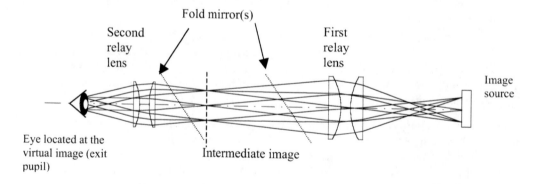

FIGURE 5.7   A pupil-forming optical design is similar to a compound microscope, binoculars, or a periscope.

The advantage is that the pupil-forming design provides more path length from the image plane to the eye. This gives the designer more freedom to insert mirrors as required to fold the optical train away from the face to a more advantageous weight and center of gravity location. The disadvantages are that the additional lenses increase the weight and cost of the HMD and that outside the exit pupil—the image of the stop — there is no imagery. This approach is typically used when the image source is large (such as a CRT) or where it is desirable to move the weight away from the front of the face such as in Figure 5.1c and Figure 5.3.

In each case, the optical design must be capable of collimating, magnifying, and relaying the image with sufficiently small amounts of residual aberrations,[39] with manual focus (if required), and with proper alignment (if a binocular system). In addition, the optical design must provide a sufficiently large exit

TABLE 5.2 Some of the Advantages and Disadvantages of Pupil-Forming and Non-Pupil-Forming Optical Designs for HMDs

	Non-pupil-forming (simple magnifier)	Pupil-forming (relayed lens design)
Advantages	Simplest optical design Fewer lenses and lighter weight Doesn't "wipe" imagery outside of eye box Less eyebox fit problems Mechanically the simplest and least expensive	Longer path length means more packaging freedom. Can move away from front of face. More lenses provide better optical correction

Disadvantages	Short path-length puts the entire display near
	  the eyes/face
	Short path-length means less packaging
	  design freedom
		More complicated optical design
		More lenses mean heavier design
		Loss of imagery outside of pupil
		Needs precision fitting, more and finer adjustments

pupil* so the user doesn't lose the image if the HMD shifts on the head, as well as providing at least 25 mm of eye relief** to allow the user to wear eyeglasses

5.2.3 Head Mounting

It is difficult to put a precise metric on the fit or comfort of an HMD, though it is always immediately evident to the wearer. Even if the HMD image quality is excellent, the user will reject it if it doesn't fit well. Fitting and sizing are especially critical in the case of a helmet-mounted display where, in addition to being comfortable, it must provide a *precision* fit for the display relative to the pilot's eyes.

We can list the most important issues for achieving a good fit with an HMD:

- The user must be able to adjust the display to see the imagery.
- The HMD must be comfortable for long duration wear without causing "hot spots."
- The HMD must not slip with sweating or under *g*-loading, vibration, or buffeting.
- The HMD must be retained during crash or ejection.
- The weight of the head-borne equipment must be minimized.
- The mass-moment-of-inertia must be minimized.
- The mass of the head-borne components should be distributed to keep the center of gravity close to that of the head alone.

The human head weighs approximately 9 to 10 lb and sits atop the spinal column. The Occipital Condyles on the base of the skull mate to the Superior Articular Facets of the first cervical vertebra, the Atlas.[40] These two small, oblong mating surfaces on either side of the spinal column are the pivot points for the head.

The center of gravity (CG) of the head is located at or about the tragion notch, the small cartilaginous flap in front of the ear. Because this is *up* and *forward* of the head/vertebra pivot point, there is a tendency for the head to tip downwards, were it not for the strong counter force exerted by the muscles running down the back of the neck — hence, when people fall asleep they "nod off." Adding mass to the head in the form of an HMD can move the CG (now HMD + head) away from this ideal location. High vibration or buffeting, ejection, parachute opening, or crash will greatly exacerbate the effect of this extra weight and

---

*The exit pupil is found only in pupil-forming designs such as the SIM EYE (Figure 5.1c), the IHADSS (Figure 5.3), and the HIDSS (Figure 5.4). In non-pupil-forming designs of Figures 5.1a and 5.1b, it is more nearly correct to refer to a *viewing eyebox*, because there is a finite unvignetted viewing area.

**There are some differences in terminology usually relating to the writing of specifications. In the classical optical design, the eye relief is the distance along the optical axis from the last optical surface to the exit pupil. In an HMD with angled combiners, eye relief should be measured from the eye to the closest point of the combiner, whether it is on the optical axis or not.

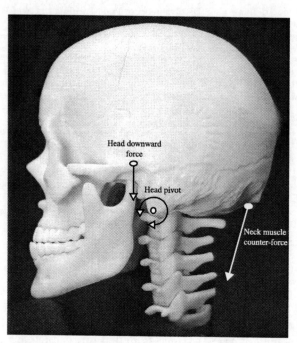

FIGURE 5.8   The human head and neck with the center of gravity located near the tragion notch and the pivot point located at the Occipital Condyles.

displaced CG, with effects that can range from fatigue and neck strain to serious or mortal injury.[41] Designers can mitigate the impact of the added head-borne hardware by first minimizing the mass of the HMD, followed by an optimization of the *location* of the mass to restore the head + HMD CG location to that of the head alone.

This is supported by the extensive biomechanics research at the U.S. Army's Aeromedical Research Labs. Figure 5.9 gives a weight vs. CG curve in the vertical direction, where the area under the curve is considered crash safe for a helicopter environment. The second graph (Figure 5.10) defines the weight/CG combination that will minimize fatigue.[12] Similar work in fixed-wing biomechanics at the Air Force's Wright-Patterson Labs has concluded that the weight of the HMD and oxygen mask cannot exceed 4 lb, and that the resulting center of gravity must also be within a specified region centered about tragion notch.[40]

Anthropometry — "the measure of Man" — is a compilation of data that define such things as the range of height for males and females, the size of our heads, and how far our eyes are apart. Used judiciously, these data can help the HMD designer achieve a proper fit, though an overreliance can be equally problematical. One of the most common mistakes made by designers is to assume a correlation between various anthropometric measurements, because almost all sizing data are *univariate* — that is, they are completely uncorrelated with other data. For example, a person who has a 95th percentile head circumference will not necessarily have a 95th percentile interpupillary distance.[42] One bivariate study did correlate head length and head breadth for male and female aviators, resulting in a rather large spread of data.[43]

There are examples where helmet and HMD developments have been less than successful as a result of an overemphasis on anthropometric data and an underemphasis on fitting, resulting in HMDs that don't fit properly (INIGHTS) or in extraneous helmet sizes (the HGU-53/P).[42]

5.3   The HMD as Part of the Visually Coupled System

In an avionics application, the HMD — be it a Helmet-Mounted Display or Helmet-Mounted Sight — is part of a Visually Coupled System (VCS) consisting of the HMD, a head tracker, and mission computer. As the pilot turns his head, the new orientation is communicated to the mission computer that updates the imagery as required. The information is always with the pilot, always ready for viewing.

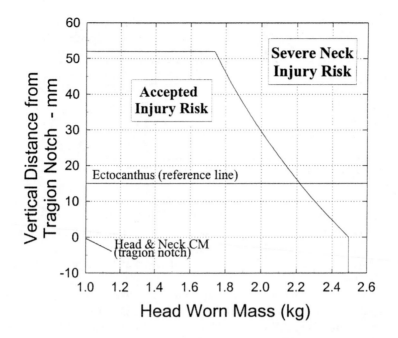

FIGURE 5.9 The USAARL weight and vertical center of gravity curve. The area under the curve is considered crash safe in helicopter environments. (Data curve courtesy of U.S. Army Aeromedical Research Labs, used with permission.)

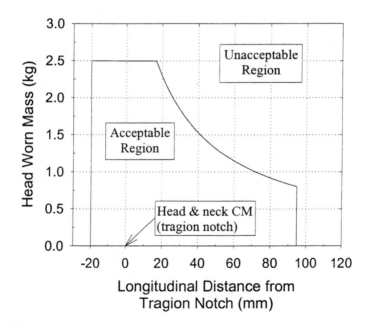

FIGURE 5.10 The USAARL weight and horizontal center of gravity curve with the area under the curve considered acceptable for fatigue in helicopter environments. (Data curve courtesy of U.S. Army Aeromedical Research Labs, used with permission.)

TABLE 5.3   The Univariate (Uncorrelated) Anthropometric Data for Key Head Features. Note the Range of Sizes for the 5th Percentile Female up to the 95th Percentile Male.[44]

Critical Head Dimensions (cm)	5% Female	95% Male
Interpupillary distance (IPD)	5.66	7.10
Head length [a]	17.63	20.85
Head width	13.66	16.08
Head circumference	52.25	59.35
Head height (ectocanthus to top of head)[a]	10.21	12.77

[a] These data are head orientation-dependent.

Early cockpit-mounted displays — Head-Down Displays — gave the pilot information on aircraft status, but required him to return his attention continuously to the interior of the cockpit. This reduced the time he could spend looking outside the aircraft. As jets got faster and the allowable reaction time for pilots got shorter, Head-Up Displays (HUD) provided the next improvement by creating a collimated, virtual image that is projected onto a combining glass located on top of the cockpit panel, in the pilot's forward line of sight.* This means the pilot does not have to redirect his attention away from the critical forward airspace or refocus his eyes to see the image. Because the imagery is collimated — it appears as though from some distant point — it can be superimposed on a distant object. This gives the pilot access to real-time geo- or aircraft-stabilized information such as compass headings, artificial horizons, or sensor imagery.

The HMD expands on this capability by placing the information in front of the pilot's eyes at all times and by linking the information to the pilot's line of sight. While the HUD provides information about only the relatively small forward-looking area of the aircraft, the HMD with head tracker can provide information over the pilot's entire field of regard, all around the aircraft with eyes- and head-out viewing. This ability to link the displayed information with the pilot's line of sight increases the area of regard over which the critical aircraft information is available. This new capability can:

- Cue the pilot's attention by providing a pointing reticle to where a sensor has located an object of interest.
- Allow the pilot to slew sensors such as FLIR for flying at night or in adverse conditions.
- Allow the pilot to aim weapons at targets that are off-boresight from the line of sight of the aircraft.
- Allow the pilot to hand-off or receive target information (or location) from a remote platform, wingman, or other crew member.
- Provide the pilot with aircraft- or geo-stabilized information.

And, in general, provide situational awareness to the pilot by giving him information about the entire space surrounding the aircraft.

One excellent example is the U.S. Army AH-64 Apache helicopter equipped with Honeywell's Integrated Helmet And Display Sighting System (IHADSS) HMD and head tracker. As the pilot moves his head in azimuth or elevation, the tracker communicates the head orientation to the servo system controlling the Pilot Night Vision System (PNVS) FLIR. The sensor follows his head movements, providing the pilot with a viewpoint as though his head were located on the nose of the aircraft. This gives the pilot the ability to "see" at night or in low light in a very intuitive and hands-off manner, similar to the way he would fly during daytime with the overlay of key flight data such as heading, altitude, and airspeed.

Studies are being conducted to find ways to squeeze even more out of the HMD in high-performance aircraft. A recent simulator study at the Naval Weapons Center used the HMD to provide "pathway in the sky" imagery to help pilots avoid threats and adverse weather.[45] Another experimental feature compensated for the loss of color and peripheral vision that accompanies g-induced loss of consciousness

---

*Head-Up Displays are discussed in Chapter 4.

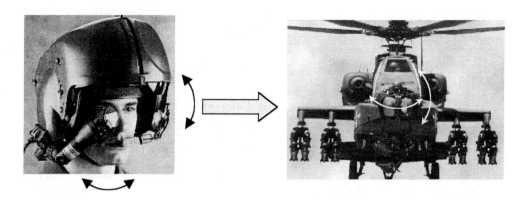

FIGURE 5.11 The linkage between the IHADSS helmet-mounted display and the Pilot's Night Vision System in the AH-64 Apache helicopter. The PNVS is slaved to the pilot's head line of sight. As he turns his head, the PNVS turns to point in the same direction.

(*g*-loc). As the pilot began to "gray-out" the symbol set was reduced down to just a few critical items, positioned closer to the pilot's central area of vision. Another study provided helicopter pilots with earth-referenced navigation waypoints overlayed on terrain and battlefield engagement areas.[46] The results showed significant improvements in navigation, landing, the ability to maintain fire sectors, and an overall reduction in pilot workload.

5.4 HMD System Considerations and Trade-Offs

As mentioned in the Introduction, good HMD design relies on a suboptimization of requirements, trading off various performance parameters and requirements. The following sections will address some of these issues.

5.4.1 Ocularity

One of the first issues to consider in an HMD is whether it should be monocular, biocular, or binocular,:

Monocular — *a single video channel viewed by a single eye*. This is the lightest, least expensive, and simplest of all three approaches. Because of these advantages, most of the current HMD systems are monocular, such as the Elbit DASH, the Vision Systems International JHMCS (Figure 5.2), and the Honeywell IHADSS (Figure 5.3). Some of the drawbacks are the potential for a laterally asymmetric center of gravity and issues associated with focus, eye dominance, binocular rivalry, and ocular-motor instability.[47,48]

Biocular — *a single video channel viewed by both eyes*. The biocular approach is more complex than the monocular design, though it stimulates both eyes, eliminating the ocular-motor instability issues associated with monocular displays. Viewing imagery with two eyes vs one has been shown to yield improvements in detection as well as providing a more comfortable viewing experience.[49,50] However, since it is now a two-eyed viewing system, the designer is subject to a much more stringent set of alignment, focus, and adjustment requirements.[51] The primary disadvantage of the biocular design is that the image source is usually located in the forehead region, making it more difficult to package. In addition, since the luminance from the single image source is split to both eyes, the brightness is cut in half.

Binocular — *each eye views an independent video channel*. This is the most complex, most expensive, and heaviest of all three options, but one which has all the advantages of a two-eyed system with the added benefit of providing partial binocular overlap (to enlarge the horizontal field of view), stereoscopic imagery, and more packaging design freedom. Examples are the Kaiser Electronics HIDSS (Figure 5.4) and the Kaiser Electro-Optics SIM EYE (Figure 5.1c). A binocular HMD is subject to the same alignment, focus, and adjustment requirements as the biocular design, but the designer can move both the optics and the image sources *symmetrically away* from the face.

TABLE 5.4 Advantages and Disadvantages of Monocular, Biocular, and Binocular HMDs

Configuration		Advantages	Disadvantages
Monocular (1 image source viewed by 1 eye)		Lightest weight Simplest to align Least expensive	Potential for asymmetric center of gravity Potential for ocular-motor instability, eye dominance, and focus issues
Biocular (1 image source viewed by both eyes)		Simple electrical interface Lightweight Inexpensive	More complex alignment than monocular Difficult to package Difficult for see-through
Binocular (2 image sources viewed by both eyes)		Stereo imagery Partial binocular overlap Symmetrical center of gravity	Most difficult to align Heaviest Most expensive

5.4.2 Field of View and Resolution

When asked about HMD requirements, users will typically want more of both field-of-view (FOV) *and* resolution. This is not surprising since the human visual system has a total field of view of 200° horizontal by 130° vertical[52] with a grating acuity of 2 min of arc[53] in the central foveal region, something that HMD designers have yet to replicate. For daytime air-to-air applications in a fixed-wing aircraft, a large FOV is probably not necessary to display the symbology shown in Figure 5.5. If it is a simple sighting reticle, the FOV can be approximately 6°. For an HMD such as the JHMCS system where the pilot will receive aircraft and weapons status information, a 20° FOV is more effective. If the HMD is intended to display sensor imagery for nighttime pilotage such as with the IHADSS (a rectangular 30° by 40° FOV), the pilot will "paint" the sky with the HMD, creating a mental map of his surroundings. The larger FOV is advantageous, because it provides peripheral cues that contribute to the pilot's sense of self-stabilization, and it lowers pilot workload by reducing the range of head movements needed to fill in the mental map.[54–56] Most night vision goggles such as the ANVIS-6 have a field of view of 40° circular, though most pilots would prefer more. The Comanche HIDSS will have a rectangular field of view of 35° by 52°.

While display resolution contributes to overall image quality, there is also a direct relationship with performance. If we examine the Johnson criteria for image recognition, we can see that the amount of resolution required is (like most HMD-related issues) task-dependent. For an object such as a tank, increased resolution will allow the pilot to Detect ("something is there"), Recognize ("it's a tank"), or Identify ("it's a T-72 tank")[57] at a particular distance.

While more of each is desirable, FOV and resolution in an HMD are linked by the relationship:

$$H = F * \mathrm{Tan}\ \Theta$$

where F is the focal length of the collimating lens. If :

- H is the size of the image source, then Θ *is the field of view*, or the apparent size of the virtual image in space.
- H is the pixel size, then Θ *is the resolution* or apparent size of the pixel in image space.

Thus, the focal length of the collimating lens *simultaneously* governs the field of view (which you want large) *and* the resolution (which you want small). For a display with a single image source, the result is either wide field of view, *or* high resolution, but *not both* at the same time.

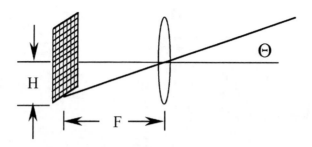

FIGURE 5.12 The focal length of the collimating lens determines the relationship between H, the size of the image source (or pixel size) and Θ, the field of view (or the resolution).

Given this $F * Tan\Theta$ invariant, there are at least four ways to increase the field of view of a display and still maintain resolution. These are (1) high-resolution area of interest, (2) partial binocular overlap, (3) optical tiling, and (4) dichoptic area of interest.[58,59] Of these, partial binocular overlap is preferable for binocular flight applications, though optical tiling is under investigation to expand the field of view of night vision goggles.[60]

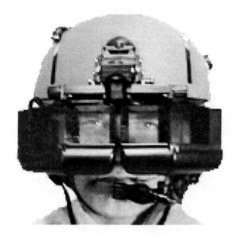

FIGURE 5.13 Proview™ 100 HMD mounted on SPH-4B helicopter helmet for a simulator application. This is an example of an optically tiled see-through HMD with 100° by 30° field of view. (Photo courtesy of Kaiser Electro-Optics, used with permission.)

Partial binocular overlap results when the two HMD optical channels are canted either inward (convergent overlap) or outward (divergent overlap). This enlarges the horizontal field of view, while maintaining the same resolution as the individual monocular channels. Partial overlap requires that two image sources and two video channels are available and that the optics and imagery are properly configured to compensate for any residual optical aberrations. Concerns have been voiced about the required minimum binocular overlap as well as the possibility that perceptual artifacts such as binocular rivalry — referred to as "luning" — may have an adverse impact on pilot performance. Although the studies found image fragmentation did place some workload on the pilot/test subjects,[61,62] all were conducted using static imagery. Several techniques have been effective in reducing the rivalry effects and their associated perceptual artifacts.[63]

It should be kept in mind that the resolution of the VCS is a product of the resolution of the HMD and of the imaging sensor. While an HMD with very high resolution may provide a high-quality image, pilotage performance may still be limited by the resolution of the imaging sensor such as the FLIR or camera. In most cases, it is preferable to match the field of view of the HMD with that of

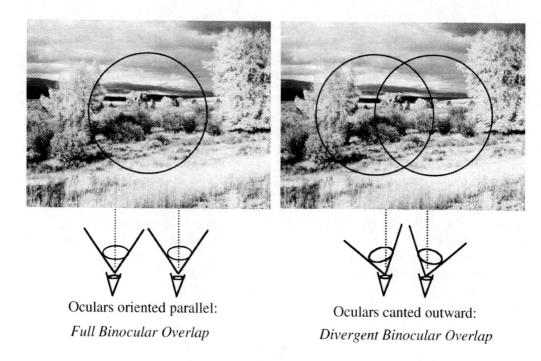

Oculars oriented parallel:

Full Binocular Overlap

Oculars canted outward:

Divergent Binocular Overlap

FIGURE 5.14   Comparison of a full binocular overlap and divergent partial binocular overlap. Note the increase in viewable imagery in the horizontal direction with the divergent overlap.

the sensor to achieve a 1:1 correspondence between sensor and display to ensure an optimum flying configuration.

5.4.3   Luminance and Contrast in High Ambient Luminance Environments

In the high ambient luminance environment of an aircraft cockpit, daylight readability of displays is a critical issue. The combining element in an HMD is similar to the combiner of a HUD, reflecting the projected imagery into the pilot's eyes. The pilot looks through the combining glass and sees the imagery superimposed on the outside world, so it cannot be 100% reflective — pilots always prefer to have as much see-through as possible. To view the HMD imagery against a bright background such as sun-lit clouds or snow, this less-than-perfect reflection efficiency means that the image source must be that much brighter. The challenge is to provide a combiner with good see-through transmission and still provide a high-luminance image. There are limitations, though, because all image sources have a luminance maximum governed by the physics of the device as well as size, weight, and power of any ancillary illumination. In addition, other factors such as the transmission of the aircraft canopy and pilot's visor must be considered when determining the required image source luminance, as shown in Figure 5.15.

The image source luminance (B_I) is attenuated before entering the eye by the transmission of the collimating optics (T_O) and the reflectance of the combiner (R_C). The pilot views the distant object through the combiner (T_C or $1 - R_C$), the protective visor (T_V), and the aircraft transparency (T_A) against the bright background (B_A). We can calculate the image source luminance for a desired contrast ratio (CR) of 1.3 using the expression:[11]

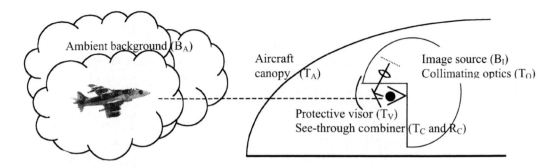

FIGURE 5.15 Contributions for determining image source luminance requirements for an HMD in an aircraft cockpit.

$$CR = \frac{B_A + B_{Display}}{B_A}$$

where we know that the display luminance to the eye is given by:

$$B_{Display} = B_I * T_O * R_C$$

and, as observed by the pilot, the background is given by:

$$B_O = T_C * T_V * T_A * B_A$$

Rewriting, we can see that:

$$CR = \frac{1 + B_I * T_O * R_C}{T_C * T_V * T_A * B_A}$$

We can substitute some nominal values for the various contributions as given in the following table:

TABLE 5.5 Contributions for the Display Luminance Calculations for Four Different HMD Configurations

		Case 1 — Clear Visor, 50% Combiner Transmission	Case 2 — dark Visor, 50% Combiner Transmission	Case 3 — Clear Visor, 80% Combiner Transmission	Case 4 — Dark Visor, 80% Combiner Transmission
Optics transmission	T_O	85%	85%	85%	85%
Combiner reflectance	R_C	50%	50%	20%	20%
Combiner transmission	T_C	50%	50%	80%	80%
Visor transmission	T_V	87%	12%	87%	12%
Aircraft canopy transmission	T_C	80%	80%	80%	80%
Ambient background luminance	B_C	10,000 fL	10,000 fL	10,000 fL	10,000 fL
Required image source luminance	B_I	**2,456 fL**	**339 fL**	**9,826 fL**	**1,355 fL**

The first two cases compare the difference when the pilot is wearing a Class 1 (clear) vs. a Class 2 (dark) visor.[64] The dark visor reduces the ambient background luminance, improving HMD image contrast against the bright clouds or snow. These first two cases are relatively simple because they assume a combiner with 50% transmission and 50% reflectance (ignoring other losses). Since pilots need more see-through, this means a reduced reflectance. Cases 3 and 4 assume this more realistic combiner configuration with both clear and dark visors, resulting in a requirement for a much brighter image source.

One of the ways to improve both see-through transmission and reflectance is to take advantage of high-reflectance holographic notch filters and V-coats. The problem is that while these special coatings reflect more of a specific display color, they transmit less of that *same* color, which can alter perceptions of cockpit display color as well as external coloration.

5.5   Summary

Head-mounted displays can provide a distinctly unique and personal viewing experience no other display technology can match. By providing the pilot with display information that is linked to head orientation, the pilot is freed from having to return his attention to the cockpit interior and is able to navigate and fly the aircraft in a more intuitive and natural manner. This is an effective means of providing a pilot with aircraft status as well as information about the surrounding airspace.

But these capabilities are not without a price. HMDs require careful attention to the complex interactions between hardware and human perceptual issues, made only more complex by the need for the HMD to provide life support in an aviation environment. Only when all factors are considered and the requirements successfully suboptimized with an understanding of the aviator's tasks and environment, will this be accomplished.

Recommended Reading

Barfield, W. and Furness, T. A., *Virtual Environments and Advanced Interface Design*, New York, Oxford University Press, 1995.

Boff, K. R. and Lincoln, J. E., Engineering Data Compendium, Human Perception and Performance, Human Engineering Division, Harry G. Armstrong Aerospace Medical Research Laboratory, Wright-Patterson Air Force Base, OH, 1988.

Kalawsky, R. A., *The Science of Virtual Reality and Virtual Environments*, New York, Addison-Wesley, 1993.

Karim, M. A., Ed., *Electro-Optical Displays*, New York, Marcel Dekker, 1992.

Lewandowski, R. J., *Helmet- and Head-Mounted Displays*, Selected SPIE papers on CD-ROM, SPIE Press, 11, 2000.

Melzer, J. E. and Moffitt, K. W., Eds., *Head-Mounted Displays: Designing for the User*, McGraw-Hill, New York, 1997.

Rash, C. E., Ed., *Helmet-Mounted Displays: Design Issues for Rotary-Wing Aircraft*, U.S. Government Printing Office, Washington, D.C., 1999.

Velger, M., *Helmet-Mounted Displays and Sights*, Norwood, MA, Artech House, 1998.

References

1. Kalawsky R. S., *The Science of Virtual Reality and Virtual Environments: A Technical, Scientific and Engineering Reference on Virtual Environments*, Addison-Wesley: Wokingham, England (1996).
2. Schmidt, G. W. and Osborn, D. B., Head-mounted display system for surgical visualization, *Proc. SPIE, Biomed. Optoelectron. Instrum.*, 2396, 345, 1995.
3. Pankratov, M. M., New surgical three-dimensional visualization system, *Proc. SPIE, Lasers in Surgery Adv. Characterization Ther. Syst.*, 2395, 143, 1995.

4. Casey, C. J., Helmet-mounted displays on the modern battlefield, *Proc. SPIE, Helmet- and Head-Mounted Displays IV*, 3689, 270, 1999.

5. Browne, M. P., Head-mounted workstation displays for airborne reconnaissance applications, *Proc. SPIE, Cockpit Displays V: Displays for Defense Applications*, 3363, 348, 1998.

6. Lacroix, M., Melzer, J., Helmet-mounted displays for flight simulators, *Proc. IMAGE VII Conf.*, Tucson Az, 12–17 June, (1994).

7. Casey, C. J. and Melzer, J. E., Part-task training with a helmet integrated display simulator system, *Proc. SPIE, Large-Screen Projection, Avionic and Helmet-Mounted Displays*, 1456, 175, 1991.

8. Thomas, M. and Geltmacher, H., Combat simulator display development, *Inf. Display*, 4&5, 23, 1993.

9. Foote, B., Design guidelines for advanced air-to-air helmet-mounted display systems, *Proc. SPIE, Helmet- and Head-Mounted Displays, III*, 3362, 94, 1998.

10. Belt, R. A., Kelley, K., and Lewandowski, R., Evolution of helmet-mounted display requirements and Honeywell HMD/HMS systems, *Proc. SPIE, Helmet- and Head-Mounted Displays III*, 3362, 373, 1998.

11. Kocian, D. F., Design considerations for virtual panoramic display (VPD) helmet systems, *AGARD Conf. Proc. No. 425*, 22-1, 1987.

12. Rash, C. E. (Ed.), *Helmet-Mounted Displays: Design Issues for Rotary-Wing Aircraft*, U.S. Government Printing Office, Washington, D.C., 1999.

13. Dornheim, M., VTAS sight fielded, shelved in 1970s, *Aviat. Week Space Technol.*, October 23, 51, 1995.

14. Dornheim, M. A. and Hughes, D., U.S. intensifies efforts to meet missile threats, *Aviat. Week Space Technol.*, October 16, 36, 1995.

15. Arbak, C., Utility evaluation of a helmet-mounted display and sight, *Proc. SPIE, Helmet-Mounted Displays*, 1116, 138, 1989.

16. Merryman, R. F. K., Vista Sabre II: integration of helmet-mounted tracker/display and high off-boresight missile seeker into F-15 aircraft, *Proc. SPIE, Helmet- and Head-Mounted Displays and Symbology Design Requirements*, 2218, 173, 1994.

17. Lake, J., NATO's best fighter is made in Russia, *The Daily Telegraph*, August, 26, 1991, p. 22.

18. Goodman, G. W., Jr., First look, first kill, *Armed Forces J. Int.*, July 2000, p. 32.

19. Braybrook, R., Looks can kill, *Armada Int.*, 4, 44, 1998.

20. Sheehy, J. B. and Wilkinson, M., Depth perception after prolonged usage of night vision goggles, *Aviat., Space Environ. Med.*, 60, 573, 1989.

21. Donohue-Perry, M. M., Task, H. L., and Dixon, S. A., Visual acuity versus field of view and light level for night vision goggles (NVGs), *Proc. SPIE, Helmet- and Head-Mounted Displays and Symbology Design Requirements*, 2218, 71, 1994.

22. Crowley, J. S., Rash, C. E., and Stephens, R. L., Visual illusions and other effects with night vision devices, *Proc. SPIE, Helmet-Mounted Displays III*, 1695, 166, 1992.

23. DeVilbiss, C. A., Ercoline, W. R., and Antonio, J. C., Visual performance with night vision goggles (NVGs) measured in U.S. Air Force aircrew members, *Proc. SPIE, Helmet- and Head-Mounted Displays and Symbology Design Requirements*, 2218, 64, 1994.

24. Gibson, J. J., *The Ecological Approach to Visual Perception*, Lawrence Erlbaum Associates, Hillsdale, NJ. 1986.

25. Sauerborn, J. P., Advances in miniature projection CRTs for helmet displays, *Proc. SPIE, Helmet-Mounted Displays III*, 1695, 102, 1992.

26. Ferrin, F. J., Selecting new miniature display technologies for head mounted applications, *Proc. SPIE, Head-Mounted Displays II*, 3058, 115, 1997.

27. Belt, R. A., Knowles, G. R., Lange, E. H., Pilney, B. J., and Girolomo, H. J., Miniature flat panels in rotary wing head mounted displays, *Proc. SPIE, Head-Mounted Displays II*, 3058, 125, 1997.

28. Urey, H., Nestorovic, N., Ng, B., and Gross, A. A., Optics designs and systems MTF for laser scanning displays, *Proc. SPIE, Helmet- and Head-Mounted Displays, IV*, 3689, 238, 1999.

29. Urey, H., Optical advantages in retinal scanning displays, *Proc. SPIE, Head- and Helmet-Mounted Displays,* 4021, 20, 2000.

30. Arbuthnot, L., Aitchison, B., Carlsen, C., King, C. N., Larsson, T., and Nguyen, T., High-luminance active matrix electroluminescent (AMEL) display for see-through symbology applications, *Proc. SPIE, Helmet- and Head-Mounted Displays, IV,* 3689, 260, 1999.

31. Semenza, P. D., The technology is pushing — but will the market pull? *Inf. Display,* 16(7), 14, 2000.

32. Rabin, J. and Wiley, R., Dynamic visual performance: comparison between helmet-mounted CRTs and LCDs, *J. SID,* 3/3, 97, 1995.

33. Gale, R., Herrmann, F., Lo, J., Metras, M., Tsaur, B., Richard, A., Ellertson, D., Tsai, K., Woodard, O., Zavaracky, M., and Presz, M., Miniature 1280 by 1024 active matrix liquid crystal displays, *Proc. SPIE, Helmet- and Head-Mounted Displays IV,* 3689, 231, 1999.

34. Woodard, O. C., Gale, R. P., Ong, H. L., and Presz, M. L., Developing the 1280 by 1024 AMLCD for the RAH-66 Comanche, *Proc. SPIE, Head- and Helmet-Mounted Displays,* 4021, 203, 2000.

35. Post, D. L., Miniature color display for airborne HMDs, *Proc. SPIE, Helmet- and Head-Mounted Displays and Symbology Design Requirements,* 2218, 2, 1994.

36. Melzer, J. E. and Moffitt, K. W., Color helmet display for the tactical environment: the pilot's chromatic perspective, *Proc. SPIE, Helmet-Mounted Displays III,* 1695, 47, 1992.

37. MacGillis, A., Flying at high G's alters pilot's perception of colors, *Natl. Defense,* 16, September 1999.

38. Task, H. L., HMD image sources, optics and visual interface, in *Head-Mounted Displays: Designing for the User,* Melzer, J. E. and Moffitt, K. W., Eds., McGraw-Hill, New York, 1997, chap. 3.

39. Fischer, R. E., Fundamentals of HMD Optics, in *Head-Mounted Displays: Designing for the User,* Melzer, J. E. and Moffitt, K. W., Eds., McGraw-Hill, New York, 1997, chap. 4.

40. Perry, C. E. and Buhrman, J. R., Biomechanics in HMDs, in *Head-Mounted Displays: Designing for the User,* Melzer, J. E. and Moffitt, K. W., Eds., McGraw-Hill, New York, 1997, chap. 6.

41. Guill, F. C. and Herd, G. R., An evaluation of proposed causal mechanisms for "ejection associated" neck injuries, *Aviat. Space Environ. Med.,* A26, July, 1989.

42. Whitestone, J. J. and Robinette, K. M., Fitting to maximize performance of HMD systems, in *Head-Mounted Displays: Designing for the User,* Melzer, J. E. and Moffitt, K. W., Eds., McGraw-Hill, New York, 1997, chap. 7.

43. Barnaba, J. M., Human factors issues in the development of helmet mounted displays for tactical, fixed-wing aircraft, *Proc. SPIE, Head-Mounted Displays II,* 3058, 2, 1997.

44. Gordon, C. C., Churchill, T., Clauser, C. E., Bradtmiller, B., McConville, J. T., Tebbetts, I., and Walker, R. A., 1988 Anthropometric Survey of U.S. Army Personnel: Summary Statistics Interim Report. U.S. Army Natick Tech. Rep. TR-89/027, Natick, MA, 1989.

45. Procter, P., Helmet displays boost safety and lethality, *Aviat. Week Space Technol.,* February 1, 1999, pg. 81.

46. Rogers, S. P., Asbury, C. N. and Haworth, L. A., Evaluation of earth-fixed HMD symbols using the PRISMS helicopter flight simulator, *Proc. SPIE, Helmet- and Head-Mounted Displays III,* 3389, 54, 1999.

47. Rash, C. E. and Verona, R. W., The human factor considerations of image intensification and thermal imaging systems, in *Electro-Optical Displays,* Karim, M. A., Ed., New York, Marcel Dekker, 1992, chap. 16.

48. Moffitt, K. W., Ocular responses to monocular and binocular helmet-mounted display configurations, *Proc. SPIE, Helmet-Mounted Displays,* 1116, 142, 1989.

49. Boff, K. R. and Lincoln, J. E., Engineering Data Compendium, Human Perception and Performance, Human Engineering Division, Harry G. Armstrong Aerospace Medical Research Laboratory, Wright-Patterson Air Force Base, OH, 1988.

50. Moffitt, K. W., Designing HMDs for viewing comfort, in *Head-Mounted Displays: Designing for the User,* Melzer, J. E. and Moffitt, K. W., Eds., McGraw-Hill, New York, 1997, chap. 5.

51. Self, H. C., Critical Tolerances for Alignment and Image Differences for Binocular Helmet-Mounted Displays, Tech. Rep. AAMRL-TR-86-019, Armstrong Aerospace Medical Research Laboratory, Wright-Patterson AFB OH, 1986.

52. U.S. Department of Defense, "MIL-HDBK-141 Optical Design," 1962.

53. Smith, G. and Atchison, D. A., *The Eye and Visual Optical Instruments*, New York, Cambridge University Press, 1997.

54. Wells, M. J., Venturino, M., and Osgood, R. K., Effect of field of view size on performance at a simple simulated air-to-air mission, *Proc. SPIE, Helmet-Mounted Displays*, 1116, 126, 1989.

55. Kasper, E. F., Haworth, L. A., Szoboszlay, Z. P., King, R. D., and Halmos, Z. L., Effects of in-flight field-of-view restriction on rotorcraft pilot head movement, *Proc. SPIE, Head-Mounted Displays II*, 3058, 34, 1997.

56. Szoboszlay, Z. P., Haworth, L. A., Reynolds, T. L., Lee, A. G., and Halmos, Z. L., Effect of field-of-view restriction on rotorcraft pilot workload and performance: preliminary results, *Proc. SPIE, Helmet- and Head-Mounted Displays and Symbology Design Requirements II*, 2465, 142, 1995.

57. Lloyd, J. M., *Thermal Imaging Systems*, Plenum Press, New York, 1975.

58. Melzer, J. E., Overcoming the field of view: resolution invariant in head-mounted displays, *Proc. SPIE, Helmet- and Head-Mounted Displays III*, 3362, 284–293, 1998.

59. Hoppe, M. J. and Melzer, J. E., Optical tiling for wide-field-of-view head-mounted displays, *Proc. SPIE, Current Developments in Optical Design and Optical Engineering VIII*, 3779, 146, 1999.

60. Jackson, T. W. and Craig, J. L., Design, development, fabrication, and safety-of-flight testing of a panoramic night vision goggle, *Proc. SPIE, Head- and Helmet-Mounted Displays IV*, 3689, 98, 1999.

61. Klymenko, V., Verona, R. W., Beasley, H. H., and Martin, J. S., Convergent and divergent viewing affect luning, visual thresholds, and field-of-view fragmentation in partial binocular overlap helmet-mounted displays, *Proc. SPIE, Helmet- and Head-Mounted Displays and Symbology Design Requirements*, 2218, 2, 1994.

62. Klymenko, V., Harding, T. H., Beasley, H. H., Martin, J. S., and Rash, C. E., Investigation of helmet-mounted display configuration influences on target acquisition, *Proc. SPIE, Head- and Helmet-Mounted Displays*, 4021, 316, 2000.

63. Melzer, J. E. and Moffitt, K., An ecological approach to partial binocular-overlap, *Proc. SPIE, Large Screen, Projection and Helmet-Mounted Displays*, 1456, 124, 1991.

64. U.S. Department of Defense, MIL-V-85374, Military specification, visors, shatter resistant, 1979.

6

Display Devices: RSD™ (Retinal Scanning Display)

Thomas M. Lippert
Microvision Inc.

6.1 Introduction

This chapter relates performance, safety, and utility attributes of the Retinal Scanning Display as employed in a Helmet-Mounted Pilot-Vehicle Interface, and by association, in panel-mounted HUD and HDD applications. Because RSD component technologies are advancing so rapidly, quantitative analyses and design aspects are referenced to permit a more complete description here of the first high-performance RSD System developed for helicopters.

Visual displays differ markedly in how they package light to form an image. The Retinal Scanning Display, or RSD depicted in Figure 6.1, is a relatively new optomechatronic device based initially on red, green, and blue diffraction-limited laser light sources. The laser beams are intensity modulated with video information, optically combined into a single, full-color pixel beam, then scanned into a raster pattern by a ROSE comprised of miniature oscillating mirrors, much as the deflection yoke of a cathode-ray tube (CRT) writes an electron beam onto a phosphor screen. RSDs are unlike CRTs in that conversion of electrons to photons occurs prior to beam scanning, thus eliminating the phosphor screen altogether along with its re-radiation, halation, saturation, and other brightness- and contrast-limiting factors. This means that the RSD is fundamentally different from other existing display technologies in that there is no planar emission or reflection surface — the ROSE creates an optical pupil directly. Like the CRT, an RSD may scan out spatially continuous (nonmatrix-addressed) information along each horizontal scan line, while the scan lines form discrete information samples in the vertical image dimension.

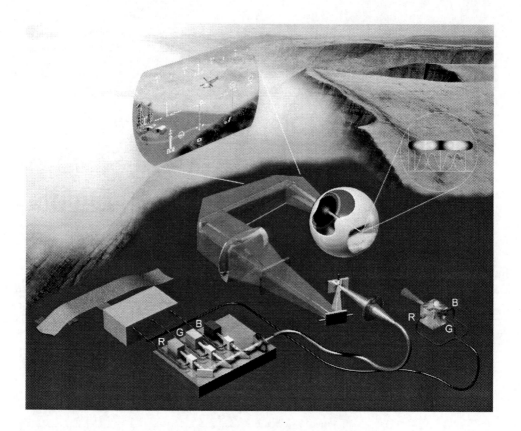

FIGURE 6.1   Functional component diagram of the RSD HMD.

6.2   An Example Avionic HMD Challenge

Consider the display engineering problem posed by Figure 6.1. An aircraft flying the contour of the earth will transit valleys as well as man-made artifacts: towers, power lines, buildings, and other aircraft. On this flight the pilot is faced with a serious visual obscurant in the form of ground fog, rendered highly opaque by glare from the sun.

The pilot's situational awareness and navigation performance are best when flying "eyes-out" the windshield, in turn requiring "eyes-out" electronic display of his own aircraft attitude and status information. Particularly under degraded visual conditions, additional imagery of obstacles (towers, the Earth, etc.) synthesized from terrain data bases and mapped into the pilot's ever-changing direction of gaze via Global Positioning System data, reduce the hazards of flight. The question has been, which technology can provide a display of adequate brightness, color, and resolution to adequately support pilotage as viewed against the harsh real-world conditions described.

For over 30 years, researchers and designers have improved the safety and effectiveness of HMDs so that mission-critical information would always be available "eyes-out" where the action is, unlike "eyes-in" traditional HDDs.[1] U.S. Army AH-64 Apache Helicopter pilots are equipped with such an HMD, enabling nap-of-the-earth navigation and combat at night with video from a visually coupled infrared imager and data computer. This particular pilot-vehicle interface has proven its reliability and effectiveness in over 1 million hours of flight and was employed with great success in the Desert Storm Campaign. Still, it lacks the luminance required for optimal grayscale display during typical daylight missions, much less the degraded conditions illustrated above.

The low luminance and contrast required for nighttime readability is relatively easy to achieve, but it is far more difficult to develop an HMD bright enough and of sufficient contrast for daylight use. The information must be displayed as a dynamic luminous transparency overlaying the real-world's complex features, colors, and motion. In order to display an image against a typical real-world daytime scene luminance of 3000 fL, the virtual display peak luminance must be about 1500 fL at the pilot's eye. And depending on the efficiency of the specific optics employed, the luminance at the display light source may need to be many times greater. The display technology that provides the best HMD solution might also provide the optimal HUD and HDD approaches.

6.3 CRTs and MFPs

Army Aviation is the U.S. military leader in deployed operational HMD systems. The Apache helicopter's monochrome green CRT Helmet Display Unit (HDU) presents pilotage FLIR (forward-looking infrared) imagery overlaid with flight symbology in a 40°(H) × 30°(V) monocular field of view (FOV). The Apache HDU was developed in the late 1970s and early 1980s using the most advanced display technology then available. The new RAH-66 Comanche Helicopter Program has expanded the display's performance requirements to include night and day operability of a monochrome green display with a binocular 52° H × 30° V FOV and at least 30° of left/right image overlap.

The Comanche's Early Operational Capability Helmet Integrated Display Sighting System (EOC HIDSS) prototype employed dual miniature CRTs. The addition of a second CRT pushed the total head-supported weight for the system above the Army's recommended safety limit. Weight could not be removed from the helmet itself without compromising safety, so even though the image quality of the dual-CRT system was good, the resulting reduction in safety margins was unacceptable.

The U.S. Army Aircrew Integrated Systems (ACIS) office initiated a program to explore alternate display technologies for use with the proven Aircrew Integrated Helmet System Program (AIHS, also known as the HGU-56/P helmet) that would meet both the Comanche's display requirements and the Army's safety requirements.

Active-matrix liquid-crystal displays (AMLCD), active-matrix electroluminescent (AMEL) displays, field-emission displays (FEDs), and organic light-emitting diodes (OLEDs) are some of the alternative technologies that have shown progress. These postage-stamp size miniature flat-panel (MFP) displays weigh only a fraction as much as the miniature CRTs they seek to replace.

AMLCD is the heir apparent to the CRT, given its improved luminance performance. Future luminance requirements will likely be even higher, and there are growing needs for greater displayable pixel counts to increase effective range resolution or FOV, and for color to improve legibility and enhance information encoding. It is not clear that AMLCD technology can keep pace with these demands.

6.4 Laser Advantages, Eye Safety

The RSD offers distinct advantages over other display technologies because image quality and color gamut are maintained at high luminances limited only by eye-safety considerations.[2,3] The light-concentrating aspect of the diffraction-limited laser beam can routinely produce source luminances that exceed that of the solar disc. Strict engineering controls, reliable safeguards, and careful certification are mandatory to minimize the risk of damage to the operator's vision.[4] Of course, these safety concerns are not limited to laser displays; any system capable of displaying extremely high luminances should be controlled, safeguarded, and certified.

Microvision's products are routinely tested and classified according to the recognized eye safety standard — the maximum permissible exposure (MPE) — for the specific display in the country of delivery. In the U.S. the applicable agency is the Center for Devices and Radiological Health (CDRH) Division of the Food and Drug Administration (FDA). The American National Standards Institute's Z136.1 reference, "The Safe Use of Lasers," provides MPE standards and the required computational procedures to assess compliance. In most of Europe the IEC 60825-1 provides the standards.

Compliance is assessed across a range of retinal exposures to the display, including single-pixel, single scan line, single video frame, 10-second, and extended-duration continuous retinal exposures. For most scanned laser displays, the worst-case exposure leading to the most conservative operational usage is found to be the extended-duration continuous display MPE. Thus, the MPE helps define laser power and scan-mirror operation-monitoring techniques implemented to ensure safe operation. Examples include shutting down the laser(s) if the active feedback signal from either scanner is interrupted and automatically attenuating the premodulated laser beam for luminance control independent of displayed contrast or grayscale.

6.5 Light Source Availability and Power Requirements

Another challenge to manufacturers of laser HMD products centers on access to efficient, low-cost lasers or diodes of appropriate collectible power (1–100 mW), suitable wavelengths (430–470, 532–580, and 607–660 nm), low video-frequency noise content (<3%), and long operating life (10,000 hr). Diodes present the most cost-effective means because they may be directly modulated up from black, while lasers are externally modulated down from maximum beam power.

Except for red, diodes still face significant development hurdles, as do blue lasers. Operational military-aviation HMDs presently require only a monochrome green, G, display which can be obtained by using a 532-nm diode-pumped solid-state (DPSS) laser with an acoustic-optic modulator (AOM). Given available AOM and optical fiber coupling efficiencies, the 1500-fL G RSD requires about 50 mW of laser beam power. Future requirements will likely include red + green, RG, and full color, RGB, display capability.

6.6 Microvision's Laser Scanning Concept

Microvision has developed a flexible component architecture for display systems (Figure 6.1). RGB video drives AOMs to impress information on Gaussian laser beams, which are combined to form full-color pixels with luminance and chromaticity determined by traditional color-management techniques. The aircraft-mounted photonics module is connected by single-mode optical fiber to the helmet, where the beam is air propagated to a lens, deflected by a pair of oscillating scanning mirrors (one horizontal and one vertical), and brought to focus as a raster format intermediate image. Finally, the image is optically collimated and combined with the viewer's visual field to achieve a spatially stabilized virtual image presentation.

The AIHS Program requires a production display system to be installed and maintained as a helicopter subsystem — designated Aircraft Retained Unit (ARU) — plus each pilot's individually fitted protective helmet, or Pilot Retained Unit (PRU). Microvision's initial concept-demonstration HMD components meet these requirements (Figure 6.2).

Microvision's displays currently employ one horizontal line-rate scanner — the Mechanical Resonant Scanner (MRS) — and a vertical refresh galvanometer. Approaches using a bi-axial microelectro-mechanical system (MEMS) scanner are under development. Also, as miniature green laser diodes become available, Microvision expects to further reduce ARU size, weight, and power consumption by transitioning to a small diode module (Figure 6.1, lower-right) embedded in the head-worn scanning engine, which would also eliminate the cost and inefficiency of the fiber optic link.

For the ACIS project, a four-beam concurrent writing architecture was incorporated to multiply by 4 the effective line rate achievable with the 16-kHz MRS employed in unidirectional horizontal writing mode. The vertical refresh scanner was of the 60-Hz saw-tooth-driven servo type for progressive line scanning. The f/40 writing beams, forming a narrow optical exit pupil (Figure 6.3), are diffraction-multiplied to form a 15-mm circular matrix of exit pupils.

The displayed resolution of a scanned-light-beam display[5] is limited by three parameters: (1) spot size and distribution as determined by cascaded scan-mirror apertures (*D*), (2) total scan-mirror deflection angles in the horizontal or vertical raster domains (*Theta*), and (3) dynamic scan-mirror flatness under normal operating conditions. Microvision typically designs to the full-width/half-maximum Gaussian spot overlap criterion, thus determining the spot count per raster line. Horizontal and vertical displayable spatial resolutions, limited by (*D*)*(*Theta*), must be supported by adequate scan-mirror dynamic flatness for the projection engine to perform at its diffraction limit. Beyond these parameters, image quality is affected by all the components common to any video projection display. Electronics, photonics, optics, and packaging tolerances are the most significant.

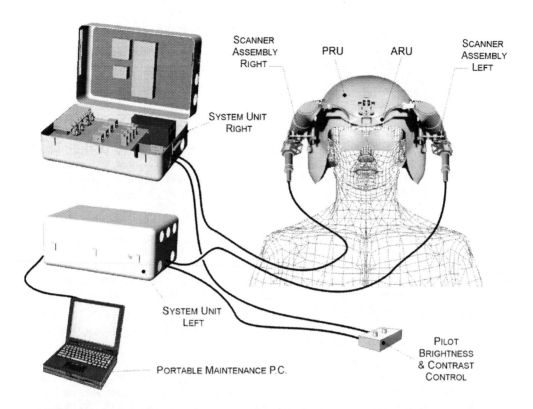

FIGURE 6.2 Microvision's RSD components meet the requirements of the AIHS HIDSS program for an HMD.

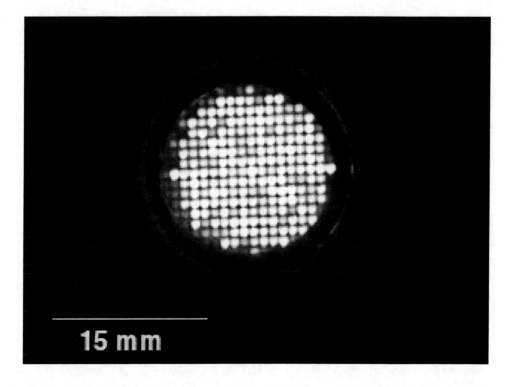

FIGURE 6.3 The far-field beamlet structure of a spot-multiplied (expanded) RSD OEP. The unexpanded 1-mm exit pupil is represented by a single central spot.

6.6.1   Government Testing of the RSD HMD Concept

Under the ACIS program, the concept version of the Microvision RSD HMD was delivered to the U.S. Army Aeromedical Research Laboratory (USAARL) for testing and evaluation in February 1999.[6]

As expected, the performance of the concept-phase system had some deficiencies when compared to the RAH-66 Comanche requirements. However, these deficiencies were few in number and the overall performance was surprisingly good for this initial development phase. Measured performance for exit pupil, eye relief, alignment, aberrations, luminance transmittance, and field-of-view met the requirements completely. The luminance output of the left and right channels — although high, with peak values of 808 and 1111 fL, respectively — did not provide the contrast values required by Comanche in all combinations of ambient luminance and protective visor. Of greatest concern was the modulation transfer function (MTF) — and the analogous Contrast Transfer Function (CTF) — exhibiting excessive rolloff at high spatial frequencies, and indicating a "soft" displayed image.

6.6.2   Improving RSD Image Quality

At the time of this writing, the second AIHS program phase is concentrating on imprving image quality. Microvision identified the sources of the luminance, contrast, and MTF/CTF deficiencies found by USAARL. A few relatively straightforward fixes such as better fiber coupling, stray light baffling, and scan-mirror edge treatment are expected to provide the luminance and low-spatial-frequency contrast improvements required to meet specification, but MTF/CTF performance at high spatial frequencies have presented a more complex set of issues.

Each image-signal-handling component in the system contributes to the overall system MTF. Although the video electronics and AOM-controller frequency responses were inadequate, they were easily remedied through redesign and component selection. Inappropriate mounting of fixed fold mirrors in the projection path led to the accumulation of several wavelengths of wave-front error and resultant image blurring. This problem, too, is readily solved.

The second class of problems pertains to the figure of the scan mirrors. Interferometer analyses of the flying spot under dynamic horizontal scanning conditions indicated excessive mirror surface deformation ($\sim$2 peak-to-peak mechanical), resulting in irregular spot growth and reduced MTF/CTF performance (Figure 6.4).

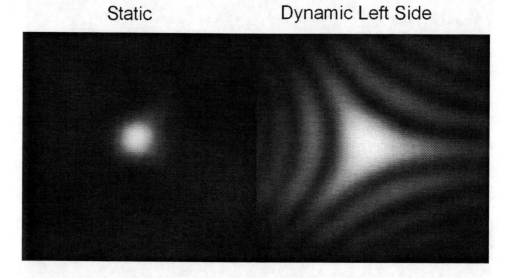

FIGURE 6.4   The effect of improved mirror design is visible in these spot (pixel) images, normalized for size but not for intensity, for scanned spots at $\sim\lambda/4$ P-P mechanical mirror deformation (left image), and $\sim2\lambda$ P-P mechanical mirror deformation (right image).

Three fast-prototyping iterations brought the mirror surface under control ($\sim\lambda/4$) to achieve acceptable spot profiles at the raster edge. Thus, the component improvements described above are expected to result in MTF/CTF performance meeting U.S. Army specification.

6.7 Next Step

The next step in the evolution of the helicopter pilot's laser HMD is the introduction of daylight-readable color. Microvision first demonstrated a color VGA format RSD HMD in 1996, followed by SVGA in 1998. Development of a 1280 × 1024-color-pixel (SXGA) binocular HMD project is being made possible by ACIS's Virtual Cockpit Optimization Program (VCOP), which begins with software-reconfigurable virtual flight simulations in 2000 and proceeds to in-flight virtual cockpit demonstrations in 2001. For these demonstrations, the aircraft's traditional control-panel instrumentation is expected to serve only an emergency backup function. Figure 6.1, with which this chapter began, represents the VCOP RGB application concept.

One configuration of the VCOP simulation/operation HMD acknowledges the limited ability of the blue component to generate effective contrast against white clouds or blue sky. Because the helmet tracker used in any visually-coupled system will "know" when the pilot is "eyes out" or "head down", the HMD may employ graphics and imaging sensor formats in daylight readable greenscale, combined with red, for "eyes out" information display across established green/yellow/red caution advisory color codes, switching to full color formats at lower luminances for "head down" displays of maps, etc.

The fundamental capabilities of the human visual system, along with ever increasing imaging sensor and digital image generation bandwidths, require HMD spatial resolutions greater than SXGA. For this reason, the US Air Force Research Laboratory has contracted Microvision Inc. to build the first known HDTV HMD (1920 × 1080 pixels in a noninterlaced 60 Hz frame refresh digital video format). The initial system will be a monocular 100-fL monochrome green fighter pilot training HMD with growth-to-daylight readable binocular color operation.

An effort of 30 years has only scratched the surface of the HMD's pilot vehicle interfacing potential. It is expected that the RSD will open new avenues of pilot-in-the-loop research and enable safer, more effective air and ground operations.

Defining Terms

Optomechatronic: Application of integrated optical, mechanical, and electronic elements for imaging and display.

Helmet-Mounted Display (HMD): Head-Up Display (HUD); Head-Down Display (HDD).

ROSE: Raster Optical Scanning Engine.

Virtual Image Projection (VIP): An optical display image comprised of parallel or convergent light bundles.

Image Viewing Zone (IVZ): The range of locations from which an entire virtual image is visible while fixating any of the image's boundaries.

Optical Exit Pupil (OEP): The aerial image formed by all compound magnifiers, which defines the IVZ.

Retinal Scanning Display (RSD): A virtual image projection display which scans a beam of light to form a visible pattern on the retina. The typical 15-mm OEP of a helmet-mounted RSD OEP permits normal helmet shifting in operational helicopter environments without loss of image. Higher-*g* environments may require larger OEPs.

Virtual Retinal Display (VRD): A subcategory of RSD specifically characterized by an optical exit pupil less than 2 mm, for Low Vision Aiding (LVA), vision testing, narrow field of view, or "agile" eye-following OEP display systems. This is the most light-efficient form of RSD.

Acknowledgments

This work was partially funded by U.S. Army Contract No. DAAH23-99-C-0072, Program Manager, Aircrew Integrated Systems, Redstone Arsenal, AL. The author wishes to express appreciation for the outstanding efforts of the Microvision Inc. design and development team, and for the guidance and support provided by the U.S. Army Aviation community, whose vision and determination have made these advances in high-performance pilotage HMD systems possible.

References

1. Rash, C. E., Ed., Helmet-Mounted Displays: Design Issues for Rotary-Wing Aircraft, U.S. Army Medical Research and Materiel Command, Fort Detrick, MD, 1999.
2. Kollin, J., A retinal display for virtual environment applications, SID Int. Symp., Digest of Technical Papers, pp. 827–828, May 1993.
3. de Wit, G. C., A Virtual Retinal Display for Virtual Reality, Doctoral Dissertation, Ponsen & Looijen BV, Wageningen, Netherlands, 1997.
4. Gross, A., Lorenson, C., and Golich, D., Eye-safety analysis of scanning-beam displays, SID Int. Symp. Digest of Technical Papers, pp. 343–345, May 1999.
5. Urey, H., Nestorovic, N., Ng, B., and Gross, A., Optics designs and system MTF for laser scanning displays, Helmet and Head Mounted Displays IV, *Proc. SPIE*, 3689, 238–248, 1999.
6. Rash, C. E., Harding, T. H., Martin, J. S., and Beasley, H. H., Concept phase evaluation of the Microvision Inc., Aircrew Integrated Helmet System, HGU-56/P, virtual retinal display. Fort Rucker, AL: U.S. Army Aeromedical Research Laboratory, USAARL Report No. 99-18, 1999.

Further Information

Microvision Inc. Website: www.mvis.com.

7

Night Vision Goggles

Dennis L. Schmickley
Boeing Helicopter Co.

7.1 Introduction

7.1.1 NVG as Part of the Avionics Suite

Visual reference to the aviator's outside world is essential for safe and effective flight. During the daylight hours and in visual meteorological conditions (VMC), the pilot relies heavily on the out-the-windshield view of the airspace and terrain for situational awareness. In addition, the pilot's visual system is augmented by the avionics which provide communication, navigation, flight control, mission, and aircraft systems information. During nighttime VMC, the pilot can improve the out-the-windshield view with the use of night vision goggles (NVG). **NVG lets the pilot see in the dark during VMC conditions!**

This chapter deals with NVG for aviation applications. There are many various nonaviation applications of NVG that are not addressed herein: NVG for personnel on the ground or underwater, and for ground vehicles and sea vehicles.

7.1.2 What Are NVG?

NVG are light image intensification (I^2) devices that amplify the night-ambient-illuminated scenes by a factor of 10^4. For this application "light" includes visual light and near infrared. The development of the microchannel plate (MCP) allowed miniature packaging of image intensifiers into a small, lightweight, helmet-mounted pair of goggles. With the NVG, the pilot views the outside scene as a green phosphor image displayed in the eyepieces.

Various terms are associated with NVG type equipment:

NVG — general term of any I^2 device, usually head-worn and binocular

I^2 — Image Intensifier type of sensor device used in NVG

ANVIS — Aviator's Night Vision Imaging System; a type of NVG designed for aviators

NVIS — Night Vision Imaging System; a general class of NVG including ANVIS

Gen II—Second-generation intensifier technology utilizing MCP and multi-alkali photocathode which enabled construction of AN/PVS-5 NVG

Gen III—Third-generation intensifier technology utilizing improved MCP and galium arsenide photocathode which enabled construction of AN/AVS-6 ANVIS

NVG HUD — Night Vision Goggle with a Head-Up Display attached

HMD — Helmet-Mounted Display; in this chapter it includes NVG HUD

PNVG — Panoramic Night Vision Goggle; usually about 100° FOV

LPNVG — Low-Profile Night Vision Goggle; usually conforms to face

AGC — Automatic gain control

7.1.3 History of NVG in Aviation

7.1.3.1 1950s

In the 1950s there was considerable and diverse research on night image intensification as reported at the Image Intensifier Symposium.[4] The applications included devices for military sensing and for astronomy and scientific research, but were not directed specifically to head-mounted pilotage devices. The U.S. Army first experimented with T-6A infrared driving binocular in helicopters in the late 1950s, according to Jenkins and Efkeman.[2] The binocular device was a near infrared (IR) converter which required an IR filtered landing light for the radiant energy, and was not satisfactory for aviation. In the late 1950s, the first continuous-channel electron multiplier research was being conducted at the Bendix Research Laboratories by George Goodrich, James Ignatowski, and William Wiley. The invention of the continuous-channel multiplier was the key step in the development of the microchannel plate (Lampton[1]).

7.1.3.2 1960s

In the early 1960s first-generation I^2 tubes were developed. The tubes allowed operation as a passive system, but the size of the three-stage I^2 tubes was too large for head-mounted applications. Passive refers to needing no active projected illumination; the system can operate using the ambient starlight illumination, thus the name "starlight scope" from the Vietnam era foot soldier's sniper scope. In the late 1960s, the production of the microchannel plates, used in the second-generation wafer technology I^2 tubes, allowed night vision devices to be packaged small enough and light enough for head-mounted applications. Thus, in the late 1960s and early 1970s the U.S. Army Night Vision and Electro-Optics Laboratory (NV&EOL) used Gen II I^2 tubes to develop NVGs for foot soldiers, and some of these NVGs were tried by aviators for night flight operations.

7.1.3.3 1970s

In 1971 the USAF began limited use of the SU-50 Electronic Binoculars. In 1973 the Army adopted the Gen II AN/PVS-5 as an "interim" NVG solution for aviators, although there were known deficiencies in low-light-level performance, weight, visual facemask obstruction, and refocusing (due to incompatibility with cockpit lighting systems). The *aviator's* night vision imaging system (ANVIS) was the first NVG developed specifically to meet the visual needs of the aviator. The NV&EOL started ANVIS development in 1976 utilizing third-generation image intensifier technology and requiring high-performance, lightweight, and improved reliability and maintainability.

7.1.3.4 1980s

Two versions of the ANVIS were introduced into military aviation:

- AN/AVS-6(V)1 for most helicopters; fits onto the helmet with a centerline mount.
- AN/AVS-6(V)2 for AH-1 Cobra only; fits onto the helmet with an offset mount.

ANVIS operation would not have been feasible or safe in the aircraft if the cockpit lighting had remained the traditional red-lighted or white-lighted incandescent illumination. In 1981 the U.S. Army released an Aeronautical Design Standard, ADS-23,[5] to establish baseline requirements for development of cockpit lighting to be compatible with ANVIS. In 1986 the Joint Aeronautical Commanders Group (JACG) released a Tri-Service specification, MIL-L-85762,[7] which defined standards for designing and measuring ANVIS-compatible lighting. GEC-Marconi introduced a Gen III projected view NVG, called the "Cat's Eye" for use in the AV-8 Harrier.

An updated MIL-L-85762A[8] was released in 1988 in which it defined NVIS as a general term (replacing the specific ANVIS term) and expanded the lighting requirements to accommodate various type NVIS. The controversial utilization of the AN/PVS-5 continued in aviation pending full fielding of ANVIS. Based upon a series of nighttime accidents often involving NVGs, a Congressional Hearing was convened (1989) to review the safety and appropriateness of NVGs in military helicopters. ANVIS was deemed necessary.

7.1.3.5 1990s

Head-up flight information symbology was desired, along with the out-the-window view, within the NVG. Integrating the symbology and imagery resulted in a new type of helmet-mounted display (HMD) referred to as the "NVG HUD". Two types of NVG HUDs were placed in service:

- AN/AVS-7 NVG HUD was installed on CH-47D and HH-60 aircraft.
- Optical Display Assembly (ODA) NVG HUD was installed on OH-58D.

NVG-compatible cockpit lighting was incorporated in high-speed fixed-wing aircraft, but an additional requirement evolved for NVG to be safe during pilot ejection. The AN/AVS-9 (model F4949) was developed for the USAF for ejection capability. In an effort to provide a greater field of view (FOV) than the normal 40° for NVG, the USAF developed a Panoramic Night Vision Goggle (PNVG) to provide about 100° FOV. Several other development programs attempted to reduce the size of the large protrusive goggle optics. Versions of the Low Profile Night Vision Goggle (LPNVG) folded the optics to fit conformally around face. Several integrated helmet development programs incorporated integral I^2 devices and electronic projected display systems. In the early 1990s, several civilian helicopter operators expressed interest in utilizing NVG. Ongoing investigations into the use of NVG in civil aviation delved into applications, safety, and FAA certification.

7.2 Fundamentals

7.2.1 Theory of Operation

An **image intensifier** is an electronic device that amplifies light energy. Light energy, photons, enter into the I^2 device through the objective lens and are focused onto a photocathode detector that is receptive to both visible and near-infrared radiation. Generation III devices use gallium arsenide as the detector. Due to the photoelectric effect, the photons striking the photocathode emit a current of electrons. Because

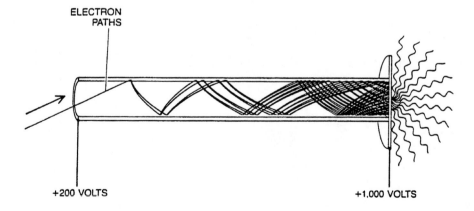

FIGURE 7.1 Electron amplification in a microchannel.[1]

the emitted electrons scatter in random directions, a myriad of parallel tubes (channels) is required to provide separation and direction of the electron current to assure that the final image will have sharp resolution. Each channel amplifier is microscopic — about 15 μm in diameter. A million or so microchannels are bundled in a wafer-shaped array about the diameter of a quarter. The wafer is called a microchannel plate (MCP). The thickness of the MCP, which is the length of the channels, is about 0.25 in. Each channel is an electric amplifier. A bias potential of about 1000 V is established along the tube,

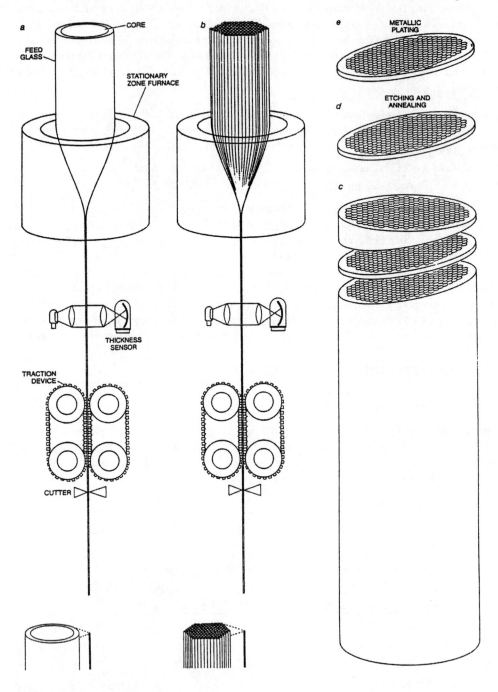

FIGURE 7.2   Double glass draw for MCP manufacture[1].

and each electron produced by the photoelectric effect accelerates through the tube toward the anode. When an electron strikes other electrons in the coated channel, they are knocked free and continue down the tube hitting other electrons in a cascade effect. The result of this multiplication of electrons is a greatly amplified signal. The amplified stream of electrons finally hits a phosphor-type fluorescent screen which, in turn, emits a large number of photons creating an image.

The microchannel plate is a solid-state light amplifier. The intensity of the image is a product of the original signal strength (i.e., the number of photons in the night scene) and the amplification gain within the channel. The fine resolution of the total image is a product of the pixel size from the MCP array and the focusing optics.

The manufacture of MCPs requires complex processes which are dependent on a two-draw glass reduction technique. A concentric tube of an outer feed glass and an inner core glass is drawn into a fine fiber about 1 mm in diameter. Then a bundle of thousands of the fibers is draw to form a multiple fiber about 50 mm in diameter. The core glass is etched out leaving a matrix of hollow glass tubes. Wafer sections are sliced, and the wafers are plated with the metallic coatings necessary for the signal amplification.

The finished product is an NVG which contains an MCP packaged inside an optical housing. The housing will contain objective lens and eyepieces appropriate for the NVG's utilization. For aviators using the NVG for pilotage, a one-to-one magnification is required. The pilot's perceived NVG image of the outside world must be equal to the actual size of the unaided-eye image of the outside real world to provide natural motion and depth perception. The image is displayed to the observer on an energized viewing screen at about 1 footLambert (fL). Screens may be the P20 or P25 phosphors. The light amplification may be 2000 or more, and to prevent phosphor damage, an automatic gain control (AGC) circuit limits the gain in high ambient conditions.

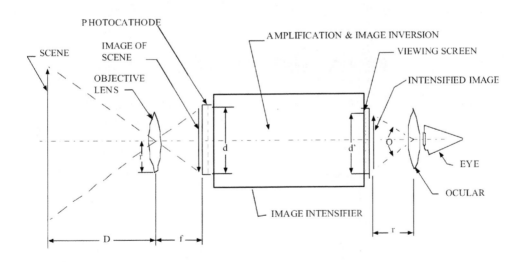

FIGURE 7.3 Typical NVIS image intensifier tube and optics.

7.2.2 I^2 Amplification of the Night Scene

Second-generation image intensifiers utilize multi-alkali photocathodes that are sensitive in the visible and near-IR bandwidth of 400–900 nm. Gen II utilization is generally limited to a minimum of quarter-moon or clear sky illumination (10^{-3} to 10^{-4} fc).

Third-generation image intensifiers utilize galium arsenide (GaAs) photocathodes which are more sensitive than Gen II and have a bandwidth of 600–900 nm. Gen III NVIS can be used in starlight and overcast conditions (10^{-4} to 10^{-5} fc).

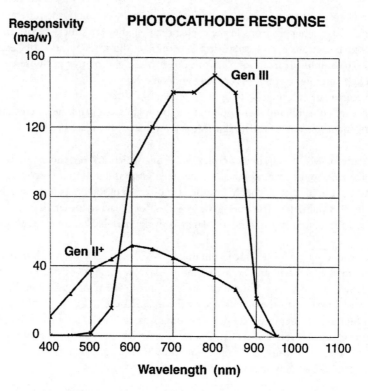

FIGURE 7.4   Photocathode sensitivity.

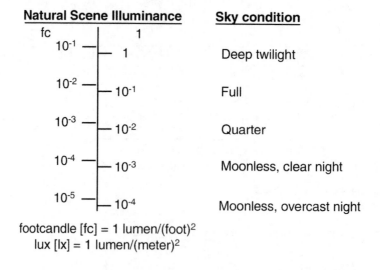

FIGURE 7.5   Illumination from the night sky.

7.2.3   NVG Does Not Work without Compatible Lighting!

NVG lighting compatibility is required for effective NVG use by pilots. If the cockpit lighting is not compatible and it emits energy with spectral wavelengths within the sensitivity range of the night vision goggles, the lighting will be amplified by the NVG and will overpower the amplification of the lower illumination in the outside visual scene.

Compatibility can be defined as a lighting system that does not render the NVG useless or hamper the crew's visual tasks (with or without NVG).

NVIS compatibility permits a crew member to observe outside scenes through vision goggles while maintaining necessary lighted information in the crew station. The Gen III NVIS are insensitive to blue/green light, so the cockpit lighting can be modified with blue cutoff filtering to reduce emitted energy in the red and near-IR regions to achieve compatibility. The complementary minus-blue coatings on the NVIS objective lens provide a sharp cutoff filter to block any red or near-IR light. Blue-green lighting allows external viewing through the ANVIS and internal viewing of the instruments by using the "look-around" technique. The ANVIS look-around design allows the pilot visual access (with unaided eyes) into the blue-green lighted cockpit without head movement. NVIS compatibility requirements are defined by MIL-L-85762.

MIL-L-85762 lighting requirements, and by default the various NVIS, have been categorized into Types and Classes to match the appropriate cockpit lighting system depending on the type of NVIS being used in the aircraft. The original issue of MIL-L-85762 was based on recommendations for ANVIS compatibility (Schmickley[7]) and addressed lighting only for ANVIS (Type I, Class A). MIL-L-86762A added Type II and Class B NVIS. The USAF is in the process of defining a Class C NVIS. A rationale was published to aid manufacturers and evaluators on interpreting the requirements (Reetz[9]).

Type I: Type I lighting components are those lighting components that are compatible with Direct View Image NVIS. Direct View Image NVIS is defined as any NVIS using Generation III image intensifier tubes which displays the intensified image on a phosphor screen in the user's direct line of sight — such as the ANVIS.

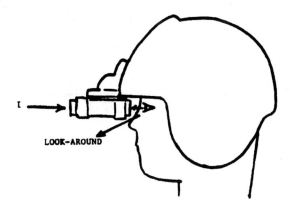

FIGURE 7.6 Type I (direct view). ANVIS with "look-around" vision into the cockpit.

Type II: Type II lighting components are those lighting components that are compatible with Projected Image NVIS. Projected Image NVIS is defined as any NVIS using Generation III image intensifier tubes which projects the intensified image on a see-through medium that reflects the image into the user's direct line of sight — such as the Cat's Eyes.

Class A: Class A lighting components are those lighting components that are compatible with NVIS using a 625-nm minus-blue objective lens filter which results in an NVIS sensitivity lens as shown in the figure below. (The standard AN/AVS-6 ANVIS are equipped with 625-nm minus-blue filters.)

Class B: Class B lighting components are those lighting components that are compatible with NVIS using a 665-nm minus-blue objective lens as shown in the figure below. Class B lighting allows red and yellow colors in cockpit displays, but the consequence is a reduced Gen III NVIS sensitivity to the outside visual scene. The 665-nm minus-blue filter reduces the NVIS sensitivity by 8 to 10% of the Class A NVIS in moonless conditions.

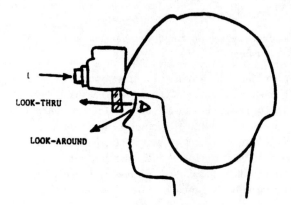

FIGURE 7.7 Type II (projected image). Cat's Eye with "look-through" outside viewing and "look-around" vision into the cockpit.

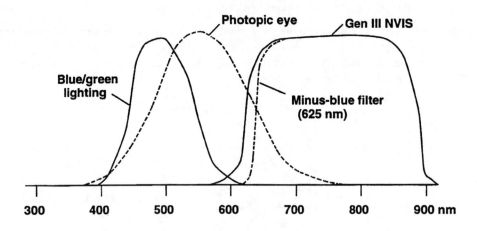

FIGURE 7.8 Typical Class A blue-green lighting and 625-nm minus-blue coating on NVIS.

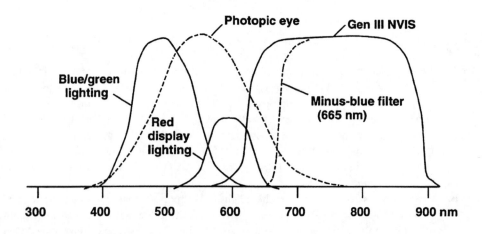

FIGURE 7.9 Typical Class B lighting allows blue-green, yellow, and red with 665-nm minus-blue coating on NVIS.

7.2.4 I² Integration into Aircraft

The integration of NVIS into an aircraft crew station usually requires very little modification with respect to the crew compartment space (volume). The primary aircraft requirements are

1. Adequate helmet and NVIS motion envelope;
2. Acceptable visual fields of view (FOV) and windshield transparency in the NVIS range;
3. Compatible cockpit lighting and displays;
4. Compatible interior (cabin) and exterior lighting.

The alerting quality of warning/caution/advisory color coding can be diminished with NVIS compatible (Class A) cockpit lighting. Audio or voice warning messages may be considered to augment the alerting characteristics.

The NVIS is normally a self-contained standalone sensor that is powered by small batteries. The cost of a typical NVIS unit is $10,000, whereas the cost of an aircraft-mounted IR sensor system is 10 to 20 times that amount. The integration of an NVG HUD requires more modification to the aircraft than the NVIS.

Incorporation of NVIS produces some advantages and some disadvantages for the aircraft and missions. **NVIS advantages usually outweigh disadvantages.** The advantages are

- NVIS allows 24-hour VFR operations (pilots say: "I'd rather fly with them.")
- Enhanced situation awareness; pilots can see the terrain.

The disadvantages are

- Limited instantaneous FOV which requires deliberate head movement;
- Neck strain and fatigue (due to increased helmet weight & increased head movement);
- Cost of equipment (NVIS + compatible lighting);
- Pilot training; currency; proficiency;
- Not useful in IMC weather or fog;
- Safety — if there is inadequate training or overexpectations of system capability.

There are known limitations of the NVIS imposed by the limited FOV. Training is required to emphasize the required head motion scanning to compensate for the FOV. Depth perception is sometimes reported as a major deficiency, although it is most likely that inadequate motion perception cues due to limited peripheral vision are a contributor to this perception.

Military training programs have been implemented to exploit the capabilities of the NVIS sensor for various types of covert missions, and to improve safety and situation awareness. Curricula have been developed "…*to assure that there is an appropriate balance of training realism and flight safety.*" Training programs include visual aids, laboratory, and simulation to cover:

- Theory of I² operation;
- FOV, FOR, adjustment;
- Moon, weather, ambient conditions;
- Different visual scans, head motion.

7.3 Applications and Examples

7.3.1 Gen III and AN/AVS-6 ANVIS

To aid night flying, in the 1980s the Army developed the Aviator's Night Vision Imaging System (ANVIS) which is a third-generation (Gen III) NVG. The ANVIS is designated as AN/AVS-6. ANVIS is lightweight (a little over 1 lb) and mounts on the pilot's helmet. The 25-mm eye relief allows the pilot to see around the eyepieces for viewing the instruments in the cockpit. The Gen III response characteristics are more sensitive than Gen II and the spectral range covers 600 to 900 nm. This spectral range takes advantage of the night sky illumination in the red and IR. Luminance gain is 2000 or greater. The FOV is 40° circular and the resolution is about 1 cy/mr. The total weight is 1.2 to 1.3 lb.

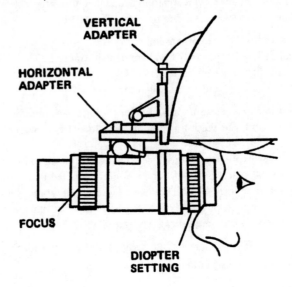

FIGURE 7.10   ANVIS adjustments.[17]

There are several adjustment features on the ANVIS to accommodate each pilot's needs:

• Inter-ocular adjustment
• Tilt adjustment
• Vertical adjustment
• Eye relief (horizontal) adjustment
• Focus adjustment
• Diopter adjustment

The pilot can also flip up the ANVIS to a "helmet stow" position. The mount has a break-away feature in case of a high *g* load or crash.

Early production models of ANVIS units produced system luminance gains of 2000 fL/fL. With improvements in manufacturing techniques and yields, and with increased photocathode sensitivities, newer units have system gains of over 5000. The Army procured large lot quantities of AN/AVS-6 through "omnibus" purchase orders. Omni IV and Omni V AN/AVS-6 have system luminance gains of 5500. The luminance gains of the intensifiers may be 10,000 to 70,000, depending on the ambient illumination being amplified, but with optics and system throughput losses, the overall system gains are 5000+. Presently, the two major suppliers in the U.S. for AN/AVS-6 are ITT and Litton, and the Army splits the procurement of the Omni lots. Adaptations and improved versions of the AN/AVS-6 include the AN/AVS-8 with a 45° FOV, and the AN/AVS-9 which has a front-mounted battery to allow use in ejection seats. The AN/AVS-9 also has a "leaky green" sensitivity to allow viewing of the HUD symbology.

7.3.2   Gen II and AN/PVS-5 NVG

The generation II AN/PVS-5 is outdated and is not now recommended for aviators. The AN/PVS-5 is discussed here because it was the most common device allowing night flying with NVG aided vision. The AN/PVS-5A provided Army ground forces with enhanced night vision capability. Later, pilots used the NVG to fly helicopters. Tests indicated that pilots using NVG could fly lower and faster than pilots without NVG, and concluded that NVG provided considerable improvement over unaided, night-adapted vision.

The AN/PVS-5A weighs 2 lbs and has a full face mask. Wearing these NVG requires the pilot to make all visual observations via the NVG, including cockpit instrument scanning. The pilot must move his head and refocus the lens to read the instruments. Annoyance, discomfort, and fatigue result from these restrictions.

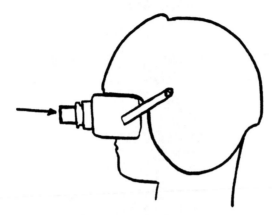

FIGURE 7.11 AN/PVS-5A NVG with full face mask.

The spectral range of the Gen II NVG is from 350 to 900 nm which includes the entire visual spectrum (380–760) plus some near-IR coverage. Most 1970s cockpits had red incandescent lamp lighting which had large red and IR emissions. The NVG's automatic gain control (AGC) shuts down the NVG in the presence of large amounts of radiant energy in the goggles' range. Therefore, the use of Gen II NVG requires that all visible lighting must be reduced below the pilot's visual threshold in order that the lighting does not degrade the NVG operation. Commonly, this is accomplished by extinguishing the lights or using a "superdim" setting. Under these conditions, crew members without NVG cannot read the cockpit instruments. Crew members with NVG must refocus from outside viewing to read the instruments. Research in the U.S. and U.K. on shared-aperatures and shared-lens attempted to provide viewing of the cockpit instruments with the NVG. Modifications to the face mask to provide peripheral and in-cockpit vision produced the "cut-away" mask.

The utilization of AN/PVS-5 NVG in aviation was controversial. The incorporation of NVG into aviation somewhat repeated the development of aviation itself, with a period of trial and error incorporation, sometimes with inadequate or inappropriate equipment, producing some pioneering breakthroughs and some accidents. In the 1980s there were nighttime accidents often involving NVGs. The *Orange County Register* published a lengthy investigative article because several of the helicopter crashes took place within the county.[14] A congressional hearing was convened to review the safety and appropriateness of NVGs in military helicopters.[15] The necessity of NVGs for night flight operations was confirmed along with an emphasis on better equipment and training. A review of AN/PVS-5 and AN/AVS-6 testing concluded both were acceptable.[16] Since that time, AN/AVS-6 ANVIS has become the preferred device for aviators.

7.3.3 Cat's Eyes

The "Cat's Eye" is a Type II (projected image) Gen III NVIS made by GEC-Marconi, and is standard in the AV-8 series of Harrier aircraft. The weight is slightly over 1 lb. The two optical combiner lenses have the I^2 image displayed for out-of-the-cockpit viewing. The combiner has see-through capability to view the aircraft's HUD. When the pilot is looking at the HUD, the I^2 imagery is automatically turned off to allow visibility of the HUD symbology. The combiner glass see-through transmission is <30%. The Cat's Eye has a 25-mm eye relief to allow look-under for cockpit instrument viewing.

7.3.4 NVG HUD

Systems termed "NVG HUD" have been produced that add head-up display (HUD) symbology onto the displayed night vision imagery provided by the NVG. Usually the HUD portion is a CRT image projected onto a combiner glass mounted in front of one of the NVG objective lens. The symbology displayed is aircraft information (attitude, altitude, airspeed, navigation data, etc.) that is generated in a processor box integrated to the aircraft systems.

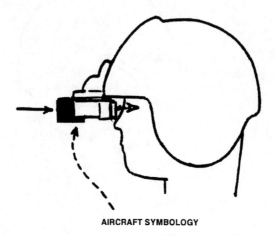

AIRCRAFT SYMBOLOGY

FIGURE 7.12 Aircraft symbologgy.

7.3.5 ANVIS HUD

Honeywell produces the Optical Display Assembly (ODA). The OH-58D that integrates the ODAS also is one of the few aircraft that provides aircraft power for the NVG (instead of self-contained battery power). Elbit produces the AN/AVS-7 NVG HUD. *Note*: The AN/AVS-7 (NVG HUD) should not be confused with AN/PVS-7 single-tube NVG for ground troops.

	Weight on NVG	I^2 FOV	Application
AN/AVS-7	0.25 lb	33° H × 24° V	CH-47, HH-60
ODA	? lb	40°	OH-58D

7.3.6 Panoramic NVG

Panoramic NVG (PNVG) have been developed for the USAF to provide an increased instantaneous FOV of the image. Night Vision Corporation developed the PNVG using four AN/PVS-7 image tubes. The four tubes produce a combined overlapping FOV of 100°.

	Weight	I^2 FOV	Type
PNVG	1.25 lb	100° H × 40° V	Direct view optics

7.3.7 Low Profile NVG

Low Profile NVG (LPNVG) have been developed for several reasons: to improve the head-borne c.g., to allow visors, and to reduce possible injury caused by the protrusion of the longer I^2 tubes. The depth is 2 to 3 in. compared to 5 to 6 in. for other NVIS. ITT developed the Modular, Ejection-Rated, Low profile, Imaging for Night (MERLIN) Aviator Goggle for use by pilots in high-performance fixed-wing aircraft. Litton produces the AN/AVS-502 LPNVG for multi-role missions (parachute operations, weapons firing). Canadian Air Forces approved the AN/AVS-502 for flight engineers in the cabin where head clearance and winds are issues. Systems Research Laboratories (SRL) developed the Eagle Eye™ for fixed-wing and multi-role.

	Weight	I^2 FOV	Type
MERLIN	1.8 lb	35°	See-through optics
AN/AVS-502	1.5 lb	40°	See-through optics
Eagle Eye™	1.2 lb	40°	See-through optics

7.3.8 Integrated I^2 Systems

I^2 sensors can be incorporated in avionics suites in several ways besides standalone NVG on a crew member's helmet. One method is to incorporate an I^2 sensor on-board an IR sensor pod to provide video imagery of either I^2 or IR to the crew.

Several integrated helmet designs and future helmet concepts are integrating I^2 devices along with CRT, LCD, and LED helmet-mounted displays.

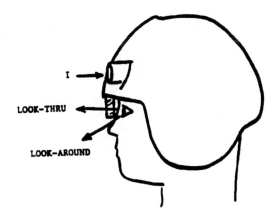

FIGURE 7.13 Integrated helmet.

7.3.9 Testing and Maintaining the NVG

NVIS manufacturers also supply testing and servicing equipment. Examples are the ANV-126 NVG Test Set, TS-6 Night Vision Device Test Set Kit, TS-4348/UV Night Vision Device Assessor, and the TS-10 Night Vision Leak Test and Purge Kit.

7.3.10 Lighting Design Considerations

NVIS-compatible aircraft interior lighting is essential to allow night flying with NVIS. Interior lighting consists of primary lighting (instrument and control panels), secondary lighting (task lights, area lights, floodlights), signals (warning, caution, advisory), and electronic displays.

The key specification that defines NVIS-compatible lighting is MIL-L-85762A. This specification is unique in that it specifies two independent characteristics for the lighting system:

1. **Luminance and chromaticity requirements** for visual (unaided eye) viewing in a dark cockpit, and
2. **Radiance requirements** for limiting any NVIS interference.

The luminance levels remain approximately the same as traditional red and white lighting systems in all previous aircraft. The chromaticity requirements generally produce a blue-green lighted cockpit. Four lighting colors for aviation have been defined in MIL-L-85762A (where u' and v' are 1976 UCS chromaticity coordinates of the defined color):

NVIS Green A — The color for primary, secondary, and advisory lighting. The chromaticity limits are within a circle of radius .037 with the center at $u' = .131, v' = .623$.

NVIS Green B — The color for special lighting components needing saturated color (monochromatic) for contrast. The chromaticity limits are within a circle of radius .057 with the center at $u' = .131, v' = .623$.

NVIS Yellow — The color for master caution and warning signals in Class A cockpits. The chromaticity limits are within a circle of radius .083 with the center at $u' = .274, v' = .622$.

NVIS Red — The color for warning signals in Class B cockpits. The chromaticity limits are within a circle of radius .060 with the center at $u' = .450, v' = .550$.

Chromaticity and luminance requirements for various types of cockpit lighting and displays and cabin lighting are listed in Table VIII of MIL-L-85762A.

NVG compatibility is not assured with proper chromaticity coordinates alone. Lights with different spectral compositions can appear visually as the same color. Similar visual colors are called metamers. But all colored lights used in the NVG cockpit must have filtering to block almost all the energy in the 600- to 900-nm range. The "NVIS-visible" portion of the lighting emission is to be limited per the NVIS radiance definition: NVIS radiance (NR) is the integral of the curve generated by multiplying the spectral radiance of the light source by the relative spectral response of the NVIS.

"Formula 14a" of MIL-L-85762A is used to calculate the NVIS radiance of Class A lighting equipment, and "Formula 14b" is for the NVIS radiance of Class B equipment.

$$\text{NVIS radiance (NR}_A) = G(\lambda)_{max} \int_{450}^{930} G_A(\lambda)SN(\lambda)d\lambda \qquad \text{(Formula 14a)}$$

$$\text{NVIS radiance (NR}_B) = G(\lambda)_{max} \int_{450}^{930} G_B(\lambda)SN(\lambda)d\lambda \qquad \text{(Formula 14b)}$$

where:

$G_A(\lambda)$	= relative NVIS response of Class A equipment
$G_B(\lambda)$	= relative NVIS response of Class B equipment
$G(\lambda)_{max}$	= 1 ma/w
$N(\lambda)$	= spectral radiance of lighting component (w/cm^2 sr nm)
S	= scaling factor
dl	= 5 nm

For example, to be compatible, a Class A lighting system requirement is to have the blue-green primary lighting not exceed 1.7×10^{-10} NR$_A$ when the lighting produces 0.1 fL luminance. If the lighting component is actually greater than 0.1 fL when it is measured, the scaling factor S scales the NR to 0.1 fL.

For cockpits where red or multicolor displays are desired, a similar equation for NR$_B$ applies to assure Class B compatibility. Note that "Class B" NVIS must be utilized with a Class B cockpit.

NR requirements for various types of cockpit lighting and displays and cabin lighting are listed in Table IX of MIL-L-85762A.

All other aircraft lighting, not just the cockpit lighting, must be made compatible with NVG. This includes stray light from the aircraft's interior cabin, the aircraft's exterior lighting system, and any

Chromaticity Requirements (from Table VIII, MIL-L-85762A)

Lighting Component(s)	TYPE 1 Class A					TYPE 1 Class B					TYPE II Class A					TYPE II Class B				
	u_i'	v_i'	r	Cd/m² (fL)	NVIS Color	u_i'	v_i'	r	Cd/m² (fL)	NVIS Color	u_i'	v_i'	r	Cd/m² (fL)	NVIS Color	u_i'	v_i'	r	Cd/m² (fL)	NVIS Color
Primary	.088	.543	.037	0.343 (0.1)	Green A						.088	.543	.037	0.343 (0.1)	Green A					
Secondary	.088	.543	.037	0.343 (0.1)	Green A						.088	.543	.037	0.343 (0.1)	Green A					
Illuminated Controls	.088	.543	.037	0.343 (0.1)	Green A						.088	.543	.037	0.343 (0.1)	Green A					
Compartment lighting	.088	.543	.037	0.343 (0.1)	Green A			Same			.088	.543	.037	0.343 (0.1)	Green A			Same		
Utility, work, and inspection	.088	.543	.037	0.343 (0.1)	Green A			as			.088	.543	.037	0.343 (0.1)	Green A			as		
Caution and advisory signals	.088	.543	.037	0.343 (0.1)	Green A			Class A			.088	.543	.037	0.343 (0.1)	Green A			Class A		
Jump lights	.088	.543	.037	17.2 (5.0)	Green A						.088	.543	.037	17.2 (5.0)	Green A					
	.274	.622	.083	51.5 (15.0)	Yellow						.274	.622	.083	51.5 (15.0)	Yellow					
Special lighting components where increased display emphasis by highly saturared (monochromatic) color is necessary, or adequate display light readability cannot be achieved with "GREEN A"	.131	.623	.057	0.343 (0.1)	Green B						.131	.623	.057	0.1	Green B					
Warning signal	.274	.622	.083	51.5 (15.0)	Yellow	.274	.622	.083	51.5 (15.0)	Yellow	.274	.622	.083	51.5 (15.0)	Yellow	.274	.622	.083	51.5 (15.0)	Yellow
	.450	.550	.060	51.5 (15.0)	Red	.450	.550	.060	51.5 (15.0)	Red	.450	.550	.060	51.5 (15.0)	Red	.450	.550	.060	51.5 (15.)	Red
Master Caution signal	.274	.622	.083	51.5 (15.0)	Yellow			Same as Class A		Yellow	.274	.622	.083	51.5 (15.0)	Yellow			Same as Class A		Yellow

Note: u_i' and v_i' = 1976 UCS chromaticity coordinates of the center point of the specified color area; r = radius of the allowable circular area on the 1976 UCS chromaticity diagram for the specified color;   fL = footLamberts;   Cd/m² = candela/(meter)².

NVIS Radiance Requirements (from Table IX, MIL-L-85762-A)

Lighting Components	TYPE 1						TYPE II					
	Class A			Class B			Class A			Class B		
	Not Less than: (NR_A)	Not Greater than: (NR_A)	fL	Not less than: (NR_B)	Not Greater than: (NR_B)	fL	Not Less than: (NR_A)	Not Greater than: (NR_A)	fL	Not Less than: (NR_B)	Not Greater than: (NR_B)	fL
Primary	—	1.7×10^{-10}	0.1		Same as Class A (see Note)		—	1.7×10^{-10}	0.1		Same as Class A (see Note)	
Secondary	—	1.7×10^{-10}	0.1				—	1.7×10^{-10}	0.1			
Illuminated Controls	—	1.7×10^{-10}	0.1				—	1.7×10^{-10}	0.1			
Compartment	—	1.7×10^{-10}	0.1				—	1.7×10^{-10}	0.1			
Utility, work and inspection lights	—	1.7×10^{-10}	0.1				—	$1.7 - 10^{-10}$	0.1			
Caution and advisory lights	—	1.7×10^{-10}	0.1				—	$1.7 - 10^{-10}$	0.1			
Jump lights	1.7×10^{-8}	5.0×10^{-8}	5.0	1.6×10^{-8}	4.7×10^{-8}	5.0	—	5.0×10^{-8}	5.0	—	4.7×10^{-8}	5.0
Warning signal	5.0×10^{-8}	1.5×10^{-7}	15.0	4.7×10^{-8}	1.4×10^{-7}	15.0	—	1.5×10^{-7}	15.0	—	1.4×10^{-7}	15.0
Master Caution Signal	5.0×10^{-8}	1.5×10^{-7}	15.0	4.7×10^{-8}	1.4×10^{-7}	15.0	—	1.5×10^{-7}	15.0	—	1.4×10^{-7}	15.0
Emergency Exit Lighting	5.0×10^{-8}	1.5×10^{-7}	15.0	4.7×10^{-8}	1.4×10^{-7}	15.0	—	1.5×10^{-7}	15.0	—	1.4×10^{-7}	15.0

	NR_A	NR_B	fL	NR_A	NR_B	fL	NR_A	NR_B	fL	NR_A	NR_B	fL
Electronic and electro-optical displays (monochromatic)	—	1.7×10^{-10}	0.5	—	1.6×10^{-10}	0.5	—	1.7×10^{-10}	0.5	—	1.6×10^{-10}	0.5
Electronic and electro-optical displays multicolor — White	—	2.3×10^{-9}	0.5	—	2.2×10^{-9}	0.5	—	2.3×10^{-9}	0.5	—	2.2×10^{-9}	0.5
— MAX	—	1.2×10^{-8}	0.5	—	1.1×10^{-8}	0.5	—	1.2×10^{-8}	0.5	—	1.1×10^{-8}	0.5
HUD systems	1.7×10^{-9}	5.1×10^{-9}	5.0	1.6×10^{-9}	4.7×10^{-9}	0.5	—	1.7×10^{-9}	5.0	—	1.6×10^{-9}	5.0

NR_A = NVIS radiance requirements for Class A equipment.
NR_B = NVIS radiance requirements for Class B equipment.
fL = footLamberts.

Note: For these lighting components, Class B equipment shall meet all Class A requirements of this specification. The relative NVIS response data for Class A equipment, $G_A(\lambda)$, shall be substituted for $G_B(\lambda)$ to calculate NVIS radiance.

external lights such as runway or shipboard lights. Often, exterior lights on military aircraft are extinguished to provide covertness. If exterior lights are required during NVG operations, they usually are in one of two categories:

- **Visible and NVG compatible** — such as electroluminescent formation lights which are green visible strips and are not degrading to pilots who are using NVG.
- **Invisible and NVG usable** — covert IR lights that provide illumination for the pilot using NVG or allow signaling or alerting to the pilot operating with NVG.

The cabin and cargo compartment interior lighting must be made NVG compatible if the aft crew uses NVG or if the cabin lighting is seen from the crew station. The cabin compartment in the HH-60Q "Medevac" helicopter requires white lighting for the medical personnel to attend to patients. The cabin has blackout curtains to protect the NVG compatibility of the crew station and to block any visual signature to the outside world.

7.3.11   Types of Filters/Lighting Sources

Aircraft lighting systems use various types of illuminating sources and lamps: incandescent, electroluminescent, fluorescent, light-emitting diode (LED), liquid crystal display (LCD), and cathode ray tube (CRT). Cockpit lighting can usually be modified by adding blue or blue-green glass filters. Glass filter companies and suppliers such as Schott, Corning, Wamco, Hoffman Engineering, and Kopp have produced usable filters. Usually, plastic filtering has not worked with incandescent sources since IR is transmitted freely, but Korry has developed a moldable plastic composition for NVG-compatible products. Manufacturers of filters, measurement equipment, exterior lighting (Grimes, Oxley, Luminescent Systems Inc., et al.) and interior lighting (Control Products Corp., Korry, Oppenheimer, IDD, Eaton, et al.) can be found through organizations involved in aircraft lighting such as ALI and SAE.

7.3.12   Evaluating Aircraft Lighting

A qualitative method of evaluating the NVG compatibility of the overall cockpit is available. The method is a field evaluation that should be conducted on a clear, moonless night with the aircraft parked in a secluded area away from disturbing light sources. A standard tri-bar resolution target board (e.g., USAF 1951), with patterns consisting of three horizontal and three vertical bar pairs arranged in decreasing size, is mounted in front of the aircraft. The resolution pattern is illuminated by the ambient starlight environment. The pilot (or observer) wears the NVG and views the resolution pattern while looking through the windshield. With all the aircraft/cockpit lighting extinguished, the pilot first determines the smallest resolvable line pair that is observed. Then, as each lighting zone or display is turned on, the pilot continues to report the smallest resolvable line pair. Lighting zones and displays are activated individually and then simultaneously. If the lighting and displays have no effect on the minimum resolvable pattern observed, then the cockpit is considered to be compatible with the NVG because there is no impact on goggle performance. Visually observed reflections from the lighting in the canopy or windshield can also be evaluated for NVG compatibility. Compatibility usually is demonstrated if the reflections are not apparent when viewed through the NVG.

7.3.13   Measurement Equipment

Laboratory measurements of the aircraft lighting components are obtained to quantify the following photometric and radiometric characteristics of the light output:

Luminance
Chromaticity
NVIS Radiance

Laboratory measurements use the guidelines of MIL-L-85762A to provide quantitative data to verify that the lighting components are NVG compatible. Units of radiometric measures are consistent with

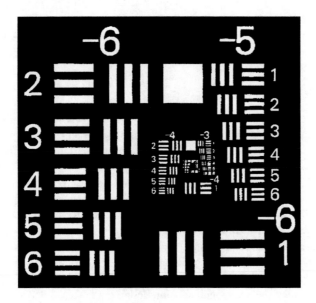

FIGURE 7.14   USAF 1951 Resolution Target.

terms in other electromagnetic radiant energy applications. The measurements based upon the visual eye response of the average human observer are termed photometric measurements. Luminance of lighted cockpit control panel and display presentation is frequently called "brightness." The color of the light is also necessary in defining the visual characteristic of the lighted presentations, and spectroradiometric measurements determine the chromaticity to quantitatively define the color. The typical chromaticity coordinates used are from the 1976 CIE UCS system.

Radiometric Term	Unit	Photometric Term	SI Unit	English Unit
Radiant flux	Watt	Luminous flux	Lumen	Lumen
Radiant	Watt/steradian	Luminous	Candela	Candela
Intensity	Watt/steradian/m^2	Intensity	Cd/m^2	FootLambert
Radiance	Watt/m^2	Luminance	Lux	Footcandle
Irradiance		Illuminance		

Radiant energy for the NVIS-weighted response is measured by a radiometer with very low energy sensitivity. The data is used to calculate the "NVIS Radiance" (as defined in MIL-L-85762) to determine the compatibility with the pilot's NVIS device. Some companies that manufacture photometric, radiometric, and spectroradiometric measurement equipment that can determine visual and NVIS characteristics are:

- Optronic Laboratories, Orlando, FL   (http://www.olinet.com/)
- Photo Research, Chatsworth, CA   (http://www.photoresearch.com/)
- Instrument Systems, Ottawa, Ontario   (http://www.instrumentsystems.com/)
- Gamma Scientific, San Diego, CA   (http://www.gamma-sci.com/)

7.3.14   Nighttime Illumination — Moon Phases

Flight planning requires knowledge of current weather conditions and the geography and topography along the route of the flight plan. For night flights when using NVG, night sky illumination, including the moon's phase and position at various times, is very important in planning the NVG flight. Astronomical data is available to determine times of sunrise, sunset, moonrise, moonset, and twilights. Data is also found for positions of the sun and moon, and on moon phase and illumination. The US Naval

Observatory offers a web version of the Multi-year Interactive Computer Almanac (MICA) on the Observatory's Astronomical Applications web site, *http://aa.usno.navy.mil/AA/*. A DOS or Mac version of the MICA Interactive Astronomical Almanac can also be ordered: NTIS Order No. PB93-500163, 5285 Port Royal Rd., Springfield, VA 22161.

7.3.15   NVG in Civil Aviation

NVG have application to civil aviation. The NVG enhances night VFR situation awareness and obstacle avoidance by allowing direct vision of the horizon, terrain, shadows, and other aircraft. The use of NVG does not require the operation to be covert. While NVG were primarily developed for military applications, NVG are being used in a variety of civilian situations requiring increased night viewing and safe night flying conditions. The forestry service uses NVG, not only to increase the safety in night fire-fighting operations, but also to find hot spots not readily seen by the unaided eye. Emergency Medical Services (EMS) helicopters utilize NVG for navigating into remote rescue sites. Civilian and commercial use of NVG in aircraft, land vehicles, and ships is growing.

The SAE G-10 Aerospace Behavioral Engineering Technology Committee, Vertical Flight Subcommittee, has been assessing human factors issues associated with NVG for application to civil aviation.

The SAE A-20 Aircraft Lighting Committee has prepared the following Aerospace Recommended Practices (ARP) documents to allow general aviation design guidance similar to military specifications and standards which defined NVG-compatible lighting:

ARP4168 — This SAE ARP recommends considerations for light sources for designing NVG-compatible lighting.

ARP4169 — This SAE ARP describes the functions and characteristics of NVG filters used in NVG compatible lighting.

ARP4967 — This SAE ARP covers design considerations for NVIS-compatible panels (also known as "integrally illuminated information panels" or "lightplates"). Panels may utilize incandescent, electroluminescent (EL), or light-emitting diode (LED) sources that are filtered to meet requirements specified in MIL-L-85762.

ARP4392 — This SAE ARP describes the recommended performance levels for NVIS-compatible aircraft exterior lighting equipment. Category I lights are compatible to be viewed by NVIS. Category II lights are illuminators to allow NVIS viewing of the surroundings. The "lights" may not be in the visible spectrum.

The FAA has conducted several studies and requested recommendations for civil application of NVG (Green[19–22]). The primary emerging philosophy for the incorporation of NVG into civil aviation is that "NVG do not enable flight". The use of NVG will not enable any mode of flight which cannot be flown visually within the framework of the existing regulatory authority.

Because civil aviation does not have the regimented control of pilots and aircraft as in the military, there is a danger to the public if untrained operators fly in ill-equipped, unregulated, and noncompatible aircraft. Therefore, minimum civil regulations and standards must be imposed. The future integration of NVG use in civil aviation will depend on the following key issues:

1. Limiting the I^2 device to Gen III;
2. Modification of cockpit lighting;
3. Modification of interior lighting;
4. Modification of exterior lighting;
5. Establishing training programs;
6. Updating FARs 61, 91, 135, et al.

Civil aviation should limit the Night Vision Device to Generation III ANVIS. The military experience has demonstrated that an NVG made for aviators is necessary. The third-generation sensor is preferred for starlight sensitivity. Gen II NVG with 625-nm minus-blue filters will work with MIL-L-85762A compatible lighting, but the filters reduce Gen II effectiveness. Without MIL-L-85762A lighting, the NVG Automatic Gain Control (AGC) can give a false sense of compatibility.

Cockpit lighting for civil aviation will have to be NVG compatible. All nighttime lighting requires NVG-compatible filtering. That normally includes control panel lightplates, numeric display read-outs, Warning/Caution/Advisory (W/C/A) legends, floodlights, flashlights, and electronic displays (CRTs, LCDs, LEDs). The MIL-L-85762A approach yields best compatibility results. An integral approach yields better lighting, although existing equipment can be modified with add-on bezels or filters. These add-ons can block viewing or reduce daylight readability.

Color coding of W/C/A legends (if red warning lights are utilized) and use of multicolor electronic displays (e.g., weather radar) must be limited to the use of Class B NVG with a 665-nm minus-blue filter.

Cabin and interior lighting for civil aviation will have to be NVG compatible. The cabin and cargo compartment interior lighting must be made NVG compatible, or else the compartment and lighting must be shielded from the cockpit. If the compartment is not isolated from the cockpit, then the passengers and crew must not operate carry-on lighting sources that are not NVG compatible. The carry-on equipment may include radios, television, computers, recorders, CD players, cellular phones, and flashlights. Also, smoking should be prohibited because smoking produces a noncompatible glow.

Exterior lighting for civil aviation will have to be NVG compatible. At present, NVG exterior lighting, including the ARP4392 exterior lighting, is not compliant with the Federal Aviation Regulations (FAR) for "see and be seen" navigation and anticollision lights necessary for civil aviation VFR flight. Invisible (covert) lighting will not be allowed as the only lighting for civil aviation. It will be necessary to develop and approve standards for exterior lighting which will be

- **Visible** (blue-green) to other aircraft VFR pilots not using NVG;
- **Visible and NVG compatible** (not degrading) to other aircraft VFR pilots who are using NVG; and
- **NVG compatible** (not degrading) to allow the pilot of the aircraft to operate with NVG.

New training systems will have to be established to support NVG use in civil aviation. Civilian pilots utilizing NVG will have to have minimum ground and flight training similar to that developed within the military. The basic ground training will include the theory of I^2 device, NVG limitations, NVG adjustments, nighttime moon and starlight illumination, FOV, and different visual scan and head motion techniques.

The FAA will have to establish certification and standards of NVG use in civil aviation. In order to allow NVG utilization in civil aviation, the FAA will have to modify regulations for pilot certification and ratings (FAR 61), equipment and flight rules (FAR 91), operating limitations (FAR 135), and airworthiness standards for various aircraft types (FAR 27, 29, etc.). Authority to operate with NVG may be documented through FAR, Special Federal Aviation Regulation (SFAR), Advisory Circular (AC), Type Certificate (TC), Supplemental Type Certificate (STC), Technical Standards Orders (TSO), Kinds of Operations List (KOL), and Proposed Master Minimum Equipment List (PMMEL).

References

1. "The Microchannel Image Intensifier," Michael Lampton, *Scientific American*, Vol. 245, No. 5, November 1981, pp 62–71.
2. "Development of an Aviator's Night Vision Imaging System (ANVIS)," Albert Efkeman and Donald Jenkins, presented at SPIE Int. Tech. Symp. Exhibit, July 28–August 1, 1980, San Diego, CA.
3. TC 1-204 Night Flight Techniques and Procedures, U.S. Army.
4. Image Intensifier Symposium (Proceedings), U.S. Army Engineer Research and Development Laboratories, October 1958.
5. "Aircrew Station Lighting for Compatibility with Night Vision Goggle Use," ADS-23, Aeronautical Design Standard, U.S. Army Aviation Research and Development Command, May 1981.
6. "Aircraft Lighting Requirements for Aviator's Night Vision Imaging System (ANVIS) Compatibility," Dennis L. Schmickley, Rep. No. NADC-83032-60, Naval Air Development Center, April 1983.

7. "Lighting, Aircraft, Interior, Aviator's Night Vision Imaging System (ANVIS) Compatible," MIL-L-85762, Military Specification, January 1986.
8. "Lighting, Aircraft, Interior, Night Vision Imaging System (NVIS) Compatible", MIL-L-85762A, Military Specification, August 1988.
9. "Rationale Behind the Requirements Contained in Military Specifications MIL-L-85762 and MIL-L-85762A," Ferdinand Reetz, III, Rep. No. NADC-87060-20, Naval Air Development Center, September 1987.
10. "Lighting, Aircraft, Interior, Night Vision Imaging System (NVIS) Compatible," ASC/ENFC 96-01, Interface Document, March 1996.
11. "Rationale Behind the Requirements Contained in ASC/ENFC 96-01 Lighting, Aircraft, Interior, Night Vision Imaging System (NVIS) Compatible and Military Specification MIL-L-85762," James C. Byrd, Wright Patterson AFB, April 1996.
12. "Night Lighting and Night Vision Goggle Compatibility," Alan R. Pinkus, AGARD Lecture Series No. 156, Advisory Group for Aerospace Research and Development, North Atlantic Treaty Organization, April 1988.
13. "Aviator's Night Vision Imaging System AN/AVS-6(V)1, AN/AVS-6(V)2," MIL-A-49425(CR), Military Specification, November 1989.
14. "Death in the Dark," Edward Humes, *The Orange County Register,* CA, December 4, 1988.
15. "Night Vision Goggles," Hearing before the Investigations Subcommittee of the Committee on Armed Services, House of Representatives, held March 21, 1989, U.S. Government Printing Office, Washington, D.C.
16. "Review of Testing Performed on AN/PVS-5 and AN/AVS-6 Aviation Night Vision Goggles," Office of the Director, Operational Test and Evaluation, June 1989.
17. "Helicopter Flights with Night Vision Goggle — Human Factors Aspects," Michael S. Brickner, NASA Technical Memorandum 101039, March 1989.
18. "Review of the use of NVG in Flight Training," rep. for the Deputy Secretary of Defense, July 1989.
19. "Rotorcraft Night Vision Goggle Evaluation," David L. Green, Rep. DOT/FAA/RD-19/11.
20. "Civil Use of Night Vision Devices — Evaluation Pilot's Guide Part I," David L. Green, Rep. FAA/RD-94/18, July 1994.
21. "Civil Use of Night Vision Devices — Evaluation Pilot's Guide Part II," David L. Green, Rep. FAA/RD-94/19, July 1994.
22. "Assessment of Night Vision Goggle Workload — Flight Test Engineer's Guide," David L. Green, Rep. FAA/RD-94/20, July 1994.
23. "Night Vision Goggle (NVG) Filters," SAE Aerospace Recommended Practice ARP4169, February 1989.
24. "Night Vision Goggle (NVG) Compatible Light Sources," SAE Aerospace Recommended Practice ARP4168, February 1989.
25. "Lighting, Aircraft Exterior, Night Vision Imaging System (NVIS) Compatible," SAE Aerospace Recommended Practice ARP4392, June 1993.
26. "Night Vision Imaging Systems (NVIS) Integrally Illuminated Information Panels," SAE Aerospace Recommended Practice ARP4967, March 1995.

Further Information

- "IESNA Lighting Handbook," published by The Illuminating Engineering Society of North America, 120 Wall Street, NYC, NY 10005. (http://www.iesna.org/)
- US Army Night Vision & Electronics Directorate (NVESD), Ft. Belvoir, VA
- Aerospace Lighting Institute, Clearwater, FL (http://www.aligodfrey.com/)
- Commission Internationale de l'Eclairage (International Commission on Illumination). (*http://www.ping.at/cie/*)
- Society of Automotive Engineers [A-20 and G-10 committees]. (http://www.sae.org/)

8

Speech Recognition and Synthesis

Douglas W. Beeks
Rockwell Collins

8.1   Introduction

The application of speech recognition (SR) in aviation is rapidly evolving and moving toward more common use on future flightdecks. The concept of using SR in aviation is not new. The use of speech recognition and voice control (VC) has been researched for more than 20 years, and many of the proposed benefits have been demonstrated in varied applications. Continuing advances in computer hardware and software are making the use of voice control applications on the flightdeck more practical, flexible, and reliable. There is little argument that the easiest and most natural and ideal way for a human to interact with a computer is by direct voice input (DVI).

While speech recognition has improved over the past several years, speech recognition has not reached the level of capability and reliability of one person talking to another. Using SR and DVI in a flightdeck atmosphere likely brings to mind thoughts of the computer on board the starship Enterprise from the science fiction classic *Star Trek*, or possibly of the HAL9000 computer from the movie *2001: A Space Odyssey*. The expectation of a voice control system like the computer on the Enterprise and the HAL9000 computer, is that it be highly reliable, work in adverse and stressful conditions, be transparent to the user, and understand its users accurately without having to tailor their individual speech and vocabulary to suit the system. Current speech recognition and voice control systems are not able to achieve this level of performance expectations, although the ability and flexibility of speech recognition and its application to voice control has increased over the past few years. Whether or not a speech recognition system will ever be able to function to the level of one person speaking to another remains to be seen.

The current accuracy rate of speech recognition is in the lower to mid 90% range. Some speaker-dependent systems, and generally those with small vocabularies, have shown accuracy rates into the upper 90% range. While at first glance that might sound good, consider that with a 90% accuracy rate, 1 in 10 words will be incorrectly recognized. Also consider that this 90% and greater accuracy may be under ideal conditions; many times this high accuracy rate is achieved in a controlled and sterile lab environment. Under actual operating conditions, including cockpit noise, random noises, bumps and thumps, multiple people talking at once, etc. the accuracy rate of speech recognition systems can erode significantly.

Currently, several military applications are planning on using SR to provide additional methods to support the Man-Machine Interface (MMI) to reduce the workload on the pilot in advanced aircraft. Boeing is incorporating SR into the new Joint Strike Fighter, and the Eurofighter Typhoon is also adding SR capabilities to its aircraft. Numerous aviation companies worldwide are conducting research and studies into how the available SR technology can be incorporated into current equipment designs and designs of the future for both the civilian and military marketplace. Speech recognition technology will likely be first used in military applications, with the technology working its way into civil aviation by the year 2005.

8.2   How Speech Recognition Works: A Simplistic View

Speech recognition is based on statistical pattern matching. One of the more common methods of speech recognition based on pattern matching uses Hidden Markov Modeling (HMM) comprising two types of pattern models, the acoustical model and the language model. Which of the two models will be used, and in some cases both will be required, depends on the complexity of the application. Complex speech recognition applications, such as those supporting continuous or connected speech recognition, will use a combination of the acoustical and language models.

In a simple application using only the acoustical model, the application will process the uttered word into phonemes, which are the fundamental part of speech. These phonemes are converted to a digital format. This digital format, or pattern, is then matched against stored patterns by the speech processor in search of a match from a stored database of word patterns. From the match, the phoneme, and word can be identified.

In a more complex method, the speech processor will convert the utterance to a digital signal by sampling the voice input at some rate, commonly 16 kHz. The required acoustical signal processing can be accomplished using several techniques. Some commonly used techniques are Linear Predictive Coding (LPC) cochlea modeling, Mel Frequency Cepstral Coefficients (MFCC), and others. For this example, the sampled data is converted to the frequency domain using a fast-Fourier transformation. The transformation will analyze the stored data at 1/30th to 1/100th of a second (3.3 ms to 100 ms) intervals, and convert the value into the frequency domain. The resulting graph from the converted digital input will be compared against a database of known sounds. From these comparisons, a value known as a feature number will be determined.

The feature numbers will be used to reference a phoneme found using that feature number. This, ideally, would be all that is required to identify a particular phoneme, however, this will not work for a number of reasons. Background noises, the user not pronouncing a word the same way every time, and the sound of a phoneme will vary, depending on the surrounding phonemes that may add variance to the sound being processed. To overcome problems of variability of the different phonemes, the phonemes are assigned to more than one feature number. Since the speech input was analyzed at an interval of 1/30th to 1/100th of a second and a phoneme or sound may last from 500 ms to 2 s, many feature numbers may be assigned to a particular sound. By using statistical analysis of these feature numbers and the probability that any one sound may contain those feature numbers, the probability of that sound being a particular phoneme can be determined.

To be able to recognize words and complete utterances, the speech recognizer must also be able to determine the beginning and the end of a phoneme. The most common method to determine the beginning and endpoint is by using the Hidden Markov Models (HMM) technique. The HMM is a state transition model and will use probabilities of feature numbers to determine the likelihood of transitioning from one state to another. Each phoneme is represented by a HMM. The English language is made up of 45 to 50 phonemes. A sequence of HMM will represent a word. This would be repeated for each word in the vocabulary. While the system can now recognize phonemes, phonemes do not always sound the same, depending on the phoneme preceding and following it. To address this problem, phonemes are placed in groups of three, called tri-phones, and as an aid in searching, similar sounding tri-phones are grouped together.

From the information obtained from the HMM state transitions, the recognizer is able to hypothesize and determine which phoneme likely was spoken, and then by referring this to a lexicon, the recognizer is able to determine the word that likely was spoken.

This is an overly simplified definition of the speech recognition process. There are numerous adaptations of the HMM technique and other modeling techniques. Some of these techniques are neural networks (NNs), dynamic time warping (DTW), and combinations of techniques.

8.2.1 Types of Speech Recognizers

There are two types of speech recognizers, speaker-dependent and speaker-independent.

8.2.1.1 Speaker-Dependent Systems

Speaker-dependent recognition is exactly that, speaker dependent. The system is designed to be used by one person. To operate accurately, the system will need to be "trained" to the user's individual speech patterns. This is sometimes referred to as "enrollment" of the speaker with the system. The speech patterns for the user will be recorded and patterned from which a template will be created for use by the speech recognizer. Because of the required training and storage of specific speech templates, the performance and accuracy of the speaker-dependent speech recognition engine will be tied to the voice patterns of a specific registered user. Speaker-dependent recognition, while being the most restrictive, is the most accurate, with accuracy rates in the mid to upper 90% range. For this reason, past research and applications for cockpit applications have opted to use speaker-dependent recognition.

The major drawback of this system is that it is dedicated to a single user, and that it must be trained prior to its use. Many applications will allow the speech template to be created elsewhere prior to use on the hosting system. This can be done at separate training stations prior to using the target system by transferring the created user voice template to the target system. If more than one user is anticipated, or if the training of the system is not desirable, a speaker-independent system might be an option.

8.2.1.2 Speaker-Independent Recognizers

Speaker-independent recognition systems are independent of the user. This type of system is intended to allow multiple users to access a system using voice input. Examples of speaker-independent systems are directory assist programs and an airline reservation system with a voice input driven menu system. Major drawbacks with a speaker-independent system, in addition to increased complexity and difficult implementation, are its lower overall accuracy rate, higher system overhead, and slower response time. The impact of these drawbacks continues to lessen with increased processor speeds, faster hardware, and increased data storage capabilities.

A variation of the speaker-independent system is the speaker-adaptive system. The speaker-adaptive system will adapt to the speech pattern, vocabulary, and style of the user. Over time, as the system adapts to the users' speech characteristics, the error rate of the system will improve, exceeding that of the independent recognizer.

8.2.2 Vocabularies

A vocabulary is a list of words that are valid for the recognizer. The size of a vocabulary for a given speech recognition system affects the complexity, processing requirements, and the accuracy of that system. There are no established definitions for how large a vocabulary should be, but systems using smaller vocabularies can result in better recognizer accuracy. As a general rule, a small vocabulary may contain up to 100 words, a medium vocabulary may contain up to 1000 words, a large vocabulary may contain up to 10,000 words, and a very large vocabulary may contain up to 64,000 words, and above that the vocabulary is considered unlimited. Again, this is a general rule and may not be true in all cases.

The size of a vocabulary will be dependent upon the purpose and intended function of the application. A very specific application may require only a few words and make use of a small vocabulary, while an application that would allow dictation or setting up airline reservations would require a very large vocabulary.

How can the size and contents of a vocabulary be determined? The words used by pilots are generally specific enough to require a small to medium vocabulary. Words that can or should be in the vocabulary could be determined in a number of ways. Drawing from the knowledge of how pilots would engage a desired function or task is one way. This could be done using a questionnaire or some similar survey method.

Another way to gather words for the vocabulary is to set up a lab situation and use the "Wizard of Oz" technique. This technique would have a test evaluator behind the scenes acting upon the commands given by a test subject. The test subject would have various tasks and scenarios to complete. While the test subject runs through the tasks, the words and phrases used by the subject are collected for evaluation. After running this process numerous times, the recorded spoken words and phrases will be used to construct a vocabulary list and command syntax, commonly referred to as a grammar. The vocabulary could be refined in further tests by only allowing those contained words and phrases to be valid, and have test subjects again run through a suite of tasks. Observations would be made as to how well the test subjects were able to complete the tasks using the defined vocabulary and syntax. Based on these tests, and the evaluation results, the vocabulary is modified as required.

A paper version of the evaluation process could be administered by giving the pilot a list of tasks, and then asking them to write out what commands they would use to perform the task. Following this data collection step, a second test could be generated having the pilot choose from a selected list of words and commands what he would likely say to complete the task. As a rule, pilots will tend to operate in a predictable manner, and this lends itself to a reduced vocabulary size and structured grammar.

8.2.3   Modes of Operation for Speech Recognizers

There are two modes of operation for a speech recognizer: continuous recognition, and discrete or isolated word recognition.

8.2.3.1   Continuous Recognition

Continuous speech recognition systems are able to operate on a continuous spoken stream of input in which the words are connected together. This type of recognition is more difficult to implement due to several inherent problems such as determining start and stop points in the stream and the rate of the spoken input.

The system must be able to determine the start and endpoint of a spoken stream of continuous speech. Words will have varied starting and ending phonemes depending on the surrounding phonemes. This is called "co-articulation." The rate of the spoken speech has a significant impact on the accuracy of the recognition system. The accuracy will degrade with rapid speech.

8.2.3.2   Discrete Word Recognition

Discrete or isolated word recognition systems operate on single words at a time. The system requires a pause between saying each word. The pause length will vary, and on some systems the pause length can be set to determined lengths. This type of recognition system is the simplest to perform because the endpoints are easier for the system to locate, and the pronunciation of a word is less likely to affect the pronunciation of other words (co-articulation effects are reduced). A user of this type of system will speak in a broken fashion. This system is the type most people think of in terms of a voice recognition system.

8.2.4   Methods of Error Reduction

There are no real standards by which error rates of various speech recognizers are measured and defined. Many systems claim accuracy rates in the high 90% range, but under actual usage with surrounding noise conditions, the real accuracy level may be much less. Many factors can impact the accuracy of SR systems.

Some of these factors include the individual speech characteristics of the user, the operating environment, and the design of the SR system itself.

There are four general error types impacting the performance of a SR system; these are substitution errors, insertion errors, rejection errors, and operator errors,

- Substitution errors occur when the SR system incorrectly identifies a word from the vocabulary. An example might be the pilot calling out "Tune COM one to one two four point seven" and the SR system incorrectly recognizes that the pilot spoke "Tune NAV one to one two four point seven." The SR system substituted NAV in place of COM. Both words may be defined and valid in the vocabulary, but the system selected the wrong word.

- Insertion errors may occur when some source of sound other than a spoken word is interpreted by the system as valid speech. Random cockpit noise might at some time be identified as a valid word to the SR system. The use of noise-canceling microphones and PTT can help to reduce this type of error.

- Rejection errors occur when the SR system fails to respond to the user's speech, even if the word or phrase was valid.

- Operator errors occur when the user is attempting to use words or phrases that are not identifiable to the SR system. A simple example might be calling out "change the radio frequency to one one eight point six" instead of "Tune COM one to one one point eight six," which is recognized by the vocabulary.

When designing a speech recognition application, several design goals and objective should be kept in mind:

- **Limitations of the hardware and the software** — Keep in mind the limitations of the hardware and the software being used for the application. Will the system need to have continuous recognition and discrete word recognition? Will the system need to be speaker independent, or will the reduced accuracy in using a speaker-independent recognizer be acceptable. Will the system be able to handle the required processing in an acceptable period of time? Will the system operate acceptably in the target environment?

- **Safety** — Will using SR to interface with a piece of equipment compromise safety? Will an error in recognition have a serious impact on the safety of flight? If the SR system should fail, is there an alternate method of control for that application?

- **Train the system in the environment in which it is intended to be used** — As discussed earlier, a SR system that has a 99% accuracy in the lab, may be frustrating and unusable in actual cockpit conditions. The speech templates or the training of the SR system needs to be done in the actual environment, or in as similar an environment as possible.

- **Don't try to use SR for tasks that don't really fit** — The problem with a new tool, like a new hammer, is that everything becomes a nail to try out that new hammer. Some tasks are natural candidates for using SR, many are not. Do not force SR onto a task if it is not appropriate for use of SR. Doing so will add significant risk and liability. Good target applications for SR include radio tuning functions, navigation functions, FMS functions, and display mode changes. Bad target applications for SR would be things that can affect the safety of flight, in short, anything that will kill you.

- **Incorporate error correction mechanisms** — Have the system repeat, using either voice synthesis or through a visual display, what it interprets, and allow the pilot to accept or reject this recognition. Allow the system to be able to recognize invalid recognition. If the recognizer interprets that it heard the pilot call out an invalid frequency, it should recognize it as invalid and possibly query the pilot to repeat, or prompt the pilot by saying or displaying that the frequency is invalid.

- **Provide feedback of the SR system's activities** — Allow the user to interact with the SR system. Have the system speak, using voice synthesis, or display what it is doing. This will allow the user to either accept or reject the recognizer interpretation. This may also serve as a way to prompt a user for more data that may have been left out of the utterance. "Tune COM 1 to...." After a delay, the system might query the user for a frequency: "Please select frequency for COM1." If the user selects some repeated command, the system may repeat back the command as it is executed: "Tuning COM 1 to ...."

8.2.4.1 Reduced Vocabulary

One way to dramatically increase the accuracy of a SR system is to reduce the number of words in a vocabulary. In addition to the reduction in words, the words should be carefully chosen to weed out words that sound similar.

Use a trigger phrase to gain the attention of the recognizer. The trigger phrase might be as simple as "computer ... " followed by some command. In this example, "computer" is the trigger phrase and alerts the recognizer that a command is likely to follow. This can be used with a system that is always on-line and listening.

Speech recognition errors can be reduced using a noise-canceling microphone. The flightdeck is not the quiet, sterile place a lab or a desktop might be. There are any number of noises and chatter that could interfere with the operation of speech recognition. Like humans, a recognizer can have increased difficulty in understanding commands in a noisy environment. In addition to the use of noise-canceling microphones, the use of high-quality omnidirectional microphones will offer further reduction in recognition errors. Using push-to-talk (PTT) microphones will help to reduce the occurrence of insertion errors as well as recognition errors.

8.2.4.2 Grammar

Grammar definition plays an important role in how accurate a SR application may be. It is used to not only define which words are valid to the system, but what the command syntax will be. A grammar notation frequently used in speech recognition is Context Free Grammar (CFG). A sample of a valid command in CFG is

$$\langle start \rangle \ = \ tune(COM \,|\, NAV) \ radio$$

This definition would allow valid commands of "tune COM radio," and "tune NAV radio." Word order is required, and words cannot be omitted. However, the grammar can be defined to allow for word order and omitted words.

8.3 Recent Applications

Though speech recognition has been applied to various flightdeck applications over the past 20 years, limitations in both hardware and software capability have kept the use of speech recognition from serious contention as a flightdeck tool. Even though there have been several notable applications of speech recognition in the recent past, and there are several current applications of speech recognition in the cockpit of military aircraft, it will likely be several more years before the civilian market will see such applications reach the level of reliability and pilot acceptance to see them commonly available.

In the mid 1990s, NASA performed experiments using speech recognition and voice control on an OV-10A aircraft. The experiment involved 12 pilots. The speech recognizer used for this study was an ITT VRS-1290 speaker-dependent system. The vocabulary used in this study was small, containing 54 words. The SR system was tested using the 12 pilots under three separate conditions: on the ground, 1g conditions, and 3g conditions. There was no significant difference in SR system performance found between the three conditions. The accuracy rates for the SR system under these three test conditions was 97.27% in hangar conditions, 97.72 under 1g conditions, and 97.11% under 3g conditions.[3]

A recent installation that is now in production is a military fighter, the Eurofighter, Typhoon. This aircraft will be the first production aircraft with voice interaction as a standard OEM configuration with speech recognition modules (SRMs). The speech recognizer is speaker dependent, and sophisticated enough to recognize continuous speech. The supplier of the voice recognition system for this aircraft is Smiths Industries. In addition, the system has received general pilot acceptance. Since the system is speaker dependent, the pilot must train the speech recognizer to his unique voice patterns prior to its use. This is done at ground-based, personal computer (PC) support stations. The PC is used to create a voice template for a specific pilot. The created voice template is then transferred to the aircraft prior to flight, via a data loader. Specifications for the recognizer include a 250-word vocabulary, a 200-ms response time, continuous speech recognition, and an accuracy rate of 95–98%.[2]

Another recent application of speech recognition technology is in the Joint Strike Fighter (JSF) being developed by Boeing and BAe Systems. Continuous speech recognition is being integrated into the cockpit. The speech recognition system will provide selected cockpit controls sole operation by using voice commands. The JSF speech recognition system will be used to allow the pilot to avoid the distraction of selected manual tasks while remaining focused on more critical aspects of the mission. The supplier of the speech recognition system for this aircraft is ITT Industries' Voxware (formerly VERBEX) voice recognition system. The Voxware system was chosen for this application due its recognized and previously proven ability to perform in a noisy cockpit environment.[1]

8.4   Flightdeck Applications

The use of speech recognition, the enabling technology for voice control, should not be relied on as the sole means of control or entering data and commands. Speech recognition is more correctly defined as an assisted method of control; and should have reversionary controls in place if the operation and performance of the SR system is no longer acceptable. It is not a question of whether voice control will find its way into mainstream aviation cockpits, but a question of when and to what degree. As the technology of SR continues to evolve, care must be exercised so that SR does not become a solution looking for a problem to solve. Not all situations will be good choices for the application of SR. In a high workload atmosphere, such as the flightdeck, the use of SR could be a logical choice for use in many operations, leading to a reduction in workload and heads-down time.

Current speech recognition systems are best assigned to tasks that are not in themselves critical to the safety of flight. In time, this will change as the technology evolves. The thought of allowing the speech recognition system to gain the ability to directly impact flight safety brings to mind an example that occurred at a speech recognition conference several years ago. While a speech recognition interface on a PC was being discussed and demonstrated before an audience, a member of the audience spoke out "format C: return," or something to that effect. The result was the main drive on the computer was formatted, erasing its contents. Normally an event such as this impacts no one's safety, however, if such unrestricted control were allowed on an aircraft, there would be serious results.

Some likely applications for voice control on the flightdeck are navigation functions; communications functions such as frequency selection, toggling of display modes, checklist functions, etc.

8.4.1   Navigation Functions

For navigation functions, SR could be used as a method of entering waypoints and inputting FMS data. Generally, most tasks requiring the keyboard to be used to enter data into the FMS would make good use of a SR system. This would allow time and labor savings in what is a repetitive and time consuming task. Another advantage of using SR is that the system is able to reduce confusion and guide the user by requesting required data. The use of SR with FMS systems is being evaluated and studied by both military and civilian aviation.

8.4.2   Communication Functions

For communication functions, voice control could be used to tune radio frequencies by calling out that frequency. For example, "Tune COM1 to one one eight point seven." The SR system would interpret this utterance, and would place the frequency into stand-by. The system may be designed to have the SR system repeat the recognized frequency back through a voice synthesizer to the pilot for confirmation prior to the frequency being placed into standby. The pilot would then accept the frequency and make it active or reject it. This would be done with a button press to activate the frequency. Another possible method of making a frequency active would be to do this by voice alone. This does bring about some added risk, as the pilot will no longer be physically making the selection. This could be done by a simple, "COM one Accept" to accept the frequency, but leave it in pre-select. Reject the frequency by saying, "COM one Reject," and to activate the frequency by saying, "COM one activate."

The use of SR would also allow a pilot to query systems, such as by requesting a current frequency setting; "What is COM one?" The ASR system could then respond with the current active frequency and possible the pre-select. This response could be by voice or by display. Other possible options would be to have the SR respond to ATC commands by moving the command frequency change to the pre-select automatically. Having done this, the pilot would only have to command "Accept," "Activate," or "Reject." The radio would never on its own, place a frequency from standby to active mode.

With the use of a GPS position-referenced database, a pilot might only have to call out "Tune COM one Phoenix Sky Harbor Approach." By referencing the current aircraft location to a database, the SR systems could look up the appropriate frequency and place it into pre-select. The system might respond back with, "COM one Phoenix Sky Harbor Approach at one two oh point seven." The pilot would then be able to accept and activate the frequency without having to know the correct frequency numbers or having to dial the frequency into the radio. Clearly a time-saving operation. Possible drawbacks are out-of-date radio frequencies in the database or no frequency listing. This can be overcome by being able to call out specific frequencies if required. "Tune COM one to one two oh point seven."

8.4.3   Checklist

The use of speech recognition is almost a natural for checklist operations. The pilot may be able to command the system with "configure for take-off." This could lead to the system bringing up an appropriate checklist for take-off configuration. The speech system could call out the checklist items as they occur and the pilot, having completed and verified the task, could press a button to accept and move on to the next task. It may be possible to allow a pilot to verbally check-off a task, vs. a button selection; however, that does bring about an opportunity for a recognition error.

Defining Terms

Accuracy: Generally, accuracy refers to the percentage of times that a speech recognizer will correctly recognize a word. This accuracy value is determined by dividing the number of times that the recognizer correctly identifies a word by the number of words input into the SR system.

Continuous speech recognition: The ability of the speech recognition system to accept a continuous, unbroken stream of words and recognize it as a valid phrase.

Discrete word recognition: This refers to the ability of a speech recognizer to recognize a discrete word. The words must be separated by a gap or pause between the previous word and successive words. The pause will typically be 150 ms or longer. The use of such a system is characterized by "choppy" speech to ensure the required break between words.

Grammar: This is a set of syntax rules determining valid commands and vocabulary for the SR system. The grammar will define how words may be ordered and what commands are valid. The grammar definition structure most commonly used is known as "context free grammar" or CFG.

Isolated word recognition: The ability of the SR system to recognize a specific word in a stream of words. Isolated word recognition can be used as a "trigger" to place the SR system into an active standby mode, ready to accept input.

Phonemes: Phonemes are the fundamental parts of speech. The English language is made up from 45 to 50 individual phonemes.

Speaker Dependent: This type of system is dependent upon the speaker for operation. The system will be trained to recognize one person's speech patterns and acoustical properties. This type of system will have a higher accuracy rate than a speaker-independent system, but is limited to one user.

Speaker Independent: A speaker-independent system will operate regardless of the speaker. This type of system is the most desirable for a general use application, however the accuracy rate and response rate will be lower than the speaker-dependent system.

Speech Synthesis: The use of an artificial means to create speech-like sounds.

Text to Speech: A mechanism or process in which text is transformed into digital audio form and output as "spoken" text. Speech synthesis can be used to allow a system to respond to a user verbally.

Tri-Phones: These are groupings of three phonemes. The sound a phoneme makes can vary depending on the phoneme ahead of it and after it. Speech recognizers use tri-phones to better determine which phoneme has been spoken based upon the sounds preceding and following it.

Verbal Artifacts: These are words or phrases, spoken with the intended command that have no value content to the command. This is sometimes referred to simply as garbage when defining a specific grammar. Grammars may be written to allow for this by disregarding and ignoring these utterances, for example, the pilot utterance, "uhhhhhmmmmmmmm, select north up mode." The "uhhhh-hmmmmmmm" would be ignored as garbage.

Vocabulary: The vocabulary a speech recognition system is made up of the words or phrases that the system is to recognize. Vocabulary size is generally broken into four sizes; small, with tens of words, medium with a few hundred words, large with a few thousand words, very large with up to 64,000 words, and unlimited. When a vocabulary is defined, it will contain words that are relative, and specific to the application.

References

1. Boeing JSF to feature voice-recognition technology, [On-Line]. Available: www.boeing.com/news/releases/2000/news_release_000222o.htm.
2. The Eurofighter Typhoon Speech Recognition Module, [On-Line]. Available: www.smithsind-aerospace.com/PRODS/CIS/Voice.htm
3. Williamson, David T., Barry, Timothy P., and Liggett, Kristen K., Flight test results of ITT VRS-1290 in NASA OV10A. Pilot-Vehicle Interface Branch (WL/FIGP), WPAFB, OH.

Bibliography

Anderson, Timothy R., Applications of speech-based control, in *Proc. Alternative Control Technologies: Human Factors Issues*, 14-15 Oct., 1998, Wright-Patterson AFB, OH, (ISBN 92-837-1003-7).

Anderson, Timothy R., The technology of speech-based control, in *Proc. Alternative Control Technologies: Human Factors Issues*, 14-15 Oct., 1998, Wright-Patterson AFB, OH, (ISBN 92-837-1003-7).

Bekker, M. M., "A comparison of mouse and speech input control of a text-annotation system," Faculty of Industrial Design Engineering, Delft University of Technology, Jaffalaan 9, 2628 BX Delft, The Netherlands.

Boeing, JSF to feature voice-recognition technology, [On-Line]. Available: www.boeing.com/news/releases/2000/news_release_000222o.htm.

Eurofighter Typhoon Speech Recognition Module, Available: www.smithsind-aerospace.com/PRODS/CIS/Voice.htm.

Hart, Sandra G., Helicopter human factors, in *Human Factors in Aviation*, Wiener, Earl L. and Nagel, David C., Eds., Academic Press, San Diego, 1988, chap. 18.

Hopkin, V. David, Air traffic control, in *Human Factors in Aviation*, Wiener, Earl L. and Nagel, David C., Eds., Academic Press, San Diego, 1988, chap. 19.

Jones, Dylan M., Frankish, Clive R., and Hapeshi, K., Automatic Speech Recognition in Practice, Behav. Inf. Technol., 2, 109–122, 1992.

Leger, Alain, Synthesis and expected benefits analysis, in *Proc. Alternative Control Technologies: Human Factors Issues*, 14-15 Oct., 1998, Wright-Patterson AFB, OH, (ISBN 92-837-1003-7).

Rood, G. M., Operational rationale and related issues for alternative control technologies, in *Proc. Alternative Control Technologies: Human Factors Issues*, 14-15 Oct., 1998, Wright-Patterson AFB, OH, (ISBN 92-837-1003-7).

Rudnicky, Alexander I. and Hauptmann, Alexander G., Models for evaluating interaction protocols in speech recognition, School of Computer Science, Carnegie Mellon University, Pittsburgh, PA, 1991.

Wickens, Christopher D. and Flach, John M., Information processing, in *Human Factors in Aviation*, Wiener, Earl L. and Nagel, David C., Eds., Academic Press, San Diego, 1988, chap. 5.

Williamson, David T., Barry, Timothy P., and Liggett, Kristen K., Flight test results of ITT VRS-1290 in NASA OV10A. Pilot-Vehicle Interface Branch (WL/FIGP), WPAFB, OH.

Williges, Robert C., Williges, Beverly H., and Fainter, Robert G., Software interfaces for aviation systems, in *Human Factors in Aviation*, Wiener, Earl L. and Nagel, David C., Eds., Academic Press, San Diego, 1988, chap. 14.

Further Information

There are numerous sources for additional information on speech recognition. A search of the Internet on "speech recognition" will yield many links and information sources. The list will likely contain companies and corporations that deal primarily in speech recognition products. Some of these companies include:

• Analog Devices	(800) 262-5643	www.analog.com
• AT&T Adv Speech Products Group	(800) 592-8766	www.att.com/aspg
• Brooktrout Technology	(617) 449-4100	www.techspk.com
• Dialogic	(201) 993-3000	www.dialogic.com
• Dragon Systems	(800) 825-5897	www.dragonsys.com
• Entropic Cambridge Research Labs	(202) 547-1420	www.entropic.com
• IBM Speech Products	(800) 825-5263	www.software.ibm.com/is/voicetype
• Kurzweil Applied Intelligence	(617) 883-5151	www.kurzweil.com
• Lernout & Hauspie	(617) 238-0960	www.lhs.com
• Nuance Communications	(415) 462-8200	www.nuance.com
• Oki Semiconductor	(408) 720-1900	www.oki.com
• Philips Speech Processing	(516) 921-9310	www.speech.be.philips.com
• PureSpeech	(617) 441-0000	www.speech.com
• Sensory	(408) 744-1299	www.SensoryInc.com
• Smith Industries	(610) 296-5000	www.smithsind-aerospace.com/
• Speech Solutions	(800) 773-3247	www.speechsolutions.com
• Texas Instruments	(800) 477-8924 x 4500	www.ti.com

9

Human Factors Engineering and Flight Deck Design

Kathy H. Abbott
Federal Aviation Administration

9.1 Introduction

This chapter briefly describes Human Factors Engineering and considerations for civil aircraft flight deck design. The motivation for providing the emphasis on the Human Factor is that the operation of future aviation systems will continue to rely on humans in the system for effective, efficient, and safe operation. Pilots, mechanics, air traffic service personnel, designers, dispatchers, and many others are the basis for successful operations now and for the foreseeable future. There is ample evidence that failing to adequately consider humans in the design and operations of these systems is at best inefficient and at worst unsafe.

This becomes especially important with the continuing advance of technology. Technology advances have provided a basis for past improvements in operations and safety and will continue to do so in the future. New alerting systems for terrain and traffic avoidance, data link communication systems to augment voice-based radiotelephony, and new navigation systems based on Required Navigation Performance are just a few of the new technologies being introduced into flight decks.

Often, such new technology is developed and introduced to address known problems or to provide some operational benefit. While introduction of new technology may solve some problems, it often introduces others. This has been true, for example, with the introduction of advanced automation.[1,2] Thus, while new technology can be part of a solution, it is important to remember that it will bring issues that may not have been anticipated and must be considered in the larger context (equipment design, training, integration into existing flight deck systems, procedures, operations, etc.). These issues are especially important to address with respect to the human operator.

The chapter is intended to help avoid vulnerabilities in the introduction of new technology and concepts through the appropriate application of Human Factors Engineering in the design of flight decks. The chapter first introduces the fundamentals of Human Factors Engineering, then discusses the flight deck design process. Different aspects of the design process are presented, with an emphasis on the incorporation of Human Factors in flight deck design and evaluation. To conclude the chapter, some additional considerations are raised.

9.2 Fundamentals

This section provides an overview of several topics that are fundamental to the application of Human Factors Engineering (HFE) in the design of flight decks. It begins with a brief overview of Human Factors, then discusses the design process. Following that discussion, several topics that are important to the application of HFE are presented: the design philosophy, the interfaces and interaction between pilots and flight decks, and the evaluation of the pilot/machine system.

9.2.1 Human Factors Engineering

It is not the purpose of this section to provide a complete tutorial on Human Factors. The area is quite broad and the scientific and engineering knowledge about human behavior and human performance, and the application of that knowledge to equipment design (among other areas), is much more extensive than could possibly be cited here.[3–8] Nonetheless, a brief discussion of certain aspects of Human Factors is desirable to provide the context for this chapter.

For the purposes of this chapter, Human Factors and its engineering aspects involve the application of knowledge about human capabilities and limitations to the design of technological systems.[9] Human Factors Engineering also applies to training, personnel selection, procedures, and other topics, but those topics will not be expanded here.

Human capabilities and limitations can be categorized in many ways, with one example being the SHEL model.[6] This conceptual model describes the components *Software, Hardware, Environment, and Liveware*. The SHEL model, as described in Reference 6, is summarized below.

The center of the model is the human, or *Liveware*. This is the hub of Human Factors. It is the most valuable and most flexible component of the system. However, the human is subject to many limitations, which are now predictable in general terms. The "edges" of this component are not simple or straight, and it may be said that the other components must be carefully matched to them to avoid stress in the system and suboptimal performance. To achieve this matching, it is important to understand the characteristics of this component:

- **Physical size and shape** — In the design of most equipment, body measurements and movement are important to consider at an early stage. There are significant differences among individuals, and the population to be considered must be defined. Data to make design decisions in this area can be found in anthropometry and biomechanics.

- **Fuel requirements** — The human needs fuel (e.g., food, water, and oxygen) to function properly. Deficiencies can affect performance and well-being. This type of data is available from physiology and biology.

- **Input characteristics** — The human has a variety of means for gathering input about the world around him or her. Light, sound, smell, taste, heat, movement, and touch are different forms of information perceived by the human operator; for effective communication between a system and the human operator, this information must be understood to be adequately considered in design. This knowledge is available from biology and physiology.

- **Information processing** — Understanding how the human operator processes the information received is another key aspect of successful design. Poor human-machine interface or system design that does not adequately consider the capabilities and limitations of the human information processing system can strongly affect the effectiveness of the system. Short- and long-term memory limitations are factors, as are the cognitive processing and decision-making processes used. Many human errors can be traced to this area. Psychology, especially cognitive psychology, is a major source of data for this area.

- **Output characteristics** — Once information is sensed and processed, messages are sent to the muscles and a feedback system helps to control their actions. Information about the kinds of forces that can be applied and the acceptable direction of controls are important in design decisions. As another example, speech characteristics are important in the design of voice communication systems. Biomechanics and physiology provide this type of information.

- **Environmental tolerances** — People, like equipment, are designed to function effectively only within a narrow range of environmental conditions such as temperature, pressure, noise, humidity, time of day, light, and darkness. Variations in these conditions can all be reflected in performance. A boring or stressful working environment can also affect performance. Physiology, biology, and psychology all provide relevant information on these environmental effects.

It must be remembered that humans can vary significantly in these characteristics. Once the effects of these differences are identified, some of them can be controlled in practice through selection, training, and standardized procedures. Others may be beyond practical control and the overall system must be designed to accommodate them safely. This *Liveware* is the hub of the conceptual model. For successful and effective design, the remaining components must be adapted and matched to this central component.

The first of the components that requires matching to the characteristics of the human is *Hardware*. This interface is the one most generally thought of when considering human-machine systems. An example is designing seats to fit the sitting characteristics of the human. More complex is the design of displays to match the human's information processing characteristics. Controls, too, must be designed to match the human's characteristics, or problems can arise from, for example, inappropriate movement or poor location. The user is often unaware of mismatches in this liveware-hardware interface. The natural human characteristic of adapting to such mismatches masks but does not remove their existence. Thus this mismatch represents a potential hazard to which designers should be alerted.

The second interface with which Human Factors Engineering is concerned is that between Liveware and Software. This encompasses the nonphysical aspects of the systems such as procedures, manual and checklist layout, symbology, and computer programs. The problems are often less tangible than in the Liveware-Hardware interface and more difficult to resolve.

One of the earliest interfaces recognized in flying was between the human and the environment. Pilots were fitted with helmets against the noise, goggles against the airstream, and oxygen masks against the altitude. As aviation matured, the environment became more adapted to the human (e.g., through pressurized aircraft). Other aspects that have become more of an issue are disturbed biological rhythms and related sleep disturbances because of the increased economic need to keep aircraft, and the humans that operate them, flying 24 hours a day. The growth in air traffic and the resulting complexities in operations are other aspects of the environment that are becoming increasingly significant now and in the future.

The last major interface described by the SHEL model is the human-human interface. Traditionally, questions of performance in flight have focused on individual performance. Increasingly, attention is being paid to the performance of the team or group. Pilots fly as a crew; flight attendants work as a team; maintainers, dispatchers, and others operate as groups; therefore, group dynamics and influences are important to consider in design.

The SHEL model is a useful conceptual model, but other perspectives are important in design as well. The reader is referred to the references cited for in-depth discussion of basic human behavioral considerations, but a few other topics are especially relevant to this chapter and are discussed here: usability, workload, and situation awareness.

9.2.1.1 Usability

The usability of a system is very pertinent to its acceptability by users; therefore, it is a key element to the success of a design. Nielsen[10] defines usability as having multiple components:

- Learnability — the system should be easy to learn
- Efficiency — the system should be efficient to use
- Memorability — the system should be easy to remember

- Error — the system should be designed so that users make few errors during use of the system, and can easily recover from those they do make
- Satisfaction — the system should be pleasant to use, so users are subjectively satisfied when using it.

This last component is indicated by subjective opinion and preference by the user. This is important for acceptability, but it is critical to understand that there is a difference between subjective preference and performance of the human-machine system. In some cases, the design that was preferred by the user was not the design that resulted in the best performance. This illustrates the importance of both subjective input from representative end users and objective performance evaluation.

9.2.1.2 Workload

In the context of the commercial flight deck, workload is a multidimensional concept consisting of: (1) the duties, amount of work, or number of tasks that a flight crew member must accomplish; (2) the duties of the flight crew member with respect to a particular time interval during which those duties must be accomplished; and/or (3) the subjective experience of the flight crew member while performing those duties in a particular mission context. Workload may be either physical or mental.[11]

Both overload (high workload, potentially resulting in actions being skipped or executed incorrectly or incompletely) and underload (low workload, leading to inattention and complacency) are worthy of attention when considering the effect of design on human-machine performance.

9.2.1.3 Situation Awareness

This can be viewed as the perception on the part of a flight crew member of all the relevant pieces of information in both the flight deck and the external environment, the comprehension of their effects on the current mission status, and the projection of the values of these pieces of information (and their effect on the mission) into the near future.[11]

Situation awareness has been cited as an issue in many incidents and accidents, and can be considered as important as workload. As part of the design process, the pilot's information requirements must be identified, and the information display must be designed to ensure adequate situation awareness. Although the information is available in the flight deck, it may not be in a form that is directly usable by the pilot, and therefore of little value.

Another area that is being increasingly recognized as important is the topic of organizational processes, policies and practices.[12] It has become apparent that the influence of these organizational aspects is a significant, if latent, contributor to potential vulnerabilities in design and operations.

9.2.2 Flight Deck Design

The process by which commercial flight decks are designed is complex, largely unwritten, variable, and nonstandard.[11] That said, Figure 9.1 is an attempt to describe this process in a generic manner. It represents a composite flight deck design process based on various design process materials. The figure is not intended to exactly represent the accepted design process within any particular organization or program; however, it is meant to be descriptive of generally accepted design practice. (For more detailed discussion of design processes for pilot-system integration and integration of new systems into existing flight decks, see References 13 and 14.)

The figure is purposely oversimplified. For example, the box labeled "Final Integrated Design" encompasses an enormous number of design and evaluation tasks, and can take years to accomplish. It could be expanded into a figure of its own that includes not only the conceptual and actual integration of flight deck components, but also analyses, simulations, flight tests, certification and integration based on these evaluations.

Flight deck design necessarily requires the application of several disciplines, and often requires trade-offs among those disciplines. Human Factors Engineering is only one of the disciplines that should be part of the process, but it is a key part of ensuring that the flight crew's capabilities and limitations are considered. Historically, this process tends to be very reliant on the knowledge and experiences of individuals involved in each program.

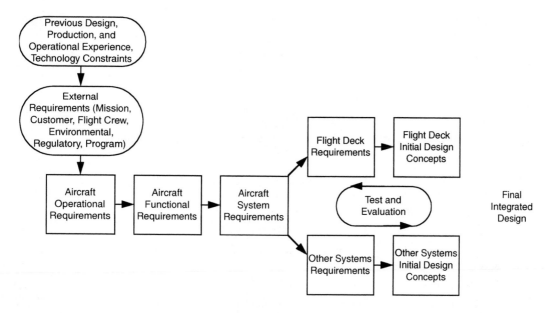

FIGURE 9.1 Simplified representation of the flight deck design process (from NASA TM 109171).

Human-centered or user-centered design has been cited as a desirable goal. That is, design should be focused on supporting the human operator of the system, much as discussed above on the importance of matching the hardware, software, and environment to the human component. A cornerstone of human-centered design is the design philosophy.

9.2.2.1 Flight Deck Design Philosophy

The design philosophy, as embodied in the top-level philosophy statements, guiding principles, and design guidelines, provides a core set of beliefs used to guide decisions concerning the interaction of the flight crew with the aircraft systems. It typically deals with issues such as allocation of functions between the flight crew and the automated systems, levels of automation, authority, responsibility, information access and formatting, and feedback, in the context of human use of complex, automated systems.[1,11]

The way pilots operate airplanes has changed as the amount of automation and the automation's capabilities have increased. Automation has both provided alternate ways of accomplishing pilot tasks performed on previous generations of airplanes and created new tasks. The increased use of and flight crew reliance on flight deck automation makes it essential that the automation act predictably with actions that are well understood by the flight crew. The pilot has become, in some circumstances, a supervisor or manager of the automation.

Moreover, the automation must be designed to function in a manner that directly supports flight crews in performing their tasks. If these human-centered design objectives are not met, the flight crew's ability to properly control or supervise system operation is limited, leading to confusion, automation surprises, and unintended airplane responses.

Each airplane manufacturer has a different philosophy regarding the implementation and use of automation. Airbus and Boeing are probably the best-known for having different flight deck design philosophies. However, there is general agreement that the flight crew is and will remain ultimately responsible for the safety of the airplane they are operating.

Airbus has described its automation philosophy as:

- Automation must not reduce overall aircraft reliability, it should enhance aircraft and systems safety, efficiency, and economy
- Automation must not lead the aircraft out of the safe flight envelope and it should maintain the aircraft within the normal flight envelope

- Automation should allow the operator to use the safe flight envelope to its full extent, should this be necessary due to extraordinary circumstances
- Within the normal flight envelope, the automation must not work against operator inputs, except when absolutely necessary for safety

Boeing has described its philosophy as follows:

- The pilot is the final authority for the operation of the airplane
- Both crew members are ultimately responsible for the safe conduct of the flight
- Flight crew tasks, in order of priority, are safety, passenger comfort, and efficiency
- Design for crew operations based on pilot's past training and operational experience
- Design systems to be error tolerant
- The hierarchy of design alternatives is simplicity, redundancy, and automation
- Apply automation as a tool to aid, not replace, the pilot
- Address fundamental human strengths, limitations, and individual differences — for both normal and nonnormal operations
- Use new technologies and functional capabilities only when:
  - They result in clear and distinct operational or efficiency advantages, and
  - There is no adverse effect to the human-machine interface

One of the significant differences between the design philosophies of the two manufacturers is in the area of envelope protection. Airbus' philosophy has led to the implementation of what has been described as "hard" limits, where the pilot can provide whatever control inputs he or she desires, but the airplane will not exceed the flight envelope. In contrast, Boeing has "soft" limits, where the pilot will meet increasing resistance to control inputs that will take the airplane beyond the normal flight envelope, but can do so if he or she chooses. In either case, it is important for the pilot to understand what the design philosophy is for the airplane being flown.

Other manufacturers may have philosophies that differ from Boeing and Airbus. Different philosophies can be effective if each is consistently applied in design, training, and operations, and if each supports flight crew members in flying their aircraft safely. To ensure this effectiveness, it is critical that the design philosophy be documented explicitly and provided to the pilots who will be operating the aircraft, the trainers, and the procedure developers.

9.2.2.2   Pilot/Flight Deck Interfaces

The layout, controls, displays and amount of automation in flight decks have evolved tremendously in commercial aviation.[15,16] What is sometimes termed the "classic" flight deck, which includes the B-727, the DC-10, and early series B-747, is typically characterized by dedicated displays, where one piece of data is generally shown on a dedicated gage or dial as the form of display. These aircraft are relatively lacking in automation. A representative "classic" flight deck is shown in Figure 9.2. All of these aircraft are further characterized by the relative simplicity of their autopilot, which offers one or a few simple modes in each axis. In general, a single instrument indicates the parameter of a single sensor. In a few cases, such as the Horizontal Situation Indicator, a single instrument indicates the "raw" output of multiple sensors. Regardless, the crew is generally responsible for monitoring the various instruments and realizing when a parameter is out of range. A simple caution and warning system exists, but it covers only the most critical system failures.

The first generation of "glass cockpit" flight decks, which include the B-757/767, A-310, and MD-88, receive their nickname due to their use of cathode ray tubes (CRTs). A representative first-generation "glass cockpit" flight deck is shown in Figure 9.3. A mix of CRTs and instruments was used in this generation of flight deck, with instruments used for primary flight information such as airspeed and altitude. A key innovation in this flight deck was the "map display" and its coupling to the Flight Management System (FMS). This enabled the crew to program their flight plan into a computer and see their planned track along the ground, with associated waypoints, on the map display. Accompanying the introduction of the map

FIGURE 9.2   Representative "classic" flight deck (DC-10).

FIGURE 9.3   Representative first-generation "glass cockpit" (B-757) flight deck.

FIGURE 9.4   Representative second-generation "glass cockpit" (Airbus A320) flight deck.

display and FMS were more complex autopilots (added modes from the FMS and other requirements). This generation of aircraft also featured the introduction of an integrated Caution and Warning System, usually displayed in a center CRT with engine information. A major feature of this Caution and Warning System was that it prioritized alerts according a strict hierarchy of "warnings" (immediate crew action required), "cautions" (immediate crew awareness and future action required), and "advisories" (crew awareness and possible action required).[17]

The second generation of "glass cockpit" flight decks, which include the B-747-400, A-320/330/340, F-70/100, MD-11, and B-777, are characterized by the prevalence of CRTs (or LCDs in the case of the B-777) on the primary instrument panel. A representative second-generation "glass cockpit" flight deck is shown in Figure 9.4. CRT/LCDs are used for all primary flight information, which is integrated on a few displays. In this generation of flight deck, there is some integration of the FMS and autopilot — certain pilot commands can be input into either the FMS or autopilot and automatically routed to the other.

There are varying levels of aircraft systems automation in this generation of flight deck. For example, the MD-11 fuel system can suffer certain failures and take corrective action — the crew is only notified if they must take some action or if the failure affects aircraft performance. The caution and warning systems in this generation of flight decks are sometimes accompanied by synoptic displays that graphically indicate problems. Some of these flight decks feature fly-by-wire control systems — in the case of the A-320/330/340, this capability has allowed the manufacturer to tailor the control laws such that the flying qualities of these various size aircraft appear similar to pilots. The latest addition to this generation of flight deck, the B-777, has incorporated "cursor control" for certain displays, allowing the flight crew to use a touchpad to interact with "soft buttons" programmed on these displays.

FIGURE 9.5 Gulfstream GV flight deck.

Of note is the way that this flight deck design evolution affects the manner in which pilots access and manage information. Figure 9.2 illustrates the flight deck with dedicated gages and dials, with one display per piece of information. In contrast, the flight deck shown in Figure 9.4 has even more information available, and the pilot must access it in entirely different manner. Some of the information is integrated in a form that the pilot can more readily interpret (e.g., moving map displays). Other information must be accessed through pages of menus. The point is that there has been a fundamental change in information management in the flight deck, not through intentional design but through introduction of technology, often for other purposes.

An example is shown in Figure 9.5 from the business aircraft community illustrating that the advanced technology discussed here is not restricted to large transport aircraft. In fact, new technology is quite likely to be more quickly introduced into these smaller, sophisticated aircraft.

Major changes in the flight crew interface with future flight decks are expected. While it is not known exactly what the flight decks of the future will contain or how they will function, some possible elements may include:

- Sidestick control inceptors, interconnected and with tailorable force/feel, preferably "backdriven" during autopilot engagement.
- Cursor control devices, which the military has used for many years, but the civil community is just starting to use (e.g., in the Boeing 777).
- Multifunction displays.
- Management of subsystems through displays and control-display units.
- "Mode-less" flight path management functions.
- Large, high-resolution displays having multiple signal sources (computer-generated and video).

- Graphical interfaces for managing certain flight deck systems.
- High-bandwidth, two-way datalink communication capability embedded in appropriate flight deck systems
- Replacement of paper with "electronic flight bags."
- Voice interfaces for certain flight deck systems.

These changes will continue to modify the manner in which pilots manage information within the flight deck, and the effect of such changes should be explicitly considered in the flight deck design process.

9.2.2.3 Pilot/Flight Deck Interaction

Although it is common to consider the pilot interfaces to be the only or primary consideration in human factors in flight deck design, the interaction between the pilot(s) and the flight deck must also be considered. Some of the most visible examples of the importance of this topic, and the consequences of vulnerabilities in this area, are in the implementation of advanced automation.

Advanced automation (sophisticated autopilots, autothrust, flight management systems, and associated displays and controls) has provided large improvements in safety (e.g., through reduced pilot workload in critical or long-range phases of flight) and efficiency (improved precision of flying certain flight paths). However, vulnerabilities have been identified in the interaction between the flight crews and modern systems.[2]

For example, on April 26, 1994, an Airbus A300–600 operated by China Airlines crashed at Nagoya, Japan killing 264 passengers and flight crew members. Contributing to the accident were conflicting actions taken by the flight crew and the airplane's autopilot. During complex circumstance, the flight crew attempted to stay on glide slope by commanding nose-down elevator. The autopilot was then engaged, and because it was still in go-around mode, commanded nose-up trim. A combination of an out-of-trim condition, high engine thrust, and retracting the flaps too far led to a stall. The crash provided a stark example of how a breakdown in the flight crew/automation interaction can affect flight safety. Although this particular accident involved an A300–600, other accidents, incidents, and safety indicators demonstrate that this problem is not confined to any one airplane type, airplane manufacturer, operator, or geographical region.

A lesson to be learned here is that design of the interaction between the pilot and the systems must consider human capabilities and limitations. A good human-machine interface is necessary but may not be sufficient to ensure that the system is usable and effective. The interaction between the pilot and the system, as well as the function of the system itself, must be carefully "human engineered."

9.2.3 Evaluation

Figure 9.1 showed test and evaluation (or just evaluation, for the remainder of the discussion) as an integral part of the design process. Because evaluation is (or should be) such an important part of design, some clarifying discussion is appropriate here. (See Reference 18 for a more detailed discussion of the evaluation issues that are summarized below.)

Evaluation often is divided into verification (the process of demonstrating that the system works as designed) and validation (the process of assessing the degree to which the design achieves the system objectives of interest). Thus, validation goes beyond asking whether the system was built according to the plan or specifications; it determines whether the plan or specifications were correct for achieving the system objectives.

One common use of the term "evaluation" is as a synonym of "demonstration." That is, evaluation involves turning on the system and seeing if it basically resembles what the designer intended. This does not, however, provide definitive information on safety, economy, reliability, maintainability, or other concerns that are generally the motivation for evaluation.

It is not unusual for evaluation to be confused with demonstration, but they are not the same. In addition, there are several different types and levels of evaluation that are useful to understand. For example, **formative** evaluation is performed during the design process. It tends to be informal and subjective, and its results should be viewed as hypotheses, not definitive results. It is often used to evaluate requirements. In contrast, **formal** evaluation is planned during the design but performed with a prototype to assess the performance of the human/machine system. Both types of evaluations are required, but the rest of this discussion focuses on formal evaluation.

Another distinction of interest in understanding types of evaluation is the difference between **absolute** vs. **comparative** evaluations. **Absolute** evaluation is used when assessing against a standard of some kind. An example would be evaluating whether the pilot's response time using a particular system is less than some prespecified number. **Comparative** evaluation compares one design to another, typically an old design to a new one. Evaluating whether the workload for particular tasks in a new flight deck is equal to or less than in an older model is an example comparative evaluation. This type of evaluation is often used in the airworthiness certification of a new flight deck, to show its acceptability relative to an older, already certified flight deck. It may be advantageous for developers to expand an absolute evaluation into a comparative evaluation (through options within the new system) to assess system sensitivities.

Yet another important distinction is between **objective** vs. **subjective** evaluation. **Objective** evaluation measures the degree to which the objective criteria (based on system objectives) have been met. **Subjective** evaluation focuses on users' opinions and preferences. Subjective data are important but should be used to support the objective results, not replace them.

Planning for the evaluation should proceed in parallel with design rather than after the design is substantially completed. Evaluation should lead to design modification, and this is most effectively done in an iterative fashion.

Three basic issues, or levels of evaluation, are worth considering. The first is **compatibility**. That is, the physical presentation of the system must be compatible with human input and output characteristics. The pilot has to be able to read the displays, reach the controls, etc. Otherwise, it doesn't matter how good the system design is; it will not be usable.

Compatibility is important but not sufficient. A second issue is **understandability**. That is, just because the system is compatible with human input-output capabilities and limitations does not necessarily mean that it is understandable. The structure, format, and content of the pilot-machine dialogue must result in meaningful communication. The pilot must be able to interpret the information provided, and be able to "express" to the system what he or she wishes to communicate. For example, if the pilot can read the menu, but the options available are meaningless, that design is not satisfactory.

A designer must ensure that the design is both compatible and understandable. Only then should the third level of evaluation be addressed: that of **effectiveness**. A system is effective to the extent that it supports a pilot or crew in a manner that leads to improved performance, results in a difficult task being made less difficult, or enables accomplishing a task that otherwise could not have been accomplished. Assessing effectiveness depends on defining measures of performance based on the design objectives. Regardless of these measures, there is no use in attempting to evaluate effectiveness until compatibility and understandability are ensured.

Several different methods of evaluation can be used, ranging from static paper-based evaluations to in-service experience. The usefulness and efficiency of a particular method of evaluation naturally depends on what is being evaluated. Table 9.1 shows the usefulness and efficiency of several methods for each of the levels of evaluation.

As can be seen from this discussion, evaluation is an important and integral part of successful design.

9.3 Additional Considerations

9.3.1 Standardization

Generally, across manufacturers, there is a great deal of variation in existing flight deck systems design, training, and operation. Because pilots often operate different aircraft types, or similar aircraft with different equipage, at different points in time, another way to avoid or reduce errors is standardization of equipment, actions, and other areas.[19]

TABLE 9.1   Methods of Evaluation[18]

Method	Levels of Evaluation		
	Compatibility	Understandability	Effectiveness
Paper Evaluation: Static	Useful and Efficient	Somewhat Useful but Inefficient	Not Useful
Paper Evaluation: Dynamic	Useful and Efficient	Somewhat Useful but Inefficient	Not Useful[a]
Part-Task Simulator: "Canned" Scenarios	Useful but Inefficient	Useful and Efficient	Marginally Useful but Efficient[a]
Part-Task Simulator: Model Driven	Useful but Inefficient	Useful and Efficient	Somewhat Useful and Efficient
Full-Task Simulator	Useful but Very Inefficient	Useful but Inefficient	Useful but Somewhat Inefficient
In-Service Evaluation	Useful but Extremely Inefficient	Useful but Very Inefficient	Useful but Inefficient

[a] Can be effective for formative evaluation.

It is not realistic (or even desirable) to think that complete standardization of existing aircraft will occur. However, for the sake of the flight crews who fly these aircraft, appropriate standardization of new systems/technology/operational concepts should be pursued, as discussed below.

Appropriate standardization of procedures/actions, system layout, displays, color philosophy, etc. is generally desirable, because it has several potential advantages, including:

- Reducing potential for crew error/confusion due to negative transfer of learning from one aircraft to another;
- Reducing training costs, because you only need to train once; and
- Reducing equipment costs because of reduced part numbers, inventory, etc.

A clear example of standardization in design and operation is the Airbus A320/330/340 commonality of flight deck and handling qualities. This has advantages of reduced training and enabling pilots to easily fly more than one airplane type.

If standardization is so desirable, why is standardization not more prevalent? There are concerns that inappropriate standardization, rigidly applied, can be a barrier to innovation, product improvement, and product differentiation. In encouraging standardization, known issues should be recognized and addressed.

One potential pitfall of standardization that should be avoided is to standardize on the lowest common denominator. Another question is to what level of design prescription should standardization be done, and when does it take place? From a human performance perspective, consistency is a key factor. The actions and equipment may not be exactly the same, but should be consistent. An example where this has been successfully applied is in the standardization of alerting systems,[16] brought about by the use of industry-developed design guidelines. Several manufacturers have implemented those guidelines into designs that are very different in some ways, but are generally consistent from the pilot's perspective.

There are several other issues with standardization. One of them is related to the introduction of new systems into existing flight decks. The concern here is that the new system should have a consistent design/operating philosophy with the flight deck into which it is being installed. This point can be illustrated by the recent introduction of a warning system into modern flight decks. In introducing this new system, the question arose whether the display should automatically be brought up if an alert occurs (replacing the current display selected by the pilot). One manufacturer's philosophy is to bring the display up automatically when an alert occurs; another manufacturer's philosophy is to alert the pilot, then have the pilot select the display when desired. This is consistent with the philosophy of that flight deck of providing the pilot control over the management of displays. The trade-off between standardization across aircraft types (and manufacturers) and internal consistency with flight deck philosophy is very important to consider and should probably be done on a case-by-case basis.

The timing of standardization, especially with respect to introduction of new technology, is also critical.[4] It is desirable to deploy new technology early, because some problems are only found in the actual operating environment. However, if we standardize too early, then there is a risk of standardizing on a design that has not accounted for that critical early in-service experience. We may even unintentionally standardize a design that is error inducing. However, attempt to standardize too late and there may already be so many variations that no standard can be agreed upon. It is clear that standardization must be done carefully and wisely.

9.3.2 Error Management

Human error, especially flight crew error, is a recurring theme and continues to be cited as a primary factor in a majority of aviation accidents.[2,20] It is becoming increasingly recognized that this issue must be taken on in a systematic way, or it may prove difficult to make advances in operations and safety improvements. However, it is also important to recognize that human error is also a normal by-product of human behavior, and most errors in aviation do not have safety consequences. Therefore, it is important for the aviation community to recognize that error cannot be completely prevented and that the focus should be on error management.

In many accidents where human error is cited, the human operator is blamed for making the error; in some countries the human operator is assigned criminal responsibility, and even some U.S. prosecutors seem willing to take similar views. While the issue of personal responsibility for the consequences of one's actions is important and relevant, it also is important to understand why the individual or crew made the error(s). In aviation, with very rare exceptions, flight crews (and other humans in the system) do not intend to make errors, especially errors with safety consequences. To improve safety through understanding of human error, it may be more useful to address errors as *symptoms* rather than *causes* of accidents. The next section discusses understanding of error and its management, then suggests some actions that might be constructive.

Human error can be distinguished into two basic categories: (a) those which presume the intention is correct, but the action is incorrect, (including *slips* and *lapses*), and (b) those in which the intention is wrong (including *mistakes* and *violations*).[21–23]

Slips are where one or more incorrect actions are performed, such as in a substitution or insertion of an inappropriate action into a sequence that was otherwise good. An example would be setting the wrong altitude into the mode selector panel, even when the pilot knew the correct altitude and intended to enter it.

Lapses are the omission of one or more steps of a sequence. For example, missing one or more items in a checklist that has been interrupted by a radio call.

Mistakes are errors where the human did what he or she intended, but the planned action was incorrect. Usually mistakes are the result of an incorrect diagnosis of a problem or a failure to understand the exact nature of the current situation. The plan of action thus derived may contain very inappropriate behaviors and may also totally fail to rectify a problem. For example, a mistake would be shutting down the wrong engine as a result of an incorrect diagnosis of a set of symptoms.

Violations are the failure to follow established procedures or performance of actions that are generally forbidden. Violations are generally deliberate (and often well-meaning), though an argument can be made that some violation cases can be inadvertent. An example of a violation is continuing on with a landing even when weather minima have not been met before final approach. It should be mentioned that a "violation" error may not necessarily be in violation of a regulation or other legal requirement.

Understanding differences in the types of errors is valuable because management of different types may require different strategies. For example, training is often proposed as a strategy for preventing errors. However, errors are a normal by-product of human behavior. While training can help reduce some types of errors, they cannot be completely trained out. For that reason, errors should also be addressed by other means, and considering other factors, such as the consequences of the error or whether the effect of the error can be reversed. As an example of using design to address known potential errors, certain switches in the flight deck have guards on them to prevent inadvertent activation.

Error management can be viewed as involving the tasks of error avoidance, error detection, and error recovery.[23] Error avoidance is important, because it is certainly desirable to prevent as many errors as possible. Error detection and recovery are important, and in fact it is the safety consequences of errors that are most critical.

It seems clear that experienced pilots have developed skills for performing error management tasks. Therefore, it is possible that design, training, and procedures can directly support these tasks, if we get a better understanding of those skills and tasks. However, the understanding of those skills and tasks is far from complete.

There are a number of actions that should be taken with respect to dealing with error, some of them in the design process:

Stop the blame that inhibits in-depth addressing of human error, while appropriately acknowledging the need for individual and organizational responsibility for safety consequences. The issue of blaming the pilot for errors has many consequences, and provides a disincentive to report errors.

Evaluate errors in accident and incident analyses. In many accident analyses, the reason an error is made is not addressed. This typically happens because the data are not available. However, to the extent possible with the data available, the types of errors and reasons for them should be addressed as part of the accident investigation.

Develop a better understanding of error management tasks and skills that can support better performance of those tasks. This includes:

- Preventing as many errors as possible through design, training, procedures, proficiency, and any other intervention mechanism;
- Recognizing that it is impossible to prevent all errors, although it is certainly important to prevent as many as possible; and
- Addressing the need for error management, with a goal of error tolerance in design, training, and procedures.

System design and associated flight crew interfaces can and should support the tasks of error avoidance, detection, and recovery. There are a number of ways of accomplishing this, some of which are mentioned here. One of these ways is through user-centered design processes that ensure that the design supports the human performing the desired task. An example commonly cited is the navigation display in modern flight decks, which integrates information into a display that provides information in a manner directly usable by the flight crew. This is also an example of a system that helps make certain errors more detectable, such as entering an incorrect waypoint. Another way of contributing to error resistance is designing systems that cannot be used or operated in an unintended way. An example of this is designing connectors between a cable and a computer such that the only place the cable connector fits is the correct place for it on the computer; it will not fit into any other connector on the computer.

9.3.3 Integration with Training/Qualification and Procedures

To conclude, it is important to point out that flight deck design should not occur in isolation. It is common to discuss the flight deck design separately from the flight crew qualification (training and recency of experience), considerations, and procedures. And yet, flight deck designs make many assumptions about the knowledge and skills of the pilots who are the intended operators of the vehicles. These assumptions should be explicitly identified as part of the design process, as should the assumptions about the procedures that will be used to operate the designed systems. Design should be conducted as part of an integrated, overall systems approach to ensuring safe, efficient, and effective operations.

References

1. Billings, Charles E., *Aviation Automation: The Search for a Human-Centered Approach,* Lawrence Erlbaum Associates, 1997.
2. Federal Aviation Administration, The Human Factors Team Report on: The Interfaces Between Flightcrews and Modern Flight Deck Systems, July 1996.
3. Sanders, M. S. and McCormick, E. J., *Human Factors in Engineering and Design,* 7th ed., New York: McGraw-Hill, 1993.
4. Norman, Donald A., *The Psychology of Everyday Things,* also published as *The Design of Everyday Things,* Doubleday, 1988.
5. Wickens, C. D., *Engineering Psychology and Human Performance,* 2nd ed., New York: Harper Collins College, 1991.
6. Hawkins, F., *Human Factors in Flight,* 2nd ed., Avebury Aviation, 1987.
7. Bailey, R. W., *Human Performance Engineering: A Guide for System Designers,* Englewood Cliffs, NJ: Prentice-Hall,1982.
8. Chapanis, A., *Human Factors in Systems Engineering,* New York: John Wiley & Sons, 1996.
9. Cardosi, K. and Murphy, E. Eds., Human Factors in the Design and Evaluation of Air Traffic Control Systems, DOT/FAA/RD-95/3, 1995.
10. Nielsen, Jakob, *Usability Engineering,* New York: Academic Press, 1993.
11. Palmer, M. T., Roger, W. H., Press, H. N., Latorella, K. A., and Abbott, T. S., NASA Tech. Memo., 109171, January 1995.
12. Reason, J., *Managing the Risks of Organizational Accidents,* Ashgate Publishing, 1997.
13. Society of Automotive Engineers, *Pilot-System Integration,* Aerospace Recommended Practice (ARP) 4033, 1995.
14. Society of Automotive Engineers, *Integration Procedures for the Introduction of New Systems to the Cockpit,* ARP 4927, 1995
15. Sexton, G., Cockpit: Crew Systems Design and Integration, in Wiener, E. and Nagel, D., Eds., *Human Factors in Aviation,* San Diego, CA: Academic Press, 1988.
16. Arbuckle, P. D., Abbott, K. H., Abbott, T. S., and Schutte, P. C., Future Flight Decks, 21st Congr. Int. Council Aeronautical Sci., Paper Number 98-6.9.3, September, 1998.
17. Federal Aviation Administration, Aircraft Alerting Systems Standardization Study, Volume II: Aircraft Alerting Systems Design Guidelines, FAA Rep. No. DOT/FAA/RD/81–38, II, 1981.
18. Electric Power Research Institute, Rep. NP-3701: Computer-Generated Display System Guidelines, Vol. 2: Developing an Evaluation Plan, September 1984.
19. Abbott, K., Human Error and Aviation Safety Management, *Proc. Flight Saf. Found. 52nd Int. Air Saf. Semin.,* November 8–11, 1999.
20. Boeing Commercial Airplane Group, *Statistical Summary of Commercial Jet Aircraft Accidents, World Wide Operations 1959–1995,* April 1996.
21. Reason, J. T., *Human Error,* New York: Cambridge University Press, 1990.
22. Hudson, P.T.W., van der Graaf, G.C., and Verschuur, W.L.G., Perceptions of Procedures by Operators and Supervisors, Paper SPE 46760, *HSE Conf. Soc. Pet. Eng.,* Caracas, 1998.
23. Hudson, P.T.W., Bending the Rules. II. Why do people break rules or fail to follow procedures? and, What can you do about it?
24. Wiener, Earl L., Intervention Strategies for the Management of Human Error, *Flight Safety Digest,* February 1995.

10

Batteries

David G. Vutetakis
Douglas Battery Co.

10.1 Introduction

The battery is an essential component of almost all aircraft electrical systems. Batteries are used to start engines and auxiliary power units, to provide emergency backup power for essential avionics equipment, to assure no-break power for navigation units and fly-by-wire computers, and to provide ground power capability for maintenance and preflight checkouts. Many of these functions are mission critical, so the performance and reliability of an aircraft battery is of considerable importance. Other important requirements include environmental ruggedness, a wide operating temperature range, ease of maintenance, rapid recharge capability, and tolerance to abuse.

Historically, only a few types of batteries have been found to be suitable for aircraft applications. Until the 1950s, vented lead-acid (VLA) batteries were used exclusively [Earwicker, 1956]. In the late 1950s, military aircraft began converting to vented nickel-cadmium (VNC) batteries, primarily because of their superior performance at low temperature. The VNC battery subsequently found widespread use in both military and commercial aircraft [Fleischer, 1956; Falk and Salkind, 1969]. The only other type of battery used during this era was the vented silver-zinc battery, which provided an energy density about three times higher than VLA and VNC batteries [Miller and Schiffer, 1971]. This battery type was applied to several types of U.S. Air Force fighters (F-84, F-105, and F-106) and U.S. Navy helicopters (H-2, H-13, and H-43) in the 1950s and 1960s. Although silver-zinc aircraft batteries were attractive for reducing weight and size, their use has been discontinued due to poor reliability and high cost of ownership.

In the late 1960s and early 1970s, an extensive development program was conducted by the U.S. Air Force and Gulton Industries to qualify sealed nickel-cadmium (SNC) aircraft batteries for military and commercial applications [McWhorter and Bishop, 1972]. This battery technology was successfully demonstrated on a Boeing KC-135, a Boeing 727, and a UH-1F helicopter. Before the technology could be transitioned into production, however, Gulton Industries was taken over by SAFT and a decision was made to terminate the program.

In the late 1970s and early 1980s, the U.S. Navy pioneered the development of sealed lead-acid (SLA) batteries for aircraft applications [Senderak and Goodman, 1981]. SLA batteries were initially applied to the AV-8B and F/A-18, resulting in a significant reliability and maintainability (R&M) improvement compared with VLA and VNC batteries. The Navy subsequently converted the C-130, H-46, and P-3 to SLA batteries. The U.S. Air Force followed the Navy's lead, converting numerous aircraft to SLA batteries, including the A-7, B-1B, C-130, C-141, KC-135, F-4, and F-117 [Vutetakis, 1994]. The term "High Reliability, Maintenance-Free Battery," or HRMFB, was coined to emphasize the improved R&M capability of sealed-cell aircraft batteries. The use of HRMFBs soon spun off into the commercial sector, and numerous commercial and general aviation aircraft today have been retrofitted with SLA batteries.

In the mid-1980s, spurred by increasing demands for HRMFB technology, a renewed interest in SNC batteries took place. A program to develop advanced SNC batteries was initiated by the U.S. Air Force, and Eagle-Picher Industries was contracted for this effort [Flake, 1988; Johnson et al., 1994]. The B-52 bomber was the first aircraft to retrofit this technology. SNC batteries also have been developed by ACME for several aircraft applications, including the F-16 fighter, Apache AH-64 helicopter, MD-90, and Boeing 777 [Anderman, 1994].

A recent development in aircraft batteries is the "low maintenance" or "ultra-low maintenance" nickel-cadmium battery [Scardaville and Newman, 1993]. This battery is intended to be a direct replacement of conventional VNC batteries, avoiding the need to replace or modify the charging system. Although the battery still requires scheduled maintenance for electrolyte filling, the maintenance frequency can be decreased significantly. This type of battery has been under development by SAFT and more recently by Marathon. Limited flight tests have been performed by the U.S. Navy on the H-1 helicopter. Application of this technology to commercial aircraft is also being pursued.

Determining the most suitable battery type and size for a given aircraft type requires detailed knowledge of the application requirements (load profile, duty cycle, environmental factors, and physical constraints) and the characteristics of available batteries (performance capabilities, charging requirements, life expectancy, and cost of ownership). With the various battery types available today, considerable expertise is required to size, select, and prepare technical specifications. The information contained in this chapter will provide general guidance for original equipment design and for upgrading existing aircraft batteries. More detailed information can be found in the sources listed at the end of the chapter.

10.2 General Principles

10.2.1 Battery Fundamentals

Batteries operate by converting chemical energy into electrical energy through electrochemical discharge reactions. Batteries are composed of one or more cells, each containing a **positive electrode, negative electrode, separator,** and **electrolyte**. Cells can be divided into two major classes: primary and secondary. Primary cells are not rechargeable and must be replaced once the reactants are depleted. Secondary cells are rechargeable and require a DC charging source to restore reactants to their fully charged state. Examples of primary cells include carbon-zinc (Leclanche or dry cell), alkaline-manganese, mercury-zinc, silver-zinc, and lithium cells (e.g., lithium-manganese dioxide, lithium-sulfur dioxide, and lithium-thionyl chloride). Examples of secondary cells include lead-lead dioxide (lead-acid), nickel-cadmium, nickel-iron, nickel-hydrogen, nickel-metal hydride, silver-zinc, silver-cadmium, and lithium-ion. For aircraft applications, secondary cells are the most prominent, but primary cells are sometimes used for powering critical avionics equipment (e.g., flight data recorders).

Batteries are rated in terms of their **nominal voltage** and **ampere-hour capacity**. The voltage rating is based on the number of cells connected in series and the nominal voltage of each cell (2.0 V for lead-acid and 1.2 V for nickel-cadmium). The most common voltage rating for aircraft batteries is 24 V. A 24-V lead-acid battery contains 12 cells, while a 24-V nickel-cadmium battery contains either 19 or 20 cells (the U.S. military rates 19-cell batteries at 24 V). Voltage ratings of 22.8, 25.2, and 26.4 V are also common with nickel-cadmium batteries, consisting of 19, 20, or 22 cells, respectively. Twelve-volt lead-acid batteries, consisting of six cells in series, are also used in many general aviation aircraft.

The ampere-hour (Ah) capacity available from a fully charged battery depends on its temperature, rate of discharge, and age. Normally, aircraft batteries are rated at room temperature (25°C), the **C-rate** (1-hour rate), and beginning of life. Military batteries, however, often are rated in terms of the end-of-life capacity, i.e., the minimum capacity before the battery is considered unserviceable. Capacity ratings of aircraft batteries vary widely, generally ranging from 3 to 65 Ah.

The maximum power available from a battery depends on its internal construction. High rate cells, for example, are designed specifically to have very low internal impedance as required for starting turbine engines and auxiliary power units (APUs). Unfortunately, no universally accepted standard exists for defining the peak power capability of an aircraft battery. For lead-acid batteries, the peak power typically is defined in terms of the cold-cranking amperes, or **CCA** rating. For nickel-cadmium batteries, the peak power rating typically is defined in terms of the current at maximum power, or **Imp** rating. These ratings are based on different temperatures ($-18°C$ for CCA, 23°C for Imp), making it difficult to compare different battery types. Furthermore, neither rating adequately characterizes the battery's initial peak current capability, which is especially important for engine start applications. More rigorous peak power specifications have been included in some military standards. For example, MIL-B-8565/15 specifies the initial peak current, the current after 15 s, and the capacity after 60 s, during a 14-V constant voltage discharge at two different temperatures (24 and $-26°C$).

The **state-of-charge** of a battery is the percentage of its capacity available relative to the capacity when it is fully charged. By this definition, a fully charged battery has a state-of-charge of 100% and a battery with 20% of its capacity removed has a state-of-charge of 80%. The **state-of-health** of a battery is the percentage of its capacity available when fully charged relative to its rated capacity. For example, a battery rated at 30 Ah, but only capable of delivering 24 Ah when fully charged, will have a state-of-health of $24/30 \times 100 = 80\%$. Thus, the state-of-health takes into account the loss of capacity as the battery ages.

10.3 Lead-Acid Batteries

10.3.1 Theory of Operation

The chemical reactions that occur in a lead-acid battery are represented by the following equations:

$$\text{Positive electrode: } PbO_2 + H_2SO_4 + 2H^+ + 2e^- \underset{\text{charge}}{\overset{\text{discharge}}{\rightleftarrows}} PbsO_4 + 2H_2O \qquad (1)$$

$$\text{Negative electrode: } Pb + H_2SO_4 \underset{\text{charge}}{\overset{\text{discharge}}{\rightleftarrows}} PbSO_4 + 2H^+ + 2e^- \qquad (2)$$

$$\text{Overall cell reaction: } PbO_2 + pb + 2H_2SO_4 \underset{\text{charge}}{\overset{\text{discharge}}{\rightleftarrows}} 2PbSO_4 + 2H_2O \qquad (3)$$

As the cell is charged, the sulfuric acid (H_2SO_4) concentration increases and becomes highest when the cell is fully charged. Likewise, when the cell is discharged, the acid concentration decreases and becomes most dilute when the cell is fully discharged. The acid concentration generally is expressed in terms of specific gravity, which is weight of the electrolyte compared to the weight of an equal volume of pure water.

The cell's specific gravity can be estimated from its open circuit voltage using the following equation:

$$\text{Specific Gravity (SG)} = \text{Open Circuit Voltage (OCV)} - 0.84 \qquad (4)$$

There are two basic cell types: vented and recombinant. Vented cells have a flooded electrolyte, and the hydrogen and oxygen gases generated during charging are vented from the cell container. Recombinant cells have a starved or gelled electrolyte, and the oxygen generated from the positive electrode during charging diffuses to the negative electrode where it recombines to form water by the following reaction:

$$\text{Pb} + \text{H}_2\text{SO}_4 + 1/2\text{O}_2 \rightarrow \text{PbSO}_4 + \text{H}_2\text{O} \qquad (5)$$

The recombination reaction suppresses hydrogen evolution at the negative electrode, thereby allowing the cell to be sealed. In practice, the recombination efficiency is not 100% and a resealable valve regulates the internal pressure at a relatively low value, generally below 10 psig. For this reason, sealed lead-acid cells are often called "valve-regulated lead-acid" (VRLA) cells.

10.3.2   Cell Construction

Lead-acid cells are composed of alternating positive and negative plates, interleaved with single or multiple layers of separator material. Plates are made by pasting active material onto a grid structure made of lead or lead alloy. The electrolyte is a mixture of sulfuric acid and water. In flooded cells, the separator material is porous rubber, cellulose fiber, or microporous plastic. In recombinant cells with starved electrolyte technology, a glass fiber mat separator is used, sometimes with an added layer of microporous polypropylene. Gell cells, the other type of recombinant cell, are made by absorbing the electrolyte with silica gel that is layered between the electrodes and separators.

10.3.3   Battery Construction

Lead-acid aircraft batteries are constructed using injection-molded, plastic **monoblocs** that contain a group of cells connected in series. Monoblocs typically are made of polypropylene, but ABS is used by at least one manufacturer. Normally, the monobloc serves as the battery case, similar to a conventional automotive battery. For more robust designs, monoblocs are assembled into a separate outer container made of steel, aluminum, or fiberglass-reinforced epoxy. Cases usually incorporate an electrical receptacle for connecting to the external circuit with a quick connect/disconnect plug. Two generic styles of receptacles are common: the "Elcon style" and the "Cannon style." The Elcon style is equivalent to military type MS3509. The Cannon style has no military equivalent, but is produced by Cannon and other connector manufacturers. Batteries sometimes incorporate thermostatically controlled heaters to improve low temperature performance. The heater is powered by the aircraft's AC or DC bus. Figure 10.1 shows an assembly drawing of a typical lead-acid aircraft battery; this particular example does not incorporate a heater.

10.3.4   Discharge Performance

Battery performance characteristics usually are described by plotting voltage, current, or power vs. discharge time, starting from a fully charged condition. Typical discharge performance data for SLA aircraft batteries are illustrated in Figures 10.2 and 10.3. Figure 10.4 shows the effect of temperature on the capacity when discharged at the C-rate. Manufacturers' data should be obtained for current information on specific batteries of interest.

10.3.5   Charge Methods

Constant voltage charging at 2.3 to 2.4V per cell is the preferred method of charging lead-acid aircraft batteries. For a 12-cell battery, this equates to 27.6 to 28.8 V which generally is compatible with the voltage available from the aircraft's 28-V DC bus. Thus, lead-acid aircraft batteries normally can be charged by direct connection to the DC bus, avoiding the need for a dedicated battery charger. If the voltage regulation on the DC bus is not controlled sufficiently, however, the battery will be overcharged or undercharged causing premature failure. In this case, a regulated voltage source may be necessary to achieve acceptable battery life. Some aircraft use voltage regulators that compensate, either manually or automatically, for the battery temperature by increasing the voltage when cold and decreasing the voltage when hot.

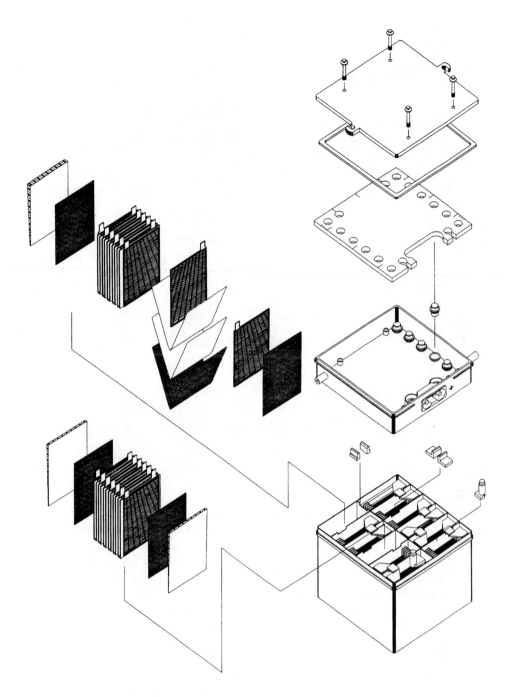

FIGURE 10.1 Assembly drawing of a lead-acid aircraft battery.

Adjusting the charging voltage in this manner has the beneficial effect of prolonging the battery's service life at high temperature and achieving faster recharge at low temperatures.

10.3.6 Temperature Effects and Limitations

Lead-acid batteries generally are rated at 25°C (77°F) and operate best around this temperature. Exposure to low ambient temperatures results in performance decline, whereas exposure to high ambient temperatures results in shortened life.

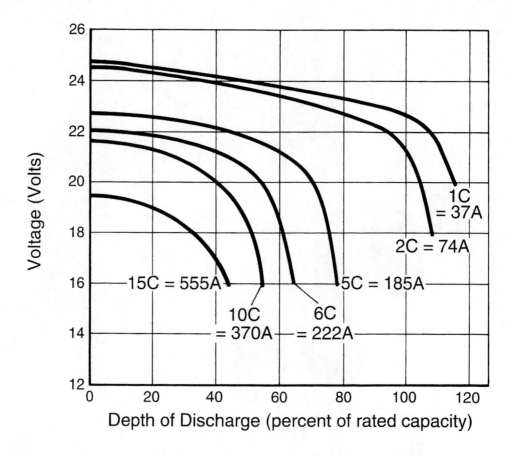

FIGURE 10.2   Discharge curves at 25°C for a 24 V/37 Ah SLA aircraft battary.

The lower temperature limit is dictated by the freezing point of the electrolyte. The electrolyte freezing point varies with acid concentration, as shown in Table 10.1. The minimum freezing point is a chilly 70°C (−95°F) at a specific gravity (SG) of 1.30. Since fully charged batteries have SGs in the range of 1.28 to 1.33, they are not generally susceptible to freezing even under extreme cold conditions. However, when the battery is discharged, the SG drops and the freezing point rises. At low SG, the electrolyte first will turn to slush as the temperature drops. This is because the water content freezes first, gradually raising the SG of the remaining liquid so that it remains unfrozen. Solid freezing of the electrolyte in a discharged battery requires temperatures well below the slush point; a practical lower limit of −30°C is often specified. Solid freezing can damage the battery permanently (i.e., by cracking cell containers), so precautions should be taken to keep the battery charged or heated when exposed to temperatures below −30°C.

The upper temperature limit is generally in the range of 60 to 70°C. Capacity loss is accelerated greatly when charged above this temperature range due to vigorous gassing and/or rapid grid corrosion. The capacity loss generally is irreversible when the battery is cooled.

10.3.7   Service Life

The service life of a lead-acid aircraft battery depends on the type of use it experiences (e.g., rate, frequency, and depth of discharge), environmental conditions (e.g., temperature and vibration), charging method, and the care with which it is maintained. Service lives can range from 1 to 5 years, depending on the application. Table 10.2 shows representative life cycle data as a function of the depth of discharge. Manufacturers' data should be consulted for specific batteries of interest.

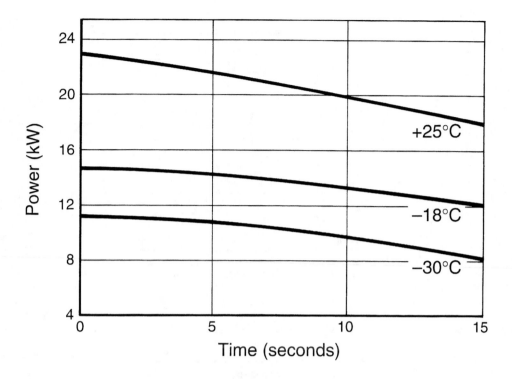

FIGURE 10.3 Maximum power curves (12 V Discharge) for a 24 V/37 Ah SLA battery.

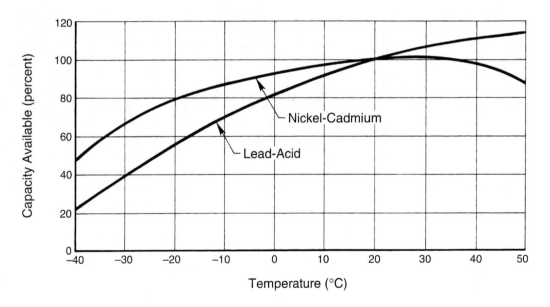

FIGURE 10.4 Capacity vs. temperature for aircraft batteries at the C-rate.

10.3.8 Storage Characteristics

Lead-acid batteries always should be stored in the charged state. If allowed to remain in the discharged state for a prolonged time period, the battery becomes damaged by "sulfation." Sulfation occurs when lead sulfate forms into large, hard crystals, blocking the pores in the active material. The sulfation creates

TABLE 10.1   Freezing Points of Sulfuric Acid-Water Mixtures

Specific Gravity at 15° C	Cell OCV (Volts)	Battery OCV (Volts)	Freezing Point	
			(°C)	(°F)
1.000	1.84	22.08	0	+32
1.050	1.89	22.68	−3	+26
1.100	1.94	23.28	−8	+18
1.150	1.99	23.88	−15	+5
1.200	2.04	24.48	−27	−17
1.250	2.09	25.08	−52	−62
1.300	2.14	25.68	−70	−95
1.350	2.19	26.28	−49	−56
1.400	2.24	26.88	−36	−33

TABLE 10.2   Cycle Life Data for SLA Aircraft Batteries

Depth of Discharge (% of Rated Capacity)	Number of Cycles to End of Life
10	2000
30	670
50	400
80	250
100	200

Source: Hawker Energy Products.

a high impedance condition that makes it difficult for the battery to accept recharge. The sulfation may or may not be reversible, depending on the discharge conditions and specific cell design. The ability to recovery from deep discharge has been improved in recent years by electrolyte additives, such as sodium sulfate.

VLA batteries normally are supplied in a dry, charged state (i.e., without electrolyte), which allows them to be stored almost indefinitely (i.e., 5 years or more). Once activated with electrolyte, periodic charging is required to overcome the effect of self-discharge and to prevent sulfation. The necessary charging frequency depends on the storage temperature. At room temperature (25°C), charging every 30 days is typically recommended. More frequent charging is necessary at higher temperatures (e.g., every 15 days at 35°C), and less frequent charging is necessary at low temperatures (e.g., every 120 days at 10°C).

SLA batteries can be supplied only in the activated state (i.e., with electrolyte), so storage provisions are more demanding compared with dry charged batteries. As in the case of activated VLA batteries, periodic charging is necessary to overcome the effects of self-discharge and to prevent sulfation. The rate of self-discharge of SLA batteries varies widely from manufacturer to manufacturer, so the necessary charging frequency also varies widely. For example, recommended charging frequencies can range from 3 to 24 months.

10.3.9   Maintenance Requirements

Routine maintenance of lead-acid aircraft batteries is required to assure airworthiness and to maximize service life. For vented-cell batteries, electrolyte topping must be performed on a regular basis to replenish the water loss that occurs during charging. Maintenance intervals are typically 2 to 4 months. A capacity test or load test usually is included as part of the servicing procedure. For sealed-cell batteries, water replenishment obviously is unnecessary, but periodic capacity measurements generally are recommended. Capacity check intervals can be based either on calendar time (e.g., every 3 to 6 months after the first year) or operating hours (e.g., every 100 hours after the first 600 hours). Refer to the manufacturer's maintenance instructions for specific batteries of interest.

10.3.10 Failure Modes and Fault Detection

The predominant failure modes of lead-acid cells are summarized as follows:

- Shorts caused by growth on the positive grid, shedding or mossing of active material, or mechanical defects protruding from the grid, manifested by inability of the battery to hold a charge (rapid decline in open circuit voltage).
- Loss of electrode capacity due to active material shedding, excessive grid corrosion, sulfation, or passivation, manifested by low capacity and/or inability to hold voltage under load.
- Water loss and resulting cell dry-out due to leaking seal, repeated cell reversals, or excessive overcharge (this mode applies to sealed cells or to vented cells that are improperly maintained), manifested by low capacity and/or inability to hold voltage under load.

Detection of these failure modes is straightforward if the battery can be removed from the aircraft. The battery capacity and load capability can be measured directly and the ability to hold a charge can be inferred by checking the open circuit voltage over time. However, detection of these failure modes while the battery is in service is more difficult. The more critical the battery is to the safety of the aircraft, the more important it becomes to detect battery faults accurately. A number of on-board detection schemes have been developed for critical applications, mainly for military aircraft [Vutetakis and Viswanathan, 1995].

10.3.11 Disposal

Lead, the major constituent of the lead-acid battery, is a toxic (poisonous) chemical. As long as the lead remains inside the battery container, no health hazard exists. Improper disposal of spent batteries can result in exposure to lead, however. Environmental regulations in the U.S. and abroad prohibit the disposal of lead-acid batteries in landfills or incinerators. Fortunately, an infrastructure exists for recycling the lead from lead-acid batteries. The same processes used to recycle automotive batteries are used to recycle aircraft batteries. Federal, state, and local regulations should be followed for proper disposal procedures.

10.4 Nickel-Cadmium Batteries

10.4.1 Theory of Operation

The chemical reactions that occur in a nickel-cadmium battery are represented by the following equations:

$$\text{Positive electrode: } 2NiOOH + 2H_2O + 2e^- \underset{\text{charge}}{\overset{\text{discharge}}{\rightleftarrows}} 2Ni(OH)_2 + 2(OH)^- \qquad (6)$$

$$\text{Negative electrode: } Cd + 2(OH)^- \underset{\text{charge}}{\overset{\text{discharge}}{\rightleftarrows}} Cd(OH)_2 + 2e^- \qquad (7)$$

$$\text{Overall cell reaction: } 2NiOOH + Cd + 2H_2O \underset{\text{charge}}{\overset{\text{discharge}}{\rightleftarrows}} 2Ni(OH)_2 + Cd(OH)_2 \qquad (8)$$

There are two basic cell types: vented and recombinant. Vented cells have a flooded electrolyte, and the hydrogen and oxygen gases generated during charging are vented from the cell container. Recombinant cells have a starved electrolyte, and the oxygen generated from the positive electrode during charging diffuses to the negative electrode where it recombines to form cadmium hydroxide by the following reaction:

$$Cd + H_2O + 1/2O_2 \longrightarrow Cd(OH)_2 \qquad (9)$$

The recombination reaction suppresses hydrogen evolution at the negative electrode, thereby allowing the cell to be sealed. Unlike valve-regulated lead-acid cells, recombinant nickel-cadmium cells are sealed with a high-pressure vent that releases only during abusive conditions. Thus, these cells remain sealed under normal charging conditions. However, provisions for gas escape must still be provided when designing battery cases since abnormal conditions may be encountered periodically (e.g., in the event of a charger failure that causes an overcurrent condition).

10.4.2   Cell Construction

The construction of nickel-cadmium cells varies significantly, depending on the manufacturer. In general, cells feature alternating positive and negative plates with separator layers interleaved between them, a potassium hydroxide (KOH) electrolyte of approximately 31% concentration by weight (specific gravity 1.30), and a prismatic cell container with the cell terminals extending through the cover. The positive plate is impregnated with nickel hydroxide and the negative plate is impregnated with cadmium hydroxide. The plates differ according to manufacturer with respect to the type of the substrate, type of plaque, impregnation process, formation process, and termination technique. The most common plate structure is made of nickel powder sintered onto a substrate of perforated nickel foil or woven screens. At least one manufacturer (ACME) uses nickel-coated polymeric fibers to form the plate structure. Cell containers typically are made of nylon, polyamide, or steel. One main difference between vented cells and sealed (recombinant) cells is the type of separator. Vented cells use a gas barrier layer to prevent gases from diffusing between adjacent plates. Recombinant cells feature a porous separator system that permits gas diffusion between plates.

10.4.3   Battery Construction

Nickel-cadmium aircraft batteries generally consist of a steel case containing identical, individual cells connected in series. The number of cells depends on the particular application, but generally 19 or 20 cells are used. The end cells of the series are connected to the battery receptacle located on the outside of the case. The receptacle is usually a two-pin, quick disconnect type; both Cannon and Elcon styles commonly are used. Cases are vented by means of vent tubes or louvers to allow escape of gases produced during overcharge. Some battery designs have provisions for forced air cooling, particularly for engine start applications. Thermostatically controlled heating pads sometimes are employed on the inside or outside of the battery case to improve low-temperature performance. Power for energizing the heaters normally is provided by the aircraft's AC or DC bus. Temperature sensors often are included inside the case to allow regulation of the charging voltage. In addition, many batteries are equipped with a thermal switch that protects the battery from overheating if a fault develops or if battery is exposed to excessively high temperatures. A typical aircraft battery assembly is shown in Figure 10.5.

10.4.4   Discharge Performance

Typical discharge performance data for VNC aircraft batteries are illustrated in Figures 10.6 and 10.7. Discharge characteristics of SNC batteries are similar to VNC batteries. Figure 10.4 shows the effect of temperature on discharge capacity at the C-rate. Compared with lead-acid batteries, nickel-cadmium batteries tend to have more available capacity at low temperature, but less available capacity at high temperature. Manufacturers' data should be consulted for current information on specific batteries of interest.

10.4.5   Charge Methods

A variety of methods are employed to charge nickel-cadmium aircraft batteries. The key requirement is to strike an optimum balance between overcharging and undercharging, while achieving full charge in the required time frame. Overcharging results in excessive water loss (vented cells) or heating (sealed cells). Undercharging results in capacity fading. Some overcharge is necessary, however, to overcome coulombic inefficiencies associated with the electrochemical reactions. In practice, recharge percentages on the aircraft generally range between 105 and 120%.

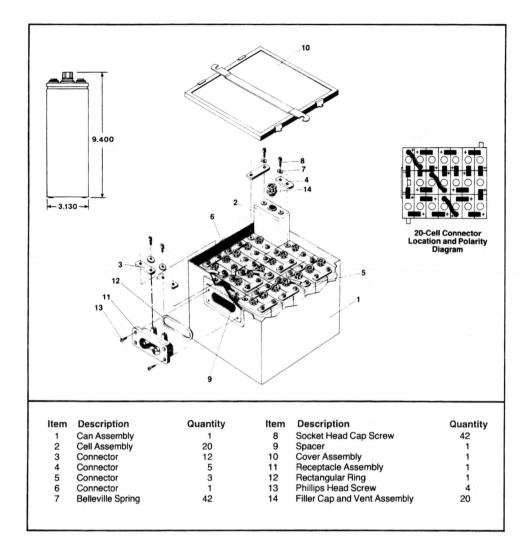

Item	Description	Quantity	Item	Description	Quantity
1	Can Assembly	1	8	Socket Head Cap Screw	42
2	Cell Assembly	20	9	Spacer	1
3	Connector	12	10	Cover Assembly	1
4	Connector	5	11	Receptacle Assembly	1
5	Connector	3	12	Rectangular Ring	1
6	Connector	1	13	Phillips Head Screw	4
7	Belleville Spring	42	14	Filler Cap and Vent Assembly	20

FIGURE 10.5 Assembly drawing of a nickel-cadmium aircraft battery.

For vented-cell batteries, common methods of charging include constant potential, constant current, or pulse current. Constant potential charging is the oldest method and normally is accomplished by floating a 19-cell battery on a 28-V DC bus. The constant current method requires a dedicated charger and typically uses a 0.5 to 1.5 C-rate charging current. Charge termination is accomplished using a temperature-compensated voltage cutoff (VCO). The VCO temperature coefficient is typically $(-)$ 4mV/°C. In some cases, two constant current steps are used, the first step at a higher rate (e.g., C-rate), and the second step at a lower rate (e.g., 1/3 to 1/5 of the C-rate). This method is more complicated, but results in less gassing and electrolyte spewage during overcharge. Pulse current methods are similar to the constant current methods, except the charging current is pulsed rather that constant.

For sealed-cell batteries, only constant current or pulse current methods should be used. Constant potential charging can cause excessive heating, resulting in thermal runaway. Special attention must be given to the charge termination technique in sealed-cell batteries, because the voltage profile is relatively flat as the battery becomess fully charged. For example, it may be necessary to rely on the battery's temperature rise rather than voltage rise as the signal for charge termination.

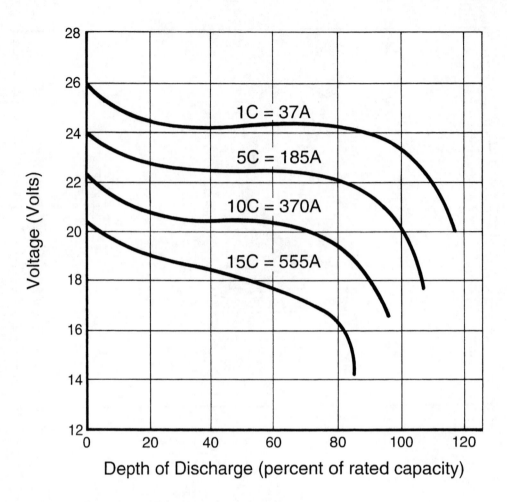

FIGURE 10.6   Discharge curves at 25°C for a 24 V/37 Ah VNC aircraft battery.

10.4.6   Temperature Effects and Limitations

Nickel-cadmium batteries, like lead-acid batteries, normally are rated at room temperature (25°C) and operate best around this temperature. Exposure to low ambient temperatures results in performance decline, and exposure to high ambient temperatures results in shortened life.

The lower temperature limit is dictated by the freezing point of the electrolyte. Most cells are filled with an electrolyte concentration of 31% KOH, which freezes at −66°C. Lower concentrations will freeze at higher temperatures, as shown in Table 10.3. The KOH concentration may become diluted over time as a result of spillage or carbonization (reacting with atmospheric carbon dioxide), so the freezing point of a battery in service may not be as low as expected. As in the case of dilute acid electrolytes, slush ice will form well before the electrolyte freezes solid. For practical purposes, a lower operating temperature limit of −40°C often is quoted.

The upper temperature limit is generally in the range of 50 to 60°C; significant capacity loss occurs when batteries are operated (i.e., repeated charge/discharge cycles) above this temperature range. The battery capacity often is recoverable, however, when the battery is cooled to room temperature and subjected to several deep discharge cycles.

10.4.7   Service Life

The service life of a nickel-cadmium aircraft battery depends on many factors, including the type of use it experiences (e.g., rate, frequency, and depth of discharge), environmental conditions (e.g., temperature

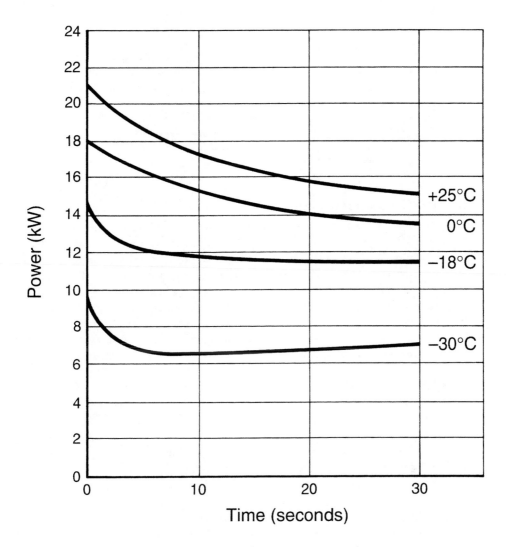

FIGURE 10.7   Maximum power curves (12 V discharge) for a 24 V/37 Ah VNC aircraft battery.

TABLE 10.3   Freezing Points of KOH-Water Mixtures

Concentration (Weight %)	Specific Gravity at 15°C	Freezing Point	
		(°C)	(°F)
0	1.000	0	+32
5	1.045	−3	+27
10	1.092	−8	+18
15	1.140	−15	+5
20	1.118	−24	−11
25	1.239	−38	−36
30	1.290	−59	−74
31	1.300	−66	−87
35	1.344	−50	−58

and vibration), charging method, and the care with which it is maintained and reconditioned. Thus, it is difficult to generalize the service life that can be expected. All things being equal, the service life of a nickel-cadmium battery is inherently longer than that of a lead-acid battery. Representative cycle life data for an SNC battery are listed in Table 10.4.

TABLE 10.4   Cycle Life Data for SNC Aircraft Batteries

Depth of Discharge (% of Rated Capacity)	Number of Cycles to End of Life
30	7500
50	4500
60	3000
80	1500
100	1000

Source: ACME Electric Corporation.

10.4.8 Storage Characteristics

Nickel-cadmium batteries can be stored in any state of charge and over a broad temperature range (i.e., -65 to 60°C). For maximum shelf life, however, it is best to store batteries between 0 and 30°C. Vented-cell batteries normally are stored with the terminals shorted together. Shorting of sealed-cell batteries during storage is not recommended, however, since it may cause cell venting and/or cell reversal.

When left on open circuit during periods of non-operation, nickel-cadmium batteries will self-discharge at a relatively fast rate. As a rule of thumb, the self-discharge rate of sealed cells is approximately 1%/day at 20°C (when averaged over 30 days), and the rate increases by 1%/day for every 10°C rise in temperature (e.g., 2%/day at 30°C, 3%/day at 40°C, etc.). The self-discharge rate is somewhat less for vented cells. The capacity lost by self-discharge usually is recoverable when charged in the normal fashion.

10.4.9 Maintenance Requirements

Routine maintenance of nickel-cadmium aircraft batteries is required to assure airworthiness and to maximize service life. Maintenance intervals for vented-cell batteries in military aircraft are typically 60 to 120 days. Maintenance intervals for commercial aircraft can be as low as 100 and as high as 1000 flight hours, depending on the operating conditions. Maintenance procedures include capacity checks, cell equalization (deep discharge followed by shorting cell terminals for at least 8 h), isolating and replacing faulty cells (only if permitted; this practice generally is not recommended), cleaning to remove corrosion and carbonate build-up, and electrolyte adjustment.

For sealed-cell batteries, maintenance requirements are much less demanding. Electrolyte adjustment is unnecessary, and the extent of corrosion is greatly reduced. However, some means of assuring airworthiness is still necessary, such as periodic capacity measurement. Manufacturers' recommendations should be followed for specific batteries of interest.

10.4.10 Failure Modes and Fault Detection

The predominant failure modes of nickel-cadmium cells are summarized as follows:

- Shorts caused by cadmium migration through the separator, swelling of the positive electrode, degradation of the separator, or mechanical defects protruding from the electrode. Manifested by inability of the battery to hold a charge (soft shorts) or dead cells (hard shorts).
- Water loss and resulting cell dry-out due to leaking seal, repeated cell reversal, or excessive overcharge (this mode applies to sealed cells or to vented cells that are improperly maintained). Manifested by low capacity and/or inability to hold voltage under load.
- Loss of negative (cadmium) electrode capacity due to passivation or active material degradation. Manifested by low capacity and/or inability to hold voltage under load. Usually reversible by deep discharge followed by shorting cell terminals, or by "reflex" charging (pulse charging with momentary discharge between pulses).
- Loss of positive (nickel) electrode capacity due to swelling or active material degradation. Manifested by low capacity that is nonrestorable.

As discussed under lead-acid batteries, detection of these failure modes is relatively straightforward if the battery can be removed from the aircraft. For example, the battery capacity and load capability can be directly measured and compared against pass/fail criteria. The occurrence of soft shorts (i.e., a high impedance short between adjacent plates) is more difficult to detect, but often can be identified by monitoring the end-of-charge voltage of individual cells.

Detection of these failure modes while the battery is in service is more difficult. As in the case of lead-acid batteries, a number of on-board detection schemes have been developed for critical applications [Vutetakis and Viswanathan, 1995]. The more critical the battery is to the safety of the aircraft, the more important it becomes to detect battery faults accurately.

10.4.11 Disposal

Proper disposal of nickel-cadmium batteries is essential because cadmium is a toxic (carcinogenic) chemical. In the U.S. and abroad, spent nickel-cadmium batteries are considered to be hazardous waste, and their disposal is strictly regulated. Several metallurgical processes have been developed for reclaiming and recycling the nickel and cadmium from nickel-cadmium batteries. These processes can be used for both vented and sealed cells. Federal, state, and local regulations should be followed for proper disposal procedures.

10.5 Applications

Designing a battery for a new aircraft application or for retrofit requires a careful systems engineering approach. To function well, the battery must be interfaced carefully with the aircraft's electrical system. The battery's reliability and maintainability depends heavily on the type of charging system to which it is connected; there is a fine line between undercharging and overcharging the battery. Many airframe manufacturers have realized that it is better to prepare specifications for a "battery system" rather than having separate specifications for the battery and the charger. This approach assures that the charging profile is tuned correctly to the specific characteristics of the battery and to the aircraft's operational requirements.

10.5.1 Commercial Aircraft

A listing of commercial aircraft batteries available from various manufacturers is given in Table 10.5. Sizes range from 12 V/6.5 Ah to 24 V/65 Ah. The table includes VLA, SLA, and VNC type batteries. SNC batteries are not included, but are available on a limited basis from several manufacturers (ACME, SAFT, and Eagle-Picher).

In general, the aircraft battery must be sized to provide sufficient emergency power to support flight essential loads in the event of failure of the primary power system. FAA regulations impose a minimum emergency power requirement of 30 min on all commercial airplanes. Some airlines impose a longer emergency requirement, such as 40 or 60 min due to frequent bad weather on their routes or for other reasons. The emergency requirement for Extended Twin Operation (ETOPS) imposed on two-engine aircraft operating over water is a full 90 min, although 60 min is allowed with operating restrictions. The specified emergency power requirement may be satisfied by batteries or other backup power sources, such as a ram air turbine. If a ram air turbine is used, a battery still is required for transient fill-in. Specific requirements pertaining to aircraft batteries can be found in the Federal Aviation Regulations (FAR), Sections 25.1309, 25.1333, 25.1351, and 25.1353. FAA Advisory Circular No. 25.1333-1 describes specific methods to achieve compliance with applicable FAR sections. For international applications, Civil Aviation Authority (CAA) and Joint Airworthiness Authority (JAA) regulations should be consulted for additional requirements.

When used for APU or engine starting, the battery must be sized to deliver short bursts of high power, as opposed to the lower rates required for emergency loads. APU start requirements on large commercial aircraft can be particularly demanding; for instance, the APU used on the Boeing 757 and 767 airplanes has a peak current requirement of 1200 A [Gross, 1991]. The load on the battery starts out very high to

TABLE 10.5 Commercial Aircraft Batteries

RATING[a]	CONCORDE	CONCORDE SLA	TELEDYNE VLA	TELEDYNE SLA	HAWKER SLA	MARATHON VNC	SAFT VNC
12V/6.5Ah					SBS-15		615
12V/10Ah							10.15
12V/15Ah				G-30s			
12V/18Ah	CB-25	RG-25	G-25 G-25M		SBS-30		
12V/23Ah	CB-35 CB-35M	RG-35	G-35 G-35M	G-35S			
12V/25Ah					SBS-40		
12V/37Ah					SBS-60		
12V/65Ah	CB12-88		G-88				
13.2V/36Ah						CA-138 SP-138	40153 40253
13.2V/40Ah						CA-13 CA-13-1	40152
13.2V/42Ah						CA-130	40353
22.8V/3Ah						CA-13 CA-125 MA-300H	19V03KHB
22.8V/5.5Ah						CA-51 CA-53 CA-54 MA-500H	
22.8V/6.5Ah							605
22.8V/7Ah						CA-7 CA-10N	19V07L
22.8V/12Ah						CA-515A/B	1201
22.8V/13Ah						CA-101 CA-103 CA-106 CA-154	12101

Rating							
22.8V/14Ah							1277
22.8V/15Ah							1277-1
22.8V/20Ah						CA-20H	12277
22.8V/22Ah						CA-21H	23175
22.8V/23Ah							19V023KHP 2353-1
22.8V/24Ah						CA-4 CA-9 MA-11 CA-24A/B CA-27 CA-272-7 KCA-727 CA-737 CA-5 KA-5h MA-5 CA-747 CA-88A/B MA-2-1	
22.8V/40Ah					9750B0818	MA-300	20V03KHB
22.8V/60Ah							
22.8V/65Ah							
24V/3Ah							
24V/5Ah	CB24-9		G-240				
24V/8Ah	CB24-9M		G-241				
24V/10Ah	CB24-11 CB24-11M	RG-24-11M	G-242 G-243 GE-54C GE-54E	G-242S G-243s	9750R0817 9750R0819 9750R0824 9750G082		
24V/14Ah	CB24-40E	RG-400E	G-640C			CA-154-5	
24V/14Ah			G-640E				

(continued)

TABLE 10.5 Commercial Aircraft Batteries (Continued)

RATING[a]	CONCORDE	CONCORDE SLA	TELEDYNE VLA	TELEDYNE SLA	HAWKER SLA	MARATHON VNC	SAFT VNC
24V/15Ah							2.10.15
							1656
							1656-1
							16156
							16256
							16356
24V/16Ah							2.10.16.1
							1600
							1606
							1666-1
							16150
24V/17Ah						CA-170A	1658
						SP-170A	1756
						CA-176	
						SP-176	
						CA-1700	
						SP-1700	
						CA-1717	
						CA-1735	
						SP-1735	
						CA-1751	
						SP-1751	
						CA-1752	
						CA-1753	
						SP-1753	
24V/18Ah			G-244		9750D0730		
			G-245		9750D0734		
			G-641		9750D0738		
					9750D0740		
					9750D0741		
					9750D0742		
					9750D0744		
					9750D0745		
					9750S0746		
					9750S0775		

24V/19Ah	CB24-20	G-246		
24V/20Ah		G-247	CA-20H-20 CA-21H-20 KTCA-21H-20	
24V/22Ah	CB24-3151 CB24-3151-1 CB24-3151E	GE-51C GE-51E		2026 2376 2376-1 2376-2 2376-5 20126 23176 23186 23376 23476 23576 23676
24V/23Ah				2371 2371-1 2506 2506-1 23180 23390 23396 23491 25106 25106-2
24V/24Ah			CA-4-20 CA-9-20 CA-27-20 CA-91-20 TCA-94A CA-727-9 CA-727-20	

(continued)

TABLE 10.5 Commercial Aircraft Batteries (Continued)

RATING[a]	CONCORDE	CONCORDE SLA	TELEDYNE VLA	TELEDYNE SLA	HAWKER SLA	MARATHON VNC	SAFT VNC
24V/25Ah	CB24-39C CB24-39E			G-639ES	9750T0639 9750E0640 9750E0647 9750E0650	KCA-727-20 CA-900 SP-900 TSP-900AT CA-910 SP-910 CA-930A SP-930A	
24V/25Ah					9750E0658 9750E0660 9750Y0662 9750T0663 9750T0667 9750E0750 9750E0751		2500 25201
24V/26Ah		RG-390E	G-639C G-639E				2378 23178
24V/27Ah							2778 2778-2 2778-4
24V/28Ah			GE-50C GE-50E				
24V/31Ah	CB24-3150 CB24-3150-1 CB24-3150E					SP-280	
24/35Ah 24V/36Ah				G-6381ES	9752D0736 9752H0754	CA-401 SP-401 CA-538 TCA-380 TSP-380	2.10.35.A 4006A 4006A-1 40100A 40206 40306

Capacity					
24V/37Ah		G-638E	9750F0530		4076
		G-638C	9750F0531		4076-1
			9750F0532		4076-2
			9750F0539		4076-5
			9750F0540		4076-9
			9750V0546		40176
					40176-4
					40176-7
					40376
					40576
					40676
					40876
					4079
					A4079
					40109-1
					40209
					A40209
24V/40Ah	RG-380E/40A				
	RG-380E/40B				
24V/40Ah				CA-5-20	400A1
				CA-5H-20	4000A1-1
				MA-5-20	4579
				CA-14	40776
				CA-16	401076
				CA-16L	401176
				CA-16L-2	40100-1
				CA-376	4050A1
				SP-376	4050A1-1
				CA-400	4071
				SP-400	4071-1
				TCA-406	4071-2
				CA-420	4080
				SP-420	
				SP-420L	
				CA-430	
				CA-440	
				SP-440	
				KTCA-747	

(continued)

TABLE 10.5   Commercial Aircraft Batteries (Continued)

RATING[a]	CONCORDE	CONCORDE SLA	TELEDYNE VLA	TELEDYNE SLA	HAWKER SLA	MARATHON VNC	SAFT VNC
24V/43Ah			G-63381C G-6381E				4078 4078-4 4078-7 40208 40208-1 40208-2 40378
24V/44Ah		RG-380E/40					
24V/45Ah	CB24-380C CB24-380E						
24V/48Ah	CB24-382E						
24V/50Ah							21931 21932
24V/65Ah						MA-2	
26.4V/77Ah							22V07L
26.4V/13Ah						CA-121	
26.4V/50Ah							5103

[a] Voltage rating is based on 1.2 V per cell for nickel-cadmium and 2.0 V per cell for lead-acid. Capacity rating is based on the one-hour rate.

deliver the in-rush current to the motor, then falls rapidly as the motor develops back electromotive force (EMF). Within 30 to 60 s, the load drops to zero as the APU ignites and the starter cutoff point is reached. The worst-case condition is starting at altitude with a cold APU and a cold battery; normally, a lower temperature limit of -18°C is used as a design point. A rigorous design methodology for optimizing aircraft starter batteries was developed by Evjen and Miller [1971].

When nickel-cadmium batteries are used for APU or engine starting applications, FAA regulations require the battery to be protected against overheating. Suitable means must be provided to sense the battery temperature and to disconnect the battery from the charging source if the battery overheats. This requirement originated in response to numerous instances of battery thermal runaway, which usually occurred when 19-cell batteries were charged from the 28-volt DC bus. Most instances of thermal runaway were caused by degradation of the cellophane gas barrier, thus allowing gas recombination and resultant cell heating during charging. Modern separator materials (e.g., Celgard) have greatly reduced the occurrence of thermal runaway as a failure mode of nickel-cadmium batteries, but the possibility still exists if the electrolyte level is not properly maintained.

10.5.2 Military Aircraft

A listing of commonly used military aircraft batteries is provided in Table 10.6. This listing includes only those batteries that have been assigned a military part number based on an approved military specification; nonstandard batteries are not included. Detailed characteristics and performance capabilities can be found by referring to the applicable military specifications. A number of nonstandard battery designs have been proliferated in the military due to the unique form, fit, and/or functional requirements of certain aircraft. Specifications for these batteries normally are obtainable only from the aircraft manufacturer. Specific examples of battery systems used in present-day military aircraft were recently described by Vutetakis [1994].

Defining Terms

Ampere-hour capacity: The quantity of stored electrical energy, measured in ampere-hours, that the battery can deliver from its completely charged state to its discharged state. The dischargeable capacity depends on the rate at which the battery is discharged; at higher discharge rates the available capacity is reduced.

C-rate: The discharge rate, in amperes, at which a battery can deliver 1 h of capacity to a fixed voltage endpoint (typically 18 or 20 V for a 24-V battery). Fractions or multiples of the C-rate also are used. C/2 refers to the rate at which a battery will discharge its capacity in 2 h; 2C is twice the C-rate or that rate at which the battery will discharge its capacity in 0.5 h. This rating system helps to compare the performance of different sizes of cells.

CCA: The numerical value of the current, in amperes, that a fully charged lead-acid battery can deliver at $-18°C$ (0°F) for 30 s to a voltage of 1.2 V per cell (i.e., 14.4 V for a 24-V battery). In some cases, 60 s is used instead of 30 s. CCA stands for cold cranking amperes.

Electrolyte: An ionically conductive, liquid medium that allows ions to flow between the positive and negative plates of a cell. In lead-acid cells, the electrolyte is a mixture of sulfuric acid (H_2SO_4) and deionized water. In nickel-cadmium cells, the electrolyte is a mixture of potassium hydroxide (KOH) dissolved in deionized water.

Imp: The numerical value of the current, in amperes, delivered after 15 s during a constant voltage discharge of 0.6 V per cell (i.e., at 12 V for a 24-V battery). The Imp rating normally is based on a battery temperature of 23°C (75°F), but manufacturers generally can supply Imp data at lower temperatures as well.

Monobloc: A group of two or more cells connected in series and housed in a one-piece enclosure with suitable dividing walls between cell compartments. Typical monoblocs come in 6-V, 12-V, or 24-v configurations. Monoblocs are commonly used in lead-acid batteries, but rarely used in nickel-cadmium aircraft batteries.

TABLE 10.6   Military Aircraft Batteries

Military Part No	Type	Rating^a (Ah)	Max. Wt. (lb)	Applications	Notes
				MIL-B-8565 Series	
D8565/1-1	SNC	2.0 (26 V)	8.6	AV-8A/C, CH-53E, MH-53E	Contains integral charger.
D8565/2-1	VNC	30	88.0	OV-10D	Superceded by M81757/12-1. MS3509 connector.
D8565/3-3	SLA	15	47.4	V-22(EMD)	MS27466T17B6S connector.
D8565/4-1	SLA	7.5	26.0	F/A-18A/B/C/D, CH-46D/E, HH-46A, UH-46A/D, F-117A	Equivalent to D8565/5-2, except uses MS3509 connector.
D8565/5-1	SLA	30	80.2	C-1A, SP-2H, A-3B, KA-3B, RA-3B, ERA-3B, NRA-3B, UA-3B, P-3A/B/C, EP-3A/B/E, RP-3A, VP-3A, AC-130A/H/U, C-130A/B/E/F/H, DC-130A, EC-130EH/G/Q, HC-130H/N/P, KC-130H/N/R/T, LC-130F/H/R, LC-130F/H/R, MC-130E/H, NC-130A/B/H, WC-130E/H, C-18A/B, EC-18B/D, C-137B/C, EC-137D, E-8A, TS-2A, US-2A/B, T-28B/C, QT-33A, MH-53J, MH-60G	
D8565/5-2	SLA	30	80.2	Same as D8565/5-1 (for aircraft equipped with Cannon style mating connector)	Equivalent to D8565/5-1, except uses Cannon connector.
D8565/6-1	SLA	1.5	6.4	V-22A, CV-22A, CH-47E	MS27466715B5S connector.
D8565/7-1	SLA	24	63.9	AV-8B, TAV-8B, VH-60A, V-22A, CV-22A	MS3509 connector.
D8565/7-2	SLA	24	63.9	Same as D8565/7-1	Replacement for D8565/7-1 with higher rate capability.
D8565/8-1	SLA	15	43.0	T-45A	Cannon connector.
D8565/9-1	SLA	24	63.0	T-34B/C, U-6A	MS3509 connector.
D8565/9-2	SLA	24	63.0	None identified	Cannon connector.
D8565/10-1	VNC	35	85.0	AH-1W	MS3509 connector. Equipped with temperature sensor.
D8565/11-1	SLA	10	34.8	F-4D/E/G, C-141B, MH-60E, NC-141A, YF-22A	Equivalent to D8565/11-2, except uses MS3509 connector.
D8565/11-2	SLA	10	34.8	None identified	Equivalent to D8565/11-1, except uses Cannon connector.
D8565/12-1	SLA	35	90.0	None identified	MS3509 connector.
D8565/13-1	SLA	10	31.0	Carousel IV, LTN-72 Inertial Navigation Systems (INS)	ARINC 1/2 ATR case.
D8565/14-1	SLA	15	45.2	F-18E/F	D38999/24YG11SN connector
D8565/15-1	SLA	35	90.0	C/KC-135 series	MS3509 connector.
				MIL-B-8565 Specials	
MS3319-1	VNC	0.75	3.5	HH-2D, SH-2D/F	MS3106-12S-3P connector.
MS3337-2	SNC	0.40	4.0	F-4s	Obsolete.

MS3346-1	VNC	2.5	10.0	A-7D/E, TA-7C	Obsolete.
MS3487-1	VNC	18	50.0	AH-1G	Equivalent to BB-649A/A.
MS17334-2	SNC	0.33	3.5	E-1B, EA-6B, US-2D	MS3106R14S-7P connector.
				MIL-B-83769 Series	
M83769/1-1	VLA	31	80.0	Same as D8565/5-1 (for aircraft equipped with Cannon style mating connector)	Supercedes AN3150. Equivalent to BB-638/U. Interchangeable with D8565/5-2 (Cannon connector).
M83769/2-1	VLA	18	56.0	AC-130H/U, NU-1B, U-6A	Supercedes AN3151. Equivalent to BB-639/U. Interchangeable with D8565/9-2 (Cannon connector).
M83769/3-1	VLA	8.4	34.0	C-141B, NC-141A	Supercedes AN3154. Equivalent to BB-640/U. Interchangeable with D8565/11-2 (Cannon connector).
M83769/4-1	VLA	18	55.0	T34B/C	Supercedes MS18045-41. Interchangeable with D8565/9-1 (MS3509 connector).
M83769/5-1	VLA	31	80.0	Same as D8565/5-1	Supercedes MS18045-42. Interchangeable with D8565/5-1 (MS3509 connector).
M83769/6-1	VLA	31	80.0	Same as D8565/5-1 (for aircraft equipped with Cannon style mating connector)	For ground use only. Equivalent to M83769/1-1 when filler caps are replaced with aerobatic vent plugs. Equipped with Cannon connector.
M83769/7-1	VLA	54(12V)	80.0	C-117D, C-118B, VC-118, C-131F, NC-131H, T-33B	Supercedes MS90379-1. Equipped with threaded terminals.
				MIL-B-81757 Series (Tri-Service)	
M81757/7-2	VNC	10	34.0	CH-46A/D/E/F, HH-46A, UH-46A/D, U-8D/F	Replaceable cells. Superceds MS24496-1 and MS24496-2.
M81757/7-3	VNC	10	34.0	Same as M81757/7-2	Nonreplaceable cells. Supercedes MS18045-44, MS18045-48 and MS90221-66W.
M81757/8-4	VNC	20	55.0	C-2A, T-2C, T-39A/B/D, OV-10A	Replaceable cells. Supercedes MS24497-3, MS24497-5, and M81757/8-2.

(continued)

TABLE 10.6 Military Aircraft Batteries (Continued)

Military Part No	Type	Rating[a] (Ah)	Max. Wt. (lb)	Applications	Notes
M81757/8-5	VNC	20	55.0	Same as M81757/8-4	Nonreplaceable cells. Supercedes MS90365-1, MS90365-2, MS90321-68W, MS90321-77, MS90321-78W, MS18045-45, MS18048-49, and M81757/8-3.
M81757/9-2	VNC	30	80.0	CT-39A/E/G, NT-39A, TC-4C, HH-1K, TH-1L, UH-1E/H/L/N, AH-1J/T, LC-130F/R, OV-1B/C/D	Replaceable cells. Supercedes MS24498-1 and MS24498-2.
N81757/9-3	VNC	30	80.0	Same as M81757/9-2	Nonreplaceable cells. Supercedes MS18045-46, MS18045-50, MS90321-75W, MS90321-69W.
M81757/10-1	VNC	6(23V)	24.0	A-6E, EA-6A, KA-6D	Nonreplaceable cells. Supercedes MS90447-2 and MS90321-84W.
M81757/11-3	VNC	20	55.0	HH-2D, SH-2D/F/G, HH-3A/E, SH-3D/G/H, UH-3A, VH-3A	Nonreplaceable cells. Supercedes MS90377-1, MS90321-79W and M81757/11-1.
M81757/11-4	VNC	20	55.0	None identified	Nonreplaceable cells with temperatur sensor. Supercedes MS90377-1, MS90321-79W and M81757/11-2.
M81757/12-1	VNC	30	88.0	OV-10D	Nonreplaceable cells, air-cooled. Supercedes D8565/2-1.
M81757/12-2	VNC	30	88.0	C-2A (REPRO), OV-10D	nonreplaceable cells, air-cooled, with temperature sensor.
M81757/13-1	VNC	30	80.0	EA-3B, ERA-3B, UA-3B	Non replaceable cells. Supercedes MS18045-75.
				MIL-B-26220 Series (U.S. Air Force)	
MS24496-1	VNC	11(C/2)	34.0	F-111A/D/E/F/G, EF-111A, FB-111A	Superceded by M81757/7-2.
MS24496-2	VNC	11(C/2)	34.0	F-4D/E/G, NF-4C/D/E, NRF-4C, RF-4C, YF-4E	Superceded by M81757/7-2.
MS24497-3	VNC	22(C/2)	55.0	None identified	Superceded by M81757/8-2.
MS24497-4	VNC	22(C/2)	60.0	B-52H	Contains integral heater.
MS24497-5	VNC	22(C/2)	55.0	B-52G, C-135, EC-135, KC-135, NC-135, NKC-135, RC-135, TC-135, TC-135, WC-135, E-4B CH-3E, NA-37B, OA-37B, OV-10A	Superceded by M81757/8-2.

Part	Type	Rate	Capacity	Application	Remarks
MS24498-1	VNC	34(C/2)	80.0	A-10A, C-20A, C-137A/B, EC-137D, OA-10A, T-37B, T-41A/B/C/D, HH-1H, UH-1N, CH-53A, MH-53J, NH-53A, TH-53A	Superceded by M81757/8-2.
MS24498-2	VNC	34(C/2)	80.0	None identified	Superceded by M81757/9-2.
MS27546	VNC	5	16.0	T-38A	Superceded by Marathon P/N 30030.
				BB-Series (U.S. Army)	
BB-432A/A	VNC	10	34.0	CH-47A/B/C, U-8F	Equivalent to M81757/7-2.
BB-432B/A	VNC	10	34.0	CH-47D	Equivalent to BB-432A/A, except includes a temperature sensor.
BB-433A/A	VNC	30	80.0	C-12C/D/F/L, OV-1D, EH-1H/X, UH-1H/V, RU-21A/B/C/H	Equivalent to M81757/9-2.
BB-434/A	VNC	20	55.0	CH-54	Equivalent to M81757/8-4.
BB-476/A	VNC	13	27.6	OH-58A/B/C	
BB-558/A	VNC	17	38.5	OH-58D	
BB-564/A	VNC	13	25.0	AH-64A	Superceded by BB-664/A.
BB-638/U	VLA	31	80.0	None identified	Equivalent to M83769/1-1.
BB-638A/U	VLA	31	80.0	None identified	Equivalent to M83769/6-1
BB-639/U	VLA	18	56.0	None identified	Equivalent to M83769/2-1.
BB-640/U	VLA	8.4	34.0	None identified	Equivalent to M83769/3-1.
BB-649A/A	VNC	18	50.0	AH-1E/F/P/S	Equivalent to MS3487-1.
BB-664/A	VNC	13	27.0	A-64A	
BB-678A/A	VNC	13	24.8	OH-6A	
BB-693A/U	VNC	30	83.0	Vulcan	
BB-708/U	VNC	5.5	15.0	OV-1D (Mission Gear Equipment)	
BB-716/A	VNC	5.5	17.5	EH-60A, HH-60H/J, SH-60B/F, UH-60A	

[a]Capacity rating is based on the one-hour rate unless otherwise noted. Voltage rating is 24 V unless otherwise noted.

Negative electrode: The electrode from which electrons flow when the battery is discharging into an external circuit. Reactants are electrochemically oxidized at the negative electrode. In the lead-acid cell, the negative electrode contains spongy lead and lead sulfate ($PbSO_4$) as the active materials. In the nickel-cadmium cell, the negative electrode contains cadmium and cadmium hydroxide ($Cd(OH)_2$) as the active materials.

Nominal voltage: The characteristic operating voltage of a cell or battery. The nominal voltage is 2.0 V for lead-acid cells and 1.2 V for nickel-cadmium cells. These voltage levels represent the approximate cell voltage during discharge at the C-rate under room-temperature conditions. The actual discharge voltage depends on the state-of-charge, state-of-health, discharge time, rate, and temperature.

Positive electrode: The electrode to which electrons flow when the battery is discharging into an external circuit. Reactants are electrochemically reduced at the positive electrode. In the lead-acid cell, the positive electrode contains lead dioxide (PbO_2) and lead sulfate ($PbSO_4$) as the active materials. In the nickel-cadmium cell, the positive electrode contains nickel oxyhydroxide ($NiOOH$) and nickel hydroxide ($Ni(OH)_2$) as the active materials.

Separator: An electrically insulating material that is used to prevent metallic contact between the positive and negative plates in a cell, but permits the flow of ions between the plates. In flooded cells, the separator includes a gas barrier to prevent gas diffusion and recombination of oxygen. In sealed cells, the separator is intended to allow gas diffusion to promote high recombination efficiency.

State-of-charge: The available capacity of a battery divided by the capacity available when fully charged, normally expressed on a percentage basis. Sometimes referred to as "true state-of-charge."

State-of-health: The available capacity of a fully charged battery divided by the rated capacity of the battery, normally expressed on a percentage basis. Sometimes referred to as "apparent state-of-charge." Can also be used in a more qualitative sense to indicate the general condition of the battery.

References

Anderman, M., 1994. "Ni-Cd Battery for Aircraft; Battery Design and Charging Options". *Proc. 9th Annu. Battery Conf. Appl. Adv.*, California State University, Long Beach, pp. 12–19.

Earwicker, G. A. 1956. "Aircraft Batteries and their Behavior on Constant-Potential Charge," in *Aircraft Electrical Engineering*, G. G. Wakefield, Ed., pp.196–224. Royal Aeronautical Society, U.K.

Evjen, J. M. and Miller, L. D., Jr., 1971. "Optimizing the Design of the Battery-Starter/Generator System," SAE Paper 710392.

Flake, R. A., 1988. "Overview on the Evolution of Aircraft Battery Systems Used in Air Force Aircraft," SAE Paper 881411.

Fleischer, A., 1956. "Nickel-Cadmium Batteries," *Proc. 10th Annu. Battery Res. Dev. Conf.*, pp. 37–41.

Falk, S. U. and Salkind, A. J., 1969. *Alkaline Storage Batteries*, pp. 466–472, John Wiley & Sons, New York, NY.

Gross, S. 1991. "Requirements for Rechargeable Airplane Batteries." *Proc. 6th Annu. Battery Conf. Appl. Adv.*, California State University, Long Beach.

Johnson, Z., Roberts, J., and Scoles, D., 1994. "Electrical Characterization of the Negative Electrode of the USAF 20-Year-Life Maintenance-Free Sealed Nickel-Cadmium Aircraft Battery over the Temperature Range −40°C to +70°C," *Proc. 36th Power Sources Conf.*, Cherry Hill, NJ, pp. 292–295.

McWhorter, T. A. and Bishop, W. S., 1972. "Sealed Aircraft Battery with Integral Power Conditioner," *Proc. 25th Power Sources Symp.*, Cherry Hill, NJ, pp. 89–91.

Miller, G. H. and Schiffer, S. F., 1971. "Aircraft Zinc-Silver Oxide Batteries," in *Zinc-Silver Oxide Batteries*, A. Fleischer, Ed., pp. 375–391, John Wiley & Sons, New York, NY.

Scardaville, P. A. and Newman, B. C., 1993. "High Power Vented Nickel-Cadmium Cells Designed for Ultra Low Maintenance," *Proc. 8th Annu. Battery Conf. Appl. Adv.*, California State University, Long Beach.

Senderak, K. L. and Goodman, A. W., 1981. "Sealed Lead-Acid Batteries for Aircraft Applications,"*Proc. 16th IECEC*, pp. 117–122.

Vutetakis, D. G., 1994. "Current Status of Aircraft Batteries in the U.S. Air Force," *Proc. 9th Annu. Battery Conf. Appl. Adv.,* California State University, Long Beach, pp. 1–6.

Vutetakis, D. G. and Viswanathan, V. V., 1995. "Determining the State-of-Health of Maintenance-Free Aircraft Batteries," *Proc. 10th Annu. Battery Conf. Appl. Adv.,* California State University, Long Beach, pp. 13–18.

Further Information

The *Handbook of Batteries and Fuel Cells* by David Linden contains extensive technical data on all battery types, and several chapters are devoted to lead-acid and nickel-cadmium batteries. The second edition of this handbook was published recently (McGraw-Hill, 1995). Engineering handbooks for nickel-cadmium batteries have been published by several battery manufacturers, including General Electric, Gates Energy (now Hawker Energy), and SAFT. An SAE specification for vented nickel-cadmium aircraft batteries, Aerospace Standard AS-8033, was published in 1981 and reaffirmed in 1988.

The following technical manuals are published by the Department of Defense and provide detailed operation and servicing instructions for aircraft batteries:

- NAVAIR 17-15BAD-1, Naval Aircraft and Naval Aircraft Support Equipment Storage Batteries. Request for this document should be referred to Commanding Officer, Naval Air Technical Services Facility, 700 Robbins Avenue, Philadelphia, PA 19111.
- T.O. 8D2-3-1, Aircraft Nickel-Cadmium Storage Batteries. Request for this document should be referred to Sacramento ALC/TILBE, McClellan AFB, CA 95652-5990.
- T.O. 8D2-1-31, Aircraft Storage Batteries (Lead-Acid Batteries). Request for this document should be referred to Sacramento ALC/TILBE, McClellan AFB, CA 95652-5990.
- T.M. 11-6140-203-23, Maintenance Manual for Aircraft Nickel-Cadmium Batteries. Requests for this document should be referred to CECOM, ATTN: AMSEL-LC-LM-LT, Fort Monmouth, NJ 07703.

The following companies manufacture aircraft batteries and may be contacted for technical assistance and pricing information:

Nickel-Cadmium Batteries

ACME Electric Corporation
Aerospace Division
528 W. 21st Street
Tempe, Arizona 85282
Phone (602) 894-6864

Eagle-Picher Industries, Inc.
3820 South Hancock Expressway
Colorado Springs, Colorado 80931
Phone (303) 392-4266

Marathon Power Technologies Company
8301 Imperial Drive
Waco, Texas 76712
Phone (817) 776-0650

SAFT America Inc.
711 Industrial Boulevard
Valdosta, Georgia 31601
Phone (912) 247-2331

Lead-Acid Batteries

Concorde Battery Corporation
2009 San Bernardino Road
West Covina, California 91790
Phone (818) 813-1234

Hawker Energy Products Ltd
Stephenson Street
Newport, Gwent NP9OXJ
United Kingdom
Phone (011) 441-633-277673

Teledyne Battery Products
840 West Brockton Avenue
Redlands, California 92374
Phone (909) 793-3131

11

Boeing B-777: Fly-By-Wire Flight Controls

Gregg F. Bartley
Boeing

11.1   Introduction

Fly-By-Wire (FBW) Primary Flight Controls have been been used in military applications such as fighter airplanes for a number of years. It has been a rather recent development to employ them in a commercial transport application. The 777 is the first commercial transport manufactured by Boeing which employees a FBW Primary Flight Control System. This chapter will examine a FBW Primary Flight Control System using the specific system on the 777 as an example. It must be kept in mind while reading this chapter that this is only a single example of what is currently in service in the airline industry. There are several other airplanes in commercial service made by other manufacturers that employ a different architecture for their FBW flight control system than described here.

A FBW flight control system has several advantages over a mechanical system. These include:

- Overall reduction in airframe weight.
- Integration of several federated systems into a single system.
- Superior airplane handling characteristics.
- Ease of maintenance.

- Ease of manufacture.
- Greater flexibility for including new functionality or changes after initial design and production.

11.2   System Overview

Conventional primary flight controls systems employ hydraulic actuators and control valves controlled by cables that are driven by the pilot controls. These cables run the length of the airframe from the cockpit area to the surfaces to be controlled. This type of system, while providing full airplane control over the entire flight regime, does have some distinct drawbacks. The cable-controlled system comes with a weight penalty due to the long cable runs, pulleys, brackets, and supports needed. The system requires periodic maintenance, such as lubrication and adjustments due to cable stretch over time. In addition, systems such as the yaw damper that provide enhanced control of the flight control surfaces require dedicated actuation, wiring, and electronic controllers. This adds to the overall system weight and increases the number of components in the system.

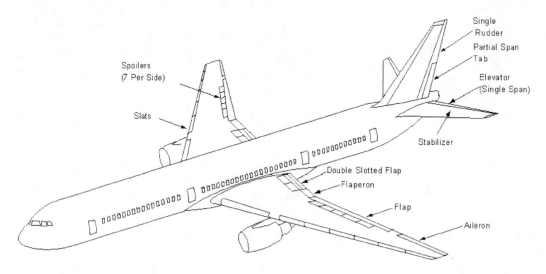

FIGURE 11.1   The Primary Flight Control System on the Boeing 777 is comprised of the outboard ailerons, flaperons, elevator, rudder, horizontal stabilizer, and the spoiler/speedbrakes.

   In a FBW flight control system, the cable control of the primary flight control surfaces has been removed. Rather, the actuators are controlled electrically. At the heart of the FBW system are electronic computers. These computers convert electrical signals sent from position transducers attached to the pilot controls into commands that are transmitted to the actuators. Because of these changes to the system, the following design features have been made possible:

- Full-time surface control utilizing advanced control laws. The aerodynamic surfaces of the 777 have been sized to afford the required airplane response during critical flight conditions. The reaction time of the control laws is much faster than that of an alert pilot. Therefore, the size of the flight control surfaces could be made smaller than those required for a conventionally controlled airplane. This results in an overall reduction in the weight of the system.
- Retention of the desirable flight control characteristics of a conventionally controlled system and the removal of the undesirable characteristics. This aspect is discussed further in the section on control laws and system functionality.
- Integration of functions such as the yaw damper into the basic surface control. This allows the separate components normally used for these functions to be removed.
- Improved system reliability and maintainability.

11.3   Design Philosophy

The philosophy employed during the design of the 777 Primary Flight Control System maintains a system operation that is consistent with a pilot's past training and experience. What is meant by this is that however different the actual system architecture is from previous Boeing airplanes, the presentation to the pilot is that of a conventionally controlled mechanical system. The 777 retains the conventional control column, wheel, and rudder pedals, whose operation are identical to the controls employed on other Boeing transport aircraft. The flight deck controls of the 777 are very similar to those of the Boeing 747-400, which employs a traditional mechanically controlled Primary Flight Control System.

Because the system is controlled electronically, there is an opportunity to include system control augmentation and envelope protection features that would have been difficult to provide in a conventional mechanical system. The 777 Primary Flight Control System has made full use of the capabilities of this architecture by including such features as:

- Bank angle protection
- Turn compensation
- Stall and overspeed protection
- Pitch control and stability augmentation
- Thrust asymmetry compensation

More will be said of these specific features later. What should be noted, however, is that none of these features limit the action of the pilot. The 777 design utilizes *envelope protection* in all of its functionality rather than *envelope limiting*. Envelope *protection* deters pilot inputs from exceeding certain predefined limits but does not prohibit it. Envelope *limiting* prevents the pilot from commanding the airplane beyond set limits. For example, the 777 bank angle protection feature will significantly increase the wheel force a pilot encounters when attempting to roll the airplane past a predefined bank angle. This acts as a prompt to the pilot that the airplane is approaching the bank angle limit. However, if deemed necessary, the pilot may override this protection by exerting a greater force on the wheel than is being exerted by the backdrive actuator. The intent is to inform the pilot that the command being given would put the airplane outside of its normal operating envelope, but the ability to do so is not precluded. This concept is central to the design philosophy of the 777 Primary Flight Control System.

11.4   System Architecture

11.4.1   Flight Deck Controls

As noted previously, the 777 flight deck utilizes standard flight deck controls; a control column, wheel, and rudder pedals that are mechanically linked between the Captain's and First Officer's controls. This precludes any conflicting input between the Captain and First Officer into the Primary Flight Control System. Instead of the pilot controls driving quadrants and cables, as in a conventional system, they are attached to electrical transducers that convert mechanical displacement into electrical signals.

A gradient control actuator is attached to the two control column feel units. These units provide the tactile feel of the control column by proportionally increasing the amount of force the pilot experiences during a maneuver with an increase in airspeed. This is consistent with a pilot's experience in conventional commercial jet transports.

Additionally, the flight deck controls are fitted with what are referred to as "backdrive actuators." As the name implies, these actuators backdrive the flight deck controls during autopilot operation. This feature is also consistent with what a pilot is used to in conventionally controlled aircraft and allows the pilot to monitor the operation of the autopilot via immediate visual feedback of the pilot controls that is easily recognizable.

11.4.2 System Electronics

There are two types of electronic computers used in the 777 Primary Flight Control System: the Actuator Control Electronics (ACE), which is primarily an analog device, and the Primary Flight Computer (PFC), which utilizes digital technology. There are four ACEs and three PFCs employed in the system. The function of the ACE is to interface with the pilot control transducers and to control the Primary Flight Control System actuation with analog servo loops. The role of the PFC is the calculation of control laws by converting the pilot control position into actuation commands, which are then transmitted to the ACE. The PFC also contains ancillary functions, such as system monitoring, crew annunciation, and all the Primary Flight Control System onboard maintenance capabilities.

Four identical ACEs are used in the system, referred to as L1, L2, C, and R. These designations correspond roughly to the left, center, and right hydraulic systems on the airplane. The flight control functions are distributed among the four ACEs. The ACEs decode the signals received from the transducers used on the flight deck controls and the primary surface actuation. The ACEs convert the transducer position into a digital value and then transmit that value over the ARINC 629 data busses for use by the PFCs. There are three PFCs in the system, referred to as L, C, and R. The PFCs use these pilot control and surface positions to calculate the required surface commands. At this time, the command of the automatic functions, such as the yaw damper rudder commands, are summed with the flight deck control commands, and are then transmitted back to the ACEs via the same ARINC 629 data busses. The ACEs then convert these commands into analog commands for each individual actuator.

11.4.3 ARINC 629 Data Bus

The ACEs and PFCs communicate with each other, as well as with all other systems on the airplane, via triplex, bi-directional ARINC 629 Flight Controls data busses, referred to as L, C, and R. The connection from these electronic units to each of the data busses is via a stub cable and an ARINC 629 coupler. Each coupler may be removed and replaced without disturbing the integrity of the data bus itself.

11.4.4 Interface to Other Airplane Systems

The Primary Flight Control System transmits and receives data from other airplane systems by two different pathways. The Air Data and Inertial Reference Unit (ADIRU), Standby Attitude and Air Data Reference Unit (SAARU), and the Autopilot Flight Director Computers (AFDC) transmit and receive data on the ARINC 629 flight controls data busses, which is a direct interface to the Primary Flight Computers. Other systems, such as the Flap Slat Electronics Unit (FSEU), Proximity Switch Electronics Unit (PSEU), and Engine Data Interface Unit (EDIU) transmit and receive their data on the ARINC 629 systems data busses. The PFCs receive data from these systems through the Airplane Information Management System (AIMS) Data Conversion Gateway (DCG) function. The DCG supplies data from the systems data busses onto the flight controls data busses. This gateway between the two main sets of ARINC 629 busses maintains separation between the critical flight controls busses and the essential systems busses but still allows data to be passed back and forth.

11.4.5 Electrical Power

There are three individual power systems dedicated to the Primary Flight Control System, which are collectively referred to as the Flight Controls Direct Current (FCDC) power system. An FCDC Power Supply Assembly (PSA) powers each of the three power systems. Two dedicated Permanent Magnet Generators (PMG) on each engine generate AC power for the FCDC power system. Each PSA converts the PMG alternating current into 28 V DC for use by the electronic modules in the Primary Flight Control System. Alternative power sources for the PSAs include the airplane Ram Air Turbine (RAT), the 28-V DC main airplane busses, the airplane hot battery buss, and dedicated 5 Ah FCDC batteries. During flight, the PSAs draw power from the PMGs. For on-ground engines-off operation or for in-flight failures of the PMGs, the PSAs draw power from any available source.

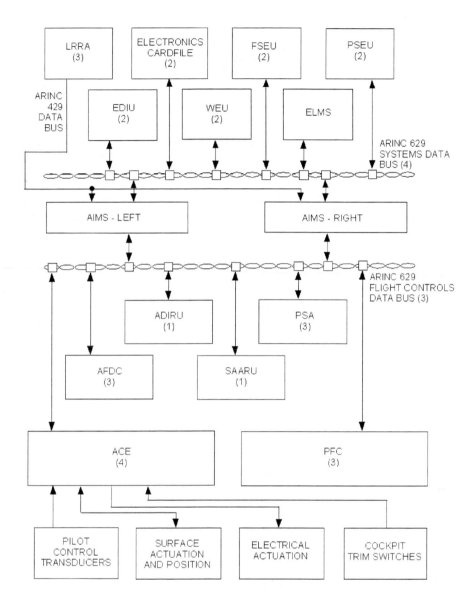

FIGURE 11.2 Block diagram of the electronic components of the 777 Primary Flight Control System, as well as the interfaces to other airplane systems.

11.5 Control Surface Actuation

11.5.1 Fly-by-Wire Actuation

The control surfaces on the wing and tail of the 777 are controlled by hydraulically powered, electrically signaled actuators. The elevators, ailerons, and flaperons are controlled by two actuators per surface, the rudder is controlled by three. Each spoiler panel is powered by a single actuator. The horizontal stabilizer is positioned by two parallel hydraulic motors driving the stabilizer jackscrew.

The actuation powering the elevators, ailerons, flaperons, and rudder have several operational modes. These modes, and the surfaces that each are applicable to, are defined below.

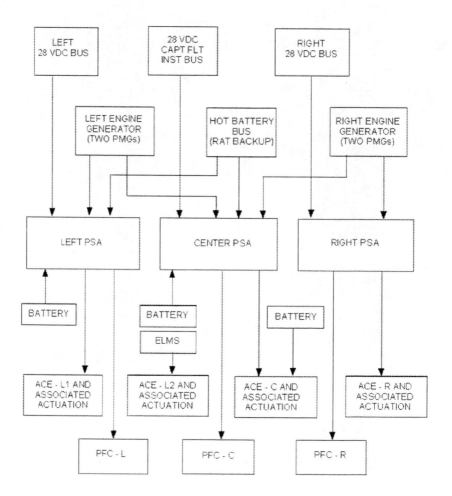

FIGURE 11.3 Block diagram of the 777 Fly-By-Wire Power Distribution System.

Active—Normally, all the actuators on the elevators, ailerons, flaperons, and rudder receive commands from their respective ACEs and position the surfaces accordingly. The actuators will remain in the active mode until commanded into another mode by the ACEs.

Bypassed—In this mode, the actuator does not respond to commands from its ACE. The actuator is allowed to move freely, so that the redundant actuator(s) on a given surface may position the surface without any loss of authority, i.e., the actuator in the active mode does not have to overpower the bypassed actuator. This mode is present on the aileron, flaperon, and rudder actuators.

Damped—In this mode, the actuator does not respond to the commands from the ACE. The actuator is allowed to move, but at a restricted rate which provides flutter damping for that surface. This mode allows the other actuator(s) on the surface to continue to operate the surface at a rate sufficient for airplane control. This mode is present on elevator and rudder actuators.

Blocked—In this mode, the actuator does not respond to commands from the ACE, and it is not allowed to move. When both actuators on a surface (which is controlled by two actuators) have failed, they both enter the "Blocked" mode. This provides a hydraulic *lock* on the surface. This mode is present on the elevator and aileron actuators.

An example using the elevator surface illustrates how these modes are used. If the inboard actuator on an elevator surface fails, the ACE controlling that actuator will place the actuator in the "Damped" mode. This allows the surface to move at a limited rate under the control of the remaining operative outboard actuator. Concurrent with this action, the ACE also arms the "Blocking" mode on the outboard actuator on the same surface. If a subsequent failure occurs that will cause the outboard actuator to be placed in the "Damped" mode by its ACE, both actuators will then be in the "Damped" mode and have their "Blocking" modes armed. An elevator actuator in this configuration enters the "Blocking" mode, which hydraulically locks the surface in place for flutter protection.

11.5.2 Mechanical Control

Spoiler panel 4 and 11 and the alternate stabilizer pitch trim system are controlled mechanically rather than electrically. Spoilers 4 and 11 are driven directly from control wheel deflections via a control cable. The alternate horizontal stabilizer control is accomplished by using the pitch trim levers on the flight deck aisle stand. Electrical switches actuated by the alternate trim levers allow the PFCs to determine when alternate trim is being commanded so that appropriate commands can be given to the pitch control laws.

Spoiler panels 4 and 11 are also used as speedbrakes, both in the air and on the ground. The speedbrake function for this spoiler pair only has two positions: stowed and fully extended. The speedbrake commands for spoilers 4 and 11 are electrical in nature, with an ACE giving an *extend* or *retract* command to a solenoid-operated valve in each of the actuators. Once that spoiler pair has been deployed by a speedbrake command, there is no control wheel speedbrake command mixing, as there is on all the other fly-by-wire spoiler surfaces.

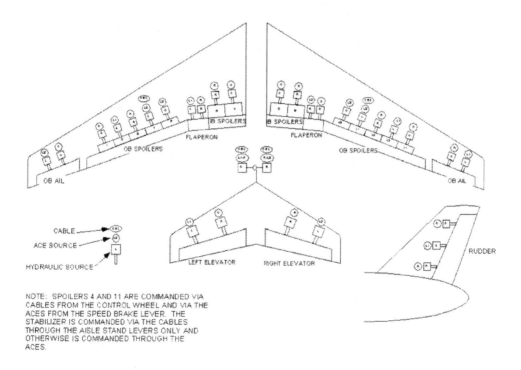

FIGURE 11.4 Schematic representation of the Boeing 777 Primary Flight Control System hydraulic power and electronic control functional distribution.

11.6 Fault Tolerance

"Fault Tolerance" is a term that is used to define the ability of any system to withstand single or multiple failures which results in either no loss of functionality or a known loss of functionality or reduced level of redundancy while maintaining the required level of safety. It does not, however, define any particular method that is used for this purpose. There are two major classes of faults that any system design must deal with. These are

- A failure which results in some particular component becoming totally inoperative. An example of this would be a loss of power to some electronic component, such that it no longer performs its intended function.
- A failure which results in some particular component remaining active, but the functionality it provides is in error. An example of this failure would be a Low Range Radio Altimeter whose output is indicating the airplane is at an altitude 500 ft above the ground when the airplane is actually 200 ft above the ground.

One method that is used to address the first class of faults is the use of redundant elements. For example, there are three PFCs in the 777 Primary Flight Control System, each with three identical computing "lanes" within each PFC. This results in nine identical computing channels. Any of the three PFCs themselves can fail totally due to loss of power or some other failure which affects all three computing lanes, but the Primary Flight Control System loses no functionality. All four ACEs will continue to receive all their surface position commands from the remaining PFCs. All that is affected is the level of available redundancy. Likewise, any single computing lane within a PFC can fail, and that PFC itself will continue to operate with no loss of functionality. The only thing that is affected is the amount of redundancy of the system. The 777 is certified to be dispatched on a revenue flight, per the Minimum Equipment List (MEL), with two computing lanes out of the nine total (as long as they are not within the same PFC channel) for 10 days and for a single day with one total PFC channel inoperative.

Likewise, there is fault tolerance in the ACE architecture. The flight control functions are distributed among the four ACEs such that a total failure of a single ACE will leave the major functionality of the system intact. A single actuator on several of the primary control surfaces may become inoperative due to this failure, and a certain number of spoiler symmetrical panel pairs will be lost. However, the pilot flying the airplane will notice little or no difference in handling characteristics with this failure. A total ACE failure of this nature will have much the same impact to the Primary Flight Control System as that of a hydraulic system failure.

The second class of faults is one that results in erroneous operation of a specific component of the system. The normal design practice to account for failures of this type is to have multiple elements doing the same task and their outputs voted or compared in some manner. This is sometimes referred to as a "voting plane." All critical interfaces into the 777 FBW Primary Flight Control System use multiple inputs which are compared by a voting plane. For interfaces that are required to remain operable after a first failure, at least three inputs must be used. For example, there are three individual Low Range Radio Altimeter (LRRA) inputs used by the PFCs. The PFCs compare all three inputs and calculates a mid-value select on the three values; i.e., the middle value LRRA input is used in all calculations which require radio altitude. In this manner, any single failure of an LRRA that results in an erroneous value will be discarded. If a subsequent failure occurs which causes the remaining two LRRA signals to disagree by a preset amount, the PFCs will throw out both values and take appropriate action in those functions which use these data.

Additionally, a voting plane scheme is used by the PFCs on themselves. Normally, a single computing lane within a PFC channel is declared as the "master" lane, and that lane is responsible for transmitting all data onto the data busses for use by the ACEs and other airplane systems. However, all three lanes are simultaneously computing the same control laws. The outputs of all three computing lanes within a single PFC channel are compared against each other. Any failure of a lane that will cause an erroneous output from that lane will cause that lane to be condemned as "failed" by the other two lanes.

Likewise, the outputs from all three PFC channels themselves are compared. Each PFC looks at its own calculated command output for any particular actuator, and compares it with the same command that was calculated by the other two PFC channels. Each PFC channel then does a mid-value select on the three commands, and that value (whether it was the one calculated by itself or by one of the other PFC channels) is then output to the ACEs for the individual actuator commands. In this manner, it is assured that each ACE receives identical commands from each of the PFC channels.

By employing methods such as those described above, it is assured that the 777 Primary Flight Control System is able to withstand single or multiple failures and be able to contain those failures in such a manner that the system remains safe and does not take inappropriate action due to those failures.

11.7 System Operating Modes

The 777 FBW Primary Flight Control System has three operating modes: Normal, Secondary, and Direct. These modes are defined below:

Normal—In the "Normal" mode, the PFCs supply actuator position commands to the ACEs, which convert them into an analog servo command. Full functionality is provided, including all enhanced performance, envelope protection, and ride quality features.

Secondary—In the "Secondary" mode, the PFCs supply actuator position commands to the ACEs, just as in the "Normal" mode. However, functionality of the system is reduced. For example, the envelope protection functions are not active in the "Secondary" mode. The PFCs enter this mode automatically from the "Normal" mode when there are sufficient failures in the system or interfacing systems such that the "Normal" mode is no longer supported. An example of a set of failures that will automatically drop the system into the "Secondary" mode is total loss of airplane air data from the ADIRU and SAARU. The airplane is quite capable of being flown for a long period of time in the "Secondary" mode. It cannot, however, be dispatched in this condition.

Direct—In the "Direct" mode, the ACEs do not process commands from the PFCs. Instead, each ACE decodes pilot commands directly from the pilot controller transducers and uses them for the closed loop servo control of the actuators. This mode will automatically be entered due to total failure of all three PFCs, failures internal to the ACEs, loss of the flight controls ARINC 629 data busses, or some combination of these failures. It may also be selected manually via the PFC disconnect switch on the overhead panel in the flight deck. The airplane handling characteristics in the "Direct" mode closely match those of the "Secondary" mode.

11.8 Control Laws and System Functionality

The design philosophy employed in the development of the 777 Primary Flight Control System control laws stresses aircraft operation consistent with a pilot's past training and experience. The combination of electronic control of the system and this philosophy provides for the feel of a conventional airplane, but with improved handling characteristics and reduced pilot workload.

11.8.1 Pitch Control

Pitch control is accomplished through what is known as a *maneuver demand* control law, which also referred to as a C*U control law. C* (pronounced "C-Star") is a term that is used to describe the blending of the airplane pitch rate and the load factor (the amount of acceleration felt by an occupant of the airplane during a maneuver). At low airspeeds, the pitch rate is the controlling factor. That is, a specific push or pull of the column by the pilot will result in some given pitch rate of the airplane. The harder the pilot pushes or pulls on the column, the faster the airplane will pitch nose up or nose down. At high airspeeds, the load factor dominates. This means that, at high airspeeds, a specific push or pull of the column by the pilot will result in some given load factor.

The "U" term in C*U refers to the feature in the control law which will, for any change in the airspeed away from a referenced trim speed, cause a pitch change to return to that referenced airspeed. For an increase in airspeed, the control law will command the airplane nose up, which tends to slow the airplane down. For a decrease in airspeed, the control law causes a corresponding speed increase by commanding the airplane nose down. This introduces an element of speed stability into the airplane pitch control. However, airplane configuration changes, such as a change in the trailing edge flap setting or lowering the landing gear, will NOT result in airplane pitch changes, which would require the pilot to re-trim the airplane to the new configuration. Thus, the major advantage of this type of control law is that the nuisance-handling characteristics found in a conventional, mechanically controlled flight control system which increase the pilot workload are minimized or eliminated, while the desirable characteristics are maintained.

While in flight, the pitch trim switches on the Captain's and First Officer's control wheels do not directly control the horizontal stabilizer as they normally do on conventionally controlled airplanes. When the trim switches are used in flight, the pilot is actually requesting a new referenced trim speed. The airplane will pitch nose up or nose down, using the elevator surfaces, in response to that reference airspeed change to achieve that new airspeed. The stabilizer will automatically trim, when necessary, to offload the elevator surface and allow it to return to its neutral surface when the airplane is in a trimmed condition. When the airplane is on the ground, the pitch trim switches do trim the horizontal stabilizer directly. While the alternate trim levers (described previously) move the stabilizer directly, even in flight, the act of doing so will also change the C*U referenced trim speed such that the net effect is the same as would have been achieved if the pitch trim switches on the control wheels had been used. As on a conventional airplane, trimming is required to reduce any column forces that are being held by the pilot.

The pitch control law incorporates several additional features. One is called landing flare compensation. This function provides handling characteristics during the flare and landing maneuvers consistent with that of a conventional airplane, which would have otherwise been significantly altered by the C*U control law. The pitch control law also incorporates Stall and Overspeed Protection. These functions will not allow the referenced trim speed to be set below a predefined minimum value or above the maximum operating speed of the airplane. They also significantly increase the column force that the pilot must hold in order to fly above or below those speeds. An additional feature incorporated into the pitch control law is turn compensation, which enables the pilot to maintain a constant altitude with minimal column input during a banked turn.

The unique 777 implementation of maneuver demand and speed stability in the pitch control laws means that:

- An established flight path remains unchanged unless the pilot changes it through a control column input, or if the airspeed changes and the speed stability function takes effect.
- Trimming is required only for airspeed changes and not for airplane configuration changes.

11.8.2   Yaw Control

The yaw control law contains the usual functionality employed on other Boeing jetliners, such as the yaw damper and rudder ratio changer (which compensates a rudder command as a function of airspeed). However, the 777 FBW rudder control system has no separate actuators, linkages, and wiring for these functions, as have been used in previous airplane models. Rather, the command for these functions are calculated in the PFCs and included as part of the normal rudder command to the main rudder actuators. This reduces weight, complexity, maintenance, and spares required to be stocked.

The yaw control law also incorporates several addition features. The gust suppression system reduces airplane tag wag by sensing wind gusts via pressure transducers mounted on the vertical tail fin and applying a rudder command to oppose the movement that would have otherwise been generated by the gust. Another feature is the wheel-rudder crosstie function, which reduces sideslip by using small amounts of rudder during banked turns.

One important feature in the yaw control is Thrust Asymmetry Compensation, or TAC. This function automatically applies a rudder input for any thrust asymmetry between the two engines which exceed approximately 10% of the rated thrust. This is intended to cancel the yawing moment associated with an engine failure. TAC operates at all airspeeds above 80 kn even on the ground during the take-off roll. It will not operate when the engine thrust reversers are deployed.

11.8.3 Roll Control

The roll control law utilized by the 777 Primary Flight Control System is fairly conventional. The outboard ailerons and spoiler panels 5 and 10 are locked out in the faired position when the airspeed exceeds a value that is dependent upon airspeed and altitude. It roughly corresponds to the airplane "flaps up" maneuvering speed. As with the yaw damper function described previously, this function does not have a separate actuator, but is part of the normal aileron and spoiler commands. The bank angle protection feature in the roll control law has been discussed previously.

11.8.4 757 Test Bed

The control laws and features discussed here were incorporated into a modified 757 and flown in the summer of 1992, prior to full-scale design and development of the 777 Primary Flight Control System. The Captain's controls remained connected to the normal mechanical system utilized on the 757. The 777 control laws were flown through the First Officer's controls. This flying testbed was used to validate the flight characteristics of the 777 fly-by-wire system, as was flown by Boeing, customer, and regulatory agency pilots. When the 777 entered into its flight test program, its handling characteristics were extremely close to those that had been demonstrated with the 757 flying testbed.

11.8.5 Actuator Force-Fight Elimination

One unique aspect of the FBW flight control system used on the 777 is that the actuators on any given surface are all fully powered at all times. There are two full-time actuators driving each of the elevator, aileron, and flaperon surfaces, just as there are three full-time actuators on the rudder. The benefit of this particular implementation is that each individual actuator was able to be sized smaller than it would have had to have been if each surface was going to be powered by a single actuator through the entire flight regime. In addition, there is not a need for any redundancy management of an active/standby actuation system. However, this does cause a concern in another area. This is a possible actuator force-fight condition between the multiple actuators on a single flight control surface.

Actuator force-fight is caused by the fact that no two actuators, position transducers, or set of controlling servo loop electronics are identical. In addition, there always will be some rigging differences of the multiple actuators as they are installed on the airplane. These differences will result in one actuator attempting to position a flight control surface in a slightly different position than its neighboring actuator. Unless addressed, this would result in a twisting moment upon the surface as the two actuators fight each other to place the surface in different positions. In order to remove this unnecessary stress on the flight control surfaces, the Primary Flight Computer control laws include a feature which "nulls out" these forces on the surfaces.

Each actuator on the 777 Primary Flight Control System includes what is referred to as a Delta Pressure, or Delta P, pressure transducer. These transducer readings are transmitted via the ACEs to the PFCs, which are used in the individual surface control laws to remove the force-fight condition on each surface. The PFCs add an additional positive or negative component to each of the individual elevator actuator commands, which results in the difference between the two Delta P transducers being zero. In this way, the possibility of any force-fight condition between multiple actuators on a single surface is removed. The surface itself, therefore, does not need to be designed to withstand these stresses, which would have added a significant amount of weight to the airplane.

11.9 Primary Flight Controls System Displays and Annunciations

The primary displays for the Primary Flight Control System on the 777 are the Engine Indication and Crew Alerting System (EICAS) display and the Multi-Function Display (MFD) in the flight deck. Any failures that require flight crew knowledge or action are displayed on these displays in the form of an English language message. These messages have several different levels associated with them, depending upon the level of severity of the failure.

Warning (Red with accompanying aural alert): A nonnormal operational or airplane system condition that requires immediate crew awareness and immediate pilot corrective compensatory action.

Caution (Amber with accompanying aural alert): A nonnormal or airplane system condition that requires immediate crew awareness. Compensatory or corrective action may be required.

Advisory (Amber with no accompanying aural alert): A nonnormal operational or airplane system condition which requires crew awareness. Compensatory or corrective action may be required.

Status (White): No Dispatch or Minimum Equipment List (MEL) related items requiring crew awareness prior to dispatch.

Also available on the MFD, but not normally displayed in flight, is the flight control synoptic page, which shows the position of all the flight control surfaces.

11.10 System Maintenance

The 777 Primary Flight Control System has been designed to keep line maintenance to a minimum, but when tasks do need to be accomplished, they are straightforward and easy to understand.

11.10.1 Central Maintenance Computer

The main interface to the Primary Flight Control System for the line mechanic is the Central Maintenance Computer (CMC) function of AIMS. The CMC uses the Maintenance Access Terminal (MAT) as its primary display and control. The role of the CMC in the maintenance of the Primary Flight Control System is to identify failures present in the system and to assist in their repair. The two features utilized by the CMC that accomplish these tasks are maintenance messages and ground maintenance tests. Maintenance messages describe to the mechanic, in simplified English, what failures are present in the system and the components possibly at fault. The ground maintenance tests exercise the system, test for active and latent failures, and confirm any repair action taken. They are also used to unlatch any EICAS and Maintenance Messages that may have become latched due to failures.

The PFCs are able to be loaded with new software through the Data Loader function on the MAT. This allows the PFCs to be updated to a new software configuration without having to take them out of service.

11.10.2 Line Replaceable Units

All the major components of the system are Line Replaceable Units (LRU). This includes all electronics modules, ARINC 629 data bus couplers, hydraulic and electrical actuators, and all position, force, and pressure transducers. The installation of each LRU has been designed such that a mechanic has ample space for component removal and replacement, as well as space for the manipulation of any required tools.

Each LRU, when replaced, must be tested to assure that the installation was accomplished correctly. The major LRUs of the system (transducers, actuators, and electronics modules) have LRU Replacement Tests that are able to be selected via a MAT pull-down menu and are run by the PFCs. These tests are user-friendly and take a minimum amount of time to accomplish. Any failures found in an LRU replacement test will result in a maintenance message, which details the failures that are present.

11.10.3 Component Adjustment

The primary surface actuators on the 777 are replaced in the same manner as on conventional airplanes. The difference is how they are adjusted. Each elevator, aileron, flaperon, and rudder actuator has what is referred to as a null adjust transducer, which is rotated by the mechanic until the actuator is positioned correctly. For example, when a rudder actuator is replaced, all hydraulic systems are depressurized except for the one that supplies power to the actuator that has just been replaced. The Null Adjust Transducer is then adjusted until the rudder surface aligns itself with a mark on the empennage, showing that the actuator has centered the rudder correctly.

The transducers used on the pilot controls are, for the most part, individual LRUs. However, there are some packages, such as the speedbrake lever position transducers and the column force transducers, which have multiple transducers in a single package. When a transducer is replaced, the Primary Flight Controls EICAS Maintenance Pages are used to adjust the transducer to a certain value at the system rig point. There are CMC-initiated LRU replacement tests which check that the component has been installed correctly and that all electrical connections have been properly mated.

11.11 Summary

The Boeing 777 fly-by-wire Primary Flight Control System utilizes new technology to provide significant benefits over that of a conventional system. These benefits include a reduction in the overall weight of the airplane, superior handling characteristics, and improved maintainability of the system. At the same time, the control of the airplane is accomplished using traditional flight deck controls, thereby allowing the pilot to fly the airplane without any specialized training when transferring from a more conventional commercial jet aircraft. The technology utilized by the 777 Primary Flight Control System has earned its way onto the airplane, and is not just technology for technology's sake.

Defining Terms

ACE: Actuator Control Electronics
ADIRU: Air Data Inertial Reference Unit
ADM: Air Data Module (Static and Total Pressure)
AFDC: Autopilot Flight Director Computer
AIMS: Airplane Information Management System
ARINC: Aeronautical Radio Inc. (Industry Standard)
C: Center
C*U: Pitch Control Law utilized in the Primary Flight Computer
CMC: Central Maintenance Computer Function in AIMS
DCGF: Data Conversion Gateway Function of AIMS
EDIU: Engine Data Interface Unit
EICAS: Engine Indication and Crew Alerting System
ELMS: Electrical Load Management System
FBW: Fly-By-Wire
FCDC: Flight Controls Direct Current (power system)
FSEU: Flap Slat Electronic Unit
L: Left
L1:   Left 1
L2:   Left 2

LRRA: Low Range Radio Altimeter
LRU: Line Replaceable Unit
MAT: Maintenance Access Terminal
MEL: Minimum Equipment List
MFD: Multi-Function Display
MOV: Motor-Operated Valve
PCU: Power Control Unit (hydraulic actuator)
PFC: Primary Flight Computer
PMG: Permanent Magnet Generator
PSA: Power Supply Assembly
R: Right
RAT: Ram Air Turbine
SAARU: Standby Attitude and Air Data Unit
TAC: Thrust Asymmetry Compensation
WEU: Warning Electronics Unit

12

Electrical Flight Controls, From Airbus A320/330/340 to Future Military Transport Aircraft: A Family of Fault-Tolerant Systems

Dominique Briere
Aerospatiale

Christian Favre
Aerospatiale

Pascal Traverse
Aerospatiale

12.1 Introduction

The first electrical flight control system for a civil aircraft was designed by Aerospatiale and installed on the Concorde. This is an analog, full-authority system for all control surfaces. The commanded control surface positions are directly proportional to the stick inputs. A mechanical back-up system is provided on the three axes.

The first generation of electrical flight control systems with digital technology appeared on several civil aircraft at the start of the 1980s with the Airbus A310 program. These systems control the slats, flaps, and spoilers. These systems were designed with very stringent safety requirements (control surface runaway must be extremely improbable). As the loss of these functions results in a supportable increase in the crew's workload, it is possible to lose the system in some circumstances.

TABLE 12.1 Incremental Introduction of New Technologies

First Flight In:	1955	1969	1972	1978–1983	1983	1987
Servo-Controls, and Artificial Feel	x	x	x	x	x	--> x
Electro-Hydraulic Actuators		x	x	x	x	--> x
Command and Monitoring Computers		x	x	x	x	--> x
Digital Computers				x	x	--> x
Trim, Yaw Damper, Protection	x	x	x	x	x	--> x
Electrical Flight Controls		x		x	x	-->x
Side-Stick, Control Laws				x		--> x
Servoed Aircraft (Auto-pilot)	x	x	x	x	x	--> x
Formal System Safety Assessment		x	x	x	x	--> x
System Integration Testing	x	x	x	x	x	--> x
	Carevelle	Concorde	A300	Flight test Concorde A300	A310, A300–600	A320

The Airbus A320 (certified in early 1988) is the first example of a second generation of civil electrical flight control aircraft, rapidly followed by the A340 aircraft (certified at the end of 1992). These aircraft benefit from the significant experience gained by Aérospatiale in the technologies used for a fly-by-wire system (see Table 12.1). The distinctive feature of these aircraft is that all control surfaces are electrically controlled and that the system is designed to be available under all circumstances.

This system was built to very stringent dependability requirements both in terms of safety (the system may generate no erroneous signals) and availability (the complete loss of the system is extremely improbable).

The overall dependability of the aircraft fly-by-wire system relies in particular on the computer arrangement (the so-called control/monitor architecture), the system tolerance to both hardware and software failures, the servo-control and power supply arrangement, the failure monitoring, and the system protection against external aggressions. It does this without forgetting the flight control laws which minimize the crew workload, the flight envelope protections which allow fast reactions while keeping the aircraft in the safe part of the flight envelope, and finally the system design and validation methods.

The aircraft safety is demonstrated by using both qualitative and quantitative assessments; this approach is consistent with the airworthiness regulation. Qualitative assessment is used to deal with design faults, interaction (maintenance, crew) faults, and external environmental hazard. For physical ("hardware") faults, both qualitative and quantitative assessments are used. The quantitative assessment covers the FAR/JAR 25.1309 requirement, and links the failure condition classification (minor to catastrophic) to its probability target.

The aim of this chapter is to describe the Airbus fly-by-wire systems from a fault-tolerant standpoint. The fly-by-wire basic principles are presented first, followed by the description of the main system features common to A320 and A340 aircraft, the failure detection and reconfiguration procedures, the A340 particularities, and the design, development, and validation procedures. Future trends in terms of fly-by-wire fault-tolerance conclude this overview.

12.2 Fly-by-Wire Principles

On aircraft of the A300 and A310 type, the pilot orders are transmitted to the servo-controls by an arrangement of mechanical components (rods, cables, pulleys, etc.). In addition, specific computers and actuators driving the mechanical linkages restore the pilot feels on the controls and transmit the autopilot commands (see Figure 12.1).

MECHANICAL FLIGHT CONTROLS

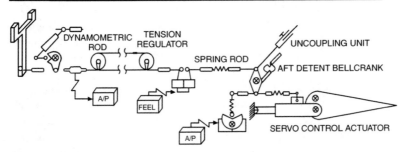

ELECTRICAL FLIGHT CONTROLS (FLY BY WIRE)

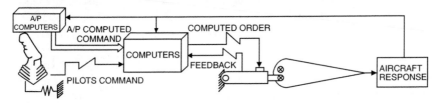

FIGURE 12.1   Mechanical and electrical flight control.

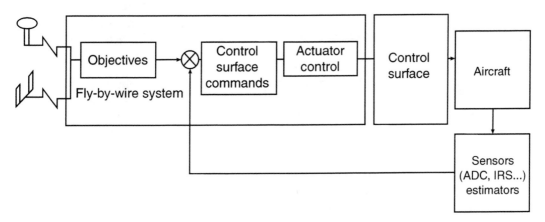

FIGURE 12.2   Flight control laws.

The term fly-by-wire has been adopted to describe the use of electrical rather than mechanical signalling of the pilot's commands to the flying control actuators. One can imagine a basic form of fly-by-wire in which an airplane retained conventional pilot's control columns and wheels, hydraulic actuators (but electrically controlled), and artificial feel as experienced in the 1970s with the Concorde program. The fly-by-wire system would simply provide electrical signals to the control actuators that were directly proportional to the angular displacement of the pilot's controls, without any form of enhancement.

In fact, the design of the A320, A321, A330, and A340 flight control systems takes advantage of the potential of fly-by-wire for the incorporation of control laws that provide extensive stability augmentation and flight envelope limiting [Favre, 1993]. The positioning of the control surfaces is no longer a simple reflection of the pilot's control inputs and conversely, the natural aerodynamic characteristics of the aircraft are not fed back directly to the pilot (see Figure 12.2).

The sidesticks, now part of a modern cockpit design with a large visual access to instrument panels, can be considered as the natural issue of fly-by-wire, since the mechanical transmissions with pulleys, cables, and linkages can be suppressed with their associated backlash and friction.

The induced roll characteristics of the rudder provide sufficient roll maneuverability of design a mechanical back-up on the rudder alone for lateral control. This permitted the retention of the advantages of the sidestick design, now rid of the higher efforts required to drive mechanical linkages to the roll surfaces.

Looking for minimum drag leads us to minimize the negative lift of the horizontal tail plane and consequently diminishes the aircraft longitudinal stability. It was estimated for the Airbus family that no significant gain could be expected with rear center-of-gravity positions beyond a certain limit. This allowed us to design a system with a mechanical back-up requiring no additional artificial stabilization.

These choices were obviously fundamental to establish the now-classical architecture of the Airbus fly-by-wire systems (Figures 12.3 and 12.4), namely a set of five full-authority digital computers controlling the three pitch, yaw, and roll axes and completed by a mechanical back-up on the trimmable horizontal stabilizer and on the rudder. (Two additional computers as part of the auto pilot system are in charge of rudder control in the case of A320 and A321 aircraft.)

Of course, a fly-by-wire system relies on the power systems energizing the actuators to move the control surfaces and on the computer system to transmit the pilot controls. The energy used to pressurize the servo-controls is provided by a set of three hydraulic circuits, one of which is sufficient to control the aircraft. One of the three circuits can be pressurized by a Ram air turbine, which automatically extends in case of an all-engine flame-out.

The electrical power is normally supplied by two segregated networks, each driven by one or two generators, depending on the number of engines. In case of loss of the normal electrical generation, an emergency generator supplies power to a limited number of fly-by-wire computers (among others). These computers can also be powered by the two batteries.

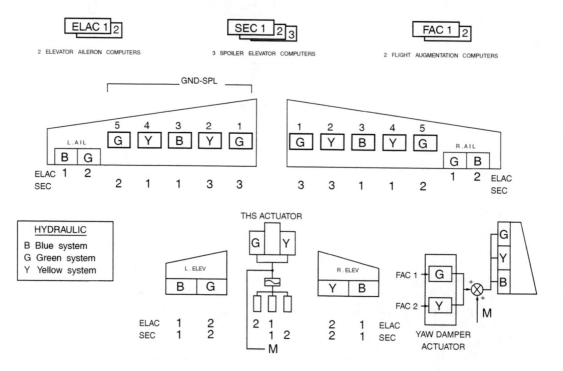

FIGURE 12.3 A320/A321 flight control system architecture.

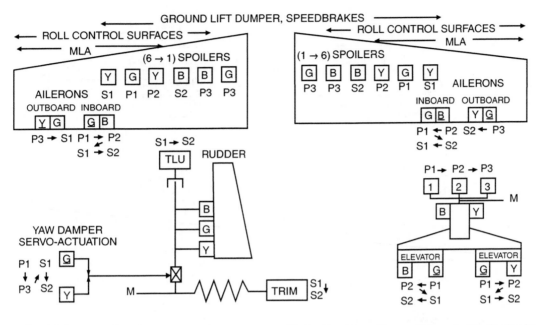

FIGURE 12.4   A330/A340 flight control system architecture.

12.3 Main System Features

12.3.1 Computer Arrangement

12.3.1.1   Redundancy

The five fly-by-wire computers are simultaneously active. They are in charge of control law computation as a function of the pilot inputs as well as individual actuator control, thus avoiding specific actuator control electronics. The system incorporates sufficient redundancies to provide the nominal performance and safety levels with one failed computer, while it is still possible to fly the aircraft safely with one single computer active.

As a control surface runaway may affect the aircraft safety (elevators in particular), each computer is divided into two physically separated channels (Figure 12.5). The first one, the control channel, is permanently monitored by the second one, the monitor channel. In case of disagreement between control and monitor, the computer affected by the failure is passivated, while the computer with the next highest priority takes control. The repartition of computers, servo-controls, hydraulic circuit, and electrical bus bars and priorities between the computers are dictated by the safety analysis including the engine burst analysis.

12.3.1.2 Dissimilarity

Despite the nonrecurring costs induced by dissimilarity, it is fundamental that the five computers all be of different natures to avoid common mode failures. These failures could lead to the total loss of the electrical flight control system.

Consequently, two types of computers may be distinguished:

  2 ELAC (elevator and aileron computers) and 3 SEC (spoiler and elevator computers) on A320/A321 and,

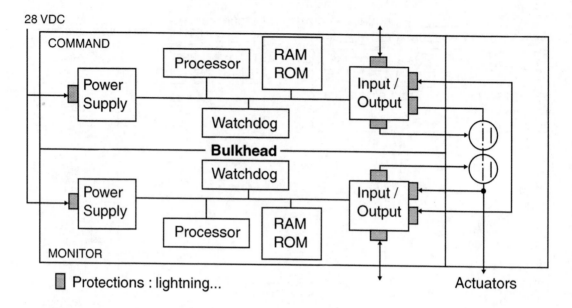

FIGURE 12.5 Command and monitoring computer architecture.

3 FCPC (flight control primary computers) and 2 FCSC (flight control secondary computers) on A330/A340.

Taking the 320 as an example, the ELACs are produced by Thomson-CSF around 68010 microprocessors and the SECs are produced in cooperation by SFENA/Aerospatiale with a hardware based on the 80186 microprocessor. We therefore have two different design and manufacturing teams with different microprocessors (and associated circuits), different computer architectures, and different functional specifications. At the software level, the architecture of the system leads to the use of four software packages (ELAC control channel, ELAC monitor channel, SEC control channel, and SEC monitor channel) when, functionally, one would suffice.

12.3.1.3 Serve-Control Arrangement

Ailerons and elevators can be positioned by two servo-controls in parallel. As it is possible to lose control of one surface, a damping mode was integrated into each servo-control to prevent flutter in this failure case. Generally, one servo-control is active and the other one is damped. In case of loss of electrical control, the elevator actuators are centered by a mechanical feedback to increase the horizontal stabilizer efficiency.

Rudder and horizontal stabilizer controls are designed to receive both mechanical and electrical inputs. One servo-control per spoiler surface is sufficient. The spoiler servo-controls are pressurized in the retracted position in case of loss of electrical control.

12.3.1.4 Flight Control Laws

The general objective of the flight control laws integrated in a fly-by-wire system is to improve the natural flying qualities of the aircraft, in particular in the fields of stability, control, and flight domain protections. In a fly-by-wire system, the computers can easily process the anemometric and inertial information as well as any information describing the aircraft state. Consequently, control laws corresponding to simple control objectives could be designed. The stick inputs are transformed by the computers into pilot control objectives which are compared to the aircraft actual state

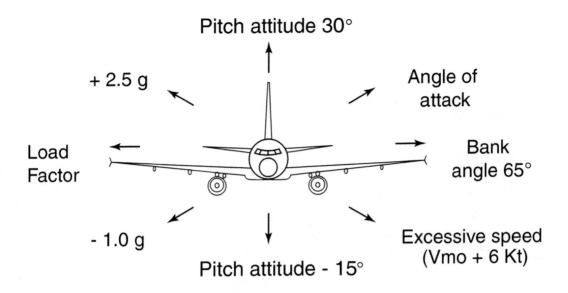

Pitch attitude 30°

+ 2.5 g

Angle of attack

Load Factor

Bank angle 65°

- 1.0 g

Pitch attitude - 15°

Excessive speed (Vmo + 6 Kt)

FIGURE 12.6 A320 flight envelope protections.

measured by the inertial and anemometric sensors. Thus, as far as longitudinal control is concerned, the sidestick position is translated into vertical load factor demands, while lateral control is achieved through roll rate, sideslip, and bank angle objectives.

The stability augmentation provided by the flight control laws improves the aircraft flying qualities and contributes to aircraft safety. As a matter of fact, the aircraft remains stable in case of perturbations such as gusts or engine failure due to a very strong spin stability, unlike conventional aircraft. Aircraft control through objectives significantly reduces the crew workload; the fly-by-wire system acts as the inner loop of an autopilot system, while the pilot represents the outer loop in charge of objective management.

Finally, protections forbidding potentially dangerous excursions out of the normal flight domain can be integrated in the system (Figure 12.6). The main advantage of such protections is to allow the pilot to react rapidly without hesitation, since he knows that this action will not result in a critical situation.

12.3.1.5 Computer Architecture

Each computer can be considered as being two different and independent computers placed side by side (see Figure 12.5). These two (sub)computers have different functions and are placed adjacent to each other to make aircraft maintenance easier. Both command and monitoring channels of the computer are simultaneously active or simultaneously passive, ready to take control.

Each channel includes one or more processors, their associated memories, input/output circuits, a power supply unit, and specific software. When the results of these two channels diverge significantly, the links between the computer and the exterior world are cut by the channel or channels which detected the failure. The system is designed so that the computer outputs are then in a dependable state (signal interrupt via relays). Failure detection is mainly achieved by comparing the difference between the control and monitoring commands with a predetermined threshold. As a result, all consequences of a single computer fault are detected and passivated, which prevents the resulting error from propagating outside of the computer. This detection method is completed by permanently monitoring the program sequencing and the program correct execution.

Flight control computers must be robust. In particular, they must be especially protected against overvoltages and undervoltages, electromagnetic aggressions, and indirect effects of lightning. They are cooled by a ventilation system but must operate correctly even if ventilation is lost.

12.3.1.6 Installation

The electrical installation, in particular the many electrical connections, also comprises a common-point risk. This is avoided by extensive segregation. In normal operation, two electrical generation systems exist without a single common point. The links between computers are limited, the links used for monitoring are not routed with those used for control. The destruction of a part of the aircraft is also taken into account; the computers are placed at three different locations, certain links to the actuators run under the floor, others overhead, and others in the cargo compartment.

12.4 Failure Detection and Reconfiguration

12.4.1 Flight Control Laws

The control laws implemented in the flight control system computers have full authority and must be elaborated as a function of consolidated information provided by at least two independent sources in agreement.

Consequently, the availability of control laws using aircraft feedback (the so-called normal laws) is closely related to the availability of the sensors. The Airbus aircraft fly-by-wire systems use the information of three air data and inertial reference units (ADIRUs), as well as specific accelerometers and rate gyros. Moreover, in the case of the longitudinal normal law, analytical redundancy is used to validate the pitch rate information when provided by a single inertial reference unit. The load factor is estimated through the pitch rate information and compared to the available accelerometric measurements in order to validate the IRS data.

After double or triple failures, when it becomes impossible to compare the data of independent sources, the normal control laws are reconfigured into laws of the direct type where the control surface deflection is proportional to the stick input. To enhance the dissimilarity, the more sophisticated control laws with aircraft feedback (the normal laws) are integrated in one type of computer, while the other type of computer incorporates the direct laws only.

12.4.2 Actuator Control and Monitor

The general idea is to compare the actual surface position to the theoretical surface position computed by the monitoring channel. When needed, the control and monitor channels use dedicated sensors to perform these comparisons. Specific sensors are installed on the servovalve spools to provide an early detection capability for the elevators. Both channels can make the actuator passive. A detected runaway will result in the servo-control deactivation or computer passivation, depending on the failure source.

12.4.3 Comparison and Robustness

Specific variables are permanently compared in the two channels. The difference between the results of the control and monitoring channels are compared with a threshold. This must be confirmed before the computer is disconnected. The confirmation consists of checking that the detected failure lasts for a sufficiently long period of time. The detection parameters (threshold, temporization) must be sufficiently "wide" to avoid unwanted disconnections and sufficiently "tight" so that undetected failures are tolerated by the computer's environment (the aircraft). More precisely, all systems tolerance (most notably sensor inaccuracy, rigging tolerances, computer asynchronism) are taken into account to prevent undue failure detection, and errors which are not detectable (within the signal and timing thresholds) are assessed in respect to their handling quality and structural loads effect.

12.4.4 Latent Failures

Certain failures may remain masked a long time after their occurrence. A typical case is a monitoring channel affected by a failure resulting in a passive state and detected only when the monitored channel itself fails. Tests are conducted periodically so that the probability of the occurrence of an undesirable event remains sufficiently low (i.e., to fulfill [FAR/JAR 25] § 25.1309 quantitative requirement). Typically, a computer runs its self-test and tests its peripherals during the energization of the aircraft, and therefore at least once a day.

12.4.5 Reconfiguration

As soon as the active computer interrupts its operation relative to any function (control law or actuator control), one of the standby computers almost instantly changes to active mode with no or limited jerk on the control surfaces. Typically, duplex computers are designed so that they permanently transmit healthy signals which are interrupted as soon as the "functional" outputs (to an actuator, for example) are lost.

12.4.6 System Safety Assessment

The aircraft safety is demonstrated using qualitative and quantitative assessments. Qualitative assessment is used to deal with design faults, interaction (maintenance, crew) faults, and external environmental hazard. For physical ("hardware") faults, both a qualitative and a quantitative assessments are done. In particular, this quantitative assessment covers the link between failure condition classification (Minor to Catastrophic) and probability target.

12.4.7 Warning and Caution

It is deemed useful for a limited number of failure cases to advise the crew of the situation, and possibly that the crew act as a consequence of the failure. Nevertheless, attention has to be paid to keep the level of crew workload acceptable. The basic rule is to get the crews attention only when an action is necessary to cope with a failure or to cope with a possible future failure. On the other hand, maintenance personnel must get all the failure information.

The warnings and cautions for the pilots are in one of the following three categories:

- Red warning with continuous sound when an immediate action is necessary (for example, to reduce airplane speed).
- Amber caution with a simple sound, such that the pilot be informed although no immediate action is needed (for example, in case of loss of flight envelope protections an airplane speed should not be exceeded).
- Simple caution (no sound), such that no action is needed (for example, a loss of redundancy).

Priority rules among these warnings and cautions are defined to present the most important message first (see also [Traverse, 1994]).

12.5 A340 Particularities

The general design objective relative to the A340 fly-by-wire system was to reproduce the architecture and principles chosen for the A320 as much as possible for the sake of commonality and efficiency, taking account of the A340 particularities (long-range four-engine aircraft).

12.5.1 System

As is now common for each new program, the computer functional density was increased between the A320 and A330/A340 programs: The number of computers was reduced to perform more functions and control an increased number of control surfaces (Figure 12.3).

12.5.2   Control Laws

The general concept of the A320 flight control laws was maintained, adapted to the aircraft characteristics, and used to optimize the aircraft performance, as follows:

- The angle of attack protection was reinforced to better cope with the aerodynamic characteristics of the aircraft.
- The dutch roll damping system was designed to survive against rudder command blocking, thanks to an additional damping term through the ailerons, and to survive against an extremely improbable complete electrical failure thanks to an additional autonomous damper. The outcome of this was that the existing A300 fin could be used on the A330 and A340 aircraft with the associated industrial benefits.
- The take-off performance could be optimized by designing a specific law that controls the aircraft pitch attitude during the rotation.
- The flexibility of fly-by-wire was used to optimize the minimum control speed on the ground (VMCG). In fact, the rudder efficiency was increased on the ground by fully and asymmetrically deploying the inner and outer ailerons on the side of the pedal action as a function of the rudder travel: the inner aileron is commanded downwards, and the outer aileron (complemented by one spoiler) is commanded upwards.
- A first step in the direction of structural mode control through fly-by-wire was made on the A340 program through the so-called "turbulence damping function" destined to improve passenger comfort by damping the structural modes excited by turbulence.

12.6   Design, Development, and Validation Procedures

12.6.1 Fly-by-Wire System Certification Background

An airline can fly an airplane only if this airplane has a type certificate issued by the aviation authorities of the airline country. For a given country, this type certificate is granted when the demonstration has been made and accepted by the appropriate organization (Federal Aviation Administration in the U.S, Joint Aviation Authorities in several European countries, etc.) that the airplane meets the country's aviation rules and consequently a high level of safety. Each country has its own set of regulatory materials although the common core is very large. They are basically composed of two parts: the requirements on one part, and a set of interpretations and acceptable means of compliance in a second part. An example of requirement is "The aeroplane systems must be designed so that the occurrence of any failure condition which would prevent the continued safe flight and landing of the aeroplane is extremely improbable" (in Federal and Joint Aviation Requirements 25.1309, [FAR/JAR 25]). An associated part of the regulation (Advisory Circular from FAA, Advisory Material — Joint from JAA 25.1309) gives the meaning and discuss such terms as "failure condition," and "extremely improbable." In addition, guidance is given on how to demonstrate compliance.

The aviation regulatory materials are evolving to be able to cover new technologies (such as the use of fly-by-wire systems). This is done through special conditions targeting specific issue of a given airplane, and later on by modifying the general regulatory materials. With respect to A320/A330/A340 fly-by-wire airplane, the following innovative topics were addressed for certification (note: some of these topics were also addressing other airplane systems):

- Flight envelope protections
- Side-stick controller
- Static stability
- Interaction of systems and structure
- System safety assessment

- Lightning indirect effect and electromagnetic interference
- Integrity of control signal transmission
- Electrical power
- Software verification and documentation, automatic code generation
- System validation
- Application-specific integrated circuit

It is noteworthy that an integration of regulatory materials is underway which is resulting in a set of four documents:

- A document on system design, verification and validation, configuration management, quality assurance [ARP 4754, 1994]
- A document on software design, verification, configuration management, quality assurance [DO178B, 1992]
- A document on hardware design, verification, configuration management, quality assurance [DO254, 2000]
- A document on the system safety assessment process [ARP 4761, 1994]

12.6.2 The A320 Experience

12.6.2.1 Design

The basic element developed on the occasion of the A320 program is the so-called SAO specification (Spécification Assistée par Ordinateur), the Aerospatiale graphic language defined to clearly specify control laws and system logics. One of the benefits of this method is that each symbol used has a formal definition with strict rules governing its interconnections. The specification is under the control of a configuration management tool and its syntax is partially checked automatically.

12.6.2.2 Software

The software is produced with the essential constraint that it must be verified and validated. Also, it must meet the world's most severe civil aviation standards (level 1 software to [D0178A, 1985]–see also [Barbaste, 1988]). The functional specification acts as the interface between the aircraft manufacturer's world and the software designer's world. The major part of the A320 flight control software specification is a copy of the functional specification. This avoids creating errors when translating the functional specification into the software specification. For this "functional" part of the software, validation is not required as it is covered by the work carried out on the functional specification. The only part of the software specification to be validated concerns the interface between the hardware and the software (task sequencer, management of self-test software inputs/outputs). This part is only slightly modified during aircraft development.

To make software validation easier, the various tasks are sequenced in a predetermined order with periodic scanning of the inputs. Only the clock can generate interrupts used to control task sequencing. This sequencing is deterministic. A part of the task sequencer validation consists in methodically evaluating the margin between the maximum execution time for each task (worst case) and the time allocated to this task. An important task is to check the conformity of the software with its specification. This is performed by means of tests and inspections. The result of each step in the development process is checked against its specification. For example, a code module is tested according to its specification. This test is, first of all, functional (black box), then structural (white box).

Adequate coverage must be obtained for the internal structure and input range. The term "adequate" does not mean that the tests are assumed as being exhaustive. For example, for the structural test of a module, the equivalence classes are defined for each input. The tests must cover the module input range taking these equivalence classes and all module branches (among other things) as a basis. These equivalence classes and a possible additional test effort have the approval of the various parties involved (aircraft manufacturer, equipment manufacturer, airworthiness authorities, designer, and quality control).

The software of the control channel is different from that of the monitoring channel. Likewise, the software of the ELAC computer is different from that of the SEC computer (the same applies to the FCPC and FCSC on the A340). The aim of this is to minimize the risk of a common error which could cause control surface runaway (control/monitoring dissimilarity) or complete shutdown of all computers (ELAC/SEC dissimilarity).

The basic rule to be retained is that the software is made in the best possible way. This has been recognized by several experts in the software field both from industry and from the airworthiness authorities. Dissimilarity is an additional precaution which is not used to reduce the required software quality effort.

12.6.2.3 System Validation

Simulation codes, full-scale simulators and flight tests were extensively used in a complementary way to design, develop, and validate the A320 flight control system (see also [Chatrenet, 1989]), in addition to analysis and peer review.

A "batch" type simulation code called OSMA (Outil de Simulation des Mouvements Avion) was used to initially design the flight control laws and protections, including the nonlinear domains and for general handling quality studies.

A development simulator was then used to test the control laws with a pilot in the loop as soon as possible in the development process. This simulator is fitted with a fixed-base faithful replica of the A320 cockpit and controls and a visual system; it was in service in 1984, as soon as a set of provisional A320 aero data, based on wind tunnel tests, was made available. The development simulator was used to develop and initially tune all flight control laws in a closed-loop cooperation process with flight test pilots.

Three "integration" simulators were put into service in 1986. They include the fixed replica of the A320 cockpit, a visual system for two of them, and actual aircraft equipment including computers, displays, control panels, and warning and maintenance equipment. One simulator can be coupled to the "iron bird" which is a full-scale replica of the hydraulic and electrical supplies and generation, and is fitted with all the actual flight control system components including servojacks. The main purpose of these simulators is to test the operation, integration, and compatibility of all the elements of the system in an environment closely akin to that of an actual aircraft.

Finally, flight testing remains the ultimate and indispensable way of validating a flight control system. Even with the current state of the art in simulation, simulators cannot yet fully take the place of flight testing for handling quality assessment. On this occasion a specific system called SPATIALL (Système Pour Acquisition et Traitement d'Informations Analogiques ARINC et Logiques) was developed to facilitate the flight test. This system allows the flight engineer to:

- Record any computer internal parameter
- Select several preprogrammed configurations to be tested (gains, limits, thresholds, etc.)
- Inject calibrated solicitations to the controls, control surfaces, or any intermediate point.

The integration phase complemented by flight testing can be considered as the final step of the validation side of the now-classical V-shaped development/validation process of the system.

12.6.3 The A340 Experience

12.6.3.1 Design

The definition of the system requires that a certain number of actuators be allocated to each control surface and a power source and computers assigned to each actuator. Such an arrangement implies checking that the system safety objectives are met. A high number of failure combinations must therefore be envisaged. A study has been conducted with the aim of automating this process.

It was seen that a tool which could evaluate a high number of failure cases, allowing the use of capacity functions, would be useful and that the possibility of modeling the static dependencies was not absolutely necessary even though this may sometimes lead to a pessimistic result. This study gave rise to a data processing tool which accepts as input an arrangement of computers, actuators, hydraulic and electrical power sources, and also specific events such as simultaneous shutdown of all engines and, therefore, a high number of power sources. The availability of a control surface depends on the availability of a certain number of these resources. This description was made using a fault tree-type support as input to the tool.

The capacity function used allows the aircraft roll controllability to be defined with regard to the degraded state of the flight control system. This controllability can be approached by a function which measures the roll rate available by a linear function of the roll rate of the available control surfaces. It is then possible to divide the degraded states of the system into success or failure states and thus calculate the probability of failure of the system with regards to the target roll controllability.

The tool automatically creates failure combinations and evaluates the availability of the control surfaces and, therefore, a roll controllability function. It compares the results to the targets. These targets are, on the one hand, the controllability (availability of the pitch control surfaces, available roll rate, etc.) and, on the other hand, the reliability (a controllability target must be met for all failure combinations where probability is greater than a given reliability target). The tool gives the list of failure combinations which do not meet the targets (if any) and gives, for each target controllability, the probability of nonsatisfaction. The tool also takes into account a dispatch with one computer failed.

12.6.3.2   Automatic programming

The use of automatic programming tools is becoming widespread. This tendency appeared on the A320 and is being confirmed on the A340 (in particular, the FCPC is, in part, programmed automatically). Such a tool has SAO sheets as inputs, and uses a library of software packages, one package being allocated to each symbol. The automatic programming tool links together the symbol's packages.

The use of such tools has a positive impact on safety. An automatic tool ensures that a modification to the specification will be coded without stress even if this modification is to be embodied rapidly (situation encountered during the flight test phase for example). Also, automatic programming, through the use of a formal specification language, allows onboard code from one aircraft program to be used on another. Note that the functional specification validation tools (simulators) use an automatic programming tool. This tool has parts in common with the automatic programming tool used to generate codes for the flight control computers. This increases the validation power of the simulations. For dissimilarity reasons, only the FCPC computer is coded automatically (the FCSC being coded manually). The FCPC automatic coding tool has two different code translators, one for the control channel and one for the monitoring channel.

12.6.3.3   System validation

The A320 experience showed the necessity of being capable of detecting errors as early as possible in the design process, to minimize the debugging effort along the development phase. Consequently, it was decided to develop tools that would enable the engineers to actually fly the aircraft in its environment to check that the specification fulfils the performance and safety objectives before the computer code exists.

The basic element of this project is the so-called SAO specification, the Aerospatiale graphic language defined to clearly specify control laws and system logics and developed for A320 program needs. The specification is then automatically coded for engineering simulation purposes in both control law and system areas.

In the control law area, OCAS (Outil de Conception Assistée par Simulation) is a real-time simulation tool that links the SAO definition of the control laws to the already-mentioned aircraft movement simulation (OSMA). Pilot orders are entered through simplified controls including side-stick and engine thrust levels. A simplified PFD (primary flight display) visualizes the outputs of the control law. The

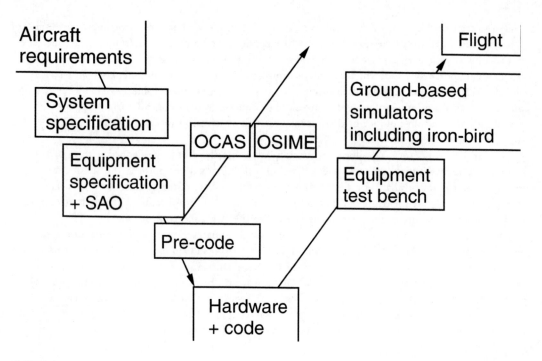

FIGURE 12.7 Validation methodology.

engineer is then in a position to physically judge by himself the quality of the control law that he has just produced, in particular with respect to law transition and nonlinear effects. In the early development phase, this very same simulation was used in the full-scale A340 development simulator with a pilot in the loop.

In the system area, OSIME (Outil de SImulation Multi Equipement) is an expanded time simulation that links the SAO definition of the whole system (control law and system logic) to the complete servo-control modes and to the simulation of aircraft movement (OSMA). The objective was to simulate the whole fly-by-wire system including the three primary computers (FCPC), the two secondary computers (FCSC), and the servo-controls in an aircraft environment.

This tool contributed to the functional definition of the fly-by-wire system, to the system validation, and to the failure analysis. In addition, the behavior of the system at the limit of validity of each parameter, including time delays, could be checked to define robust monitoring algorithms. Non-regression tests have been integrated very early into the design process to check the validity of each new specification standard.

Once validated, both in the control law and system areas using the OCAS and OSIME tools, a new specification standard is considered to be ready to be implemented in the real computers (automatic coding) to be further validated on a test bench, simulator, and on the aircraft (Figure 12.7).

12.7 Future Trends

The fly-by-wire systems developed on the occasion of the A320, A321, A340, and A330 programs now constitute an industrial standard for commercial applications and are well adapted to future military transport aircraft, thanks to the robustness of the system and its reconfiguration capabilities. What are the possible system evolutions? Among others, are the following:

1. New actuator concepts are arising. In particular, systems using both electrical and hydraulic energy within a single actuator were developed and successfully tested on A320 aircraft. This is the so-called electrical back-up hydraulic actuator or EBHA. This actuator can be used to design flight

control systems that survive the total loss of hydraulic power, which is a significant advantage for a military transport aircraft particularly in the case of battle damage.

2. The hardware dissimilarity of the fly-by-wire computer system and the experience with A320 and A340 airline operation will probably ease the suppression of the rudder and trimmable horizontal stabilizer mechanical controls of future aircraft.

3. The integration of new functions, such as structural mode control, may lead to increased dependability requirements, in particular if the loss of these functions is not allowed.

4. Finally, future flight control systems will be influenced by the standardization effort made through the IMA concept (integrated modular avionics) and by the "smart" concept where the electronics destined to control and monitor each actuator are located close to the actuator.

References

ARP 4754, 1994. *System Integration Requirements.* Society of Automotive Engineers (SAE) and European Organization for Civil Aviation electronics (EUROCAE).

ARP 4761, 1994. *Guidelines and tools for Conducting the Safety Assessment Process on Civil Airborne Systems and Equipment.* Society of Automotive Engineers (SAE) and European Organization for Civil Aviation Electronics (EUROCAE).

Barbaste, L. and Desmons, J. P., 1988. Assurance qualité du logiciel et la certification des aéronefs/ Expérience A320. ler séminaire EOQC sur la qualité des logiciels, April 1988, Brussels, pp. 135–146.

Chatrenet, D., 1989. Simulateurs A320 d'Aérospatiale: leur contribution à la conception, au développement et à la certification. *INFAUTOM 89*, Toulouse.

DO178A. 1985. *Software Considerations in Airborne Systems and Equipment Certification.* Issue A. RTCA and European Organization for Civil Aviation Electronics (EUROCAE).

DO178B, 1992. *Software Considerations in Airborne Systems and Equipment Certification.* Issue B. RTCA and European Organization for Civil Aviation Electronics (EUROCAE).

DOXXX, 1995. *Design Assurance Guidance for Complex Electronic Hardware Used in Airborne Systems.* RTCA and by European Organization for Civil Aviation Electronics (EUROCAE).

FAR/JAR 25. *Airworthiness Standards: Transport Category Airplanes.* Part 25 of "Code of Federal Regulations, Title 14, Aeronautics and Space," for the Federal Aviation Administration, and "Airworthiness Joint Aviation Requirements — large aeroplane" for the Joint Aviation Authorities.

Favre, C., 1993. Fly-by-wire for commercial aircraft — the Airbus experience. *Int. J. Control,* special issue on "Aircraft Flight Control".

Traverse, P., Brière, D., and Frayssignes, J. J., 1994. Architecture des commande de vol électriques Airbus, reconfiguration automatique et information équipage. *INFAUTOM 94*, Toulouse.

13

Navigation Systems

Myron Kayton
Kayton Engineering Co.

13.1 Introduction

Navigation is the determination of the position and velocity of the mass center of a moving vehicle. The three components of position and the three components of velocity make up a six-component *state vector* that fully describes the translational motion of the vehicle because the differential equations of motion are of second order. Surveyors now use some of the same sensors as navigators but are achieving higher accuracy as a result of longer periods of observation, a fixed location, and more complex, non-real-time data reduction.

In the usual navigation system, the state vector is derived on-board, displayed to the crew, recorded on-board, and often transmitted to the ground. Navigation information is usually sent to other on-board subsystems; for example, to the waypoint steering, engine control, communication control, and weapon-control computers. Some navigation systems, called *position-location systems*, measure a vehicle's state vector using sensors on the ground or in another vehicle (Section 13.5). These external sensors usually track passive radar returns or a transponder. Position-location systems usually supply information to a dispatch or control center.

The term *guidance* has two meanings, both of which differ from *navigation*:

1. Steering toward a destination of known position from the vehicle's present position, as measured by a navigation system. The steering equations are derived from a plane triangle for nearby destinations and from a spherical triangle for distant destinations.
2. Steering toward a destination without calculating the state vector explicitly. A guided vehicle homes on radio, infrared, or visual emissions. Guidance toward a *moving* target is usually of interest to military tactical missiles in which a steering algorithm assures impact within the maneuver and fuel constraints of the interceptor. Guidance toward a *fixed* target involves beam riding, as in the Instrument Landing System, Section 13.5.

The term *flight control* refers to the deliberate rotation of an aircraft in three-dimensions around its mass center.

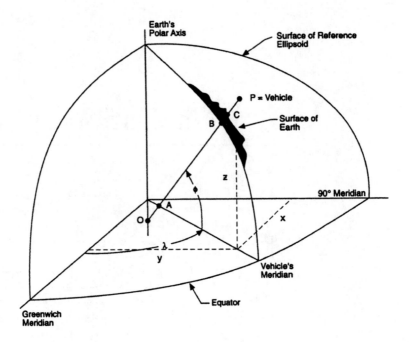

FIGURE 13.1   Latitude–longitude–altitude and x,y,z, coordinate frames. ϕ = geodetic latitude; $\overline{OP}$ is normal to the ellipsoid at B; λ = geodetic longitude; $h = \overline{BP}$ = altitude above the reference ellipsoid = altitude above mean sea level.

13.2   Coordinate Frames

Navigation is with respect to a coordinate frame of the designer's choice. For navigation within a few hundred kilometers (e.g., by helicopter), various map grids exist whose coordinates can be calculated from latitude–longitude (Fig. 13.1). NATO helicopters and land vehicles use a Universal Transverse Mercator grid. Long-range aircraft navigate relative to an earth-bound coordinate frame, the most common of which are latitude–longitude–altitude and rectangular x, y, z (Figure 13.1). Latitude–longitude–altitude coordinates are not suitable in polar regions because longitude is indeterminate. GPS does its calculations in x, y, z and may convert to latitude–longitude–altitude for readout. The most accurate world-wide reference ellipsoid is described in WGS-84, 1991. Spacecraft in orbit around the earth navigate with respect to an earth-centered, inertially nonrotating coordinate frame whose z axis coincides with the polar axis of the earth and whose x axis lies along the equator. Interplanetary spacecraft navigate with respect to a sun-centered, inertially nonrotating coordinate frame whose z axis is perpendicular to the *ecliptic* and whose x axis points to a convenient star (Battin, 1987).

13.3   Categories of Navigation

Navigation systems can be categorized as:

1. *Absolute navigation systems* that measure the state vector without regard to the path traveled by the vehicle in the past. These are of two kinds:
   - Radio systems (Section 13.5). They consist of a network of transmitters (sometimes also receivers) on the ground or in satellites. A vehicle detects the transmissions and computes its position relative to the known positions of the stations in the navigation coordinate frame. The vehicle's velocity is measured from the Doppler shift of the transmissions or from a sequence of position measurements.

- Celestial systems (Section 13.6). They measure the elevation and azimuth of celestial bodies relative to the local level and North. Electronic star sensors are used in special-purpose high-altitude aircraft and in spacecraft. Manual celestial navigation was practiced at sea for millennia (see Bowditch).

2. *Dead-reckoning navigation systems* that derive their state vector from a continuous series of measurements beginning at a known initial position. There are two kinds: those that measure vehicle heading and either speed or acceleration (Section 13.4) and those that measure emissions from continuous-wave radio stations whose signals create ambiguous "**lanes**" (Section 13.5). Dead-reckoning systems must be reinitialized as errors accumulate and if power is lost.

3. *Mapping navigation systems* that observe and recognize images of the ground, profiles of altitude, sequences of turns, or external features (Section 13.7). They compare their observations to a stored database, often on compact disc.

13.4 Dead Reckoning

The simplest dead-reckoning systems measure vehicle heading and speed, resolve speed into the navigation coordinates, then integrate to obtain position (Figure 13.2). The oldest heading sensor is the magnetic compass, a magnetized needle or electrically excited toroidal core (called a *flux* gate), as shown in Figure 13.3. It measures the direction of the earth's magnetic field to an accuracy of 2° at a steady velocity below 60° magnetic latitude. The horizontal component of the magnetic field points toward *magnetic north* or *south*. The angle from true to magnetic north is called *magnetic variation* and is stored in the computers of modern vehicles as a function of position over the region of anticipated travel (Quinn, 1996). *Magnetic deviations* caused by iron and motors in the vehicle can exceed 30° and must be compensated for in the navigation computer, using tables that account for the power-on status of subsystems.

A more complex heading sensor is the *gyrocompass*, consisting of a spinning wheel whose axle is constrained to the horizontal plane (often by a pendulum). The ship's version points north, when properly

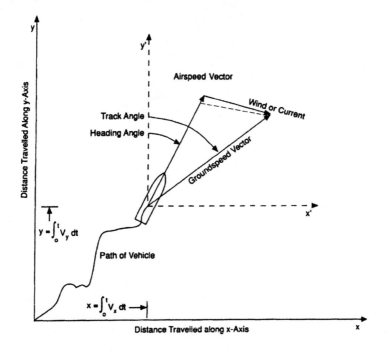

FIGURE 13.2   Geometry of dead reckoning.

FIGURE 13.3   Saturated core ("flux-gate') magnetometer, mounted on a "compass engine" board. The two orthogonal sensing coils (visible) and the drive coil, wound on the toroidal core, measure two components of the magnetic field in the plane of the toroid. (Courtesy of KVH Industries, Inc.)

compensated for vehicle motion, and exhibits errors less than a degree. The aircraft version (more properly called a *directional gyroscope*) holds any preset heading relative to earth and drifts at 50°/hr or more. Inexpensive gyroscopes (some built on silicon chips as vibrating beams with on-chip signal conditioning) are often coupled to magnetic compasses to reduce maneuver-induced errors and long-term drift.

The usual speed-sensor is a *pitot tube* that measures the dynamic pressure of the air stream from which airspeed is derived in an *air-data* computer. To compute ground speed (Figure 13.2) the velocity of the wind must be vectorially added to that of the vehicle. Hence, unpredicted wind or air current will introduce an error into the dead-reckoning computation. Most sensors are insensitive to the component of airspeed normal to their axis (*drift*). A Doppler radar measures the frequency shift in radar returns from the ground or water below the aircraft, from which speed is measured directly. Multibeam Doppler radars can measure all the components of the vehicle's velocity. Doppler radars are widely used on military helicopters.

The most precise dead-reckoning system is an *inertial navigator* in which accelerometers measure the vehicle's acceleration while gyroscopes measure the orientation of the accelerometers. An on-board computer resolves the accelerations into navigation coordinates and integrates them to obtain velocity and position. The gyroscopes and accelerometers are mounted in either of two ways:

1. Fastened directly to the airframe ("strap-down"), whereupon the sensors are exposed to the maximum angular rates and accelerations of the vehicle. This is the usual inertial navigator in 2000 (Figure 13.4). Attitude is computed by a *quaternion* algorithm (Kayton and Fried, 1997, pp. 352–356) that integrates measured angular rates in three dimensions.
2. On a servo-stabilized platform in gimbals that angularly isolate them from rotations of the vehicle. In 2000, gimballed navigators are used only on specialized, high-accuracy military aircraft. They measure attitude directly from the gimbal angles. Their instruments are in a benign angular-environment and held at a constant orientation relative to gravity.

Inertial-quality gyroscopes measure vehicle orientation within 0.1° for steering and pointing. Most accelerometers consist of a gram-sized proof-mass mounted on flexure pivots. The newest accelerometers have proof masses that are etched into silicon chips. Older gyroscopes contained metal wheels rotating in ball bearings or gas bearings. More recent gyroscopes contain rotating, vibrating rings whose frequency of oscillation measures angular rates. The newest gyroscopes are evacuated cavities or optical fibers in which counter-rotating laser beams are compared in phase to measure the sensor's angular velocity relative

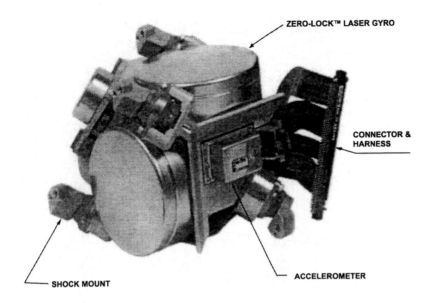

ZERO-LOCK™ LASER GYRO

CONNECTOR & HARNESS

ACCELEROMETER

SHOCK MOUNT

FIGURE 13.4 Inertial reference unit. Two laser gyroscopes (flat discs), an accelerometer, an electrical connector, and four shock mounts are visible. This unit is used in Airbuses and many military aircraft such as the F-18 and Comanche helicopter. (Courtesy of Litton Guidance and Control Systems.)

to *inertial space* about an axis normal to the plane of the beams. Vibrating hemispheres and rotating vibrating tines are the basis of some navigation-quality gyroscopes (drift rates less than 0.1 deg/h).

Fault-tolerant configurations of cleverly oriented redundant gyroscopes and accelerometers (typically four to six) detect and correct sensor failures. Inertial navigators are used aboard airliners, in most military fixed-wing aircraft, in space boosters and entry vehicles, and in manned spacecraft.

13.5 Radio Navigation

Scores of radio navigation aids have been invented and many of them have been widely deployed, as summarized in Table 13.1.

The most precise is the global positioning system (GPS), a network of 24 satellites and a half-dozen ground stations for monitoring and control. A vehicle derives its three-dimensional position and velocity from ranging signals at 1.575 GHz received from four or more satellites (U.S. military users also receive 1.227 GHz). The one-way ranging measurements depend on precise atomic clocks on the spacecraft (one part in 10E13) and on precise clocks on the aircraft (one part in 10E8) that can be calibrated to behave briefly as atomic clocks by taking redundant range measurements from satellites. The former Soviet Union deployed a similar system, called GLONASS. GPS offers better than 30-m ranging errors to civil users and 10-m ranging errors to military users. Simple receivers were available for less than $100 in 2000. GPS provides continuous worldwide navigation for the first time in history. It will make dead reckoning unnecessary on many aircraft and will reduce the cost of most navigation systems. Figure 13.5 is an artist's drawing of a GPS Block 2F spacecraft, scheduled for launch in the year 2005.

Differential GPS (DGPS) employs one or more ground stations at known locations, which receive GPS signals and transmit measured errors on a radio link to nearby ships and aircraft. DGPS improves accuracy (centimeters for fixed observers) and detects faults in GPS satellites. In the late 1990s, the United States was conducting experiments with a nationwide DGPS system of 25 to 50 ground stations. This *Wide Area Augmentation System* (WAAS) could eventually replace VORTAC (below) and Category I ILS. A denser network of DGPS stations and GPS-emulating *pseudolites*, whose stations are located at airports,

TABLE 13.1 Worldwide Radio Navigation Aids for Aviation

System	Frequency Hz	Band	Number of Stations	Number of Aeronautical Users
Loran-C/Chaika	100 kHz	LF	50	120,000
Beacon*	200–1600 kHz	MF	4000	130,000
Instrument Landing System (ILS)*	108–112 MHz / 329–335 MHz	VHF / UHF	1500	150,000
VOR*	108–118 MHz	VHF	1500	180,000
SARSAT/COSPAS	121.5 MHz / 243,406 MHz	VHF / UHF	5 satellites	200,000
JTIDS	960–1213 MHz	L	None	500
DME*	962–1213 MHz	L	1500	90,000
Tacan*	962–1213 MHz	L	850	15,000
Secondary Surveillance Radar (SSR)*	1030, 1090 MHz	L	800	250,000
GPS-GLONASS	1227, 1575 MHz	L	24 + 24 satellites	120,000
Radar Altimeter	4200 MHz	C	None	20,000
MLS*	5031–5091 MHz	C	50	100
Weather/map radar	10 GHz	X	None	10,000
Airborne Doppler radar	13–16 GHz	Ku	None	20,000
SPN-41 carrier-landing monitor	15 GHz	Ku	25	1600
SPN-42/46 carrier-landing radar	33 GHz	Ka	25	1600

*Standardized by International Civil Aviation Organization.

FIGURE 13.5 Global positioning satellite, Block 2F. L-band phased array and S-band control antennas are visible Steerable solar panels power the Spacecraft. Orbit altitude is 10,900 nmi at a 12-hour period. (Courtesy of Rockwell.)

might replace Category II and III ILS and MLS (below). In 2000, the cost, accuracy, and reliability of such a *Local Area Augmentation System* (LAAS) were still being compared with existing landing aids.

Loran is used by general aviation aircraft for en-route navigation and for nonprecision approaches to airports (in which the cloud bottoms are more than 200 feet above the runway; see Table 13.1). The 100-kHz signals are usable within 1000 nautical miles of a "chain" consisting of three or four stations. Chains cover the United States, parts of western Europe, Japan, Saudi Arabia, and a few other areas. The former Soviet Union has a compatible system called Chaika. The vehicle-borne receiver measures the difference in time of arrival of pulses emitted by two stations, thus locating the vehicle on one branch of a hyperbola (Kayton and Fried, 1997, Chapter 4.5.1). Two or more station pairs give a two-dimensional position fix at the intersection of the hyperbolas, whose typical accuracy is 0.25 nmi, limited by propagation uncertainties over the terrain between the transmitting station and the user. The measurement of 100-microsecond time differences is possible with low quality clocks (one part in 10,000) in the vehicles. Loran stations are being upgraded in 2000 so service is assured for the first decade of the third millennium. Loran will be a coarse monitor of GPS and a stand-alone navigation aid whenever GPS is deliberately taken out of service by the U.S. military. These functions may, alternatively, be provided by European or Russian navigation satellites or by private nav-com satellites. These satellite-based supplements to GPS are more accurate than Loran but are subject to the same outages as GPS: solar flares and jammers, for example.

The most widely used aircraft radio aid is VORTAC, whose stations offer three services:

1. Analog bearing measurements at 108 to 118 MHz (called VOR). The vehicle compares the phases of a rotating cardioid pattern and an omnidirectional sinusoid emitted by the ground station.
2. Pulse distance measurements (DME) at 1 GHz. The time delay for an aircraft to interrogate a VORTAC station and receive a reply is measured.
3. Tacan bearing information conveyed in the amplitude modulation of the DME replies from the VORTAC stations.

On short over-ocean flights, the inertially derived state vector drifts 1 to 2 nmi per hour. When an aircraft approaches shore, the VORTAC network updates the inertial state vector and navigation continues to the destination using VORTAC. On long over-ocean flights (e.g., trans-Pacific or polar), GPS can be used alone but is usually used with one or more inertial navigators to protect against failures.

Landing guidance throughout the western world, and increasingly in China, India, and the former Soviet Union, is with the Instrument Landing System (ILS). Transmitters adjacent to the runway create a horizontal guidance signal near 110 MHz and a vertical guidance signal near 330 MHz. Both signals are modulated such that the nulls intersect along a line in space 2.7° above the horizontal, that leads an aircraft from a distance of about 15 nmi to 50 ft above the runway. ILS gives no information about where the aircraft is located along the beam except at two or three vertical *marker beacons*. Most ILS installations are certified to the International Civil Aviation Organization's (ICAO) *Category I*, where the pilot must abort the landing if the runway is not visible at an altitude of 200 ft. Fewer than two hundred ILSs (in 2000) were certified to *Category II*, which allows the aircraft to descend to 100 ft before aborting for lack of visibility. *Category III* allows an aircraft to land at still lower weather ceilings. Category III landing aids are of special interest in western Europe, which has the worst flying weather in the developed world. Category III ILS detects its own failures and switches to a redundant channel within one second to protect aircraft that have failure while flaring-out (within 50 ft of the runway) and can no longer execute a missed approach. Once above the runway, the aircraft's bottom-mounted radar altimeter measures altitude and either the electronics or the pilot guides the flare maneuver. Landing aids are described by Kayton and Fried (1997).

Throughout the western world, civil aircraft use VOR/DME whereas military aircraft use Tacan/DME for en-route navigation. In the 1990s, China and the Commonwealth of Independent States (CIS) were installing ICAO-standard navigation aids (VOR, DME, ILS) at their international airports and along the corridors that lead to them from the borders. Overflying western aircraft navigate inertially or with GPS. Domestic flights within the CIS depended on radar tracking, nondirectional beacons, and an L-band range-angle system called *RSBN*.

It is likely that LAAS will replace or supplement ILS, which has been guaranteed to remain in service at least until the year 2010 (Federal Radionavigation Plan). The U.S. Air Force and NATO may use MLS or LAAS as a portable landing aid for tactical airstrips.

Position-location systems monitor the state vectors of many vehicles and usually display the data in a control room or dispatch center. The aeronautical bureaucracy calls them *Automatic Dependent Surveillance* (ADS) systems. Some vechicles broadcast on-board-derived position. Others derive their state vector from the ranging modulations. Table 13.1 lists *Secondary Surveillance Radars* that receive coded replies from aircraft so they can be identified by human controllers and by collision-avoidance algorithms.

A worldwide network of SARSAT-COSPAS stations monitors signals from satellite-based transponders listening on 121.5, 243, and 406 MHz, the three international distress frequencies. Software at the listening stations calculates the position of Emergency Location Transmitters carried by ships and aircraft to an accuracy of 5 to 15 km at 406 MHz or 15 to 35 km at 21.5 and 243 MHz, based on the observed Doppler-shift history. Thousands of lives have been saved world-wide, from arctic bush-pilots to tropical fishermen.

13.6   Celestial Navigation

Human navigators use sextants to measure the elevation angle of celestial bodies above the visible horizon. The peak elevation angle occurs at local noon or midnight:

$$\text{elev angle (degrees)} = 90 - \text{latitude} + \text{declination}$$

Thus at local noon or midnight, latitude can be calculated by simple arithmetic. When time became measurable at sea, with a chronometer in the 19th century and by radio in the 20th century, off-meridian observations of the elevation of two or more celestial bodies were possible at any known time of night (cloud cover permitting). These fixes were hand-calculated using logarithms, then plotted on charts. In the 1930s, hand-held sextants were built that measured the elevation of celestial bodies from an aircraft using a bubble-level reference instead of the horizon. The accuracy of celestial fixes was 3–20 miles at sea and 5–50 miles in the air, limited by the uncertainty in the horizon and the inability to make precise angular measurements on a pitching, rolling vehicle. Kayton (1990) reviews the history of celestial navigation at sea and in the air.

The first automatic star trackers were built in the late 1950s. They measured the azimuth and elevation of stars relative to a gyroscopically stabilized platform. Approximate position estimates by dead reckoning allowed the telescope to point within a fraction of a degree of the desired star. Thus, a narrow field-of-view was possible, permitting the telescope and photodetector to track stars in the daytime. An on-board computer stored the right ascension and declination of 20 to 100 stars and computed the vehicle's position. Automatic star trackers are used in long-range military aircraft and on space shuttles in conjunction with inertial navigators. Clever design of the optics and of stellar-inertial signal-processing filters achieves accuracies better than 500 ft (Kayton and Fried, 1997).

13.7   Map-Matching Navigation

As computer power grows, map-matching navigation is becoming more important. On aircraft, mapping radars and optical sensors present a visual image of the terrain to the crew. Since the 1960s, automatic map-matchers have been built that correlate the observed image to stored images, choosing the closest match to update the dead-reckoned state vector. Aircraft and cruise missiles measure the vertical profile of the terrain below the vehicle and match it to a stored profile. Matching profiles over distinctive patches of terrain, perhaps hourly, reduces the long-term drift of their inertial navigators. The profile of the terrain is measured by subtracting the readings of a baro-inertial altimeter (calibrated for altitude above sea level) and a radar altimeter (measuring terrain clearance). An on-board computer calculates the autocorrelation function between the measured profile and each of many stored profiles on possible parallel paths of the vehicle. The on-board inertial navigator usually contains a digital filter that corrects the drift of the azimuth gyroscope as a sequence of fixes is obtained. Hence the direction of flight through the stored map is known, saving the considerable computation time that would be needed to correlate for an unknown azimuth of the flight path.

The most complex mapping systems observe their surroundings, by radar or digitized video, and create their own map of the surrounding terrain. Guidance software then steers the vehicle. Optical map-matchers may be used for landings at fields that are not equipped with electronic aids.

13.8 Navigation Software

Navigation software is sometimes embedded in a central processor with other avionic-system software, sometimes confined to one or more navigation computers. The navigation software contains algorithms and data that process the measurements made by each sensor (e.g., inertial or air data). It contains calibration constants, initialization sequences, self-test algorithms, reasonability tests, and alternative algorithms for periods when sensors have failed or are not receiving information. In the simplest systems, a state vector is calculated independently from each sensor while the navigation software calculates the best estimate. Prior to 1970, the best estimate was calculated from a least squares algorithm with constant weighting functions or from a frequency-domain filter with constant coefficients. Now, a *Kalman filter* calculates the best estimate from mathematical models of the dynamics of each sensor (Kayton and Fried, 1997, Chapter 3).

Digital maps, often stored on compact disc, are carried on some aircraft and land vehicles. Terrain is displayed to the crew. Military aircraft add cultural features, hostile radar and missile functions, then superimpose their navigated position. This allows the aircraft to penetrate and escape from enemy territory. Civil operators had not significantly invested in digital databases as of 2000. Algorithms for waypoint steering and for control of the vehicle's attitude are contained in the software of the *flight management* and *flight control* subsystems.

13.9 Design Trade-Offs

The designers of a navigation system conduct trade-offs for each vehicle to determine which navigation systems to use. Tradeoffs consider the following attributes:

- *Cost*, including the construction and maintenance of transmitter stations and the purchase of on-board electronics and software. Users are concerned only with the costs of on-board hardware and software.
- *Accuracy* of position and velocity, which is specified as a circular error probable (CEP, in meters or nautical miles). The maximum allowable CEP is often based on the calculated risk of collision on a typical mission.
- *Autonomy*, the extent to which the vehicle determines its own position and velocity without external aids. Autonomy is important to certain military vehicles and to civil vehicles operating in areas of inadequate radio-navigation coverage. Classes of autonomy are described in Kayton and Fried (1997, p. 10).
- *Time delay* in calculating position and velocity, caused by computational and sensor delays.
- *Geographic coverage*. Radio systems operating below 100 kHz can be received beyond line of sight on earth; those operating above 100 MHz are confined to line of sight. On other planets, new navigation aids—perhaps navigation satellites or ground stations—will be installed,
- *Automation*. The vehicle's operator (on-board crew or ground controller) receives a direct reading of position, velocity, and equipment status, usually without human intervention. The navigator's crew station disappeared in aircraft in the 1970s. Human navigators were becoming scarce, even on ships, in the 1990s, because electronic equipment automatically selects stations, calculates waypoint steering, and accommodates failures.

References

R.H. Battin, *An Introduction to the Mathematics and Methods of Astrodynamics*, Washington: AIAA Press, 1987, 796 pp.

N. Bowditch, *The American Practical Navigator, Washington*, D.C.: U.S. Government Printing Office, 1995, 873 pp.

M. Kayton, *Navigation: Land, Sea, Air, and Space*, New York: IEEE Press, 1990, 461 pp.

M. Kayton and W.R. Fried, *Avionics Navigation Systems*, 2nd ed., New York: Wiley, 1997, 773 pp.

R.A. Minzner, *The U.S. Standard Atmosphere 1976*, NOAA Report 76-1562, NASA SP-390, 1976 or latest edition, 227 pp.

B.W. Parkinson and J.J. Spilker, Eds., *Global Positioning System, Theory and Applications*, American Institute of Aeronautics and Astronautics, 1996, 1300 pp., 2 vols.

J. Quinn, "1995 revision of joint U.S./U.K. geomagnetic field models," *J. Geomagnetism and Geo-Electricity*, 1996.

U.S. Air Force, *NAVSTAR-GPS Interface Control Document*, Annapolis, Md.: ARINC Research, 1991, 115 pp.

U.S. Government, *Federal Radionavigation Plan*, Department of Transportation, 1999, 196 pp., issued biennially.

WGS-84, U.S. Defense Mapping Agency, *World Geodetic System 1984*, Washington, D.C.: 1991.

Y. Zhao, *Vehicle Location and Navigation Systems*, Massachusetts: Artech House, 1997, 345 pp.

Further Information

AIAA Journal of Guidance and Control, bimonthly.

Commercial aeronautical standards produced by International Civil Aviation Organization (ICAO, Montreal), Aeronautical Radio, Inc. (ARINC, Annapolis, Md.), RTCA, Inc., Washington and European Organization for Civil Aviation Equipment (EUROCAE, Paris).

IEEE Transactions on Aerospace and Electronic Systems, quarterly.

Journal of Navigation, Royal Institute of Navigation (UK), quarterly.

Navigation, journal of the U.S. Institute of Navigation, quarterly.

Proceedings of the IEEE Position Location and Navigation Symposium (PLANS), biennially.

14

Navigation and Tracking

James L. Farrell

VIGIL, Inc.

14.1 Introduction

The task of navigation ("Nav") interacts with multiple avionics functions. To clarify the focus here, this chapter will not discuss tight formations, guidance, steering, minimization of fuel/noise/pollution, or managing time of arrival. The accent instead is on determining position and velocity (plus, where applicable, other variables such as acceleration, verticality, heading) with maximum accuracy reachable from whatever combination of sensor outputs are available at any time. Position can be expressed as a vector displacement from a designated point or in terms of latitude/longitude/altitude above mean sea level, above the geoid—or both. Velocity can be expressed in a locally level coordinate frame with various choices for an azimuth reference (e.g., geodetic North, Universal Transverse Mercator [UTM] grid North, runway centerline, wander azimuth with or without Earth sidereal rate torquing). In principle any set of axes could be used — such as an Earth Centered Earth Fixed (ECEF) frame for defining position by a Cartesian vector; or velocity in Cartesian coordinates or in terms of groundspeed, flight path angle, and ground track angle — in either case it is advisable to use accepted conventions.

Realization of near-optimal accuracy with any configuration under changing conditions is now routinely achievable. The method uses a means of dead-reckoning — preferably an Inertial Navigation System (INS) — which can provide essentially continuous position, velocity, and attitude in three dimensions by performing a running accumulation from derivative data. Whenever a full or partial fix is available from a nav sensor, a discrete update is performed on the entire set of variables representing the state of the nav system; the amount of reset for each state variable is determined by a weighting computation based on modern estimation. In this way, "initial" conditions applicable to the dead-reckoning device in effect are reinitialized as the "zero" time is advanced (and thus kept current) with each update. Computer-directed operations easily accommodate conditions that may arise in practice (incomplete fixes, inconsistent data rates, intermittent availability, changing measurement geometry, varying accuracies) while providing complete flexibility for backup with graceful degradation. The approach inherently combines short-term accuracy of the dead-reckoning data with the navaids' long-term accuracy. A commonly cited example of synergy offered by the scheme is a tightly coupled GPS/INS wherein the inertial information provides short-term aiding that vastly improves responsiveness of narrowband code and/or carrier tracking, while GPS information counteracts the long-term accumulation of INS error.

The goal of navigation has progressed far beyond mere determination of geographic location. Efforts to obtain double and triple "mileage" from inertial instruments, by integrating nav with sensor stabilization and flight control, are over a decade old. Older yet are additional tasks such as target designation, precision pointing, tracking, antenna stabilization, imaging sensor stabilization (and therefore transfer alignment). Digital beamforming (DBF) for array antennas (including graceful degradation to recover when some elements fail), needs repetitive data for instantaneous relative position of those elements; on deformable structures this can require multiple low-cost transfer-aligned Inertial Measuring Units (IMUs) and/or the fitting of spatial data to an aeroelastic model. The multiplicity of demands underlines the importance of integrating the operations; the rest of this chapter describes how integration should be done.

14.2   Fundamentals

To accomplish the goals just described, the best available balance is obtained between old and new information — avoiding the both extremes of undue clinging to old data and jumping to conclusions at each latest input. What provides this balance is a modern estimation algorithm that accepts each data fragment as it appears from a nav sensor, immediately weighing it in accordance with its ability to shed light on every variable to be estimated. That ability is determined by accounting for all factors that influence how much or how little the data can reveal about each of those variables: Those factors include

- Instantaneous geometry (e.g., distance along a skewed line carries implications about more than one coordinate direction),
- Timing of each measurement (e.g., distance measurements separated by known time intervals carry implications about velocity as well as position), and
- Data accuracy, compared with the accuracy of estimates existing before measurement.

Only when all these factors are taken into account are accuracy and flexibility as well as versatility maximized. To approach the ramifications gradually, consider a helicopter hovering at constant altitude, which is to be determined on the basis of repeated altimeter observations. After setting the initial *a posteriori* estimate to the first measurement $\hat{Y}$, an *a priori* estimate $\hat{x}_2^{(-)}$ is predicted for the second measurement and that estimate is refined by a second observation,

$$\hat{x}_2^{(-)} = \hat{x}_1^{(+)}; \qquad \hat{x}_2^{(+)} = \hat{x}_2^{(-)} + \frac{1}{2}z_2, \quad z_2 \triangleq \hat{Y}_2 - \hat{x}_2^{(-)} \qquad (14.1)$$

and a third observation,

$$\hat{x}_3^{(-)} = \hat{x}_2^{(+)}; \qquad \hat{x}_3^{(+)} = \hat{x}_3^{(-)} + \frac{1}{3}z_3, \quad z_3 \triangleq \hat{Y}_3 - \hat{x}_3^{(-)} \qquad (14.2)$$

and then a fourth observation,

$$\hat{x}_4^{(-)} = \hat{x}_3^{(+)}; \qquad \hat{x}_4^{(+)} = \hat{x}_4^{(-)} + \frac{1}{4}z_4, \quad z_4 \triangleq \hat{Y}_4 - \hat{x}_4^{(-)} \qquad (14.3)$$

which now clarifies the general expression for the mth observation,

$$\hat{x}_m^{(-)} = \hat{x}_{m-1}^{(+)}; \qquad \hat{x}_m^{(+)} = \hat{x}_m^{(-)} + \frac{1}{m}z_m, \quad z_m \triangleq \hat{Y}_m - \hat{x}_m^{(-)} \qquad (14.4)$$

which can be rewritten as

$$\hat{x}_m^{(+)} = \frac{m-1}{m}\hat{x}_m^{(-)} + \frac{1}{m}\hat{Y}_m, \quad m > 0 \tag{14.5}$$

Substitution of $m = 1$ into this equation produces the previously mentioned condition that the first *a posteriori* estimate is equal to the first measurement; substitution of $m = 2$, combined with that condition, yields a second *a posteriori* estimate equal to the average of the first two measurements. Continuation with $m = 3, 4, \dots$ yields the general result that, after m measurements, estimated altitude is simply the average of all measurements.

This establishes an equivalence between the *recursive* estimation formulation expressed in (14.1)–(14.5) and the *block* estimate that would have resulted from averaging all data together in one step. Since that average is widely known to be optimum when all observations are statistically equally accurate, the recursion shown here must then be optimum under that condition. For measurement errors that are sequentially independent random samples with zero mean and variance R, it is well known that the mean squared estimation error $P_m^{(+)}$ after averaging m measurements is just R/m. That is the variance of the *a posteriori* estimate (just *after* inclusion of the last observation); for the *a priori* estimate the variance $P_m^{(-)}$ is $R/(m-1)$. It is instructive to express the last equation above as a blended sum of old and new data, weighted by factors

$$\frac{R}{P_m^{(-)} + R} \equiv \frac{R/P_m^{(-)}}{1 + R/P_m^{(-)}} = \frac{m-1}{m} \tag{14.6}$$

and

$$\frac{P_m^{(-)}}{P_m^{(-)} + R} = \frac{1}{m} \tag{14.7}$$

respectively; weights depend on variances, giving primary influence to information having lower mean squared error. This concept, signified by the left-hand sides of the last two equations, is extendable to more general conditions than the restrictive (uniform variance) case considered thus far. We are now prepared to address more challenging tasks.

As a first extension, let the sequence of altimeter measurements provide repetitive refinements of estimates for both altitude x_1 and vertical velocity x_2. The general expression for the mth observation now takes a more inclusive form

$$\hat{\mathbf{x}}_m^{(-)} = \mathbf{\Phi}_m\hat{\mathbf{x}}_{m-1}^{(+)}; \quad \hat{\mathbf{x}}_m^{(+)} = \hat{\mathbf{x}}_m^{(-)} + \mathbf{W}_m z_m, \quad z_m \triangleq \hat{Y}_m - \hat{x}_{1,m}^{(-)}, \quad \mathbf{x}_m \triangleq \begin{bmatrix} x_{m,1} \\ x_{m,2} \end{bmatrix} \tag{14.8}$$

The method accommodates estimation of multiple unknowns, wherein the status of a system is expressed in terms of a *state vector* ("*state*") x, in this case a 2×1 vector containing two *state variables* ("*states*"); superscripts and subscripts continue to have the same meaning as in the introductory example, but for these states the conventions $m,1$ and $m,2$ are used for altitude and vertical velocity, respectively, at time t_m. For this dynamic case the *a priori* estimate at time t_m is not simply the previous *a posteriori* estimate; that previous state must be premultiplied by the *transition matrix*,

$$\mathbf{\Phi}_m = \begin{bmatrix} 1 & t_m - t_{m-1} \\ 0 & 1 \end{bmatrix} \tag{14.9}$$

which performs a time extrapolation. Unlike the static situation, elapsed time now matters since imperfectly perceived velocity enlarges altitude uncertainty between observations — and position measurements separated by known time intervals carry implicit velocity information (thus enabling vector estimates to be obtained from scalar data in this case). Weighting applied to each measurement is influenced by three factors:

- A sensitivity matrix $\mathbf{H}_m$ whose (i, j) element is the partial derivative of the ith component of the mth measured data vector to the jth state variable. In this scalar measurement case $\mathbf{H}_m$ is a 1×2 matrix [1 0] for all values of m.
- A covariance matrix $\mathbf{P}_m$ of error in state estimate at time t_m [the ith diagonal element = mean squared error in estimating the ith state variable and, off the diagonal, $P_{ij} = P_{ji} = \sqrt{P_{ii}P_{jj}} \times$ (correlation coefficient between i^{th} and j^{th} state variable uncertainty)].
- A covariance matrix $\mathbf{R}_m$ of measurement errors at time t_m (in this scalar measurement case $\mathbf{R}_m$ is a 1×1 "matrix," i.e., a scalar variance R_m).

Although formation of $\mathbf{H}_m$ and $\mathbf{R}_m$ follows directly from their definitions, $\mathbf{P}_m$ changes with time (e.g., recall the effect of velocity error on position error) and with measurement events (because estimation errors fall when information is added). In this "continuous-discrete" approach, uncertainty is decremented at the discrete measurement events

$$\mathbf{P}_m^{(+)} = \mathbf{P}_m^{(-)} - \mathbf{W}_m \mathbf{H}_m \mathbf{P}_m^{(-)} \tag{14.10}$$

and, between events, dynamic behavior follows a continuous model of the form

$$\dot{\mathbf{P}} = \mathbf{A}\mathbf{P} + \mathbf{P}\mathbf{A}^T + \mathbf{E} \tag{14.11}$$

where $\mathbf{E}$ acts as a forcing function to maintain positive definiteness of $\mathbf{P}$ (thereby providing stability and effectively controlling the remembrance duration—the "data window" denoted herein by T— for the estimator) while $\mathbf{A}$ defines dynamic behavior of the state to be estimated ($\dot{\mathbf{x}} = \mathbf{A}\mathbf{x}$ and $\dot{\mathbf{\Phi}} = \mathbf{A}\mathbf{\Phi}$). In the example at hand,

$$\mathbf{A} = \begin{bmatrix} 0 & 1 \\ 0 & 0 \end{bmatrix}; \qquad \begin{bmatrix} \dot{x}_1 \\ \dot{x}_2 \end{bmatrix} = \mathbf{A} \begin{bmatrix} x_1 \\ x_2 \end{bmatrix} \tag{14.12}$$

Given $\mathbf{H}_m$, $\mathbf{R}_m$, and $\mathbf{P}_m^{(-)}$ the optimal (Kalman) weighting matrix is

$$\mathbf{W}_m = \mathbf{P}_m^{(-)}\mathbf{H}_m^T(\mathbf{H}_m\mathbf{P}_m^{(-)}\mathbf{H}_m^T + R_m)^{-1} \tag{14.13}$$

which for a scalar measurement produces a vector $\mathbf{W}_m$ as the above inversion simplifies to division by a scalar (which becomes the variance R_m added to P_{11} in this example):

$$\mathbf{W}_m = \mathbf{P}_m^{(-)}\mathbf{H}_m^T/(\mathbf{H}_m\mathbf{P}_m^{(-)}\mathbf{H}_m^T + R_m) \tag{14.14}$$

The preceding (hovering helicopter) example is now recognized as a special case of this vertical nav formulation. To progress further, horizontal navigation addresses another matter, i.e., location uncertainty in more than one direction—with measurements affected by more than one of the unknowns (e.g., lines of position [LOPs] skewed off a cardinal direction such as North or East; Figure 14.1). In the classic "compass-and-dividers" approach, dead reckoning would be used to plot a running accumulation of position increments until the advent of a fix from two intersecting straight or curved LOPs. The position

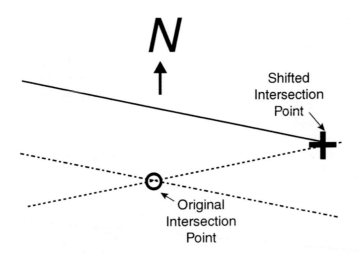

FIGURE 14.1 Nonorthogonal LOPs.

would then be reinitialized at that fixed position, from whence dead reckoning would continue until the next fix. For integrated nav we fundamentally alter that procedure as follows:

- In the reinitialization, data imperfections are taken into account. As already discussed, Kalman weighting (Equations 14.13 and 14.14) is based on accuracy of the dead-reckoning extrapolation as well as the variance of each measurement *and* its sensitivity to each state variable. An optimal balance is provided between old and new information, and the optimality inherently applies to updating of every state variable (e.g., to velocity estimates as well as position, even when only position is observed directly).
- Fixes can be incomplete. In this example, one of the intersecting LOPs may be lost. An optimal update is still provided by the partial fix data, weighted by $\mathbf{W}_m$ of (14.14).

Implications of these two alterations can be exemplified by Figure 14.1, depicting a pair of LOPs representing partial fixes, not necessarily synchronous. Each scalar measurement allows the entire state vector to be optimally updated with weighting from (14.14) in the relation,

$$\hat{\mathbf{x}}_m^{(+)} = \hat{\mathbf{x}}_m^{(-)} + \mathbf{W}_m z_m \tag{14.15}$$

where z_m is the *predicted residual* formed by subtracting the predicted measurement from the value observed at time t_m and acceptance-tested to edit out wild data points;

$$z_m = y_m + \epsilon = Y_m - \hat{Y}_M^{(-)} + \epsilon = \hat{Y}_m - \hat{Y}_m^{(-)}; \quad \hat{Y}_m^{(-)} = Y(\hat{\mathbf{x}}_m^{(-)}) \tag{14.16}$$

The measurement function $Y(\mathbf{x})$ is typically a simple analytical expression (such as that for distance from a designated point, the difference between distances from two specified station locations, GPS pseudo-range or carrier phase difference, etc.). Its partial derivative with respect to each position state is obtained by simple calculus; other components of $\mathbf{H}_m$ (e.g., sensitivity to velocity states) are zero, in which case updating of those states occurs due to dynamics from off-diagonal elements of $\mathbf{P}$ in the product $\mathbf{P}_m^{(-)}\mathbf{H}_m^\mathrm{T}$. R_m — whether constant or varying (e.g., with signal strength) — is treated as a known quantity; if not accurately known, a conservative upper bound can be used. The same is true for the covariance matrix $\mathbf{P}_0$ of error in state estimate at the time of initiating the estimation process — after which the changes are tracked by (14.10) at each measurement event, and by (14.11) between measurements — thus $\mathbf{P}$ is always available for Equations 14.13 and 14.14.

It is crucial to note that the updates are *not* obtained in the form of newly measured coordinates, as they would have been for the classical "compass-and-dividers" approach. Just as old navigators knew how to use partial information, a properly implemented modern estimator would not forfeit that capability. The example just shown provides the best updates, even with no dependably precise way of obtaining a point of intersection when motion occurs between measurements. Furthermore, even with a valid intersection from synchronized observations, the North coordinate of the intersection in Figure 14.1 would be more credible than the East. To show this, consider the consequence of a measurement error effectively raising the dashed LOP to the solid curve as shown; the North coordinate of the new inter-section point "**+**" exceeds that of point "**O**" — but by less than the East-West coordinate shift.

Unequal sensitivity to different directions is automatically taken into account via H_m —just as the dynamics of **P** will automatically provide velocity updating without explicitly forming velocity in terms of sequential changes in measurements — and just as individual values of R_m inherently account for measurement accuracy variations.

Theoretically then, usage of Kalman weighting unburdens the designer while ensuring optimum perfor-mance; no other weighing could provide lower mean squared error in the estimated value of any state. Practically, the fulfillment of this promise is realized by observing additional guidelines, some of which apply "across the board" (e.g., usage of algorithms that preserve numerical stability) while others are application dependent.

Now that a highly versatile foundation has been defined for general usage, the way is prepared for describing some specific applications. The versatility just mentioned is exhibited in the examples that follow. Attention is purposely drawn to the standard process cycle; models of dynamics and measurements are sufficient to define the operation.

14.3   Applications

Various operations will now be described, using the unified form to represent the state dynamics* with repetitive instantaneous refresh via discrete or discretized observations (fixes, whether full or partial). Finite space necessitates some limitations in scope here. First, all updates will be from position-dependent measure-ments (e.g., Doppler can be used as a source of continuous dead-reckoning data but is not considered herein for the discrete fixes). In addition, all nav reference coordinate frames under consideration will be locally level. In addition to the familiar North-East-Down (NED) and East-North-Up (ENU) frames, this includes any Wander Azimuth frame (which deviates from the geographic by only an azimuth rotation about the local vertical). Although these reference frames are not inertial (thus the velocity vector is not exactly the time integral of total acceleration as expressed in a nav frame), known kinematical adjustments will not be described in any depth here. This necessitates restricting the aforementioned data window *T* to intervals no greater than a tenth of the 84-minute Schuler period. The limitation is not very severe when considering the amount of measured data used by most modern avionics applications within a few minutes duration.

Farrell[1] is cited here for expansion of conditions addressed, INS characterization, broader error mod-eling, increased analytical development, the physical basis for that analysis, and myriad practical "*do* s and *don't* s" for applying estimation in each individual operation.

14.3.1   Position and Velocity along a Line

The vertical nav case shown earlier can be extended to the case of time-varying velocity; with accurately (not necessarily exactly) known vertical acceleration Z_V,

$$\begin{bmatrix} \dot{x}_1 \\ \dot{x}_2 \end{bmatrix} = \begin{bmatrix} 0 & 1 \\ 0 & 0 \end{bmatrix} \begin{bmatrix} x_1 \\ x_2 \end{bmatrix} + \begin{bmatrix} 0 \\ Z_V \end{bmatrix} \tag{14.17}$$

---

*A word of explanation is in order: For classical physics the term *dynamics* is reserved for the relation between forces and translational acceleration, or torques and rotational acceleration — while *kinematics* describes the relation between acceleration, velocity, and position. In the estimation field, all continuous time-variation of the state is lumped together in the term *dynamics*.

which allows interpretation in various ways. With a positive upward convention (as in the ENU reference, for example), x_1 can represent altitude above any datum while x_2 is upward velocity; a positive downward convention (NED reference) is also accommodated by simple reinterpretation. In any case, the above equation correctly characterizes actual vertical position and velocity (with true values for Z_V and all xs), and likewise characterizes *estimated* vertical position and velocity (denoted by circumflexes over Z_V and all xs). Therefore, by subtraction, it also characterizes *uncertainty in* vertical position and velocity (i.e., error in the estimate, with each circumflex replaced by a tilde $\sim$). That explains the role of this expression in two separate operations:

- Extrapolation of the *a posteriori* estimate (just after inclusion of the last observation) to the time of the next measurement, to obtain an *a priori* estimate of the state vector — which is used to predict the measurement's value. If a transition matrix can readily be formed (e.g., Equation 14.9 in the example at hand), it is sometimes, but not always, used for that extrapolation.
- Propagation of the covariance matrix from time t_{m-1} to t_m via (14.11) initialized at the *a posteriori* value $\mathbf{P}_{m-1}^{(+)}$ and ending with the *a priori* value $\mathbf{P}_m^{(-)}$. Again, an alternate form using (14.9) is an option.

After these two steps, the cycle at time t_m is completed by forming gain from (14.14), predicted residual from (14.16), update via(14.15), and decrement by (14.10).

The operation just described can be visualized in a generic pictorial representation. Velocity data in a dead-reckoning (DR) accumulation of position increments predicts the value of each measurement. The difference z between the prediction and the observed fix (symbolically shown as a discrete event depicted by the momentary closing of a switch) is weighted by position gain W_{pos} and velocity gain W_{vel} for the update. Corrected values, used for operation thereafter, constitute the basis for further subsequent corrections.

For determination of altitude and vertical velocity, the measurement prediction block in Figure 14.2 is replaced by a direct connection; altimeter fixes are compared vs. the repeatedly reinitialized accumulation of products (time increment) $\times$ (vertical velocity). In a proper implementation of Figure 14.2 time history of *a posteriori* position tracks the truth; RMS position error remains near $\sqrt{P_{11}}$. At the first measurement, arbitrarily large initial uncertainty falls toward sensor tolerance — and promptly begins rising at a rate dictated by $\sqrt{P_{22}}$. A second measurement produces another descent followed by another climb, but now at gentler slope, due to implicit velocity information gained from repeated position observations within a known time interval. With enough fix data the process approaches a quasi-static condition with $\sqrt{P_{11}}$ maintained at levels near RMS sensor error.

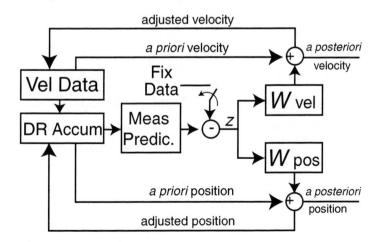

FIGURE 14.2   Position and velocity estimation.

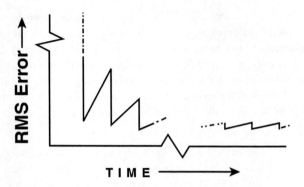

FIGURE 14.3 Time history of accuracy.

Extensive caveats, ramifications, etc. could be raised at this point; some of the more obvious ones will be mentioned here.

- In analogy with the static example, the *left* side of (14.7), with $P^{(-)}_{m11}$ substituted for $P^{(-)}_m$, implies high initial weighting followed by lighter weights as measurements accumulate. If fixes are from sensors with varying tolerance, the entire approach remains applicable; only parameter values change. The effect in Figure 14.3 would be a smaller step decrement, and less reduction in slope, when RMS fix error is larger.

- Vertical velocity can be an accumulation of products, involving instantaneous vertical acceleration which comes from data containing an accelerometer offset driven by a randomly varying error, e.g., having spectral density in conformance to **E** of (14.11). With this offset represented as a third state, another branch would be added to Figure 14.2 and an augmented form of (14.17) could define dynamics in instantaneous altitude, vertical velocity, and vertical acceleration (instead of a constant bias component, extension to exponential correlation is another common alternative);

$$\begin{bmatrix} \dot{x}_1 \\ \dot{x}_2 \\ \dot{x}_3 \end{bmatrix} = \begin{bmatrix} 0 & 1 & 0 \\ 0 & 0 & 1 \\ 0 & 0 & 0 \end{bmatrix} \begin{bmatrix} x_1 \\ x_2 \\ x_3 \end{bmatrix} + \begin{bmatrix} 0 \\ 0 \\ e \end{bmatrix} \tag{14.18}$$

Rather than ramping between fixes, position uncertainty then curves upward faster than the linear rate in Figure 14.3; curvature starts to decrease after the *third* fix. It takes longer to reach quasi-static condition, and closeness of "steady-state" $\sqrt{P_{11}}$ to RMS sensor error depends on measurement scheduling density within a data window.

- (14.12) and Figure 14.2 can also represent position and velocity estimation along another direction, e.g., North or East — or both, as developed in the next section.

14.3.2 Position and Velocity in Three-Dimensional Space

For brevity, only a succinct description is given here. First consider excursion over a meridian with position x_1 expressed as a product [latitude (*Lat*) increment] $\times$ [total radius of curvature (R_M + altitude)],

$$R_M = \frac{a_E(1 - e_E^2)}{[1 - e_E^2 \sin^2(Lat)]^{3/2}}; \quad a_E = 6378137\,\text{m}.; \; e_E^2 = (2 - f)f, f = \frac{1}{298.25722}$$

$$\tag{14.19}$$

so that, for usage of **A** in (14.12), x_2 is associated with North component V_N of velocity. North position fixes could be obtained by observing the altitude angle of Polaris (appropriately corrected for slight deviation off the North Pole). To use the formulation for travel in the East direction, the curvature radius is $(R_p + h)$,

$$R_p = a_E/\sqrt{1 - e_E^2 \sin^2(Lat)}; \quad h = \text{altitude} \tag{14.20}$$

and, while the latitude rate is $V_N/(R_M + h)$, the longitude rate is $V_E \sec(Lat)/(R_p + h)$.

Even for limited distance excursions within a data window, these spheroidal expressions would be used in kinematic state extrapolation, while our short-term ground rule allows a simplified ("flat-Earth" Cartesian) model to be used as the basis for matrix extrapolation in (14.11). The reason lies with *very* different sensitivities in Equations 14.15 and 14.16. The former is significantly less critical; a change δW would modify the *a posteriori* estimate by only the second-order product $z_m \delta W_m$. By way of contrast, small variations in an anticipated measurement (from seemingly minor model approximations) can produce an unduly large deviation in the residual — a small difference of large quantities.

Thus, for accuracy of *additive* state *vector* adjustments (such as *velocity* $\times$ $\Delta time$ products in dynamic propagation), Equations 14.19 and 14.20 properly account for path curvature and for changes in direction of the nav axes as the path progresses. At the poles, the well-known singularity in $\{\sec(Lat)\}$ of course necessitates a modified expression (e.g., Earth-centered vector).

In applying (14.12) to all three directions, a basic decision must be made at the outset. Where practical, it is desirable for axes to remain separated, which produces three uncoupled two-state estimators. An example of this form is radar tracking at long range — long enough so that, within a data window duration, the line-of-sight (LOS) direction remains substantially fixed (i.e., nonrotating). If all three axes are monitored at similar data rates and accuracies, experience has shown that even a fully coupled six-state estimator has position error ellipsoid axes aligned near the sensor's range/azimuth/elevation directions. In that case, little is lost by ignoring coupling across sensor reference axes — hence the triad of uncoupled two-state estimators, all in conformance to (14.12). To resolve vectors along cardinal directions at any time, all that is needed is the direction cosine matrix transformation between nav and sensor axes, which is always available.

When the conditions mentioned above do not hold, the reasoning needs to be revisited. If LOS direction rotates (which happens at short range), or if all three axes are *not* monitored at similar data rates, decoupling may or may not be acceptable; in any case it is suboptimal. If one axis (or a pair of axes) is *un*monitored a fully coupled six-state estimator can dramatically outperform the uncoupled triad. In that case, although (14.12) represents uncoupled dynamics for each axis, coupling comes from multiple changing projections in measurement sensitivity $\mathbf{H}$ as the sensor sight-line direction rotates.

Even the coupled formulation has a simple dynamic model in partitioned form; for a relative position vector $\mathbf{R}$ and velocity $\mathbf{V}$ driven by perturbing acceleration $\mathbf{e}$,

$$\begin{bmatrix} \dot{\mathbf{R}} \\ \dot{\mathbf{V}} \end{bmatrix} = \begin{bmatrix} 0 & I \\ 0 & 0 \end{bmatrix} \begin{bmatrix} \mathbf{R} \\ \mathbf{V} \end{bmatrix} + \begin{bmatrix} 0 \\ e \end{bmatrix} \tag{14.21}$$

where I and 0 are null and identity partitions. The next section extends these concepts.

14.3.3 Position, Velocity, and Acceleration of a Tracked Object

In this chapter it has been repeatedly observed that velocity can be inferred from position-dependent measurements separated by known time intervals. In fact, a velocity *history* can be inferred. As a further generalization of methods just shown, the position reference need not be stationary. In the example now to be described, the origin will move with a supersonic jet carrying a radar and INS. Furthermore, the object whose state is being estimated can be external, with motions that are independent of the platform carrying the sensors that provide all the measurements.

For tracking, first consider the uncoupled case already described, wherein each of three separate estimator channels corresponds to a sensor reference axis direction and each channel has three kinematically related states, representing that directional component of relative (sensor-to-tracked-object) position, relative velocity, and total (not relative) acceleration of the tracked object.* The expression used to propagate state estimates between measurements in a channel conforms to standard kinematics, i.e.,

$$
\begin{bmatrix} \hat{x}_{m1}^{(-)} \\ \hat{x}_{m2}^{(-)} \\ \hat{x}_{m3}^{(-)} \end{bmatrix} = \begin{bmatrix} 1 & t_m - t_{m-1} & \frac{1}{2}(t_m - t_{m-1})^2 \\ 0 & 1 & t_m - t_{m-1} \\ 0 & 0 & 1 \end{bmatrix} \begin{bmatrix} \hat{x}_{m-1,1}^{(+)} \\ \hat{x}_{m-1,2}^{(+)} \\ \hat{x}_{m-1,3}^{(+)} \end{bmatrix} - \begin{bmatrix} \frac{1}{2}(t_m - t_{m-1})q_m \\ q_m \\ 0 \end{bmatrix}
\tag{14.22}
$$

where q_m denotes the component, along the sensor channel direction, of the change in INS velocity during $(t_m - t_{m-1})$. In each channel, $\mathbf{E}$ of (14.11) has only one nonzero value, a spectral density related to data window and measurement error variance σ^2 by

$$
E_{33} = (20\sigma^2/T^5)/g^2 \qquad (g/\text{sec})^2/\text{Hz}
\tag{14.23}
$$

To change this to a fully coupled 9-state formulation, partition the 9×1 state vector into three 3×1 vectors $\mathbf{R}$ for relative position, $\mathbf{V}_r$ for relative velocity, and $\mathbf{Z}_T$ for the tracked object's total acceleration — all expressed in the INS reference coordinate frame. The partitioned state transition matrix is then constructed by replacing each diagonal element in (14.22) by a 3×3 identity matrix $\mathbf{I}_{33}$, each zero by a 3×3 null matrix, and multiplying each above-diagonal element by $\mathbf{I}_{33}$. Consider this transition matrix to propagate covariances as expressed in *sensor reference axes*, so that parameters applicable to a sensing channel are used in (14.23) for each measurement.

Usage of different coordinate frames for states (e.g., geographic in the example used here) and $\mathbf{P}$ (sensor axes) must of course be taken into account in characterizing the estimation process. An orthogonal triad $\mathbf{I}_b\mathbf{J}_b\mathbf{K}_b$ conforms to directions of sensor sight-line $\mathbf{I}_b$, its elevation axis $\mathbf{J}_b$, in the normal plane, and the azimuth axis $\mathbf{I}_b \times \mathbf{J}_b$ normal to both. The instantaneous direction cosine matrix $\mathbf{T}_{b/A}$ will be known (from the sensor pointing control subsystem) at each measurement time. By combination with the transformation $\mathbf{T}_{A/G}$ from geographic to airframe coordinates (obtained from INS data), the transformation from geographic to sensor coordinates is

$$
\mathbf{T}_{b/G} = \mathbf{T}_{b/A}\mathbf{T}_{A/G}
\tag{14.24}
$$

which is used to resolve position states along $\mathbf{I}_b\mathbf{J}_b\mathbf{K}_b$:

$$
\frac{1}{|\mathbf{R}|} \mathbf{T}_{b/G}\mathbf{R} = \begin{bmatrix} 1 \\ p_A \\ -p_E \end{bmatrix}
\tag{14.25}
$$

where p_A and p_E — small fractions of a radian — are departures above and to the right, respectively, of the *a priori* estimated position from the sensor sight-line (which due to imperfect control does not look exactly where the tracked object is anticipated at t_m).

---

*Usage of relative acceleration states would have sacrificed detailed knowledge of INS velocity history, characterizing ownship acceleration instead with the random model used for the tracked object. To avoid that unnecessary performance degradation the dynamic model used here, in contrast to (14.18), has a forcing function with nonzero mean.

For application of (14.16), p_A and p_E are recognized in the role of *a priori* estimated measurements — adjusting the "dot-off-the-crosshairs" azimuth ("AZ") and elevation ("EL") observations so that a full three-dimensional fix (range, AZ, EL) in this operation would be

$$
\begin{bmatrix} y_R \\ y_{AZ} \\ y_{EL} \end{bmatrix} = \begin{bmatrix} 1 & 0 & 0 \\ 0 & \dfrac{1}{|\mathbf{R}|} & 0 \\ 0 & 0 & -\dfrac{1}{|\mathbf{R}|} \end{bmatrix} \mathbf{T}_{b/G}\mathbf{R} - \begin{bmatrix} 0 \\ p_A \\ -p_E \end{bmatrix}
\tag{14.26}
$$

Since $\mathbf{R}$ contains the first three states, its matrix coefficient in (14.26) provides the three nonzero elements of $\mathbf{H}$; e.g., for scalar position observables, these are comprised of

- The top row of $\mathbf{T}_{b/G}$ for range measurements,
- The middle row of $\mathbf{T}_{b/G}$ divided by scalar range for azimuth measurements,
- The bottom row of $\mathbf{T}_{b/G}$ divided by scalar range $\times\ (-1)$, for elevation measurements.

By treating scalar range coefficients as well as the direction cosines as known quantities in this approach, *both* the dynamics *and* the observables are essentially linear in the state. This has produced success in nearly all applications within the experience of this writer. The sole need for extension arose when distances and accuracies of range data were extreme (the cosine of the angle between the sensor sight-line and range vector could not be set at unity). Other than that case, the top row of $\mathbf{T}_{b/G}$ suffices for relative position states, and also for relative velocity states when credible Doppler measurements are available.

A more thorough discourse would include a host of additional material, including radar and optical sensing considerations, sensor stabilization — with its imperfections isolated from tracking, error budgets, kinematical correction for gradual rotation of the acceleration vector, extension to multiple track files, sensor fusion, myriad disadvantages of alternative tracking estimator formulations, etc. The ramifications are too vast for inclusion here.

14.3.4 Position, Velocity, and Attitude in Three-Dimensional Space (INS Aiding)

In the preceding section, involving determination of velocity history from position measurement sequences, dynamic velocity variations were expressed in terms of an acceleration vector. For nav (as opposed to tracking of an external object) with high dynamics, the history of velocity is often tied to the angular orientation of an INS. In straight-and-level Northbound flight, for example, an unknown tilt ψ_N about the North axis would produce a fictitious ramping in the indicated East velocity V_E; in the short-term this effect will be indistinguishable from a bias n_{aE} in the indicated lateral component (here, East) of accelerometer output. More generally, velocity *vector* error will have a rate

$$
\dot{v} = \psi \times \mathbf{A} + \mathbf{n_a} = -\mathbf{A} \times \psi + \mathbf{n_a}
\tag{14.27}
$$

where bold symbols (v, $\mathbf{n}$) contain the geographic components equal to corresponding scalars denoted by italicized quantities (v,n) and $\mathbf{A}$ represents the vector, also expressed in geographic coordinates, of the total nongravitational acceleration experienced by the IMU. Combined with the intrinsic kinematical relation between v and a position vector error r, in a naw frame rotating at $\tilde{\omega}$ rad/sec, the 9-state dynamics with a time-invariant misorientation ψ can be expressed via 3×3 matrix partitions,

$$\begin{bmatrix} \dot{r} \\ \dot{v} \\ \dot{\psi} \end{bmatrix} = \begin{bmatrix} -\tilde{\omega}\times & \mathbf{I} & \mathbf{0} \\ \mathbf{0} & \mathbf{0} & (-\mathbf{A}\times) \\ \mathbf{0} & \mathbf{0} & \mathbf{0} \end{bmatrix} \begin{bmatrix} r \\ v \\ \psi \end{bmatrix} + \begin{bmatrix} \mathbf{0} \\ \mathbf{n_a} \\ \mathbf{e} \end{bmatrix} \qquad (14.28)$$

which lends itself to numerous straighforward interpretations. For brevity, these will simply be listed here:

- For strapdown systems, it is appropriate to replace vectors such as $\mathbf{A}$ and $\mathbf{n_a}$ by vectors initially expressed in vehicle coordinates and transformed into geographic coordinates, so that parameters and coefficients will appear in the form received.

- Although both $\mathbf{n_a}$ and $\mathbf{e}$ appear as forcing functions, the latter drives the highest-order state and thus exercises dominant control over the data window.

- If $\mathbf{n_a}$ and $\mathbf{e}$ contain both bias and time-varying random (noisy) components, (14.28) is easily reexpressible in augmented form, wherein the biases can be estimated along with the corrections for estimated position, velocity, and orientation. Especially for accelerometer bias elements, however, observability is often limited; therefore the usage of augmented formulations should be adopted judiciously. In fact, the number of states should in many cases be *reduced*, as in the next two examples:

  - In the absence of appreciable sustained horizontal acceleration, the azimuth element of misorientation is significantly less observable than the tilt components. In some operations this suggests replacing (14.28) with an eight-state version obtained by omitting the ninth state and deleting the last row and column of the matrix.

  - When the last *three* states are omitted — while the last three rows and columns of the matrix are deleted — the result is the fully coupled three-dimensional position and velocity estimator (14.21).

The options just described can be regarded as different modes of the standard cyclic process already described, with operations defined by dynamics and measurement models. Any discrete observation could be used with (14.28) or an alternate form just named, constituting a mode subject to restrictions that were adopted here for brevity (position-dependent observables only, with distances much smaller than Earth radius).

At this point, expressions could be given for measurements as functions of the states and their sensitivities to those state variables: (14.26) provides this for range and angle data; it is now appropriate to discuss GPS information in an integrated nav context.

14.3.5   Individual GPS Measurements as Observables

The explosive growth of navigation applications within the past decade has been largely attributed to GPS. Never before has there been nav data source of such high accuracy, reachable from any location on the Earth's surface at any time. Elsewhere in this book the reader has been shown how GPS data can be used to

- Solve for 3D position and user clock offset with pseudo-range observations received simultaneously from each of four space vehicles (SVs),

- Use local differential GPS corrections that combine, for each individual SV, compensation for propagation delays plus SV clock and ephemeris error,

- Compensate via wide-area augmentation which, though not as accurate as local, is valid for much greater separation distances between the user and reference station,

- Use differencing techniques with multiple SVs as well as multiple receivers to counteract the effects of the errors mentioned and of user clock offsets,

- Apply these methods to carrier phase as well as to pseudo-range so that, once the cycle count ambiguities are resolved, results can be accurate to within a fraction of the L-band wavelength.

Immediately we make a definite departure from custom here; each scalar GPS observable will call for direct application of (14.15). To emphasize this, results will first be described for instances of sparse measurement scheduling. Initial runs were made with real SV data, taken before the first activation of selective availability (SA) degradations, collected from a receiver at a known stationary location but spanning intervals of several hours. Even with that duration consumed for the minimum required measurements, accuracies of 1 or 2 m were obtained: not surprising for GPS with good geometery and no SA.

The results just mentioned, while not considered remarkable, affirm the point that full fixes are not at all necessary with GPS. They also open the door for drawing dependable conclusions when the same algorithms are driven by simulated data containing errors from random number generators. A high-speed aircraft simulation was run under various conditions, always with no more than one pseudo-range observation every 6 sec. (and furthermore with some gaps even in that slow data rate). Since the results are again unremarkable, only a brief synopsis suffices here:

- Estimates converged as soon as the measurements accumulated were sufficient to produce a nav solution (e.g., two asynchronous measurements from each of three noncoplanar SVs for a vehicle moving in three dimensions, with known clock state, or four SVs with all states initially unknown).
- Initial errors tended to wash out; accuracies of the estimates just mentioned were determined by measurement error levels amplified by geometry.
- Velocity errors tended toward levels proportional to a ratio (RMS measurement error)/(T), where T here represents time elapsed since the first measurement on a course leg, or the data window — whichever is smaller. The former definition of the denominator produced a transient at the onset and when speed or direction changed.
- Doppler data reduced the transient, and INS velocity aiding minimized or removed it.
- Extreme initial errors interfered with these behavioral patterns somewhat — readily traceable to usage of imprecise direction cosines — but the effects could be countered by reinitialization of estimates with *a posteriori* values and recycling the measurements.

These results mirror familiar real-world experience (including actual measurement processing by this author); they are used here to emphasize adequacy of partial fixes at low rates, which many operational systems fail to exploit.

Although the approach just described is well known (i.e., in complete conformance to the usual Kalman filter updating cycle) and the performance unsurprising, the last comment is significant. There are numerous applications wherein SV sight-lines are often obscured by terrain, foliage, buildings, or structure of the vehicle carrying the GPS receiver. In addition, there can be SV outages (whether from planned maintenance or unexpected failures), intermittently strong interference or weak signals, unfavorable multipath geometry in certain SV sight-line directions, etc., and these problems can arise in critical situations.

At the time of this writing there are widespread opportunities, prompted by genuine need, to replace loose (cascaded) configurations by tightly coupled (integrated) configurations. Accentuating the benefit is the bilateral nature of tight integration. As the tracking loops (code loop and, where activated, carrier phase track) contribute to the estimator, the updated state enhances ability to maintain stable loop operation. For a properly integrated GPS/INS, this enhancement occurs even with narrow bandwidth in the presence of rapid change. Loop response need not follow the dynamics, only the error in perceived dynamics.

It is also noted that the results just described are achievable under various conditions and can be scaled over a wide accuracy range. Sensitivity **H** of an individual SV observation contains the SV-to-receiver unit vector; when satellite observations are differenced, the sensitivity **H** contains the difference of two SV-to-receiver unit vectors. Measurements may be pseudo-ranges with SA ($\sigma > 30$ m), pseudo-ranges without SA ($\sigma < 10$ m, typically), differentially corrected pseudo-ranges ($\sigma = 1$ or 2 m), or carrier phase (with ambiguities resolved, σ at 1 cm or less). In all cases, attainable performance is determined by σ and the span of **H** for each course leg. An analogous situation holds for other navaids when used with the standard updating procedure presented herein.

14.4   Conclusion

Principles of nav system integration have been described, within the limits of space. Inevitably, some restrictions in scope were adopted; those wishing to pursue the topic in greater depth may consult the sources which follow.

References

1. Farrell, J. L., *Integrated Aircraft Navigation,* Academic Press, New York, 1976. (Now available in paperback only; 800/628-0885 or 410/647-6165.)
2. Bierman, *Factorized Methods for Discrete Sequential Estimation,* Academic Press, New York, 1977.
3. Institute of Navigation Redbooks (reprints of selected GPS papers); Alexandria, VA, 703/683-7101.
4. Brown and Hwang, *Introduction to Random Signals and Applied Kalman Filtering,* John Wiley & Sons, New York, 1996.
5. Kayton and Fried (Eds.), *Avionics Navigation Systems,* John Wiley & Sons, New York, 1997.

Further Information

1. Journal and Conference Proceedings from the Institute of Navigation, Alexandria, VA.
2. Tutorials from Conferences sponsored by the Institute of Navigation (Alexandria, VA) and the Position Location And Navigation Symposium (PLANS) of the Institute of Electrical and Elecronic Engineers (IEEE).
3. Transactions of the Institute of Electrical and Electronic Engineers (IEEE) Aerospace and Electronic Systems Society (AES).

15

Flight Management Systems

Randy Walter

Smiths Industries

15.1 Introduction

The flight management system typically consists of two units, a computer unit and a control display unit. The computer unit can be a standalone unit providing both the computing platform and various interfaces to other avionics or it can be integrated as a function on a hardware platform such as an Integrated Modular Avionics cabinet (IMA). The Control Display Unit (CDU or MCDU) provides the primary human/machine interface for data entry and information display. Since hardware and interface implementations of flight management systems can vary substantially, this discussion will focus on the functional aspects of the flight management system.

The flight management system provides the primary navigation, flight planning, and optimized route determination and en route guidance for the aircraft and is typically comprised of the following interrelated functions: navigation, flight planning, trajectory prediction, performance computations, and guidance.

To accomplish these functions the flight management system must interface with several other avionics systems. As mentioned above, the implementations of these interfaces can vary widely depending upon the vintage of equipment on the aircraft but generally will fall into the following generic categories.

- Navigation sensors and radios
  - Inertial/attitude reference systems
  - Navigation radios
  - Air data systems
- Displays
  - Primary flight and navigation
  - Multifunction
  - Engine
- Flight control system
- Engine and fuel system
- Data link system
- Surveillance systems

Figure 15.1 depicts a typical interface block diagram.

Today, flight management systems can vary significantly in levels of capability because of the various aviation markets they are intended to serve. These range from simple point to point lateral navigators to the more sophisticated multisensor navigation, optimized four-dimensional flight planning/guidance systems. The flight management system in its simplest form will slowly diminish as reduced separation airspace standards place more demands on the aircraft's ability to manage its trajectory more accurately, even though lateral-only navigators will continue to have a place in recreational general aviation.

With its current role in the aircraft, the flight management system becomes a primary player in the current and future CNS/ATM environment. Navigation within RNP airspace, data-linked clearances and weather, aircraft trajectory-based traffic management, time navigation for aircraft flow control, and seamless low-visibility approach guidance all are enabled through advanced flight management functionality.

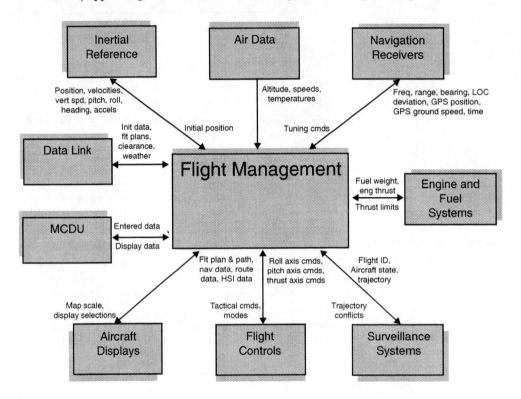

FIGURE 15.1   Typical interface block diagram.

15.2 Fundamentals

At the center of the FMS functionality is the flight plan construction and subsequent construction of the four-dimensional aircraft trajectory defined by the specified flight plan legs and constraints and the aircraft performance. Flight plan and trajectory prediction work together to produce the four-dimensional trajectory and consolidate all the relevant trajectory information into a flight plan/profile buffer. The navigation function provides the dynamic current aircraft state to the other functions. The vertical, lateral steering, and performance advisory functions use the current aircraft state from navigation and the information in the flight plan/profile buffer to provide guidance, reference, and advisory information relative to the defined trajectory and aircraft state.

- The navigation function — responsible for determining the best estimate of the current state of the aircraft.
- The flight planning function — allows the crew to establish a specific routing for the aircraft.

- The trajectory prediction function — responsible for computing the predicted aircraft profile along the entire specified routing.
- The performance function — provides the crew with aircraft unique performance information such as takeoff speeds, altitude capability, and profile optimization advisories.
- The guidance functions — responsible for producing commands to guide the aircraft along both the lateral and vertical computed profiles.

Depending on the particular implementation, the ancillary I/O, BITE, and control display functions may be included as well. Since the ancillary functions can vary significantly, this discussion will focus on the core flight management functions.

There are typically two loadable databases that support the core flight management functions. These are the navigation database which must be updated on a monthly cycle and the performance database that only gets updated if there's been a change in the aircraft performance characteristics (i.e., engine variants or structural variants affecting the drag of the aircraft).

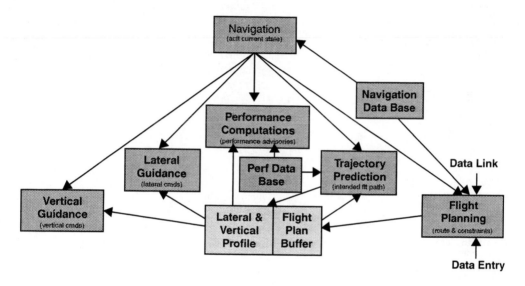

FIGURE 15.2 Flight management functional block diagram.

The navigation database contains published data relating to airports, navaids, named waypoints, airways and terminal area procedures along with RNP values specified for the associated airspace. The purpose of the navigation data base is twofold. It provides the navigation function location, frequency, elevation, and class information for the various ground-based radio navigation systems. This information is necessary to select, auto-tune, and process the data from the navigation radios (distance, bearing, or path deviation) into an aircraft position. It also provides the flight plan function with airport, airport-specific arrival, departure, and approach procedures (predefined strings of terminal area waypoints), airways (predefined enroute waypoint strings), and named waypoint information that allows for rapid route construction. A detailed description of the actual data content and format can be found in ARINC 424.

The performance database contains aircraft/engine model data consisting of drag, thrust, fuel flow, speed/altitude envelope, thrust limits, and a variety of optimized and tactical speed schedules that are unique to the aircraft. Figure 15.2 shows the interrelationships between the core functions and the databases.

15.2.1 Navigation

The navigation function within the FMS computes the aircraft current state (generally WGS-84 geodetic coordinates) based on a statistical blending of multisensor position and velocity data. The aircraft current state data usually consists of:

- Three-dimensional position (latitude, longitude, altitude)
- Velocity vector
- Altitude rate
- Track angle, heading, and drift angle
- Wind vector
- Estimated Position Uncertainty (EPU)
- Time

The navigation function is designed to operate with various combinations of autonomous sensors and navigation receivers. The position update information from the navigation receivers is used to calibrate the position and velocity data from the autonomous sensors, in effect providing an error model for the autonomous sensors. This error model allows for navigation coasting based on the autonomous sensors while maintaining a very slow growth in the EPU. If the updating from navigation aids such as DME, VOR, or GPS is temporarily interrupted, navigation accuracy is reasonably maintained, resulting in seamless operations. This capability becomes very important for operational uses such as RNAV approach guidance where the coasting capability allows completion of the approach even if a primary updating source such as GPS is lost once the approach is commenced. A typical navigation sensor complement consist of:

- Autonomous sensors
  - Inertial reference
  - Air data
- Navigation receivers
  - DME receivers
  - VOR/LOC receivers
  - GPS receivers

The use of several navigation data sources also allows cross-checks of raw navigation data to be performed to ensure the integrity of the FMS position solution.

15.2.1.1 Navigation Performance

The navigation function, to be RNP airspace compliant per DO-236, must compute an Estimated Position Uncertainty (EPU) that represents the 95% accuracy performance of the navigation solution. The EPU is computed based on the error characteristics of the particular sensors being used and the variance of the individual sensors position with respect to other sensors. The RNP for the airspace is defined as the minimum navigation performance required for operation within that airspace. It is specified by default values based on the flight phase retrieved from the navigation data base for selected flight legs or crew-entered in response to ATC-determined airspace usage. A warning is issued to the crew if the EPU grows larger than the RNP required for operation within the airspace. The table below shows the current default RNP values for the various airspace categories.

Airspace Definition	Default RNP
Oceanic — no VHF navaids within 200 nm	12.0 nm
Enroute — above 15,000 ft	2.0 nm
Terminal	1.0 nm
Approach	0.5 nm

A pictorial depiction of the EPU computation is shown below for a VOR/VOR position solution. A similar representation could be drawn for other sensors.

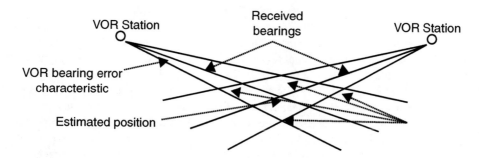

As can be seen from the diagram, the estimated position uncertainty (EPU) is dependent on the error characteristics of the particular navigation system being used as well as the geometric positioning of the navaids themselves. Other navigation sensors such as an inertial reference system have error characteristics that are time-dependent. More information pertaining to EPU and various navigation navaid system error characteristics can be found in RTCA DO-236.

15.2.1.2 Navigation Receiver Management

The various navigation receivers require different levels of FMS management to obtain a position update solution.

GPS — The GPS receiver is self-managing in that the FMS receives position, velocity, and time information without any particular FMS commands or processing. Typically, the FMS will provide an initial position interface to reduce the satellite acquire time of the receiver and some FMSs may provide an estimated time of arrival associated with a final approach fix waypoint to support the Predictive Receiver Autonomus Integrity Monitor (PRAIM) function in the GPS. More information on the GPS interface and function can be found in ARINC 743.

VHF navaids (DME/VOR/ILS) — The DME/VOR/ILS receivers must be tuned to an appropriate station to receive data. The crew may manually tune these receivers but the FMS navigation function will also auto-tune the receivers by selecting an appropriate set of stations from its stored navigation database and sending tuning commands to the receiver(s). The selection criteria for which stations to tune are

- Navaids specified within a selected flight plan procedure, while the procedure is active.
- The closest DME navaids to the current aircraft position of the proper altitude class that are within range (typically 200 nm).
- Collocated DME/VORs within reasonable range (typically 25 nm).
- ILS facilities if an ILS or LOC approach has been selected into the flight plan and is active.

Since DMEs receive ranging data and VORs receive bearing data from the fixed station location, the stations must be paired to determine a position solution as shown below:

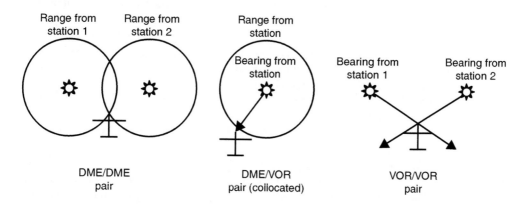

The pairing of navaids to obtain a position fix is based on the best geometry to minimize the position uncertainty (minimize the portion of EPU caused by geometric dilution of precision, GDOP). As can be seen from the figure above, the FMS navigation must process range data from DMEs and bearing data from VORs to compute an estimated aircraft position. Further, since the DME receives slant range data from ground station to aircraft, the FMS must first correct the slant range data for station elevation and aircraft altitude to compute the actual ground-projected range used to determine position. The station position, elevation, declination, and class are all stored as part of the FMS navigation data base. There are variations in the station-tuning capabilities of DME receivers. A standard DME can only accept one tuning command at a time, an agility-capable DME can accept two tuning commands at a time, and a scanning DME can accept up to five tuning commands at a time. VOR receivers can only accept one tuning command at a time.

An ILS or LOC receiver works somewhat differently in that it receives cross-track deviation information referenced to a known path into a ground station position. These facilities are utilized as landing aids and therefore are located near runways. The FMS navigation function processes the cross-track information to update the cross-track component of its estimated position. More information about DME/VOR/ILS can be found in ARINC 709, 711, and 710, respectively.

15.2.2   Flight Planning

The basis of the FMC flight profile is the route that the aircraft is to fly from the departure airport to the destination airport. The FMS flight planning function provides for the assembly, modification, and activation of this route data known as a flight plan. Route data are typically extracted from the FMC navigation data base and typically consists of a departure airport and runway, a standard instrument departure (SID) procedure, enroute waypoints and airways, a standard arrival (STAR) procedure, and an approach procedure with a specific destination runway. Often the destination arrival (or approach transition) and approach procedure are not selected until the destination terminal area control is contacted. Once the routing, along with any route constraints and performance selections, are established by the crew, the flight plan is assembled into a "buffer" that is utilized predominantly by the trajectory predictions in computing the lateral and vertical profile the aircraft is intended to fly from the departure airport to the destination airport.

The selection of flight planning data is done by the crew through menu selections either on the MCDU or navigation display or by data link from the airline's operational control. Facilities are also provided for the crew to define additional navigation/route data by means of a supplemental navigation data base. Some of the methods for the crew to create new fixes (waypoints) are listed below.

PBD Waypoints — Specified as bearing/distance off existing named waypoints, navaids, or airports.

PB/PB Waypoints — Specified as the intersections of bearings from two defined waypoints.

ATO Waypoints — Specified by an along-track offset (ATO) from an existing flight plan waypoint. The waypoint that is created is located at the distance entered and along the current flight plan path from the waypoint used as the fix. A positive distance results in a waypoint after the fix point in the flight plan while a negative distance results in a waypoint before the fix point.

Lat/Lon Waypoints — Specified by entering in the latitude/longitude coordinates of the desired waypoint.

Lat/Lon Crossing Waypoints — Created by specifying a latitude or longitude. A waypoint will be created where the active flight plan crosses that latitude or longitude. Latitude or longitude increments can also be specified, in which case several waypoints are created where the flight plan crosses the specified increments of latitude or longitude.

Intersection of Airways — Created by specifying two airways. A waypoint will be created at the first point where the airways cross.

Fix Waypoints — Created by specifying a "fix" reference. Reference information includes creation of abeam waypoints and creation of waypoints where the intersections of a specified radial or distance from the "fix" intersects the current flight plan.

Runway Extension Waypoints — Created by specifying a distance from a given runway. The new waypoint will be located that distance from the runway threshold along the runway heading.

Abeam Waypoints — If a direct-to is performed, selection of abeam points results in waypoints being created at their abeam position on the direct-to path. Any waypoint information associated with the original waypoint is transferred to the newly created waypoints.

FIR/SUA Intersection Waypoints — Creates waypoints where the current flight plan crosses FIR boundaries and Special Use Areas (SUA) that are stored in the navigation data base.

The forward field of view display system shows a presentation of the selected segments of the flight plan as the flight plan is being constructed and flown.

The crew can modify the flight plan at any time. The flight plan modification can come from crew selections or via data link from the airline operational communications or air traffic control in response to a tactical situation. An edit to the flight plan creates a modified (or temporary) version of the flight plan that is a copy of the active flight plan plus any accrued changes made to it. Trajectory predictions are performed on the modified flight plan with each edit and periodically updated, which allows the crew to evaluate the impact of the flight plan changes prior to acceptance. When the desired changes have been made to the crew's satisfaction this modified flight plan is activated by the crew.

15.2.2.1 Flight Plan Construction

Flight plans are normally constructed by linking data stored in the navigation data base. The data may include any combination of the following items:

- SID/STAR/approach procedures
- Airways
- Prestored company routes
- Fixes (en route waypoints, navaids, nondirectional beacons, terminal waypoints, airport reference points, runway thresholds)
- Crew-defined fixes (as referenced above)

These selections may be strung together using clearance language, by menu selection from the navigation data base, by specific edit actions, or data link.

Terminal area procedures (SIDs, STARs, and approaches) consist of a variety of special procedure legs and waypoints. Procedure legs are generally defined by a leg heading, course or track, and a leg termination type. The termination type can be specified in many ways such as an altitude, a distance, or intercept of another leg. More detail on the path construction for these leg types and terminators will be discussed in the trajectory predictions section. Refer to ARINC 424 specification for further detail about what data and format are contained in the NDB to represent these leg types and terminations.

AF	DME Arc to a Fix
CA	Course to an Altitude
CD	Course to a Distance
CF*	Course to a Fix
CI	Course to an Intercept
CR	Course to Intercept a Radial
DF*	Direct to a Fix
FA*	Course from Fix to Altitude
FC	Course from Fix to Distance
FD	Course from Fix to DME Distance
FM	Course from Fix to Manual Term
HA*	Hold to an Altitude
HF*	Hold, Terminate at Fix after 1 Circuit

*These leg types are recommended in DO-236 as the set that produces consistent ground tracks and the only types that should be used within RNP airspace.

HM*   Hold, Manual Termination
IF*   Initial Fix
PI   Procedure Turn
RF*   Constant Radius to a Fix
TF*   Track to Fix
VA   Heading to Altitude
VD   Heading to Distance
VI   Heading to Intercept next leg
VM   Heading to Manual Termination
VR   Heading to Intercept Radial

Many of these leg types and terminations have appeared because of the evolution of equipment and instrumentation available on the aircraft and do not lend themselves to producing repeatable, deterministic ground tracks. For example, the ground track for a heading to an altitude will not only be dependent on the current wind conditions but also the climb performance of each individual aircraft. One can readily see that to fly this sort of leg without an FMS, the crew would follow the specified heading using the compass until the specified altitude is achieved, as determined by the aircraft's altimeter. Unfortunately, every aircraft will fly a different ground track and in some cases be unable to make a reasonable maneuver to capture the following leg. For the FMS, the termination of the leg is "floating" in that the lat/lon associated with the leg termination must be computed. These nondeterministic-type legs present problems for the air traffic separation concept of RNP airspace and for this reason RTCA DO-236 does not recommend the use of these legs in terminal area airspace, where they are frequently used today. These leg types also present added complexity in the FMS path construction algorithms since the path computation becomes a function of aircraft performance. With the advent of FMS and RNAV systems, in general, the need for non-deterministic legs simply disappears along with the problems and complexities associated with them.

Waypoints may also be specified as either "flyover" or nonflyover". A flyover waypoint is a waypoint whose lat/lon position must be flown over before the turn onto the next leg can be initiated whereas a nonflyover waypoint does not need to be overflown before beginning the turn onto the next leg.

15.2.2.2   Lateral Flight Planning

To meet the tactical and strategic flight planning requirements of today's airspace, the flight planning function provides various ways to modify the flight plan at the crew's discretion.

Direct-to — The crew can perform a direct-to to any fix. If the selected fix is a downtrack fix in the flight plan, then prior flight plan fixes are deleted from the flight plan. If the selected fix is not a downtrack fix in the flight plan, then a discontinuity is inserted after the fix and existing flight plan data are preserved.

Direct/intercept — The direct/intercept facility allows the crew to select any fixed waypoint as the active waypoint and to select the desired course into this waypoint. This function is equivalent to a direct-to except the inbound course to the specified fix may be specified by the crew. The inbound course may be specified by entering a course angle, or if the specified fix is a flight plan fix, the crew may also select the prior flight plan-specified course to the fix.

Holding pattern — Holding patterns may be created at any fix or at current position. All parameters for the holding pattern are editable including entry course, leg time/length, etc.

Fixes — Fixes may be inserted or deleted as desired. A duplicate waypoint page will automatically be displayed if there is more than one occurrence of the fix identifier in the navigation database. Duplicate fixes are arranged in order starting from the closest waypoint to the previous waypoint in the flight plan.

Procedures — Procedures (SIDs, STARs, and approaches including missed approach procedures) may be inserted or replaced as desired. If a procedure is selected to replace a procedure that is in the flight plan, the existing procedure is removed and replaced with the new selection.

Airway segments — Airway segments may be inserted as desired.

Missed approach procedures — The flight planning function also allows missed approach procedures to be included in the flight plan. These missed approach procedures can either come from the navigation database where the missed approach is part of a published procedure, in which case they will be automatically included in the flight plan, or they can be manually constructed by entry through the MCDU. In either case, automatic guidance will be available upon activation of the missed approach.

Lateral offset — The crew can create a parallel flight plan by specifying a direction (left or right of path) and distance (up to 99 nm) and optionally selecting a start and/or end waypoint for the offset flight plan. The flight planning function constructs an offset flight plan, which may include transition legs to and from the offset path.

15.2.2.3 Vertical Flight Planning

Waypoints can have associated speed, altitude, and time constraints. A waypoint speed constraint is interpreted as a "cannot exceed" speed limit, which applies at the waypoint and all waypoints preceding the waypoint if the waypoint is in the climb phase, or all waypoints after it if the waypoint is in the descent phase. A waypoint altitude constraint can be of four types — "at," "at or above," "at or below," or "between." A waypoint time constraint can be of three types — "at," "after," "before," "after" and "before" types are used for en route track-crossings and the "at" type is planned to be used for terminal area flow control.

Vertical flight planning consists of selection of speed, altitude, time constraints at waypoints (if required or desired), cruise altitude selection, aircraft weight, forecast winds, temperatures, and destination barometric pressure as well as altitude bands for planned use of aircraft anti-icing. A variety of optimized speed schedules for the various flight phases are typically available. Several aircraft performance-related crew selections may also be provided. All these selections affect the predicted aircraft trajectory and guidance.

15.2.2.4 Atmospheric Models

Part of the flight planning process is to specify forecast conditions for temperatures and winds that will be encountered during the flight. These forecast conditions help the FMS to refine the trajectory predictions to provide more accurate determination of ETAs, fuel burn, rates of climb/descent, and leg transition construction.

The wind model for the climb segment is typically based on an entered wind magnitude and direction at specified altitudes. The value at any altitude is interpolated between the specified altitudes to zero on the ground and merged with the current sensed wind. Wind models for use in the cruise segment usually allow for the entry of wind (magnitude and direction) for multiple altitudes at en route waypoints. Future implementation of en route winds may be via a data link of a geographical current wind grid database maintained on the ground. The method of computing winds between waypoints is accomplished by interpolating between entries or by propagating an entry forward until the next waypoint entry is encountered. Forecast winds are merged with current winds obtained from sensor data in a method which gives a heavier weighting to sensed winds close to the aircraft and converges to sensed winds as each waypoint-related forecast wind is sequenced. The wind model used for the descent segment is a set of altitudes with associated wind vector entered for different altitudes. The value at any altitude is interpolated from these values, and blended with the current sensed wind.

Forecast temperature used for extrapolating the temperature profile is based on the International Standard Atmosphere (ISA) with an offset (ISA deviation) obtained from pilot entries and/or the actual sensed temperature.

$$Forecast\ temperature = 15 + ISA\ dev - 0.00198 \times altitude \quad altitude < 36{,}089$$
$$Forecast\ temperature = -56.5 \quad altitude > 36{,}089$$

Air pressure is also utilized in converting speed between calibrated airspeed, mach, and true airspeed.

$$\delta\ (Pressure\ ratio) = (1 - 0.0000068753 * altitude)^{5.2561} \quad altitude < 36{,}089$$
$$\delta\ (Pressure\ ratio) = 0.22336 * e^{(4.8063 * (36089 - altitude)/100{,}000)}$$

15.2.3 Trajectory Predictions

Given the flight plan, the trajectory prediction function computes the predicted four-dimensional flight profile (both lateral and vertical) of the aircraft within the specified flight plan constraints and aircraft performance limitations, based on entered atmospheric data and the crew-selected modes of operation.

The lateral path and predicted fuel, time, distance, altitude, and speed are obtained for each point in the flight plan (waypoints as well as inserted vertical breakpoints such as speed change, cross-over, level off, T/C, T/D points). The flight profile is continuously updated to account for nonforecasted conditions and tactical diversions from the specified flight plan.

To simplify this discussion, the flight path trajectory is broken into two parts — the lateral profile (the flight profile as seen from overhead) and the vertical profile (the flight profile as seen from the side). However, the lateral path and vertical path are interdependent in that they are coupled to each other through the ground speed parameter. Since the speed schedules that are flown are typically constant CAS/mach speeds for climb and descent phases, the TAS (or ground speed) increases with altitude for the constant CAS portion and mildly decreases with altitude for the constant mach portion, as shown in the following equations.

$$Mach = sqrt\ [(1/\delta\{[1 + 0.2(CAS/661.5)^2]^{3.5} - 1\} + 1)^{0.286} - 1]$$
$$TAS = 661.5 \times mach \times sqrt[\theta]$$
$$CAS = calibrated\ airspeed\ in\ knots$$
$$TAS = true\ airspeed\ in\ knots$$
$$\delta = atmospheric\ pressure\ ratio\ (actual\ temperature\ /S.L.\ std.\ temperature)$$
$$\theta = atmospheric\ temperature\ ratio\ (actual\ temperature\ /S.L.\ std.\ temperature)$$

The significance of the change in airspeed with altitude will become apparent in the construction of the lateral and vertical profile during ascending and descending flights as described in the next section. Further, since the basic energy balance equations used to compute the vertical profile use TAS, these speed conversion formulas are utilized to convert selected speed schedule values to true airspeed values.

15.2.3.1 Lateral Profile

Fundamentally, the lateral flight profile is the specified route (composed of procedure legs, waypoints, hold patterns, etc.), with all the turns and leg termination points computed by the FMS according to how the aircraft should fly them. The entire lateral path is defined in terms of straight segments and turn segments which begin and end at either fixed or floating geographical points. Computing these segments can be difficult because the turn transition distance and certain leg termination points are a function of predicted aircraft speed (as noted in the equations below), wind, and altitude, which, unfortunately, are dependent on how much distance is available to climb and descend. For example, the turn transition at a waypoint requires a different turn radius and therefore a different distance when computed with different speeds. The altitude (and therefore speed of the aircraft) that can be obtained at a waypoint is dependent upon how much distance is available to climb or desend. So, the interdependency between speed and leg distance presents a special problem in formulating a deterministic set of algorithms for computing the trajectory. This effect becomes significant for course changes greater than 45°, with the largest effect for legs such as procedure turns which require a 180° turn maneuver.

Lateral turn construction is based on the required course change and the aircraft's predicted ground speed during the turn. If the maximum ground speed that the aircraft will acquire during the required course change is known, a turn can be constructed as follows:

$$Turn\ Radius\ (ft) = (GS^2)/(g \times tan\phi)$$
$$Turn\ Arclength\ (ft) = \Delta\ Course \times Turn\ Radius$$
$$GS = maximum\ aircraft\ ground\ speed\ during\ the\ turn$$
$$g = acceleration\ due\ to\ gravity$$
$$\phi = nominal\ aircraft\ bank\ angle\ used\ to\ compute\ a\ turn.$$

For legs such as constant radius to a fix (RF) where the turn radius is specified, a different form of the equation is used to compute the nominal bank angle that must be used to perform the maneuver.

$$\phi \; = \; arctan[\,GS^2/(turn \; radius \times g)\,]$$

To determine the maximum aircraft ground speed during the turn the FMC must first compute the altitude at which the turn will take place, and then the aircraft's planned speed based on the selected speed schedule and any applicable wind at that altitude. The desired bank angle required for a turn is predetermined based on a trade-off between passenger comfort and airspace required to perform a lateral maneuver.

The basis for the lateral profile construction is the leg and termination types mentioned in the flight plan section. There are four general leg types:

- Heading (V) — aircraft heading
- Course (C) — fixed magnetic course
- Track (T) — computed great circle path (slowly changing course)
- Arc (A or R) — an arc defined by a center (fix) and a radius

There are six leg terminator types:

- Fix (F) — terminates at geographic location
- Altitude (A) — terminates at a specific altitude
- Intercept next leg (I) — terminates where leg intercepts the next leg
- Intercept radial (R) — terminates where leg intercepts a specific VOR radial
- Intercept distance (D or C) — terminates where leg intercepts a specific DME distance or distance from a fix
- Manual (M) — leg terminates with crew action

Not all terminator types can be used with all leg types. For example, a track leg can only be terminated by a fix since the definition of a track is the great circle path between two geographic locations (fixes). Likewise, arc legs are only terminated by a fix. In a general sense, heading and course legs can be graphically depicted in the same manner understanding that the difference in the computation is the drift angle (or aircraft yaw).

Figure 15.3 depicts a graphical construction for the various leg and terminator types. The basic construction is straightforward. The complexity arises from the possible leg combinations and formulating proper curved transition paths between them. For example, if a TF leg, is followed by a CF leg where the specified course to the fix does not pass through the terminating fix for the prior TF leg, then a transition path must be constructed to complete a continuous path between the legs.

In summary, the lateral flight path computed by the FMC contains much more data than straight lines connecting fixed waypoints. It is a complete prediction of the actual lateral path that the aircraft will fly under FMS control. The constructed lateral path is critical because the FMC will actually control the aircraft to it by monitoring cross-track error and track angle error, and issuing roll commands to the autopilot as appropriate.

15.2.3.2 Vertical Profile

The fundamental basis for the trajectory predictor is the numerical integration of the aircraft energy balance equations including variable weight, speed, and altitude. Several forms of the energy balance equation are used to accommodate unrestricted climb/descent, fixed gradient climb/descent, speed change, and level flight. The integration steps are constrained by flight plan-imposed altitude and speed restrictions as well as aircraft performance limitations such as speed and buffet limits, maximum altitudes,

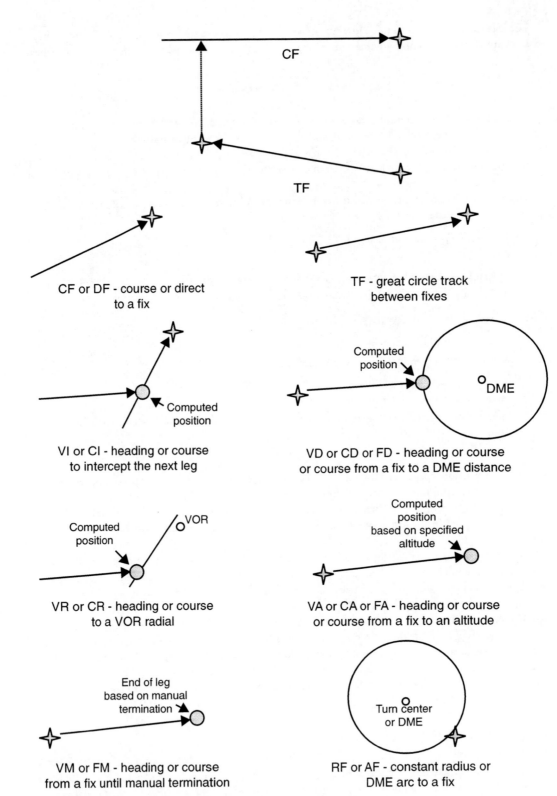

FIGURE 15.3 Basic lateral leg construction.

and thrust limits. The data that drives the energy balance equations come from the airframe/engine-dependent thrust, fuel flow, drag, and speed schedule models stored in the performance data base. Special construction problems are encountered for certain leg types such as an altitude-terminated leg because the terminator has a floating location. The location is dependent upon where the trajectory integration computes the termination of the leg. This also determines the starting point for the next leg.

The trajectory is predicted based on profile integration steps — the smaller the integration step the more accurate the computed trajectory. For each step the aircraft's vertical speed, horizontal speed, distance traveled, time used, altitude change, and fuel burned is determined based on the projected aircraft target speed, wind, drag, and engine thrust for the required maneuver. The aircraft's vertical state is computed for the end of the step and the next step is initialized with those values. Termination of an integration step can occur when a new maneuver type must be used due to encountering an altitude or speed constraint, flight phase change, or special segments such as turn transitions where finer integration steps may be prudent. The vertical profile is comprised of the following maneuver types: unrestricted ascending and descending segments, restricted ascending and descending segments, level segments, and speed change segments. Several forms of the energy balance equation are used depending on the maneuver type for a given segment of the vertical profile. Assumptions for the thrust parameter are maneuver type and flight phase dependent.

15.2.3.3 Maneuver Types

Unrestricted ascending and descending segments — The following form of the equation is typically used to compute the average vertical speed for fixed altitude steps (dh is set integration step). Using fixed altitude steps for this type of segment allows for deterministic step termination at altitude constraints. For ascending flight the thrust is general assumed to be the take-off, go-around, or climb thrust limit. For descending flight the thrust is generally assumed to be at or a little above flight idle.

$$V/S = \frac{\dfrac{(T - D)V_{ave}}{GW}}{\dfrac{T_{act}}{T_{std}} + \dfrac{V_{ave}}{g}\dfrac{dV_{true}}{dh}}$$

where:

T = Avg. thrust (lb)

D = Avg. drag (lb)

GW = A/C gross wt (lb)

T_{act} = Ambient temp (K)

T_{std} = Std. day temp (K)

V_{ave} = Average true airspeed (ft/sec)

g = 32.174 ft/sec^2

dV_{true} = Delta V_{true} (ft/sec)

dh = Desired altitude step (ft)

The projected aircraft true airspeed is derived from the pilot-selected speed schedules and any applicable airport or waypoint-related speed restrictions. Drag is computed as a function of aircraft configuration, speed, and bank angle. Fuel flow and therefore weight change is a function of the engine thrust.

Once V/S is computed for the step the other prediction parameters can be computed for the step.

$$dt = \frac{dh}{V/S}, \quad \text{where} \quad dt = \text{delta time for step}$$

$$ds = dt(V_{true} + \text{average along track wind for segment}), \quad \text{where} \quad ds = \text{delta distance for step}$$

$$dw = dt \times \text{fuel flow}(T), \quad \text{where} \quad dw = \text{delta weight for step}$$

Restricted ascending and descending segments — The following form of the equation is typically used to compute the average thrust for fixed altitude steps (dh and V/S are predetermined). Using fixed altitude steps for this type of segment allows for deterministic step termination at altitude constraints. The average V/S is either specified or computed based on a fixed flight path angle (FPA).

$$V/S_{ave} = GS_{ave} \tan FPA, \quad \text{where} \quad GS_{ave} = \text{segment ground speed (ft/sec)}.$$

The fixed *FPA* can in turn be computed based on a point to point vertical flight path determined by altitude constraints, which is known as a geometric path. With a specified *V/S* or *FPA* segment the thrust required to fly this profile is computed.

$$T = \frac{W \times V/S_{ave}}{V_{ave}} - \left(1 + \frac{V_{ave}}{g}\frac{dV_{true}}{dh}\right) + D$$

The other predicted parameters are computed as stated for the unrestricted ascending and descending segment.

Level segments — Constant-speed-level segments are a special case of the above equation. Since dV_{true} and V/S_{ave} are by definition zero for level segments, the equation simplifies to $T = D$. Level segments are typically integrated based on fixed time or distance steps so the other predicted parameters are computed as follows:

$dt = $ set integration step

and

$ds = dt(V_{true} + \text{average along track wind for segment}) \qquad ds = \text{delta distance for step}$

or

$ds = $ set integration step

and

$dt = ds/(V_{true} + \text{average along track wind for segment}) \qquad dt = \text{delta time for step}$

$dw = dt \times \text{fuel flow}(T) \qquad\qquad\qquad\qquad\qquad dw = \text{delta weight for step}$

Speed change segments — The following form of the equation is typically used for speed change segments to compute the average time for a fixed dV_{true} step. The V/S_{ave} used is predetermined based on ascending, descending, or level flight along with the operational characteristics of the flight controls or as for the case of geometric paths computed based on the required *FPA*. The thrust is assumed to be flight idle for descending flight, take-off or climb thrust limit for ascending flight, or cruise thrust limit for level flight.

$$dt = dV_{true}/g\left\{\frac{(T-D)}{GW} - \left(\frac{T_{act}}{T_{std}}\frac{V/S_{ave}}{V_{ave}}\right)\right\}$$

$$dh = V/S_{ave} \times dt$$

For all maneuver types the altitude rate, speed change, or thrust must be corrected for bank angle effects if the maneuver is performed during a turn transition. The vertical flight profile that the FMC computes along the lateral path is divided into three phases of flight: climb, cruise, and descent.

The climb phase — The climb phase vertical path, computed along the lateral path, is typically composed of the segments shown in Figure 15.4.

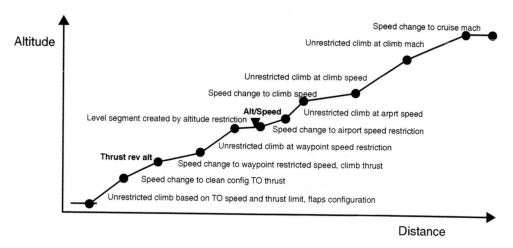

FIGURE 15.4   Typical climb profile.

In addition to these climb segments, there can also be altitude level-off segments created by altitude restrictions at climb waypoints, and additional target speed acceleration segments created by speed restrictions at climb waypoints.

The cruise phase — The cruise phase vertical path, computed along the lateral path, is very simple. It's typically composed of a climb speed to cruise speed acceleration or deceleration segment followed by a segment going to the FMC-computed top of descent. The cruise phase typically is predicted level at cruise altitude via several distance- or time-based integration steps. Unlike the climb and descent phase, the optimal cruise speeds slowly change with the changing weight of the aircraft, caused by fuel burn. If step climb or descents are required during the cruise phase, these are treated as unrestricted ascending flight and fixed V/S or FPA descents. At each step the FMC computes the aircraft's along-path speed, along-path distance traveled, and fuel burned based on the projected aircraft target speed, wind, drag, and engine thrust. The projected aircraft true airspeed is derived from the pilot-selected cruise speed schedule and applicable airport-related speed restrictions. Drag is computed as a function of aircraft speed and bank angle. For level flight, thrust must be equal to drag. Given the required thrust, the engine power setting can be computed, which becomes the basis for computing fuel burn and throttle control guidance.

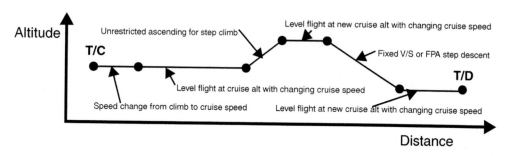

FIGURE 15.5   Typical cruise profile.

The descent phase — The descent phase vertical path, computed along the lateral path, can be composed of several vertical leg types as shown in the following figure:

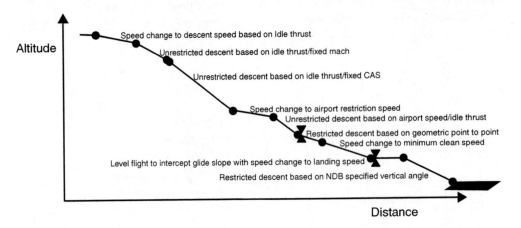

FIGURE 15.6   Typical descent profile.

In addition to these descent segments, there can also be altitude level-off segments created by altitude restrictions at descent waypoints and additional targets speed deceleration segments created by speed restrictions at descent waypoints as well as eventual deceleration to the landing speed for the selected flaps configuration.

15.2.3.4   NDB Vertical Angles

These leg types are generally used in the approach. The desired approach glide slope angle that assures obstacle clearance is packed as part of the waypoint record for the approach in the Navigation Data Base (NDB). The angle is used to compute the descent path between the waypoint associated with the angle and the first of the following to be encountered (looking backwards)

1. Next lateral waypoint with an NDB vertical angle record
2. Next "at" constraint
3. First approach waypoint

A new NDB gradient can be specified on any waypoint. This allows the flexibility to specify multiple FPAs for the approach if desired. The integration basis for this leg assumes a thrust level compatible with maintaining the selected speed schedule at the descent rate specified by the NDB angle. Decelerations that can occur along these legs because of various restrictions (both regulatory and airframe) assume performing the speed change at idle thrust at the vertical speed specified by the NDB angle. If within the region where flaps are anticipated, then the deceleration model is based on a flaps configuration performance model.

Default approach vertical angle — Generally, this leg is used in lieu of a specified NDB angle to construct a stabilized nominal glide slope between the glide slope intercept altitude (typically 1500 ft above the runway) to the selected runway. The integration basis for this leg is the same as the NDB angle.

Direct to vertical angle — This leg type provides a vertical "Dir to" capability for use in tactical situations. The path constructed is the angle defined by the current 3-D position of the aircraft and the next appropriate reference point (usually the next altitude constraint). For a pending vertical "direct to" the direct to angle is updated on a periodic basis to account for the movement of the aircraft. In determining the direct to angle the aircraft 3-D position is extrapolated to account for the amount of time required to compute the trajectory for VNAV guidance to avoid path overshoots when the trajectory is available. The integration basis for this leg assumes a thrust level compatible with maintaining the selected speed schedule at the descent rate specified by the direct to gradient. Decelerations that can occur along these descent legs because of various restrictions (both regulatory and aircraft) assume performing the speed change at idle thrust for the anticipated flaps/landing gear configuration.

Computed vertical angle — This leg type provides constant angle vertical paths between constraints that are part of the vertical flight plan. These geometric paths provide for repeatable, stabilized, partial power descent paths at lower altitudes in the terminal area. The general rules for proper construction of these paths are

- Vertical maneuvering should be minimized. This implies that a single angle to satisfy a string of altitude constraints is preferred. This can occur when "At or above" and "At or below" altitude constraints are contained in the flight plan.

- If a string of "At or above" and/or "At or below" constraints can be satisfied with an unrestricted, idle power path, then that path is preferred.

- Computed gradient paths should be checked for flyability (steeper than idle). The computation of the idle path (for the anticipated and idle with drag devices deployed) should account for a minimum deceleration rate if one is contained within the computed gradient leg.

The integration basis for this leg assumes a thrust level compatible with maintaining the selected speed schedule at the descent rate specified by the computed vertical angle. Decelerations that can occur along these descent legs because of various restrictions (both regulatory and airframe) assume performing the speed change at idle thrust limited for a maximum deceleration rate.

Constant V/S — This leg type provides a strategic, shallower-than-idle initial descent path if desired. The construction of this path is dependent on a vertical speed and intercept altitude being requested. The integration basis for this leg assumes a thrust level compatible with maintaining the selected speed schedule at the descent rate specified by the commanded v/s. Decelerations that can occur along these descent legs because of various restrictions (both regulatory and airframe) assume performing the speed change at idle thrust limited for a maximum deceleration rate.

Unrestricted descent — The unrestricted descent uses performance data to construct an energy-balanced idle descent path when not constrained by altitude constraints. The integration basis for this leg assumes maintaining the selected speed schedule at idle thrust. This results in a changing vertical speed profile. Decelerations that can occur along these descent legs because of various restrictions (both regulatory and aircraft) assume performing the speed change at a minimum vertical speed rate and idle thrust limited for a maximum deceleration rate. The minimum vertical speed can be based on energy sharing or a precomputed model. An idle thrust factor allows the operator to create some margin (shallower or steeper) in the idle path construction.

15.2.4 Performance Computations

The performance function provides the crew information to help optimize the flight or provide performance information that would otherwise have to be ascertained from the aircraft performance manual. FMSs implement a variety of these workload reduction features, only the most common functions are discussed here.

15.2.4.1 Speed Schedule Computation

Part of the vertical flight planning process is the crew selection of performance modes for each flight phase based on specific mission requirements. These performance modes provide flight profile optimization through computation of flight phase-dependent, optimized speed schedules that are used as a basis for both the trajectory prediction, generation of guidance speed targets, and other performance advisories.

The selection of a specific performance mode for each flight phase results in the computation of an optimized speed schedule, which is a constant CAS, constant mach pair, which becomes the planned speed profile for each flight phase. The altitude where the CAS and mach are equivalent is known as the crossover altitude. Below the crossover altitude the CAS portion of the speed schedule is the controlling speed parameter and above the crossover altitude the mach portion is the controlling speed. The performance parameter that is optimized is different for each performance mode selection.

Climb

- Economy (based on Cost Index) — speed that optimizes overall cost of operation (lowest cost).
- Maximum angle of climb — speed that produces maximum climb rate with respect to distance.
- Maximum rate of climb — speed that produces maximum climb rate with respect to time.
- Required time of arrival speed (RTA) — speed that optimizes overall cost of operation, while still achieving a required time of arrival at a specific waypoint.

Cruise

- Economy (based on Cost Index) — speed that optimizes overall cost of operation (lowest cost).
- Maximum endurance — speed that produces lowest fuel burn rate, maximizing endurance time.
- Long range cruise — speed that produces best fuel mileage, maximizing range.
- Required time of arrival (RTA) — speed that optimizes overall cost of operation, while still achieving a required time of arrival at a specific waypoint.

Descent

- Economy (based on Cost Index) — speed that optimizes overall cost of operation (lowest cost).
- Maximum descent rate — speed that produces maximum descent rate with respect to time.
- Required time of arrival (RTA) — speed that optimizes overall cost of operation, while still achieving a required time of arrival at a specific waypoint.

All flight phases allow a manually entered CAS/mach pair as well.

It may be noted that one performance mode that is common to all flight phases is the "economy" speed mode which minimizes the total cost of operating the airplane on a given flight. This performance mode uses a Cost Index, which is the ratio of time-related costs (crew salaries, maintenance, etc.) to fuel cost as one of the independent variables in the speed schedule computation.

$$\textbf{Cost Index (CI)} = \textbf{flight time-related cost/fuel cost}$$

The cost index allows airlines to weight time and fuel costs based on their daily operations.

15.2.4.2   Maximum and Optimum Altitudes

An important parameter for the flight crew is the optimum and maximum altitude for the aircraft/engine type, weight, atmospheric conditions, bleed air settings, and the other vertical flight planning parameters. The optimum altitude algorithm computes the most cost-effective operational altitude based solely on aircraft performance and forecasted environmental conditions. Fundamentally, the algorithm searches for the altitude that provides the best fuel mileage.

$$\textbf{Altitude that maximizes the ratio: ground speed/fuel burn rate}$$

The maximum altitude algorithm computes the highest attainable altitude based solely on aircraft performance and forecasted environmental conditions, while allowing for a specified rate of climb margin.

$$\textbf{Altitude that satisfies the equality: min climb rate} = \textbf{TAS} \times \textbf{(thrust} - \textbf{drag)/weight}$$

Optimum altitude is always limited by maximum altitude. The algorithms for these parameters account for the weight reduction caused by the fuel burn in achieving the altitudes. The speeds assumed are the selected performance modes.

Trip altitude — Another important computation that allows the crew to request an altitude clearance to optimize the flight is the recommended cruise altitude for a specified route known as trip altitude. This altitude may be different from the optimum altitude in that for short trips the optimum altitude may not be achievable because of the trip distance. This algorithm searches for the altitude that satisfies the climb and descent while preserving a minimum cruise time.

Alternate destinations — To help reduce crew workload during flight diversion operations the FMS typically provides alternate destination information. This computation provides the crew with distance, fuel, and ETA for selected alternate destinations. The best trip cruise altitude may be computed as well. The computations are based either on a direct route from the current position to the alternate or continuing to the current destination, execution of a missed approach at the destination, and then direct to the alternate. Also computed for these alternate destinations are available holding times at the present position and current fuel state vs. fuel required to alternates. Usually included for the crew convenience is the CDU/MCDU retrieval of suitable airports that are nearest the aircraft.

Step climb/descent — For longer-range flights often the achievable cruise altitude is initially lower than the optimum because of the heavy weight of the aircraft. As fuel is burned off and the aircraft weight reduced, it becomes advantageous to step climb to a higher altitude for more efficient operation. The FMS typically provides a prediction of the optimum point(s) at which a step climb/descent maneuver may be initiated to provide for more cost-effective operation. This algorithm considers all the vertical flight planning parameters, particularly the downstream weight of the aircraft, as well as entered wind data. The time and distance to the optimum step point for the specified step altitude is displayed to the crew, as well as the percent savings/penalty for the step climb/descent vs. the current flight plan. For transoceanic aircraft it is typical for the trajectory prediction function to assume that these steps will be performed as part of the vertical profile, so that the fuel predictions are more aligned with what the aircraft will fly.

Thrust limit data — To prevent premature engine maintenance/failure and continued validation of engine manufacturer's warrantees, it becomes important not to overboost the aircraft engines. The engine manufacturers specify flight phase-dependent thrust limits that the engines are designed to operate reliably within. These engine limits allow higher thrust levels when required (take-off, go-around, engine out) but lower limits for non-emergency sustained operation (climb and cruise). The thrust limits for take-off, climb, cruise, go around, and continuous modes of operation are computed based on the current temperature, altitude, speed, and type of engine/aircraft and engine bleed settings. Thrust limit data are usually represented by "curve sets" in terms of either engine RPM (N1) or engine pressure ratio (EPR), depending on the preferred engine instrumentation package used to display the actual engine thrust. The "curve sets" typically have a temperature-dependent curve and an altitude-dependent curve along with several correction curves for various engine bleed conditions. The algorithms used to compute the thrust limits vary among engine manufacturers.

Take-off reference data — The performance function provides for the computation, or entry, of V_1, V_R and V_2 take-off speeds for selected flap settings and runway, atmospheric, and weight/CG conditions. These speeds are made available for crew selection for display on the flight instruments. In addition, take-off configuration speeds are typically computed. The take-off speeds and configuration speeds are stored as data sets or supporting data sets in the performance database.

Approach reference data — Landing configuration selection is usually provided for each configuration appropriate for the operation of the specific aircraft. The crew can select the desired approach configuration and the state of that selection is made available for other systems. Selection of an approach configuration also results in the computation of a landing speed based on a manually entered wind correction for the destination runway. In addition, approach configuration speeds are computed and displayed for reference and selection for display on the flight instruments. The approach and landing speeds are stored as data sets in the performance database.

Engine-out performance — The performance function usually provides engine-out performance predictions for the loss of at least one engine. These predictions typically include:

- Climb at engine-out climb speed
- Cruise at engine-out cruise speed
- Driftdown to engine-out maximum altitude at driftdown speed
- Use of maximum continuous thrust

The engine out speed schedules are retrieved from the performance data base and the trajectory predictions are computed based on the thrust available from the remaining engines and the increased aircraft drag created by engine windmilling and aircraft yaw caused by asymmetrical thrust.

15.2.5 Guidance

The FMS typically computes roll axis, pitch axis, and thrust axis commands to guide the aircraft to the computed lateral and vertical profiles as discussed in the trajectory predictions section. These commands may change forms depending on the particular flight controls equipment installed on a given aircraft. Other guidance information is sent to the forward field of view displays in the form of lateral and vertical path information, path deviations, target speeds, thrust limits and targets, and command mode information.

15.2.5.1 Lateral Guidance

The lateral guidance function typically computes dynamic guidance data based on the predicted lateral profile described in the trajectory predictions section. The data are comprised of the classic horizontal situation information:

- Distance to go to the active lateral waypoint (DTG)
- Desired track (DTRK)
- Track angle error (TRKERR)
- Cross-track error (XTRK)
- Drift angle (DA)
- Bearing to the go to waypoint (BRG)
- Lateral track change alert (LNAV alert)

A common mathematical method to compute the above data is to convert the lateral path lat/lon point representation and aircraft current position to earth-centered unit vectors using the following relationships:

P = *earth centered unit position vector with x, y, z components*

$X = COS\ (lat)\ COS\ (lon)$

$Y = COS\ (lat)\ SIN\ (lon)$

$Z = SIN\ (lat)$

For the following vector expressions $\times$ is the vector cross product and $\cdot$ is the vector dot product. For any two position vectors that define a lateral path segment:

$N = Pst \times Pgt$	N = unit vector normal to **Pst** and **Pgt**
$Pap = N \times (Ppos \times N)$	Pgt = go to point unit position vector
	Pst = start point unit position vector
$DTGap = earth\ radius * arcCOS\ (Pgt \cdot Pap)$	Pap = along path position unit vector
$DTGpos = earth\ radius * arcCOS\ (Pgt \cdot Ppos)$	$Ppos$ = current position unit vector
$XTRK = -earth\ radius * arcCOS\ (Pap \cdot Ppos)$ (full expression)	
$XTRK = -earth\ radius * N \cdot Ppos$ (good approximation)	
$Est = Z \times P$	E_{st} = East-pointing local level unit vector
$Nth = P \times Est$	Nth = North-pointing local level unit vector

$$Z = \begin{vmatrix} 0 \\ 0 \\ 1 \end{vmatrix} \quad Z \text{ axis unit vector}$$

$$DTRK = arcTAN\ [(-N \cdot Nth_{ap})/(-N \cdot Est_{ap})]$$

$$BRG = arcTAN\ [(-N \cdot Nth_{pos})/(-N \cdot Est_{pos})]$$

$$TRKERR = DTRK - Current\ Track$$

$$DA = Current\ Track - Current\ Heading$$

LNAV Alert is set when the DTG/ground speed < 10 sec from turn initiation

The above expressions can also be used to compute the distance and course information between points that are displayed to the crew for the flight plan presentation. The course information is generally displayed as magnetic courses, due to the fact that for many years a magnetic compass was the primary heading sensor and therefore all navigation information was published as magnetic courses. This historical-based standard requires the installation of a worldwide magnetic variation model in the FMS since most of the internal computations are performed in a true course reference frame. Conversion to magnetic is typically performed just prior to crew presentation.

The lateral function also supplies data for a graphical representation of the lateral path to the navigation display, if the aircraft is so equipped, such that the entire lateral path can be displayed in an aircraft-centered reference format or a selected waypoint center reference format. The data for this display are typically formatted as lat/lon points with identifiers and lat/lon points with straight and curved vector data connecting the points. Refer to ARINC 702A for format details. In the future the FMS may construct a bit map image of the lateral path to transmit to the navigation display instead of the above format.

Lateral leg switching and waypoint sequencing — As can be seen in the lateral profile section, the lateral path is composed of several segments. Most lateral course changes are performed as "flyby" transitions. Therefore anticipation of the activation of the next vertical leg is required, such that a smooth capture of that segment is performed without path overshoot. The turn initiation criteria are based on the extent of the course change, the planned bank angle for the turn maneuver, and the ground speed of the aircraft.

$$Turn\ Radius = Ground\ Speed^2/[g * TAN\ (\phi_{nominal})]$$

$$Turn\ initiation\ Distance = Turn\ Radius/TAN\ (Course\ Change/2) + roll\ in\ distance$$

The roll in distance is selected based on how quickly the aircraft responds to a change in the aileron position. Transitions that are flyby but require a large course change (>135°) typically are constructed for a planned overshoot because of airspace considerations. Turn initiation and waypoint sequence follow the same algorithms except the course change utilized in the above equations is reduced from the actual course change to delay the leg transition and create the overshoot. The amount of course change reduction is determined by a balance in the airspace utilized to perform the overall maneuver. For "flyover" transitions, the activation of the next leg occurs at the time the "flyover" waypoint is sequenced.

The initiation of the turn transition and the actual sequence point for the waypoint are not the same for "flyby" transitions. The waypoint is usually sequenced at the turn bisector point during the leg transition.

Roll control — Based on the aircraft current state provided by the navigation function and the stored lateral profile provided by the trajectory prediction function, lateral guidance produces a roll steering command that can be engaged by the flight controls. This command is both magnitude and rate limited based on aircraft limitations, passenger comfort, and airspace considerations. The roll command is computed to track the straight and curved path segments that comprise the lateral profile. The roll control is typically a simple control law driven by the lateral cross-track error and track error as discussed in the prior subsection as well as a nominal roll angle for the planned turn transitions. The nominal roll angle is zero for straight segments but corresponds to the planned roll angle used to compute lateral transition paths to follow the curved segments.

$$Roll = xtrk\ gain \times xtrk + trk\ gain \times trk\ error + \phi_{nominal}$$

where

$$\phi_{nominal} = nominal\ planned\ roll\ angle.$$

The gain values used in this control loop are characteristic of the desired aircraft performance for a given airframe and flight controls system.

Lateral capture path construction — At the time LNAV engagement with the flight controls occurs, a capture path is typically constructed that guides the airplane to the active lateral leg. This capture path is usually constructed based on the current position and track of the aircraft if it intersects the active lateral leg. If the current aircraft track does not intersect the active lateral leg, then LNAV typically goes into an armed state waiting for the crew to steer the aircraft into a capture geometry before fully engaging to automatically steer the aircraft. Capture of the active guidance leg, is usually anticipated to prevent overshoot of the lateral path.

15.2.5.2 Vertical guidance

The vertical guidance function provides commands of pitch, pitch rate, and thrust control to the parameters of target speeds, target thrusts, target altitudes, and target vertical speeds (some FMS provide only the targets depending on the flight management/flight control architecture of the particular aircraft). Much like the lateral guidance function, the vertical guidance function provides dynamic guidance parameters for the active vertical leg to provide the crew with vertical situation awareness. Unlike the lateral guidance parameters, the vertical guidance parameters are somewhat flight phase dependent.

Flight Phase	Vertical Guidance Data
Takeoff	Take-off speeds V1, V2, VR
	Take-off thrust limit
Climb	Target speed based on selected climb speed schedule,
	flight plan speed restriction, and airframe limitations
	Target altitude intercept
	Alt constraint violation message
	Distance to top of climb
	Climb thrust limits
Cruise	Target speed based on selected cruise speed schedule,
	flight plan speed restriction, and airframe limitations
	Maximum and optimum altitude
	distance to step climb point
	Distance to top of descent
	Cruise thrust limit
	Cruise thrust target
Descent	Target speed based on selected descent speed schedule,
	flight plan speed restriction, and airframe limitations
	Target altitude intercept
	Vertical deviation
	Desired V/S
	Energy bleed-off message
Approach	Target speed based on dynamic flap configuration
	Vertical deviation
	Desired V/S
Missed	Target speed based on selected climb speed schedule,
Approach	flight plan speed restriction, and airframe limitations
	Target altitude intercept
	Alt constraint violation msg
	Distance to top of climb
	Go-around thrust limit

Vertical guidance is based on the vertical profile computed by the trajectory prediction function as described in a previous section as well as performance algorithms driven by data from the performance data base.

The mathematical representation of the vertical profile is the point type identifier, distance between points, which includes both lateral and vertical points, speed, altitude, and time at the point. Given this information, data for any position along the computed vertical profile can be computed.

$$Path\ gradient = (alt_{start} - alt_{end})/distance\ between\ points$$

Therefore the path reference altitude and desired V/S at any point is given by:

$$Path\ altitude = alt_{end} + path\ gradient * DTGap$$

$$Vertical\ deviation = current\ altitude - path\ altitude$$

$$Desired\ V/S = path\ gradient * current\ ground\ speed$$

In the same manner time and distance data to any point or altitude can be computed as well. The target speed data are usually not interpolated from the predicted vertical profile since it is only valid for on-path flight conditions. Instead, it is computed based on the current flight phase, aircraft altitude, relative position with respect to flight plan speed restrictions, flaps configuration, and airframe speed envelope limitations. This applies to thrust limit computations as well.

Auto flight phase transitions — The vertical guidance function controls switching of the flight phase during flight based on specific criteria. The active flight phase becomes the basis for selecting the controlling parameters to guide the aircraft along the vertical profile. The selected altitude is used as a limiter in that the vertical guidance will not allow the aircraft to fly through that altitude (except during approach operations where the selected altitude may be pre-set for a missed approach if required). When on the ground with the flight plan and performance parameters initialized, the flight phase is set to take-off. After liftoff, the phase will switch to climb when the thrust revision altitude is achieved. The switch from climb to cruise (level flight) phase usually occurs when the aircraft is within an altitude acquire band of the target altitude.

$$|Cruise\ altitude - current\ altitude| < capture\ gain * current\ vertical\ speed$$

The capture gain is selected for aircraft performance characteristics and passenger comfort. The switch from cruise to descent can occur in various ways. If the crew has armed the descent phase by lowering the preselected altitude below cruise altitude, then descent will automatically initiate at an appropriate distance before the computed T/D to allow for sufficient time for the engine to spool down to descent thrust levels so that the aircraft speed is coordinated with the initial pitch-over maneuvers. If the crew has not armed the descent by setting the selected altitude to a lower level, then cruise is continued past the computed T/D until the selected altitude is lowered to initiate the descent. Facilities are usually provided for the crew to initiate a descent before the computed T/D in response to ATC instructions to start descending.

Vertical leg switching — As can be seen in the vertical profile section, the vertical path is composed of several segments. Just as in the lateral domain it is desirable to anticipate the activation of the next vertical leg such that a smooth capture of that segment is performed without path overshoot. It therefore becomes necessary to have an appropriate criteria for vertical leg activation. This criteria is typically in the form of an inequality involving path altitude difference and path altitude rate difference.

$$|Path\ altitude\ (n) - path\ altitude\ (n + 1)| * capture\ gain < |desired\ V/S\ (n) - desired\ V/S\ (n + 1)|$$

The capture gain is determined based on airframe performance and passenger comfort.

Pitch axis and thrust axis control — The pitch command produced by vertical guidance is based on tracking the speed target, FMS path, or acquiring and holding a target altitude depending on the flight phase and situation. If VNAV is engaged to the flight controls an annunciation of the parameter controlling pitch is usually displayed in the crew's forward field of view.

Flight Phase	Pitch Axis Control	Thrust Axis Control	Pitch/Thrust Mode Annunciation
Take-off	None until safely off ground then same as climb	Take-off thrust limit	Vspd/TO limit
Climb and cruise climb	Capture and track speed target	Climb thrust limit	Vspd/CLB limit
Level flight	Capture and maintain altitude	Maintain speed target	Valt/CRZ limit
Unrestricted descent	Capture and track vertical path	Set to flight idle	Vpath/CRZ limit
Restricted descent and approach	Capture and track vertical path	Set to computed thrust required, then maintain speed	Vpath/CRZ limit
Descent path capture from below and cruise descent	Capture and track fixed V/S capture path	Set to computed thrust required, then maintain speed	Vpath/CRZ limit
Descent path capture from above	Capture and track upper speed limit	Set to flight idle	Vspd/CRZ limit
Missed approach	Capture and track speed target	Go-around thrust limit	Vspd/GA limit

Control strategy may vary with specific implementations of FMSs. Based on the logic in the above table, the following outer loop control algorithms are typically used to compute the desired control parameters.

Pitch axis control — The control algorithms below are representative of control loop equations that could be utilized and are by no means the only form that apply. Both simpler and more complex variations may be used.

Vspd

> **Capture**
> *Delta pitch = speed rate gain * (airspeed rate − capture rate)*
> **Track**
> *Delta pitch = (airspeed gain * airspeed error + speed rate gain * airspeed rate)/V$_{true}$*

Vpath

> **Capture**
> *V/S error = fixed capture V/S − current V/S*
> *Delta pitch = path capture gain * arcSIN (V/S error/V$_{true}$)*
> **Track**
> *Delta pitch = (VS gain * V/S error + alt error gain * alt error)/V$_{true}$*

Valt

> **Capture**
> *Capture V/S = alt capture gain * alt error*
> *V/S error = capture V/S − current V/S*
> *Delta pitch = V/S gain * * arcSIN (V/S error/V$_{true}$)*
> **Track**
> *Delta pitch = (VS gain * current V/S + alt error gain * alt error)/V$_{true}$*

Proper aircraft pitch rates and limits are typically applied before final formulation of the pitch command. Once again, the various gain values are selected based on the aircraft performance and passenger comfort.

Thrust axis control — The algorithms below are representative of those that could be utilized to determine thrust settings. Quite often the thrust setting for maintaining a speed is only used for an initial throttle setting. Thereafter the speed error is used to control the throttles.

Thrust Limit

Thrust limit = f (temp, alt, spd, engine bleed air): *stored as data sets in the performance DB*

Flight Idle

Idle thrust = f (temp, alt, spd, engine bleed air): *stored as data sets in the performance DB*

Thrust Required

$$T = \frac{W \times V/S_{ave}}{V_{ave}}\left(1 + \frac{V_{ave}}{g}\frac{dV_{true}}{dh}\right) + D$$

RTA (required time of arrival) — The required time of arrival or time navigation is generally treated as a dynamic phase-dependent speed schedule selection (refer to the performance section). From this standpoint the only unique guidance requirements are the determination of when to recompute the phase-dependent speed schedules based on time error at the specified point and perhaps the computation of the earliest and latest times achievable at the specified point.

RNAV approach with VNAV guidance — The only unique guidance requirement is the increased scale in the display of vertical deviation when the approach is initiated. The vertical profile for the approach is constructed as part of the vertical path by trajectory predictions, complete with deceleration segments to the selected landing speed (refer to the performance section).

15.3 Summary

This chapter is an introduction to the several functions that comprise a flight management system and has focused on the basic functionality and relationships that are fundamental to understanding the flight management system and its role in the operations of the aircraft. Clearly, there is a myriad of complexity in implementing each function that is beyond the scope of this publication.

The future evolution of the flight management system is expected to focus not on the core functions as described herein, but on the utilization within the aircraft and on the ground of the fundamental information produced by the flight management system today. The use of the FMS aircraft state and trajectory intent, within the aircraft and on the ground, to provide strategic conflict awareness is a significant step toward better management of the airspace. Communication of the optimized user-preferred trajectories will lead to more efficient aircraft operation. The full utilization of RNP-based navigation will increase the capacity of the airspace. Innovative methods to communicate FMS information and specify flight plan construction with the crew to make flight management easier to use are expected as well. Clearly, the FMS is a key system in moving toward the concepts embodied in CNS future airspace.

16

Synthetic Vision

Russell V. Parish
NASA Langley Research Center

Daniel G. Baize
NASA Langley Research Center

Michael S. Lewis
NASA Langley Research Center

16.1 Introduction

A large majority of the avionics systems introduced since the early days of flight (attitude indicators, radio navigation, instrument landing systems, etc.) have sought to overcome the issues resulting from limited visibility. Limited visibility is the single most critical factor affecting both the safety and capacity of worldwide aviation operations. In commercial aviation, over 30% of all fatal accidents worldwide are categorized as Controlled Flight Into Terrain (CFIT)—accidents in which a functioning aircraft impacts terrain or obstacles that the flight crew could not see. In general aviation, the largest accident category is Continued Flight into Instrument Meteorological Conditions, in which pilots with little experience continue to fly into deteriorating weather and visibility conditions and either collide with unexpected terrain or lose control of the vehicle because of the lack of familiar external cues. Finally, the single largest factor causing airport flight delays is the limited runway capacity and increased air traffic separation distances resulting when visibility conditions fall below visual flight rule operations. Now, synthetic vision technology will allow this *visibility* problem to be solved with a *visibility* solution, making every flight the equivalent of a clear daylight operation.

Initial attempts to solve the visibility problem with a visibility solution have used imaging sensors to enhance the pilot's view of the outside world. Such systems are termed "enhanced vision systems," which attempt to improve visual acquisition by enhancing significant components of the real-world scene. Enhanced vision systems typically use advanced sensors to penetrate weather phenomena such as darkness, fog, haze, rain, and/or snow, and the resulting enhanced scene is presented on a head up display (HUD), through which the outside real world may be visible. The sensor technologies involved include either active or passive radar or infrared systems of varying frequencies. These systems have been the subject of experiments for over two decades, and the military has successfully deployed various implementations. However, no sensor-based application has seen commercial aircraft success for a variety of reasons, including cost, complexity, and technical performance. Though technology advances are making radar and infrared sensors more affordable, they still suffer from deficiencies for commercial applications. High-frequency radars (e.g., 94 GHz) and infrared sensors have degraded range performance in heavy rain and certain fog types. Low-frequency (e.g., 9.6 GHz) and mid-frequency (e.g., 35 GHz) radars have improved range, but poor resolution displays. Active radar sensors also suffer from mutual interference issues with multiple users in close proximity. All such sensors yield only monochrome displays with potentially misleading visual artifacts in certain temperature or radar reflective environments.

A "synthetic vision system" is a display system in which the view of the outside world is provided by melding computer-generated airport scenes from on-board databases and flight display symbologies, with information derived from a weather-penetrating sensor (e.g., information from runway edge detection or object detection algorithms) or with actual imagery from such a sensor. These systems are characterized by their ability to represent, in an intuitive manner, the visual information and cues that a flight crew would have in daylight — Visual Meteorological Conditions (VMC). The visual information and cues are depicted based on precise positioning information relative to an onboard terrain database, and possibly includes traffic information from surveillance sources (such as TCAS, ASDE, etc.) and other hazard information (such as wind shear).

Synthetic vision displays are unlimited in range, unaffected by atmospheric conditions, and require only precise ownship location and readily available display media, computer memory, and processing to function. The rapid emergence of reliable GPS position information and precise digital terrain maps, including data from the Space Shuttle Radar Topography Mission (SRTM), make this approach capable of both true all-weather performance as well as extremely low cost, low maintenance operations. When fully implemented, successful synthetic vision technologies will be a revolutionary improvement in aviation safety and utility.

16.2 Background

Synthetic vision systems are intended to reduce accidents by improving a pilot's situation and spatial awareness during low-visibility conditions, including night and Instrument Meteorological Conditions (IMC). Synthetic vision technologies are most likely to help reduce the following types of accidents: CFIT, Loss of Control, and Runway Incursion (RI). CFIT is the number one cause of fatalities in revenue service flights, and the majority of CFIT accidents, runway incursion accidents, and GA loss of control accidents can be considered to be visibility-induced crew error, where better pilot vision would have been a substantial mitigating factor. Better pilot vision is provided by synthetic/enhanced vision display systems. These technologies will serve as a substantial mitigating factor for aircraft accidents of other types as well. Such display systems will substantially reduce the following accident precursors:

- Loss of vertical and lateral spatial awareness.
- Loss of terrain and traffic awareness on approach.
- Unclear escape or go-around path even after recognition of problem.
- Loss of attitude awareness.
- Loss of situation awareness relating to the runway environment.
- Unclear path guidance on the surface.

Many laboratory research efforts have investigated replacing the conventional attitude direction indicator or primary flight display for transport airplanes with a pictorial display to increase situation awareness as well as to increase operational capability for landing in low-visibility weather conditions. These research efforts have consistently demonstrated the advantages of pictorial displays over conventional display formats, and the technologies involved in implementing such concepts appear to become available in the near term. Over the past 5 years, a number of organizations have demonstrated synthetic vision-based flight, landings, and taxi operations in research aircraft. Digital data links and displays of the positions and paths of airborne and ground traffic have also been demonstrated.

The practical implementation tasks remaining are to define requirements for display configurations and associated human performance criteria, and to resolve human performance and technology issues relating to the development of synthetic vision concepts. These same tasks remain as well for the necessary enabling technologies and the supporting infrastructure and certification strategies. Aggressive, active participation by synthetic vision advocates with appropriate standards and regulatory groups is also required.

16.3 Applications

All aircraft categories are expected to benefit from synthetic vision applications, including general aviation aircraft, rotorcraft, business jets, and commercial transports (both cargo and passenger). The concepts will emphasize the cost-effective use of synthetic/enhanced vision displays, worldwide navigation, terrain, obstruction, and airport databases, and Global Positioning System (GPS)-derived navigation to eliminate "visibility-induced" (lack of visibility) errors for all aircraft categories.

The high-end general aviation aircraft (business jets) and commercial transports application of synthetic vision will prevent CFIT and Runway Incursion (RI) accidents by improving the pilot's situation awareness of terrain, obstacle, and airport surface operations during all phases of flight, with particular emphasis on the approach and landing phases, airport surface navigation, and missed approaches. Current accident data indicate that the majority of CFIT accidents involving transports occur during non-precision approaches. This application will require the examination of technology issues related to implementation of an infrastructure for autonomous precision guidance systems. The standards committees (RTCA SC-193 and EURO-CAE WG-44) that are developing requirements for terrain, obstacles, and airport surface databases and maintaining coordination with the Federal Aviation Administration's (FAA) Local Area Augmentation System (LAAS) and Wide Area Augmentation System (WAAS) programs, are aware of synthetic vision applications.

In the U.S., runway incursions have increased an average of 15% each year for the last 4 years. Worldwide, the only airline fatalities from 1987 to 1996 due to runway incursions occurred in the U.S. A runway incursion occurs any time a plane, vehicle, person or object on the ground creates a collision hazard with an airplane that is taking off or landing at an airport under the supervision of air traffic controllers. The FAA has established the Runway Incursion Reduction Program (RIRP) to develop surface technology at major airports to help reduce runway incursions. Future activities are planned by NASA with the FAA to integrate the RIRP surface infrastructure with the flight deck to enhance situation awareness of the airport surface and further reduce the possibility of runway incursions. Runway incursion reduction efforts will target surface surveillance, GPS-based navigation, and Cockpit Display of Traffic Information (CDTI) to improve situational awareness on the surface. Also to be considered are surface route clearance methodologies and onboard alerting strategies during surface operations (runway incursion, route deviation, and hazard detection alerting).

The application of synthetic vision to low-end general aviation (GA) aircraft will advance technologies to prevent GA CFIT and loss-of-control accidents by improving the pilot's situation awareness during up and away flight. Current accident data indicate a leading cause for GA loss-of-control accidents are due to pilot disorientation after inadvertent flight into low-visibility weather conditions. A low-cost synthetic vision display system for the low-end general aviation aircraft, which often operate in marginal VMC, will enable safe landing or transit to VMC in the event of the unplanned, inadvertent encounter of IMC, including low-ceiling and low-visibility weather conditions. It will also address loss of spatial situation awareness and unusual attitude issues. Successful synthetic vision concepts will also lower the workload and increase the safety of the demanding single-pilot GA IFR operations.

Synthetic vision applications to rotorcraft will be forced to supplement the database view of the outside world with a heavier dependence on imaging sensors, because the rotorcraft environment has requirements that exceed the current expectations for the content of available databases. For example, emergency medical service vehicles operate to and from innumerable ball fields and hospitals, and at very low altitudes amongst power and telephone wires. Hence rotorcraft applications will employ more of the features of enhanced vision, although low-cost imaging sensors for civilian applications will present significant challenges.

16.4 Concepts

Synthetic vision systems are based on precise ownship positioning information relative to an onboard terrain database, and traffic information from surveillance sources (such as TCAS, ADS-B, or TIS-B, air-to-air modes of the weather radar, ASDE, AMASS, etc.). Enhanced vision systems are based on display presentations of onboard weather-penetrating sensor data combined with some synthetic vision elements.

Block Diagram: Synthetic Vision Concept

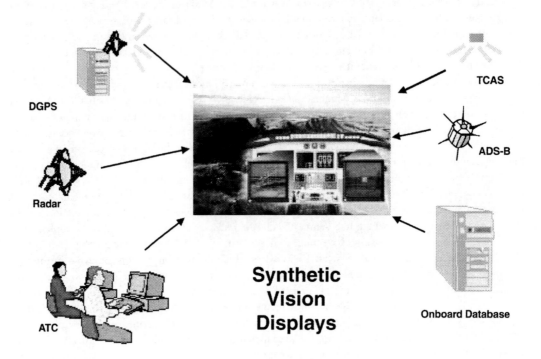

FIGURE 16.1   Possible synthetic vision system.

Figure 16.1 illustrates a top-level view of a potential high-end synthetic vision system. The specific architecture, and use of specific technology, is for illustration only. In this flight operations concept, the traffic surveillance sources of information are represented, as are the other enabling technologies of terrain/obstacle/airport databases (including curvilinear approach waypoint data), data link, and DGPS/LAAS/WAAS. Surface operations sources of surveillance information could be ASDE/AMASS (via TIS-B) or ADS-B. Controller Pilot Datalink Communications (CPDLC) may also be considered.

These system concepts address information conveyance for two separate functional levels:, tactical and strategic. Tactical concepts generally provide alerting information and may include escape guidance methodologies, while strategic concepts are more concerned with incident prevention/avoidance. These display concepts allow for presentation of three-dimensional perspective scenes with necessary and sufficient information and realism to be nearly equivalent to daylight VMC, thus increasing spatial awareness (terrain, attitude, and traffic). Symbolic information can be overlaid on these scenes to enhance situational awareness through, for example, presentation of an artificial horizon, heading, attitude indications, and pitch and/or velocity vector references. Tactical guidance capability can also be provided by highway or pathway-in-the-sky symbology for approach and landing, departure, and go-around, or other ATC-cleared routings.

16.5   Challenges

Of the technologies and issues involved in the cost-effective use of synthetic/enhanced vision displays, worldwide navigation, terrain, obstacle, and airport databases, and GPS-derived navigation to eliminate visibility-induced errors, challenges exist in the areas of human factors, avionics displays, DGPS, certification, and databases.

As with any new avionics concept, the important principles of human-centered design and other human factors considerations must be applied, as human error is a principal cause of aircraft accidents of any type. A majority of CFIT and approach, landing, takeoff, and runway incursion accidents have been attributed to visibility-induced crew error. Pilot situation and spatial awareness should be dramatically improved by synthetic vision display systems, and the maturity of the human factors discipline is such that effective designs can be confidently expected.

For avionics applications, the most significant display technology issues that have the potential for limiting the implementation of synthetic vision are

- Limited cockpit space available for display media
- Limited display capability (graphics, size, field of view, brightness, color) of current commercial aircraft
- Retrofit cost

The tactical (PFD) and strategic (Nav Display) pictorial concepts may be presented on existing form factor A, B, or D displays in a retrofit application. However, as larger displays may be more effective in presenting these concepts, other form factors will be considered for new cockpits. Other display mechanisms, such as HUDs or head-mounted systems, are not illustrated in Figure 16.1 but will be investigated. Expectations for display solutions are

- Panel-mounted for near-term applications in modern "glass" flight decks
- Head-up displays for retrofit to older "steam gauge" flight decks
- Head-mounted (and head-tracked for unlimited field-of-view) or direct in-window displays for future applications
- All display types would have guidance and other flight symbology superimposed on the terrain image

Another important enabling technology for synthetic vision systems is the DGPS infrastructure. It is anticipated that unaugmented civil-code GPS will be suitable for en route operations, and that the FAA's WAAS and/or LAAS is required for approach and landing (meter/submeter accuracy). The intent in this arena for synthetic vision system enthusiasts is to support, supplement, and complement research and modernization work currently underway by the FAA, NASA, and other governmental and private entities around the world to implement LAAS/WAAS and other precision positioning systems.

Implementation issues such as cost-effectiveness and certification, however, provide perhaps the greatest challenges to full realization of commercially viable synthetic vision systems. The successful development of a compelling business case to serve as an economic incentive over and above the safety benefits of a synthetic vision system is a significant hurdle. The leading candidate for that business case is increased operational capability in low-visibility conditions. Tactical use of synthetic vision as a replacement for today's PFD requires the certification of a flight-critical system. While certification to that level is a lengthy and expensive process, that effort is beginning. Also needed are solid, certifiable processes for database development assurance, quality assurance, and integrity assurance, and standards for the database content sources and maintenance. Efforts are also underway in this area.

Database implementation issues are equally challenging. However, the recent formulation of the joint RTCA/EUROCAE committee to develop the industry requirements for terrain, obstacle, and airport databases, indicates a desire by the world aeronautical community to solve the database issues. In line with this activity is the FAA's Notice of Proposed Rulemaking (NPRM) requiring all airplanes with turbine engines and six or more passenger seats to carry a TAWS using a computer database for providing terrain displays and warnings. To carry such technology beyond the terrain warning stage to applications of strategic planning and tactical navigation and guidance, however, compounds the database implementation issues. The most significant are

- Cost and validation of accurate high-resolution worldwide terrain, obstacle, and airport databases ($50 million by some estimates)
- Liability, maintenance, and ownership of the data

It seems clear the database implementation issues will require not only involvement of the appropriate American government agencies (FAA, NASA, NOAA, NIMA), but also ICAO and member governments' funding and sponsorship.

The capability of synthetic vision systems is limited only by the resolution and accuracy of the terrain database. Two key potential concerns with a synthetic vision approach and their mitigation strategies are as follows:

1. "How can you trust the database is correct?"

   - The terrain database will be certified to necessary standards at the start. Aircraft operations (cruise flight, approach, landing, taxi) will only be allowable to the certified fidelity of the database.
   - It will constantly improve over time. (Every clear daytime approach will be confirmation of the basic presentation.)
   - If necessary, processing of radar altimeter or existing weather radar signals may be used to "confirm" the actual terrain surface with the displayed database in real time.
   - Flight guidance cues used by the flight crew will come from the same GPS positioning as will be certifiably acceptable for instrument-only (no synthetic vision) approaches.

2. "What about obstacles or traffic that are not in the database?"

   - Airborne and ground traffic position information data-linked to the aircraft would be readily displayed
   - The proper flight path — always clear of buildings and towers — would be clearly displayed and obvious to follow.
   - The database would be updated on a regular cycle, much like today's paper approach charts.
   - Such obstacles and/or traffic are, of course, not able to be seen in today's nonsynthetic vision, low-visibility operations.
   - If necessary, additional modes may be added to the onboard weather radar to detect obstacles and traffic not in the database.
   - If necessary, imaging sensors may be added to augment the synthetic scene.

The *approximate* requirements for the database, "nested" in four layers of resolution, for example, are as follows:

Location from Airport Spacing (m)	Resolution (m)	Accuracy (m)	Grid
30 miles—Enroute	~150	~50	~500
30 − 5 miles — Approach/Departure	~50	~30	~200
5 − 0.5 miles — Landing/Takeoff	~10	~5	~50
0 miles — Surface Ops	~0.5	~1	NA
For comparison:			
Shuttle SRTM	20	8–16	30

The realization of such a database and its supporting infrastructure is somewhat dependent on the following:

- The shuttle radar topography mapping mission is expected to provide more than adequate terrain data for the enroute and approach/departure databases for approximately 80% of the earth's surface.
- The support of the National Imagery and Mapping Agency (NIMA) is considered critical for developing and releasing the worldwide enroute and approach/departure level data. International defense/security issues may limit the release of higher-resolution data.

- Local airport authorities and/or other providers (not the government) are expected to develop the landing/takeoff and surface ops databases to specified certification standards for individual airports. The safety and operational benefits to be gained by adding an airport to the evolving database are expected to be a significant motivation.

16.6   Conclusion

Synthetic vision display concepts allow for presentation of three-dimensional perspective scenes with necessary and sufficient information and realism to be equivalent to a bright, clear, sunny day, regardless of the outside weather condition, for increased spatial awareness (terrain, attitude, and traffic). Symbolic information can be overlaid on these scenes to enhance situational awareness and to provide tactical guidance capability through, for example, presentation of pathway-in-the-sky symbology. In spite of the numerous challenges and hurdles to be faced and overcome to prove synthetic vision applications practical, not just as a research demonstration, but as a viable, implementable capability, the technologies are available in the near term and the safety and operational benefits seem obvious. Solving a *visibility* problem with a *visibility* solution just plain makes sense. There is little doubt that synthetic vision-based flight *will be* the standard method for low-visibility operations in the near future.

Defining Terms

ADS-B	Automated Dependent Surveillance-Broadcast
AMASS	Airport Movement Area Safety System
ASDE	Airport Surface Detection Equipment
ATC	Air Traffic Control
CDTI	Cockpit Display of Traffic Information
CFIT	Controlled Flight into Terrain
CPDLC	Controller Pilot Datalink Communications
DGPS	Differential Global Positioning System
EuroCAE	European Organisation for Civil Aviation Equipment
FAA	Federal Aviation Administration
GA	General Aviation
GHz	Gigahertz
GPS	Global Positioning System
HUD	Head-Up Display
IFR	Instrument Flight Rules
IMC	Instrument Meteorological Conditions
LAAS	Local Area Augmentation System
NASA	National Aeronautics and Space Administration
Nav	Navigation
NIMA	National Imagery and Mapping Agency
NOAA	National Oceanic and Atmospheric Administration
NPRM	Notice of Proposed Rulemaking
PFD	Primary Flight Display
RI	Runway Incursion
RIRP	Runway Incursion Reduction Program
RTCA	Requirements and Technical Concepts for Aviation
SC	Special Committee
SRTM	Space Shuttle Radar Topography Mission
TAWS	Terrain Awareness Warning Systems
TCAS	Traffic Alert and Collision Avoidance System
TIS-B	Traffic Information Services — Broadcast

U.S.	United States
VMC	Visual Meteorological Conditions
WAAS	Wide Area Augmentation System
WG	Working Group

Further Information

NASA's Aviation Safety Program, Synthetic Vision Project: http://avsp.larc.nasa.gov/

17
Enhanced Situation Awareness

Barry C. Breen
Honeywell

17.1 Enhanced Ground Proximity Warning System

The Enhanced Ground Proximity Warning System (EGPWS)* is one of the newest systems becoming standard on all military and civilian aircraft. Its purpose is to help provide situational awareness to terrain and to provide predictive alerts for flight into terrain. This system has a long history of development and its various modes of operation and warning/advisory functionality reflect that history:

- Controlled Flight Into Terrain (CFIT) is the act of flying a perfectly operating aircraft into the ground, water, or a man-made obstruction. Historically, CFIT is the most common type of fatal accident in worldwide flying operations.

- Analysis of the conditions surrounding CFIT accidents, as evidenced by early flight recorder data, Air Traffic Control (ATC) records, and experiences of pilots in Controlled Flight Towards Terrain (CFTT) incidents, have identified common conditions which tend to precede this type of accident.

- Utilizing various onboard sensor determinations of the aircraft current state, and projecting that state dynamically into the near future, the EGPWS makes comparisons to the hazardous conditions known to precede a CFIT accident. If the conditions exceed the boundaries of safe operation, an aural and/or visual warning/advisory is given to alert the flight crew to take corrective action.

---

*There are other synonymous terms used by various government/industry facets to describe basically the same equipment. The military (historically at least) and at least one non-U.S. manufacturer refer to GPWS and EGPWS as Ground Collision Avoidance Systems (GCAS), although the military is starting to use the term EGPWS more frequently. The FAA, in its latest regulations concerning EGPWS functionality, have adopted the term Terrain Awareness Warning System (TAWS).

17.2   Fundamentals of Terrain Avoidance Warning

The current state of the aircraft is indicated by its position relative to the ground and surrounding terrain, attitude, motion vector, accelerations vector, configuration, current navigation data, and phase of flight. Depending upon operating modes (see next section) required or desired, and EGPWS model and complexity, the input set can be as simple as GPS position and pressure altitude or fairly large including altimeters, air data, flight management data, instrument navigation data, accelerometers, inertial references, etc. (see Figure 17.1.)

The primary input to the "classic" GPWS (nonenhanced) is the Low Range Radio (or Radar) Altimeter (LRRA), which calculates the height of the aircraft above the ground level (AGL) by measuring the time it takes a radio or radar beam directed at the ground to be reflected back to the aircraft. Imminent danger of ground collision is inferred by the relationship of other aircraft performance data relative to a safe height above the ground. With this type of system, level flight toward terrain can only be implied by detecting rising terrain under the aircraft; for flight towards steeply rising terrain, this may not allow enough time for corrective action by the flight crew.

The EGPWS augments the classic GPWS modes by including in its computer memory a model of the earth's terrain and man-made objects, including airport locations and runway details. With this digital terrain elevation and airports database, the computer can continuously compare the aircraft state vector to a virtual three-dimensional map of the real world, thus predicting an evolving hazardous situation much in advance of the LRRA-based GPWS algorithms.

The EGPWS usually features a colored or monochrome display of terrain safely below the aircraft (shades of *green* for terrain and *blue* for water is standard). When a potentially hazardous situation exists,

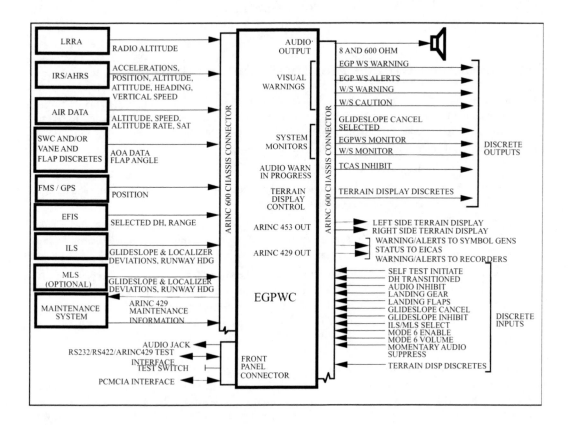

FIGURE 17.1   Typical air transport EGPWS installation.

the EGPWS alerts the flight crew with aural and/or visual warnings. Advisory (informative) situational awareness information may consist simply of an aural statement, for example, "One Thousand," as the aircraft height AGL passes from above to below 1000 ft. Cautionary alerts combine an informative aural warning, e.g., "Too Low Flaps" for flying too low and slow without yet deploying landing flaps, with a *yellow* visual alert. The cautionary visual alert can be an individual lamp with a legend such as "GPWS" or "TERRAIN;" it can be a yellow text message displayed on an Electronic Flight Instrument System (EFIS) display; or, in the case of the enhanced alert with a display of surrounding terrain, the aural "CAUTION TERRAIN" or "TERRAIN AHEAD," accompanied by both a *yellow* lamp and a rendering of the hazardous terrain on the display in *bright yellow*.

When collision with terrain is imminent and immediate drastic recovery action must be taken by the flight crew to avert disaster, the standard aural alert is a loud, commanding "PULL UP" accompanied by a *red* visual alert. Older aircraft with no terrain display utilize a single red "pull up" lamp; modern EFIS-equipped aircraft put up the words PULL UP in bright red on the Primary Flight Display (PFD). On the display of surrounding terrain, usually integrated on the EFIS Horizontal Situation Indicator, the location of hazardous terrain turns *bright red*.*

17.3 Operating Modes

The various sets of hazardous conditions that the EGPWS monitors and provides alerts for are commonly referred to as **Modes**.** These are described in detail in the following paragraphs.

Modes 1 through 4 are the original classic GPWS modes, first developed to alert the pilots to unsafe trajectory with respect to the terrain. The original analogue computer model had a single red visual lamp and a continuous siren tone as an aural alert for all modes. Aircraft manufacturer requirements caused refinement to the original modes, and added the voice "Pull Up" for Modes 1 through 4 and a new Mode 5 "Glideslope". Mode 6 was added with the first digital computer models about the time of Boeing 757/767 aircraft introduction; and Mode 7 was added when windshear detection became a requirement in about 1985.***

The latest addition to the EGPWS are the Enhanced Modes: Terrain Proximity Display, Terrain Ahead Detection, and Terrain Clearance Floor. For many years, pilot advocates of GPWS requested that Mode 2 be augmented with a display of approaching terrain. Advances in memory density, lower costs, increased computing power, and the availability of high-resolution maps and Digital Terrain Elevation Databases (DTED) enabled this advancement. Once displayable terrain elevation database became a technical and economic reality, the obvious next step was to use the data to *look ahead* of the aircraft path and predict terrain conflict well before it happened, rather than waiting for the downward-looking sensors.

Combining the DTED with a database of airport runway locations, heights, and headings allows the final improvement — warnings for normal landing attempts where there is no runway.

---

*Note that all of the EGPWS visual indication examples in this overview discussion are consistent with the requirements of FAR 25.1322.

**The EGPWS modes described here are the most common for commercial and military transport applications. Not discussed here are more specialised warning algorithms, more closely related to terrain-following technology, that have been developed for military high-speed low-altitude operations. These are more related to advanced terrain database guidance, which is outside the scope of enhanced situation awareness function.

***Though not considered CFIT, analysis of windshear-related accidents has resulted in the development of reactive windshear detection algorithms. At the request of Boeing, their specific reactive windshear detection algorithm was hosted in the standard commercial GPWS, about the same time the 737-3/4/500 series aircraft was developed. By convention this became Mode 7 in the GPWS. The most common commercially available EGPWS computer contains a Mode 7 consisting of both Boeing and non-Boeing reactive windshear detection algorithms, although not all aircraft installations will use Mode 7. There also exist "standalone" reactive windshear detection computers; and some aircraft use only predictve wind shear detection, which is a function of weather radar.

17.3.1   Mode 1 — Excessive Descent Rate

The first ground proximity mode warns of an excessive descending barometric altitude rate near the ground, regardless of terrain profile. The original warning was a straight line at 4000 ft/min barometric sinkrate, enabled at 2400 ft AGL, just below the altitude at which the standard commercial radio altimeters came into track (2500 ft AGL). This has been refined over the years to a current standard for Mode 1 consisting of two curves, an outer cautionary alert and a more stringent inner warning boundary. Exceeding the limits of the outer curve results in the voice alert "Sinkrate;" exceeding the inner curve results in the voice alert "Pull Up."

Figure 17.2 illustrates the various Mode 1 curves, including the current standard air transport warnings, the DO-161A minimum warning requirement, the original curve and the Class B TSO C151

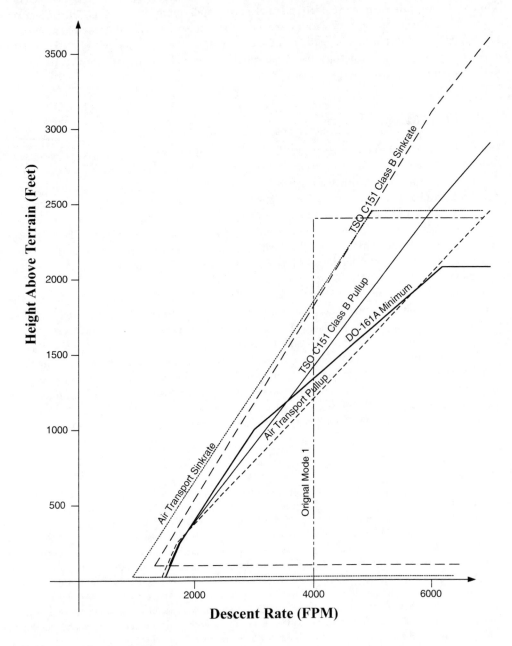

FIGURE 17.2   Mode 1 warning curves.

curves for 6 to 9 passenger aircraft and general aviation. Note that the Class B curves, which use GPS height above the terrain database instead of radio altitude are not limited to the standard commercial radio altimeter range of 2500 ft AGL.

17.3.2   Mode 2 — Excessive Closure Rate

Rate-of-change of radio altitude is termed the closure rate, with the positive sense meaning that the aircraft and the ground are coming closer together. When the closure rate initially exceeds the Mode 2 warning boundary, the alert "Terrain Terrain" is given. If the warning condition persists, the voice is changed to a "Pull Up" alert.

Closure rate detection curves are the most difficult of the classic GPWS algorithms to design. Tall buildings, towers, trees, and rock escarpments in the area of final approach can cause sharp spikes in the computed closure rate. Modern Mode 2 algorithms employ complex filtering of the computed rate, with varying dynamic response dependent upon phase of flight and aircraft configuration. The Mode 2 detection algorithm is also modified by specific problem areas by using latitude, longitude, heading, and selected runway course — a technique in the EGPWS termed "Envelope Modulation."

Landing configuration closure rate warnings are termed Mode 2B; cruise and approach configurations are termed Mode 2A. Figure 17.3 illustrates some of the various Mode 2A curves, including the original first Mode 2 curve, the current standard air transport Mode 2A "Terrain-Terrain-Pull-Up" warning curve, and the DO-161A nominal Mode 2A warning requirement. Note that the Class B TSO C151 EGPWS does not use a Radio Altimeter and therefore has no Mode 2.

17.3.3   Mode 3 — Accelerating Flight Path Back into the Terrain after Take-off

Mode 3 (Figure 17.4) is active from liftoff until a safe altitude is reached. This mode warns for failure to continue to gain altitude. The original Mode 3, still specified in DO-161A as Mode 3A, produced warnings for any negative sinkrate after take-off until 700 ft of ground clearance was reached. The mode has since been redesigned (designated 3B in DO-161A) to allow short-term sink after take-off but detect a trend to a lack of climb situation. The voice callout for Mode 3 is "Don't Sink." This take-off mode now remains active until a time-integrated ground clearance value is exceeded; thus allowing for a longer protection time with low-altitude noise abatement maneuvering before climb-out.

Altitude loss is computed by either sampling and differentiating altitude MSL or integrating altitude rate during loss of altitude. Because a loss is being measured, the altitude can be a corrected or uncorrected pressure altitude, or an inertial or GPS height. Typical Mode 3 curves are linear, with warnings for an 8-ft loss at 30 ft AGL, increasing to a 143-ft loss at 1500 ft AGL.

17.3.4   Mode 4 — Unsafe Terrain Clearance Based on Aircraft Configuration

The earliest version of Mode 4 was a simple alert for descent below 500 ft with the landing gear up. Second generations of Mode 4 added additional alerting at lower altitudes for flaps not in landing position. The warning altitude for flaps was raised to the 500-ft level for higher descent rates. There are three of these types of Mode 4 curves still specified as alternate minimum performance requirements in DO-161A (see Figure 17.5).

Modern Mode 4 curves are airspeed-enhanced, rather than descent rate alone, and for high airspeeds will give alerts at altitudes up to 1000 ft AGL.

Currently, EGPWS Mode 4 has three types of alerts based upon height AGL, mach/airspeed, and aircraft configuration, termed Modes 4A, 4B, and 4C (Figure 17.6). Two of the curves (4A, 4B) are active during cruise until full landing configuration is achieved with a descent "close to the ground" — typically 700 ft for a transport aircraft. Mode 4C is active on take-off in conjunction with the previously described

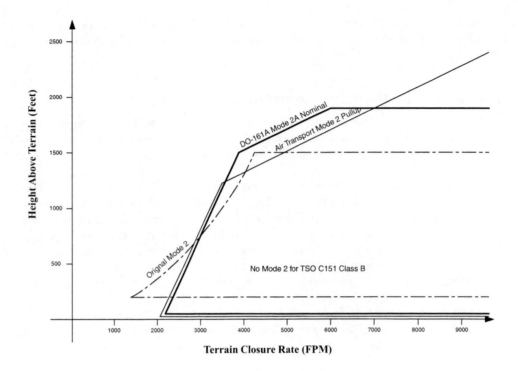

FIGURE 17.3   Mode 2 curves.

Mode 3. All three alerts are designed with the intent to warn of flight "too close to the ground" for the current speed/configuration combination. At higher speeds, the alert commences at higher AGL and the voice alert is always "Too Low Terrain." At lower speeds, Mode 4A warning is "Too Low Gear" and the Mode 4B warning is "Too Low Flaps."

Mode 4C compliments Mode 3, which warns on an absolute loss of altitude on climb-out, by requiring a continuous gain in height above the terrain. If the aircraft is rising, but the terrain under is also rising, Mode 4C will alert "Too Low Terrain" on take-off if sufficient terrain clearance is not achieved prior to Mode 3 switching out.

17.3.5   Mode 5 — Significant Descent Below the ILS Landing Glide Path Approach Aid

This Mode warns for failure to remain on an instrument glide path on approach. Typical warning curves alert for 1.5 to 2.0 dots below the beam, with a wider divergence allowed at lower altitudes. The alerts and warnings are only enabled when the crew is flying an ILS approach, as determined by radio frequency selections and switch selection. Most installations also include separate enable switch and a warning cancel for crew use when flying some combination of visual and or other landing aids and deviation from the ILS glide path is intentional. Although the mode is typically active from 1000 ft AGL down to about 30 ft, allowance in the design of the alerts must also be made for beam capture from below, and level maneuvering between 500 and 1000 ft without nuisance alerting.

Figure 17.7 illustrates the Mode 5 warnings for a typical jet air transport. When the outer curve is penetrated, the voice message "Glideslope" is repeated at a low volume. If the deviation below the beam increases or altitude decreases, the repetition rate of the voice is increased. If the altitude/deviation combination falls within the inner curve, the voice volume increases to the equivalent of a warning message and the repetition rate is at maximum.

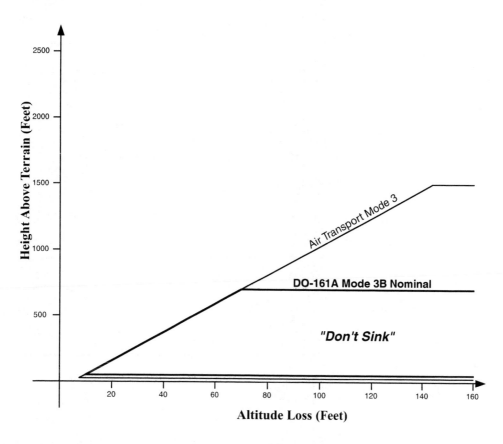

FIGURE 17.4 Mode 3 curves.

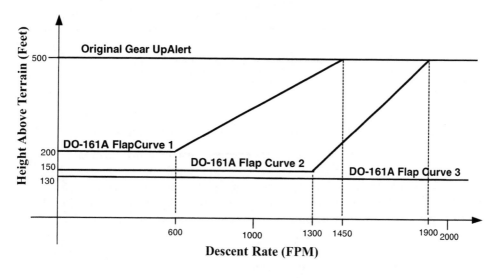

FIGURE 17. 5 Old GPWS Mode 4 curves.

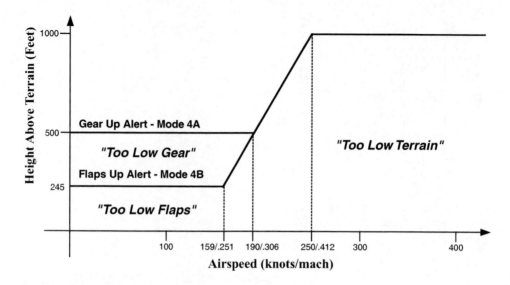

FIGURE 17.6   EGPWS Mode 4.

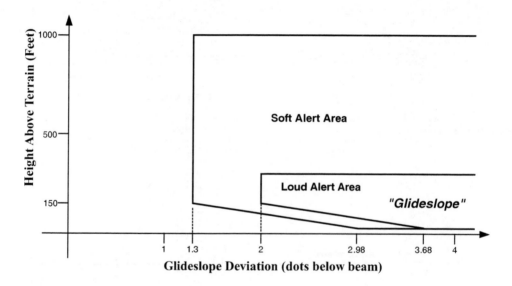

FIGURE 17.7   EGPWS Mode 5.

17.3.6   Mode 6 — Miscellaneous Callouts and Advisories

The first application of this Mode consisted of a voice alert added to the activation of the decision height discrete on older analog radio altimeters. This voice alert is "Minimums" or "Decision Height," which adds an extra level of awareness during the landing decision point in the final approach procedure. Traditionally, this callout would be made by the pilot not flying (PNF). Automating the callout frees the PNF from one small task enabling him to more easily monitor other parameters during the final approach.

This mode has since been expanded as a "catch all" of miscellaneous aural callouts requested by air transport manufacturers and operators, many of which also were normally an operational duty of the PNF (see Figure 17.8). In addition to the radio altitude decision height, callouts are now available at barometric minimums, at an altitude approaching the decision height or barometric minimums, or at various

Callout Voice	Description
Radio Altimeter	Activates at 2500 feet as radio altimeter comes into track
Twenty five hundred	(alternate to Radio Altimeter)
One Thousand	Activates at 1000 feet AGL
Five Hundred (smart)	Activates at 500 feet AGL for non-precision approaches only
One Hundred	Activates at 100 feet AGL
Fifty	50 feet AGL
Forty	40 feet AGL
Thirty	30 feet AGL
Twenty	20 feet AGL
Ten	10 feet AGL
Approaching Minimums	100 feet above the selected decision height
Minimums	At pilot selected decision height – may be AGL or barometric
Decision Height	(alternate to Minimums)

FIGURE 17.8   Examples of EGPWS Mode 6 callouts.

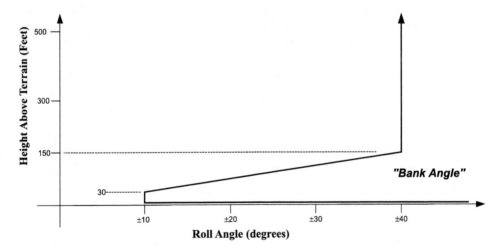

FIGURE 17.9   EGPWS Mode 6 overbank (excessive roll) alert.

combinations of specific altitudes. There are also "smart callouts" available, that only call the altitude for nonprecision approaches (ILS not tuned). The EGPWS model used on Boeing aircraft will also callout for V_1 on take-off and give aural "engine out" warnings. Finally, included in the set of Mode 6 callouts are warnings of overbanking (excessive roll angle).

17.3.7   Mode 7 — Flight into Windshear Conditions

Windshear is a sudden change in wind direction and/or windspeed over a relatively short distance in the atmosphere and can have a detrimental effect on the performance of an aircraft. The magnitude of a windshear is defined precisely in engineering terms by the sum of the rate of change of horizontal wind, and the vertical wind divided by the true airspeed of the aircraft:

$$ F = -\left(\frac{w_{\text{wind}}}{V_A} + \frac{\dot{u}_{\text{wind}}}{g} \right) $$

where

  F is expressed in units of g and is positive for increasing energy windshears
  w_{wind} = vertical wind velocity (fps), positive for downdrafts
  $\dot{u}_{wind} = \dfrac{du_{wind}}{dt}$ = rate of change of horizontal wind velocity
  V_A = true airspeed (fps)
  g = gravitational acceleration, 32.178 fps^2

There are various techniques for computing this windshear factor from onboard aircraft sensors (air data, inertial accelerations, etc.). The EGPWS performs this computation and alerts the crew with the aural "Windshear, Windshear, Windshear," when the factor exceeds predefined limits as required by TSO C117a.

17.3.8   Envelope Modulation

Early GPWS equipment was plagued by false and nuisance warnings, causing pilots to distrust the equipment when actual hazardous conditions existed. Many approach profiles and radar vectoring situations violated the best-selected warning curve designs. Even as GPWS algorithms were improved, there still existed some approaches that required close proximity to terrain prior to landing.

Modern GPWS equipment adapts to this problem by storing a table of known problem locations and providing specialized warning envelope changes when the aircraft is operating in these areas. This technique is known as GPWS envelope modulation.

An example exists in the southerly directed approaches to Glasgow Scotland, Runway 23. The standard approach procedures allow an aircraft flying level at 3000 ft barometric altitude to pass over mountain peaks with heights above 1700 ft when approaching this runway. At nominal airspeeds the difference in surrounding terrain height will generate closure rates well within the nominal curve of Figure 17.3. With the envelope modulation feature the GPWS, using latitude, longitude, and heading, notes that the aircraft is flying over this specific area and temporarily lowers the maximum warning altitude for Mode 2 from 2450 ft to the minimum 1250 ft AGL. This eliminates the nuisance warning while at the same time providing the minimum required DO-161A protection for inadvertent flight closer to the mountain peaks on the approach path.

17.3.9   "Enhanced Modes"

The enhanced modes provide terrain and obstacle awareness beyond the normal sensor-derived capabilities of the standard GPWS. Standard GPWS warning curves are deficient in two areas, even with the best designs. One area is immediately surrounding the airport; which is where a large majority of CFIT accidents occur. The other is flight directly into precipitous terrain, for which little or no Mode 2 warning time may occur.

The enhanced modes solve these problems by making use of a database of terrain and obstacle spot heights and airport runway locations arranged in a grid addressed by latitude and longitude. This combined terrain/airports/obstacle database — a virtual world within the computer — provides the ability to track the aircraft position in the real world given accurate x-y-z position combined with the aircraft velocity vector.

This database technique allows three improvements which overcome the standard GPWS modes shortcomings: terrain proximity display, terrain ahead alerting, and terrain clearance floor.

17.3.9.1   Terrain Proximity Display

The terrain proximity display is a particular case of a horizontal (plan view) moving map designed to enhance vertical and horizontal situational awareness. The basic display is based upon human factors studies recommending a minimum of contours and minimum of coloring. The display is purposely compatible with existing three-color weather radar displays, allowing economical upgrade of existing equipment.

Terrain well below the flight path of the aircraft is depicted in shades of green, brighter green being closer to the aircraft and sparse green-to-black for terrain far below the aircraft. Some displays additionally allow water areas to be shown in cyan (blue). Terrain in the proximity of the aircraft flight path, but posing no immediate danger (it can be easily flown over or around) is depicted in shades of yellow. Terrain well above the aircraft (nominally more than 2000 ft above flight level), toward which continued safe flight is not possible, is shown in shades of red.

17.3.9.2   Terrain Ahead Alerting

Terrain (and/or obstruction) alerting algorithms continually compare the state of the aircraft flight to the virtual world and provide visual and/or aural alerts if impact is possible or probable. Two levels of alerting are provided, a cautionary alert and a hard warning. The alerting algorithm design is such that, for a steady approach to hazardous terrain, the cautionary alert is given much in advance of the warning alert. Typical design criteria may try to issue caution up to 60s in advance of a problem and a warning within 30s.

Voice alerts for the cautionary alert are "Caution, Terrain" or "Terrain Ahead." For the warnings on turboprop and turbojet aircraft, the warning aural is "Terrain Terrain Pullup" or "Terrain Ahead Pullup," with the pullups being repeated continuously until the aircraft flight path is altered to avoid the terrain.

In conjunction with the aural alerts, yellow and red lamps may be illuminated, such as with the standard GPWS alerts. The more compelling visual alerts are given by means of the Terrain Awareness Display. Those areas that meet the criteria for the cautionary alert are illuminated in a bright yellow on the display. If the pullup alert occurs, those areas of terrain where an immediate impact hazard exists are illuminated in bright red. When the aircraft flight path is altered to avoid the terrain, the display returns to the normal terrain proximity depictions as the aural alerts cease.

17.3.9.3   Terrain Clearance Floor

The standard Modes 2 and 4 are desensitized when the aircraft is put in landing configuration (flaps down and/or gear lowered) and thus fail to alert for attempts at landing where there is no airport. Since the EGPWS database contains the exact position of all allowable airport runways, it is possible to define an additional alert, a terrain clearance floor, at all areas where there are no runways. When the aircraft descends below this floor value, the voice alert "Too Low Terrain" is given. This enhanced mode alert is also referred to as premature descent alert.

17.4   EGPWS Standards

ARINC 594 — Ground Proximity Warning System: This is the first ARINC characteristic for Ground Proximity Warning Systems and defines the original analog interfaced system. It applies to the original model (MkI and MkII) GPWS systems featuring Modes 1–5, manufactured by Sundstrand Data Control, Bendix, Collins, Litton and others. It also applies to the AlliedSignal (Honeywell) MkVII digital GPWS, which featured Modes 1–7 and a primarily analog interface for upgrading older models.

ARINC 743 — Ground Proximity Warning System: This characteristic applies to primarily digital (per ARINC 429 DITS) interfaced Ground Proximity Warning Systems, such as the AlliedSignal/Honeywell MkV series, which was standard on all newer Boeing aircraft from the 757/767 up through the introduction of the 777.

ARINC 762 — Terrain Avoidance and Warning System: This characteristic, still in draft form at the time of this writing, is an update of ARINC 743 applicable to the primarily digital interfaced (MkV) Enhanced GPWS.

ARINC 562 — Terrain Avoidance and Warning System: This proposed ARINC characteristic will be an update of ARINC 594, applicable to the primarily analog interfaced (MkVII) Enhanced GPWS.

RTCA DO-161A — Minimum Performance Standards, Airborne Ground Proximity Warning System: This 1976 document still provides the minimum standards for the classic GPWS Modes 1–5. It is required by both TSO C92c and the new TSO C151 for EGPWS (TAWS).

TSO C92c — Ground Proximity Warning, Glideslope Deviation Alerting Equipment: This TSO covers the classic Modes 1– 6 minimum performance standards. It basically references DO-161A and customizes and adds features of the classic GPWS which were added subsequent to DO-161A, including voice callouts signifying the reason for the alert/warnings, Mode 6 callouts, and bank angle alerting.

CAA Specification 14 (U.K.) — Ground Proximity Warning Systems: This is the United Kingdom CAA standard for Modes 1–5 and also specifies some installation requirements. As with the U.S. TSOs, Spec 14 references DO-161A and customizes and augments features of the classic GPWS which are still required for U.K. approvals. Most notably, the U.K. version of Mode 5 is less stringent and requires a visual indication of Mode 5 cancellation. Spec 14 also requires that a stall warning inhibit the GPWS voice callouts, a feature which is found only on U.K.-certified installations.

TSO C117a — Airborne Windshear Warning and Escape Guidance Systems for Transport Airplanes: This TSO defines the requirements for EGPWS Mode 7, reactive low level windshear detection.

TSO C151a — Terrain Awareness and Warning System (TAWS): This new TSO supersedes TSO C92c for certain classes of aircraft being required to feature the enhanced modes. It also extends coverage down to smaller aircraft, in anticipation of further rulemaking requiring GPWS type equipment. It describes two classes of TAWS equipment, the standard EGPWS becomes Class A. For smaller aircraft with limited equipment, a new Class B set of requirements are created that add a subset of the DO-161A modes that can be accomplished soley with a source of three-dimensional position and an airports and terrain database.

FAR 121.360 — This rule requires GPWS on most "for revenue" passenger aircraft, including air transport, charters, and regional airlines.

TAWS NPRM — This proposed rulemaking would replace FAR 121.360 and also modify Parts 135 and 91 to require GPWS per TSO C92c or EGPWS per TSO C151 Class A or B on all turbine-powered aircraft of 6 or 10 seats or more. The final form of the proposed rule is fluid at the time of this writing but is expected to released in March of 2000.

Further Information

1. *Controlled Flight Into Terrain, Education and Training Aid* — This joint publication of ICAO, Flight Safety Foundation, and DOT/FAA consists of two loose-leaf volumes and an accompanying video tape. It is targeted toward the air transport industry, containing management, operations, and crew training information, including GPWS. Copies may be obtained by contacting the Flight Safety Foundation, Alexandria, Virginia.

2. *DOT Volpe NTSC Reports on CFIT and GPWS* — These may be obtained from the USDOT and contain accident analyses, statistics, and studies of the effectivity of both the classic and enhanced GPWS warning modes. There are a number of these reports which were developed in response to NTSB requests. Of the two most recent reports, the second one pertains to the Enhanced GPWS in particular:

   - Spiller, David — Investigation of Controlled Flight Into Terrain (CFIT) Accidents Involving Multi-engine Fixed-wing Aircraft Operating Under Part 135 and the Potential Application of a Ground Proximity Warning System (Cambridge, MA: U.S. Department of Transportation, Volpe National Transportation Systems Center) March 1989.

   - Phillips, Robert O. — Investigation of Controlled Flight Into Terrain Aircraft Accidents Involving Turbine Powered Aircraft with Six or More Passenger Seats Flying Under FAR Part 91 Flight Rules and the Potential for Their Prevention by Ground Proximity Warning Systems (Cambridge, MA: U.S. Department of Transportation, Volpe National Transportation Systems Center) March 1996.

18

TCAS II

Steve Henely
Rockwell Collins

18.1   Introduction

The Traffic Alert and Collision Avoidance System (TCAS) provides a solution to the problem of reducing the risk of midair collisions between aircraft. TCAS is a family of airborne systems that function independently of ground-based air traffic control (ATC) to provide collision avoidance protection. The TCAS concept makes use of the radar beacon transponders carried by aircraft for ground ATC purposes and provides no protection against aircraft that do not have an operating transponder.

TCAS I provides proximity warning only, to aid the pilot in the visual acquisition of potential threat aircraft. TCAS II provides traffic advisories and resolution advisories (recommended evasive maneuvers) in a vertical direction to avoid conflicting traffic. Development of TCAS III, which was to provide traffic advisories and resolution advisories in the horizontal as well as the vertical direction, was discontinued in favor of emerging systems such as the ADS-B system discussed elsewhere in this book. This chapter will focus on TCAS II.

Based on a congressional mandate (Public Law 100-223), the Federal Aviation Administration (FAA) issued a rule effective February 9, 1989 that required the equipage of TCAS II on airline aircraft with more than 30 seats by December 30, 1991. Public Law 100-223 was later amended (Public Law 101-236) to permit the FAA to extend the deadline for TCAS II fleetwide implementation to December 30, 1993. In December of 1998 the FAA released a Technical Standard Order (TSO) that approved Change 7, resulting in the DO-185A TCAS II requirement. Change 7 incorporates software enhancements to reduce the number of false alerts. TCAS equipage on aircraft with 30 or more seats has been mandated in India, Argentina, Germany, Australia, and Hong Kong by the year 2000.

18.2   Components

TCAS II consists of the Mode S/TCAS Control Panel, the Mode S transponder, the TCAS computer, antennas, traffic and resolution advisory displays, and an aural annunciator. Figure 18.1 is a block diagram of TCAS II. Control information from the Mode S/TCAS Control Panel is provided to the TCAS computer via the Mode S Transponder. TCAS II uses a directional antenna, mounted on top of the aircraft. In addition to receiving range and altitude data on targets above the aircraft, this directional antenna is used to transmit interrogations at varying power levels in each of four 90° azimuth segments. An omnidirectional transmitting and receiving antenna is mounted at the bottom of the aircraft to provide

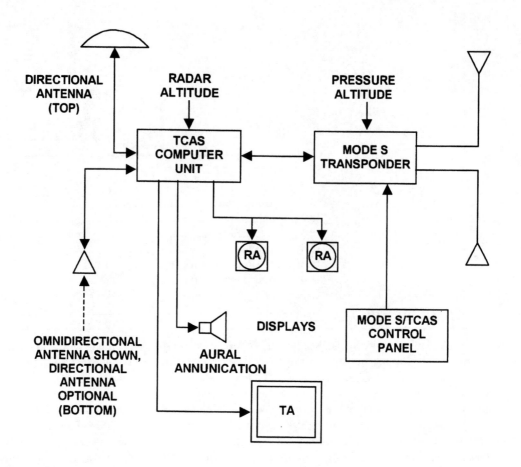

FIGURE 18.1   TCAS II block diagram.

TCAS with range and altitude data from traffic that is below the aircraft. TCAS II transmits transponder interrogations on 1030 MHz and receives transponder replies on 1090 MHz.

The Traffic Advisory (TA) display depicts the position of the traffic relative to the TCAS aircraft to assist the pilot in visually acquiring threatening aircraft. The Resolution Advisory (RA) can be displayed on a standard Vertical Speed Indicator (VSI), modified to indicate the vertical rate that must be achieved to maintain safe separation from threatening aircraft. When an RA is generated, the TCAS II computer lights up the appropriate display segments and RA compliance is accomplished by flying to keep the VSI needle out of the red segments. On newer aircraft, the RA display function is integrated into the Primary Flight Display (PFD). Displayed traffic and resolution advisories are supplemented by synthetic voice advisories generated by the TCAS II computer.

18.3   Surveillance

TCAS listens for the broadcast transmission (squitters) which is generated once per second by the Mode S transponder and contains the discrete Mode S address of the sending aircraft. Upon receipt of a valid squitter message the transmitting aircraft identification is added to a list of aircraft the TCAS aircraft will interrogate. Figure 18.2 shows the interrogation/reply communications between TCAS systems. TCAS sends an interrogation to the Mode S transponder with the discrete Mode S address contained in the squitter message. From the reply, TCAS can determine the range and the altitude of the interrogated aircraft.

There is no selective addressing capability with Mode A/C transponders, so TCAS uses the Mode C only all-call message to interrogate these types of Mode A/C transponders at a nominal rate of once per

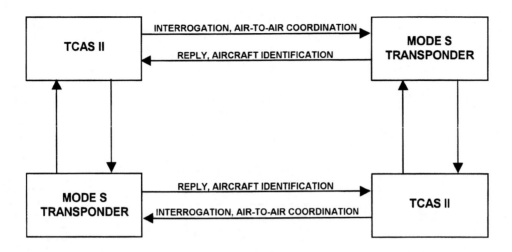

FIGURE 18.2 Interrogation/Reply between TCAS systems.

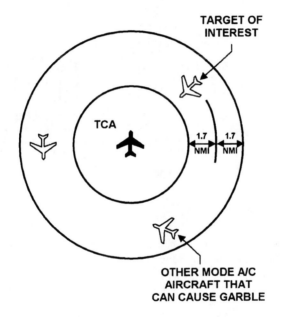

FIGURE 18.3 Synchronous garble area.

second. Mode C transponders reply with altitude data while Mode A transponders reply with no data in the altitude field. All Mode A/C transponders that receive a Mode C all-call interrogation from TCAS will reply. Since the length of the reply is 21 μs, Mode A/C-equipped aircraft within a range difference of 1.7 nmi from the TCAS will generate replies that overlap each other, as shown in Figure 18.3. These overlapping Mode A/C replies are known as synchronous garble.

Hardware degarblers can reliably decode up to three overlapping replies. The Whisper-Shout technique and directional transmissions can be used to reduce the number of transponders that reply to a single interrogation. A low power level is used for the first interrogation step in a Whisper-Shout sequence. In the second Whisper-Shout step, a suppression pulse is first transmitted at a slightly lower level than the first interrogation, followed 2 μs later by an interrogation at a slightly higher power level than the first interrogation. The Whisper-Shout procedure shown in Figure 18.4 reduces the possibility of garble

TABLE 18.1 Sensitivity Level Selection Based on Altitude

| Altitude (in Feet) | Sensitivity Level | Tau Values (in Seconds) | |
		TA	RA
0–1,000 AGL	2	20	N.A.
1,000–2,350 AGL	3	25	15
2,350–5,000 MSL	4	30	20
5,000–10,000 MSL	5	40	25
10,000–20,000 MSL	6	45	30
20,000–42,000 MSL	7	48	35
Greater than 42,000 MSL	7	48	35

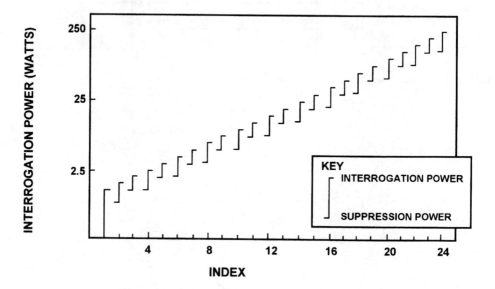

FIGURE 18.4 Whisper-Shout interrogation.

by suppressing most of the transponders that had replied to the previous interrogation, but eliciting replies from an additional group of transponders that did not reply to the previous interrogation. Directional interrogation transmissions further reduce the number of potential overlapping replies.

18.4 Protected Airspace

One of the most important milestones in the quest for an effective collision avoidance system is the development of the range/range rate (tau). This concept is based on time-to-go, rather than distance-to-go, to the closest point of approach. Effective collision avoidance logic involves a trade-off between providing the necessary protection with the detection of valid threats while at the same time avoiding false alarms. This trade-off is accomplished by controlling the sensitivity level, which determines the tau, and therefore the dimensions of the protected airspace around each TCAS-equipped aircraft.

The pilot can select three modes of TCAS operation: STANDBY, TA-ONLY, and AUTOMATIC. These modes are used by the TCAS logic to determine the sensitivity level. When the STANDBY mode is selected, the TCAS equipment does not transmit interrogations. Normally, the STANDBY mode is used when the aircraft is on the ground. In TA-ONLY mode, the equipment performs all of the surveillance functions and provides TAs but not RAs. The TA-ONLY mode is used to avoid unnecessary distractions while at low altitudes and on final approach to an airport. When the pilot selects AUTOMATIC mode, the TCAS logic selects the sensitivity level based on the current altitude of the aircraft. Table 18.1 shows the altitude thresholds at which TCAS automatically changes its sensitivity level selection and the associated tau values for altitude-reporting aircraft.

The boundary lines depicted in Figure 18.5 show the combinations of range and range rate that would trigger a TA with a 40s tau and an RA with a 25s tau. These TA and RA values correspond to sensitivity level 5 from Table 18.1. As shown in Figure 18.5, the boundary lines are modified at close range to provide added protection against slow closure encounters.

18.5   Collision Avoidance Logic

The collision avoidance logic functions are shown in Figure 18.6. This description of the collision avoidance logic is meant to provide a general overview. There are many special conditions relating to particular geometry, thresholds, and equipment configurations that are not covered in this description. Using surveillance reports, the collision avoidance logic tracks the slant range and closing speed of each target to determine the time in seconds until the closest point of approach. If the target is equipped with an altitude-encoding transponder, collision avoidance logic can project the altitude of the target at the closest point of approach.

A range test must be met and the vertical separation at the closest point of approach must be within 850 ft for an altitude-reporting target to be declared a potential threat and a traffic advisory to be generated. The range test is based on the RA tau plus approximately 15 s. A non-altitude-reporting target is declared a potential threat if the range test alone shows that the calculated tau is within the RA tau threshold associated with the sensitivity level being used.

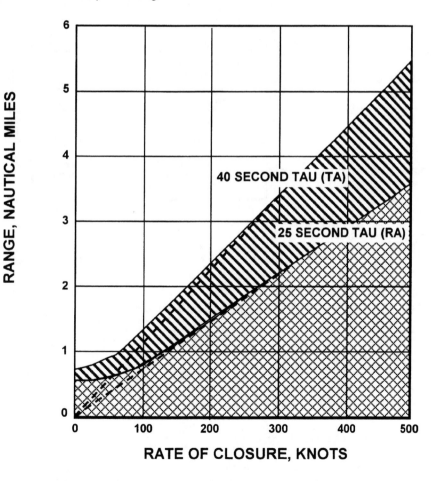

FIGURE 18.5   TA/RA Tau values for sensitivity level 5.

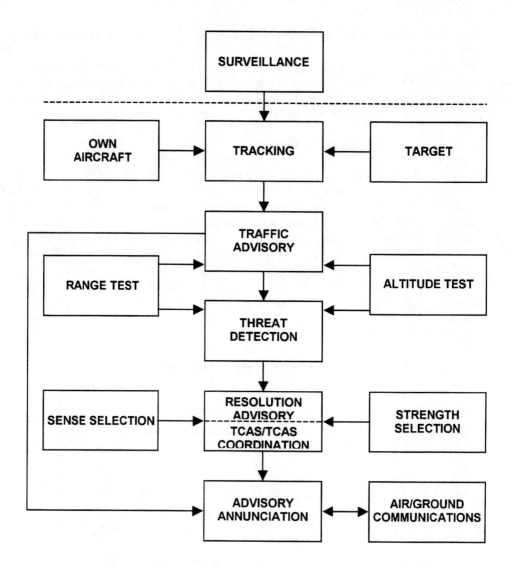

FIGURE 18.6 CAS logic functions.

A two-step process is used to determine the type of resolution advisory to be selected when a threat is declared. The first step is to select the sense (upward or downward) of the resolution advisory. Based on the range and altitude tracks of the potential threat, the collision avoidance logic models the potential threat's path to the closest point of approach and selects the resolution advisory sense that provides the greater vertical separation. The second resolution advisory step is to select the strength of the resolution advisory. The least disruptive vertical rate maneuver that will achieve safe separation is selected. Possible resolution advisories are listed in Table 18.2.

In a TCAS/TCAS encounter, each aircraft transmits Mode S coordination interrogations to the other to ensure the selection of complementary resolution advisories. Coordination interrogations contain information about an aircraft's intended vertical maneuver.

18.6   Cockpit Presentation

The traffic advisory display can either be a dedicated TCAS display or a joint-use weather radar and traffic display (see Figure 18.10). In some aircraft, the traffic advisory display will be an electronic flight

TABLE 18.2　Resolution Advisories

Upward Sense	Type	Downward Sense
Increase Climb to 2500 fpm	Positive	Increase Descent to 2500 fpm
Reversal to Climb	Positive	Reversal to Descend
Maintain Climb	Positive	Maintain Descent
Crossover Climb	Positive	Crossover Descend
Climb	Positive	Descend
Don't Descend	Negative vsl	Don't Climb
Don't Descend >500 fpm	Negative vsl	Don't Climb >500 fpm
Don't Descend >1000 fpm	Negative vsl	Don't Climb >1000 fpm
Don't Descend >2000 fpm	Negative vsl	Don't Climb >2000 fpm

Note: Any combination of climb and descent restrictions may be given simultaneously (normally in multi-aircraft encounters); fpm = feet per minute; vsl = vertical speed limit.

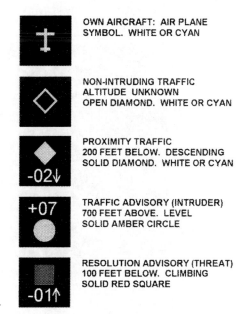

OWN AIRCRAFT: AIR PLANE SYMBOL. WHITE OR CYAN

NON-INTRUDING TRAFFIC ALTITUDE UNKNOWN OPEN DIAMOND. WHITE OR CYAN

PROXIMITY TRAFFIC 200 FEET BELOW. DESCENDING SOLID DIAMOND. WHITE OR CYAN

TRAFFIC ADVISORY (INTRUDER) 700 FEET ABOVE. LEVEL SOLID AMBER CIRCLE

RESOLUTION ADVISORY (THREAT) 100 FEET BELOW. CLIMBING SOLID RED SQUARE

FIGURE 18.7　Standardized symbology for TA display.

instrument system (EFIS) or flat panel display that combines traffic and resolution advisory information on the same display. Targets of interest on the traffic advisory display are depicted in various shapes and colors as shown in Figure 18.7.

The pilot uses the resolution advisory display to determine whether an adjustment in aircraft vertical rate is necessary to comply with the resolution advisory determined by TCAS. This determination is based on the position of the vertical speed indicator needle with respect to the lighted segments. If the needle is in the red segments, the pilot should change the aircraft vertical rate until the needle falls within the green "fly-to" segment. This type of indication is called a corrective resolution advisory. A preventive resolution advisory is when the needle is outside the red segments and the pilot should simply maintain the current vertical rate. The green segment is lit only for corrective resolution advisories. Resolution advisory display indications corresponding to typical encounters are shown in Figure 18.8.

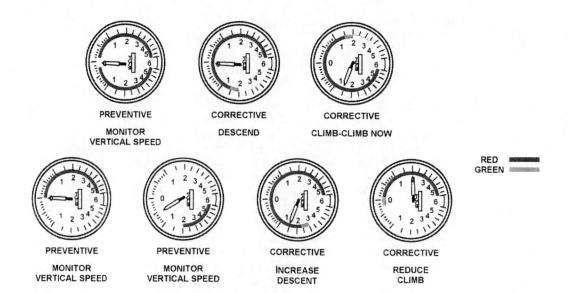

PREVENTIVE
MONITOR
VERTICAL SPEED

CORRECTIVE
DESCEND

CORRECTIVE
CLIMB-CLIMB NOW

RED
GREEN

PREVENTIVE
MONITOR
VERTICAL SPEED

PREVENTIVE
MONITOR
VERTICAL SPEED

CORRECTIVE
INCREASE
DESCENT

CORRECTIVE
REDUCE
CLIMB

FIGURE 18.8   Typical resolution advisory indications.

FIGURE 18.9   Combined traffic advisory/resolution advisory display.

Figure 18.9 shows a combined traffic advisory/resolution advisory display indicating a traffic advisory (potential threat 200 ft below), resolution advisory (threat 100 ft above) and nonthreatening traffic (1200 ft above). The airplane symbol on the lower middle section of the display indicates the location of the aircraft relative to traffic. Figure 18.10 shows an example of a joint-use weather radar and traffic display.

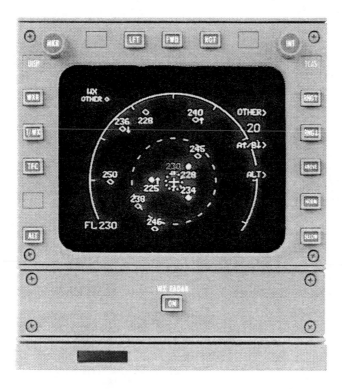

FIGURE 18.10 Joint use weather radar and traffic display.

19

Setting Requirements

Cary R. Spitzer
AvioniCon, Inc.

19.1   Requirements-Setting for Avionics Systems

Proper requirements are the *sine qua non* for building an acceptable avionics system. It is inescapable: No avionics systems can perform as expected by the customer unless the customer requirements, along with requirements from other stakeholders and relevant regulations and standards, are completely documented and understood by the avionics manufacturer. Safety, mission, cost, and certification drive the requirements.

For all aircraft, safety of flight in all possible flight regimes is the prime requirement. All aircraft without the possibility of ejection in case of an emergency typically have a probability of catastrophic failure on the order 10^{-9} per flight hour. If the crew has the potential to eject in case of an emergency the probability of failure is somewhat less demanding, but still significant.

Second only to safety, the mission of the aircraft is the principal driver of requirements. Mission requirements may be in terms of aircraft performance, ground turnaround times, or maintenance practices. Virtually every mission requirement translates into an avionics requirement in some form.

Life cycle cost is of great importance in civil aircraft and is of increasing performance in military aircraft, more specifically the emerging Joint Strike Fighter. For a typical commercial transport aircraft the acquisition cost is approximately 20 to 25% of the total life cycle cost. In military aircraft, the acquisition cost is probably a smaller fraction of the life cycle cost. It is interesting to note that for the life cycle cost for civil avionics the acquisition cost rises to 60% of the total. Avionics life cycle cost, for example, will drive, the need for built-in testing, fault tolerance, and the ratio of mean time between unscheduled removals (MTBUR) and mean time between failure (MTBF.)

Finally, certification is a major factor in avionics design. As the complexity and criticality of avionics increases so does the need for extensive certification activities. Certification issues begin with the initial definition of requirements and last until the equipment is removed from the aircraft or the aircraft is retired.

It is important to note that the requirements definition, especially after the preliminary requirements are set, is very much an iterative process between the customer and the vendor, often under the purview of a Configuration Control Board. This configuration control process, which is sometimes viewed as an intrusion on the real work, is necessary to ensure that everyone is cognizant of proposed changes to requirements and can comment on them.

Aircraft functional requirements are at the top of the requirements hierarchy. The aircraft mission is broken into phases including preflight checkout, taxi out, take-off, cruise, descent, landing, rollout, taxi in, and postflight. Additional specialized mission phases such as weather diversions, cargo drop, electronic warfare, etc. also drive the aircraft performance requirements.

Typical aircraft functional requirements include ground steering and braking, passenger comfort, navigation, communication, and environmental conditioning. Figure 19.1 shows the breakdown (function decomposition) of requirements from the aircraft level to the avionics function level. Requirements

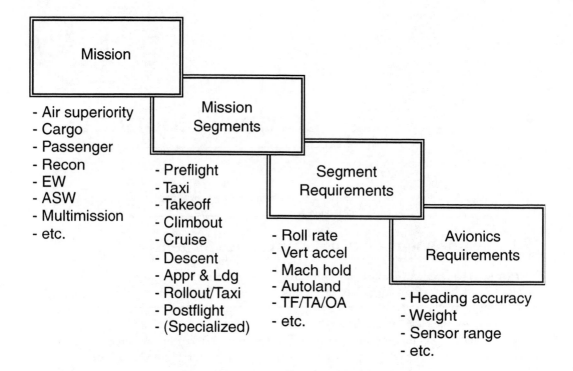

FIGURE 19.1   Function decomposition.

must be traceable and justified. In some cases every "shall" is numbered and then accounted for in test plans to demonstrate that it has been met.

An excellent example of military aircraft avionics top-level requirements is found in AFWAL-TR-1114 Architecture Specification for PAVE PILLAR Avionics. These requirements were postulated for a hypothetical high-performance fighter and guided the definition of the requirements for the F-22 Raptor (see Chapter 32). Typical PAVE PILLAR avionics requirements include:

- Two-level maintenance (flight line or depot)
- Combat turnaround time: ≤ 15 min
- Non-mission capable (for avionics): $\leq 1.2\%$
- Fault detection: 99% of all possible faults
- Fault isolation: 98% of all possible faults

Aerospace Recommended Practice (ARP) 4754 Certification Considerations for Highly-Integrated or Complex Aircraft Systems offers guidance on an aircraft-level Function Hazard Assessment (FHA) that addresses the effect on aircraft performance if a function is lost. The output of an FHA is function criticality and safety requirements, e.g., a "critical" function that must have a probability of failure of less than 10^{-9} per flight hour. This output, in turn, becomes an input to the system design and system-level FHAs. ARP 4761 System Safety Assessment, contains guidance on how to conduct the various analyses required by ARP 4754, including the well-known fault tree analysis and failure modes and effects analysis.

As requirements become more detailed the definition process becomes more amenable to automation. Computer-based processes enhance requirements traceability and validation, configuration control, and document generation. Examples of computer-based techniques include modeling and simulation (see Chapter 20), Systems Workshop, an Aerospatiale proprietary tool for use on Airbus Industrie products (see Chapter 30), and formal methods based on rigorous mathematical concepts (see Chapter 21).

DOD-HDBK-763 Human Engineering Procedures Guide is a valuable source of example techniques for determining requirements as they flow down and become more detailed. The emphasis in this handbook is, of course, on the human factors aspect, but there is a wealth of detailed general information on defining requirements.

Derived requirements are lower-level requirements that could not (or should not) be defined at the beginning of the avionics design process. Microprocessor selection is an example.

References

AFWAL-TR-87-1114, "Architecture Specification for PAVE PILLAR Avionics," January, 1987.

"Certification Considerations for Highly-Integrated or Complex Aircraft Systems," ARP 4754, SAE; 1996.

Palmer, Michael T., et al., "A Crew-Centered Flight Deck Design Philosophy for High Speed Civil Transport (HSCT) Aircraft," NASA Technical Memorandum 109171; January, 1995.

DOD-HDBK-763 Human Engineering Procedures Guide, 27 February, 1987.

"System Safety Assessment," ARP 4761, SAE; 1996.

20
Digital Avionics Modeling and Simulation

Jack Strauss
Zycad, Inc.

Terry Venema
Zycad, Inc.

Grant Stumpf
Zycad, Inc.

John Satta
Zycad, Inc.

20.1 Introduction

In order to realize unprecedented but operationally essential levels of avionics system performance, reliability, supportability, and affordability, commercial industry and the military will draw on advanced technologies, methods, and development techniques in virtually every area of aircraft design, development, and fabrication. Federated avionics architectures, integrated avionics architectures, hybrid systems architectures, and special purpose systems such as flight control systems, engine control systems, navigation systems, reconnaissance collection systems, electronic combat systems, weapons delivery systems, and communications systems all share certain characteristics, which are germane to digital systems modeling and simulation. All of these classes of avionics systems are increasing in complexity of function and design, and are making increased use of digital computer resources. Given this, commercial and military designers of new avionics systems and of upgrades to existing systems must understand, incorporate, and make use of state-of-the-art methods, disciplines, and practices of digital systems design. This chapter presents fundamentals, best practices, and examples of digital avionics systems modeling and simulation.

20.2   Underlying Principles

The fundamental principle of modeling and simulation is stated as follows. The results of most mathematical processes are either correct or incorrect. Modeling and simulation has a third possibility. The process can yield results that are correct but irrelevant [Strauss 1994]. With this startling but true realization of the possible results of modeling and simulation, it is important to understand the different perspectives that give rise to the motivation for modeling and simulation, the trade space for the development effort to include the users and systems requirements, and the technical underpinnings of the practice.

20.2.1   Historic Perspective

The past 30 years of aviation has seen extraordinary innovation in all aspects of design and manufacturing technology. Digital computing resources have been employed in all functional areas of avionics, including communication, navigation, flight controls, propulsion control, and all areas of military weapon systems. As analog, mechanical, and electrical systems have been replaced or enhanced with digital electronics, there has been an increased demand for new digital computing techniques and for higher performance digital computing resources.

Special purpose data, signal, and display processors were commonly implemented in the late 1960s and early 1970s [Swangim et al., 1989]. Special purpose devices gave way to programmable data, signal, and display processors in the early to mid 1980s. These devices were programmed at a low level—assembly language programming was common. The late 1980s and early 1990s have seen commercial and military avionics adopting the use of high-performance general-purpose computing devices programmed in high-order languages. The USAF F-22 fighter, for example, incorporates general-purpose, commercially available, microprocessors programmed in Ada. The F-22 has an operational flight program consisting of nearly one million lines of Ada code and onboard computing power on the order of 20 billion operations per second of signal processing and 500 millon intstructions per second (MIPS) of data processing. Additionally, there is increasing use of commercial off-the-shelf (COTS) products such as processor boards, storage devices, graphics displays, and fiber optic communications networks for many military and commercial avionics applications.

COTS product designers and avionics systems developers are making it a standard engineering practice to model commercial computer products and digital avionics products and systems at all levels of design abstraction. As the complexity of electronics design dramatically has increased, so too has modeling and simulation technology in both functional complexity and implementation. Complex computer-aided design (CAD) software can be several hundred thousand lines of code. These software products require advanced engineering workstation computing resources with sophisticated file and storage structures and data management schemes. Workstations capable of 50 MIPS with several-gigabyte disk drives, connected with high-speed local area networks (LANs) are common. Additionally, special purpose hardware environments, used to enhance and accelerate simulation and modeling, have increased in performance and complexity to supercomputing levels. Hardware accelerators are now capable of evaluating over a million events per second and rapid prototype equipment can reach 10^{11} events per second. At this point in history, the complexity and performance of modeling and simulation technology is equal to the digital avionics products they are used to develop.

20.2.2   Economic Perspective

For commercial system designers, a critical product development factor that has significant impact on the economic viability of any given product release is time-to-market. The first product to market generally recoups all of the nonrecurring engineering and development costs and commonly captures as much as half the total market. This is why technologies aimed at decreasing time-to-market are of increased importance to all commercial developers. Analysis is available showing the amount of development time saved as a result of digital system modeling and simulation [Donnelly, 1992]. At a lower level, cost-sensitive design has two significant issues: learning curve and packaging.

The learning curve itself is best described as an increase in productivity over time. For device manufacturing this can be measured by change in yield, the percentage of manufactured devices that survive testing. Whether it is a chip, a board, or a system, given sufficient volume, designs that have twice the yield will generally have half the cost [Hennessy and Patterson, 1990]. The design reuse inherent in digital modeling and simulation directly shortens the learning curve. Packaging at the device, board, or system level has cost implications related to fundamental design correctness and system partitioning. A case study of performance modeling for system partitioning is presented later in this text.

For military systems designers, many of the same issues affecting commercial systems designers apply, especially, as more commercial technologies are being incorporated as implementation components. Additionally, as will be shown below, mission objectives and operational requirements are the correct point of entry for modern, top-down design paradigms. However, once a development effort has been initiated, the relative cost of error discovery (shown notionally in Figure 20.1) is more expensive at each successive level of system completeness. Thus, a low-price solution which does not fully meet system requirements can turn into a high-cost problem later in the development cycle.

20.2.3 Design Perspective

Bell and Newell divided computer hardware into five levels of abstraction [Bell and Newell, 1973]. The levels are processors-memory switches (system), instruction set (algorithm), register transfer, logic, and circuit. Hardware systems at any level can also be described by their behavior, structure, and physical implementation. Walker and Thomas combined these views and proposed a model of design representation and synthesis [Walker and Thomas, 1985]. The levels are defined as the architecture level (system level), algorithmic level, functional block level (register transfer level), logic level, and circuit level. Each of these levels is defined in terms of their behavioral domain, structural domain, and physical domain. Behavioral design describes the behavior or function of the hardware using such notations as behavioral languages, register transfer languages, and logic equations. Structural design specifies functional blocks and components such as registers, gates, flip-flops, and their interconnections. Physical design describes the implementation of a design idea in a hardware platform, such as the floor plan of a circuit board and layout of transistors, standard cells, and macrocells in an integrated circuit (IC) chip.

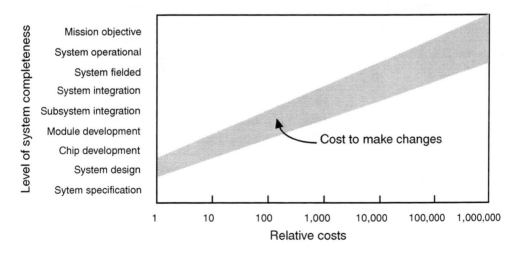

FIGURE 20.1 Relative cost of defect discovery has been shown to increase as a factor of 10 at each successive level of system completeness [Portelli et al., 1989].

TABLE 20.1 Market Factors Comparison for Commercial and Military Market Segments. Each has a different relative priority for the given criterion

Criterion	Commercial	Military
Financial Basis	Market	Budget
Development Focus	Product	Capability
Production Volume	Medium–High	Low
System Complexity	Medium	High
System Design Cycle	Short	Medium–Long
System Life Cycle	Short–Medium	Long–Very Long
Contractual Concerns	Warranty, Liability	Reliability, Mortality

Hierarchical design starts with high-level algorithmic or behavioral design. These high-level designs are converted to circuits in the physical domain. Various CAD tools are available for design entry and conversion. Digital modeling and simulation technologies and tools are directly incorporated into the process to assure correctness and completion of the design at each level and to validate the accuracy of the translation from one level to the next [Liu, 1992].

20.2.4 Market Perspective

It is generally assumed that there are two major market segments for avionics—the commercial avionics market and the military avionics market. As stated earlier, there is great similarity in the technological forces at work on both military and commercial systems. There are, however, several fundamental differences in the product development cycle and business base that are important to consider because they impact the interpretation of the cost performance analysis for modeling and simulation. These differences are summarized in Table 20.1. Consider the impact of commercial versus military production volumes on a capital investment decision for modeling and simulation technology. The relative priority of this criterion is likely to differ for commercial products as compared to military systems.

20.2.5 Requirements in the Trade Space

Technology without application, tactical, or doctrinal context is merely engineering curiosity. The development of an avionics suite, or the implementation of an upgrade to an existing set requires the careful balance of many intricate constraints. More than any other type of development, avionics has the most intricate interrelationships of constraints. The complex set of issues associated with volume, weight, power, cooling, capability, growth, reliability, and cost create the most complex engineering trades of any possible development outside the space program. The risks associated with the introduction of new technologies and the development of enabling technologies create mitigation plans that look like parallel developments. It is little wonder that avionics systems are becoming the most expensive portion of the aircraft development.

To fully exploit the dollars available for avionics development, it is necessary to invest a significant effort in an intimate understanding of the system requirements. Without a knowledge of what the pilot needs to accomplish the mission, and how each portion of the system contributes to the satisfaction of that need, it is impossible to generate appropriate trades and understand the impacts on the engineering process. Indeed, often the mission needs are vague and performance requirements are not specified. This critical feature of the development is further complicated by the fact that pilots often are unaware of detailed technical features of the requirements set, and cannot specifically identify critical system information or presentation requirements.

In the final analysis, the avionics suite is a tool used by the pilot to accomplish a task. The avionics are an extension of his senses and his capabilities. They provide orientation, perception, and function while he is attempting to complete an endlessly variable task. With this in mind, the first and most important step in the design and development of an avionics package is the development of the requirements.

20.2.6 Technical Underpinnings of the Practice

Allen defines the discipline for making predictions of system performance and capacity planning as modeling [Allen, 1994]. He further categorizes techniques in terms of rules of thumb, back-of-the-envelope calculations, statistical forecasting, analytical queuing theory modeling, simulation modeling, and benchmarking. When applied intelligently, all methods have utility. Each has specific cost and schedule impacts. For nontrivial modeling and simulation, the areas of analytical queuing theory modeling, simulation modeling, and benchmarking have the greatest information content, while rules of thumb often hold the most wisdom.

Analytical queuing theory seeks to represent the system algorithmically. The fundamental algorithm set is the M/M/1 queuing system [Klienrock, 1975]. The M/M/1 queuing system is an open system (implying an infinite stream of incoming customers or work) with one server (that which handles customers or does the work) that provides exponentially distributed service. Mathematically, the probability that the provided service will require not more than t time units is given by:

$$P[S \leq t] = 1 - e^{-t/s}$$

where S is the average service time.

For the M/M/1 queuing system, the interarrival time, that is, the time between successive arrivals, also has an exponential distribution. Mathematically, this implies that if τ describes the interarrival time, then:

$$P[\tau \leq t] = 1 - e^{-\lambda t}$$

where λ is the average arrival rate.

Therefore, with the two parameters, the average arrival rate λ and the average service time S, we completely define the M/M/1 model.

Simulation modeling is defined as driving the model of a system with suitable inputs and observing the corresponding outputs [Bratley et al., 1987]. The basic steps include construction of the model, development of the inputs to the model, measurement of the interactions within the model, formation of the statistics of the measured transactions, analysis of the statistics, and validation of the model. Simulation has the advantage of allowing more detailed modeling and analysis than analytical queuing theory, but has the disadvantage of requiring more resources in terms of development effort and computer resources to run.

Benchmarking is the process of running specialized test software on actual hardware. It has the advantage of great fidelity in that the execution time associated with a specific benchmark is not an approximation, but the actual execution time. This process is extremely useful for comparing the results of running the same benchmark on several different machines. The primary disadvantages include requiring actual hardware, which is difficult if the hardware is developmental, and unless your application is the benchmark, it is not likely to represent accurately the workload of your specific system in operation.

20.2.7 Summary Comments

Historically there have been dramatic increases in the complexity and performance of digital avionics systems. This increase in complexity has required many new tools and processes to be developed in support of avionics product design, development, and manufacture.

The commercial and military avionics marketplaces differ in many ways. Decisions made quantitatively must be interpreted in accordance with each marketplace. However, both commercial and military markets have economic forces which have driven the need for shorter development cycles. A shorter development cycle generally stresses the capabilities of design technology. Thus, both markets have similar digital system design process challenges. In fact, current policies mandate increased use of COTS products instead of ruggedized versions, and fewer military standards, generally replacing them with "best practices." As this trend continues, there exists potential for both markets to converge on common solutions for these challenges.

The most general design cycle proceeds from a concept through a design phase to a prototype test and integration phase, ending finally in release to production. The end date (be it a commercial product introduction or a military system deployment) generally does not move to the right on the schedule. Designs often take longer than scheduled due to complexity and coordination issues. The time between prototype release to fabrication and release to production, which should be spent in test and debug, gets squeezed. Ideally, lengthening the test and debug phase without compromising either design time or the production release date is desirable. Modeling and simulation does this by allowing the testing of designs before they are fabricated, when they are easier, faster, and cheaper to change.

The design process for any avionics development must begin with the development of the requirements. A full understanding of the requirements set is necessary because of the inherent constraints it places on the development and the clarity it provides as to the intended use of the system.

There are several techniques available for the analysis and prediction of a system's performance. These techniques include rules of thumb, back-of-the-envelope calculations, statistical forecasting, analytical queuing theory modeling, simulation modeling, and benchmarking. Each technique has advantages and disadvantages with respect to fidelity and cost of resources to perform. When taken together, and applied appropriately, they form the rigor base for the best practices in digital avionics systems modeling and simulation.

20.3 Best Practices

As the fundamental principle of modeling and simulation suggests, a correctly executed simulation of any model may yield results that are correct but irrelevant. Therefore, it is imperative to understand the customers, the systems, the allocation of requirements to system components, the performance of the system components, and to be able to trace component performance to system performance to customer requirements. To date, there exists no single development environment that encompasses this complete set of issues. Several discrete development environments and tools do exist. They are maturing at a rapid, although separate, rate, and tend to either complement or overlap each other. As these environments and tools, as well as the design and economic forces, continue to advance, increased use and decreased cost of entry for avionics systems designers will prevail. Presented below are summaries of best practices in requirements engineering and top-down system simulation.

20.3.1 Requirements Engineering

The process of requirement development for an avionics suite begins with the definition of the mission(s) and related requirements. Once the requirements are known, metrics can be assigned for mission utility of the various systems and subsystems involved in the development. This is essential, because the results of the requirements process are not always consistent with the engineering expectations. That is to say, the modeling and simulation process often uncovers operations and relationships that generate unexpected results. The identification and quantification of these metrics is essential to successfully navigate the complex trade space that is inherent in the development of avionics.

The effective generation of user requirements is the first step in the process. Ordinarily the user provides an initial cut at an unconstrained requirements set. The initial set of requirements needs to be validated through a simulation. Ideally, the user has correctly identified the set of unconstrained requirements that will yield the most effective utilization of the systems. If the current set of requirements yields the maximum capability, the task of the simulation process is to measure the decrease in effectiveness associated with the relaxation of each of the requirements parameters. If the current set of requirements is not optimum, then perturbation of system parameters will yield increases in unconstrained performance. In any event, the relationship between the system parameters and operational effectiveness needs to be well established before beginning any system trade activities.

At the completion of the initial assessment of the requirements, further definition of the requirements trade space requires the definition of mission scenarios for evaluation with relaxed requirement sets. The evaluation of mission performance at this level is associated with changes of requirements in association with each other. Changing the system performance parameters at various levels of requirements provides insight into the best starting point for in-depth study of engineering trades. In many instances, this evaluation is done on the basis of total avionics system costs.

Validating the requirements of avionics design by modeling and simulation depends on the validity of the simulations used. In effect, the simulations need to be calibrated to enable a valid trade study. There are several methods of validating the simulations. Many of the approved operational models are accepted as valid requirements generators, but all models need volumes of approved scenarios, threats, and concepts of operation to be valid.

The design of the avionics suite is intimately related to the design of the cockpit and the displays. In earlier designs, the capabilities of the presentations were often limited by the system elements. The integrating element was the pilot, who received most of the data generated by the avionics. Today, presentation has a significant impact on the performance of the avionics suite. In fact, the presentation requirements for the pilot can be viewed as the requirements for generation of information from the avionics. Determination of presentation necessitates simulation of some kind. This is the integration phase of the requirements with the operator.

Man-In-The-Loop (MITL) simulation is the most reliable means of determining presentation requirements. These can be part-task, or full-mission simulations that drive both presentation and capability requirements. MITL simulation is very expensive, but depending on the implementation, can yield a concise evaluation of the requirements set. There are many disciplines involved in the conduct of human factors engineering, and those references should be consulted. The essential point is that MITL simulation at some level of fidelity is required to fully validate the requirements set.

Sometimes the system requirements can be estimated with a Simulated-Man-In-The-Loop (SMITL) simulation. Working models of human performance exist at several levels of fidelity. Many of these models are adept at working with human-directed control applications, and should be evaluated if the design of the system is attempting a control implementation. The existing models are currently less than satisfying on the cognitive modeling of pilots in flight, however, several attempts are being made to develop cognitive models that will allow a much broader application of SMITL simulation.

In summary, the definition of the requirements determines the effectivity of the avionics suite. Avionics designers continue to shed the constraints imposed by technology over the years. The developments of the year 2000 and beyond will be less constrained by the technology, and the effectiveness will be determined by the application of the concepts for employment. A unique and especially potent display is as effective as a new type of circuit that improves receiver sensitivity. Avionics designs of the 21st century will be based on thorough analysis of requirements, not technological innovations.

20.3.2 Top-Down System Simulation

The Top-Down System Simulation (TDSS) is a proven risk reduction methodology that applies top-down design techniques and currently available simulation technology during development to ensure that complex systems perform correctly the first time. The benefits of applying this methodology range from lower overall cost by eliminating avoidable refabrication of hardware, to on-time schedule performance resulting from the increased visibility into hard-to-foresee integration problems. Together, these benefits frequently outweigh the initial cost of instituting a program-wide TDSS process. Its other far-reaching benefits include the resolution of specification ambiguities, validation of the hardware to specification, and implementation of manufacturing, logistics, and reprocurement documentation that is technology independent.

Until recently, the typical design process began by generating specifications to describe the system, subsystem, component, and interface requirements. Then the system design was partitioned into functions and the functions partitioned into hardware and software. During development, the hardware and software components were developed individually, and then brought together when the system was integrated. In most cases, the final integration step, when discrepancies between concept and implementation were discovered, turned out to be the schedule driver. This was true because issues had to be resolved that were unforeseen, and thus not planned for.

The TDSS methodology is designed to incorporate the best features of the typical development cycle and to add additional visibility and risk reduction to the schedule-driving integration phase. This is accomplished by using simulation techniques that allow a "virtual" integration of component models to be performed while the design is still on the drawing board and easy to change. This virtual integration eliminates the costly hardware refabrication that is frequently required when problems are not discovered until the hardware is in the lab. It also drastically reduces the time spent in the lab integrating the hardware and software, because many of the problems have already been discovered and corrected during development. Examples of this methodology include the U.S. Air Force Advanced Tactical Fighter (ATF) Demonstration and Validation (DEM-VAL) development effort. The interoperability of designs of five critical interfaces were tested through simulation. Over the five interfaces involved, the testing revealed over 200 problems, both with the designs and with the specifications on which they were based. These problems would have resulted in many iterations of hardware fabrication during the integration phase, and several of them would probably not have been detected until the system was fielded. The Air Force concluded that the application of a TDSS methodology to the DEM-VAL program alone resulted in a cost avoidance of approximately $200 million, 25 times the cost of the initiative itself.

20.3.3   TDSS Plan

A formal TDSS plan must be implemented at the very start of any development program with the agreement of all design and development participants. This plan must define the goals of TDSS on the program and the means by which these goals will be reached. For example, a specific goal may be to reduce the risk of subsystem integration through the use of modeling and simulation. The plan will define how this may be attained through the use of commercial logic simulators, off-the-shelf third-party behavioral models, and contractor design databases to perform virtual integration before the costs for hardware fabrication are incurred.

The TDSS plan will define the process, milestones, and data interchange mechanisms that will be required to achieve all of the stated TDSS goals. It will define the tasks to be performed, which Integrated Product Team (IPT) is to perform them, and to which IPT the results will be provided. In a hypothetical example, the Computer Development IPT provides design data and simulation analysis results that prove that all required functions are performed correctly and within the maximum allowable time. These results are passed to the System IPT (at higher level than the Computer Development IPT), who will use these results together with the results of other development IPTs to assess whether all components of the system, when integrated, will meet the functional requirements.

Since modeling and simulation are integral to the development process, they must not be considered as unusual or extra. Accordingly, the milestones defined in the plan are the typical development milestones, such as the Preliminary Design Review (PDR) and the Critical Design Review (CDR) for each component, board, subsystem, etc. This will ensure that the data made available by the TDSS process are used in the most effective possible manner. This means that one development phase will not be officially complete without first meeting all TDSS milestones for that phase. The intent is to prevent taking shortcuts around the process in the "rush to fabrication."

For example, the pre-CDR TDSS phase may require that the hardware be simulated at a predetermined level of abstraction prior to fabrication. The hardware cannot be released for fabrication if the TDSS results are not available or the results do not prove that the functional requirements are being met. If the plan has been agreed to up front by all concerned, then there should not be any surprises or additional effort that might increase the schedule. Even if there is additional effort required, the program will save significant refabrication and rework costs as well as integration costs.

Finally, the data formats to be exchanged between IPTs must be defined and agreed upon by representatives of the customer, the prime contractor, and appropriate subcontractors. For example, the plan might define the contents of the design database that will be supplied to the IPTs that are designing related components. So when information passes between the Computer Development IPT and the Signal Conditioning IPT, each group can effectively make use of the data. The intent is to meet the goals of the TDSS process by eliminating ambiguity and/or extra translation steps wherever possible.

It is imperative to prevent inadvertent divergence from the intent of the design. As the design proceeds from the abstract to the concrete, there are definite points at which the level of abstraction incrementally transitions from a higher level to the next lower level. Before that transition is allowed, the design at each stage of abstraction must be verified and validated.

20.3.4 TDSS Process

The digital electronic hardware development program should proceed hierarchically in the same fashion as the system is arranged. First, the overall system must be defined, modeled, and simulated. Once the system-level model is verified, it can be decomposed into subsystems (functions), each of which must be defined, modeled, simulated, and verified. The subsystems are decomposed into circuit boards containing components of various types (custom, common, standard, and COTS). Each element (subsystem, board, component) at each level of hierarchy will have its own development flow, which will resemble the development flow of the overall system. Figure 20.2 shows a typical development process. At the start of the program the system architects draft the system specification and codify the requirements while the design management team develops the overall top-down development program in the TDSS plan. At the System Requirements Review (SRR) milestone, the team agrees on what needs to be designed to meet the needs of the program. After SRR, the team finalizes the system specification and develops a series of models that describe the performance, behavior, and functions of the system. These models will be the unambiguous reference point for the design since they represent the consensus of the system architects as to what the system should do and how it should be partitioned. The models are formally approved at the System Design Review (SDR) so they can also be used to evaluate the technical soundness of architectures proposed by potential subcontractors in response to a Requests for Proposal.

Once the system is partitioned into subsystems, detailed design work begins. Using the SDR approved models, the team develops Machine Executable Specification (MES) versions. These MESs are distributed to the members of the hardware design team, who use them to construct and verify architectural and performance models of the major subsystems. The subsystem models will be integrated to create performance and architectural models of the entire system. The system architects will review these models and verify their correctness with respect to the requirements. The hardware designers will use the system and subsystem models and the appropriate MESs to begin the preliminary design work. They will define the system at further levels of hierarchical detail, including boards/LRUs, interfaces, modules, and components. They will build gradually to fully functional behavioral models of the system. In order to pass PDR successfully, they must verify that the behavioral models produce the same results as the system performance and architectural models. Once this is accomplished, detailed design work can begin.

The process repeats at the structural level, where the hardware designers are building gate-level models of new components and integrating them with behavioral or structural models of existing or COTS equipment. Once again, the exit criteria for CDR is that these detailed models produce the results that match those produced by the PDR models. Once that is verified, hardware can be fabricated with a high degree of confidence. The hardware must be verified against the models. If there is a discrepancy, the hardware must be made to match the model, not the other way around. The hardware is a technology-specific implementation of the design and, as stated previously, the models are the design.

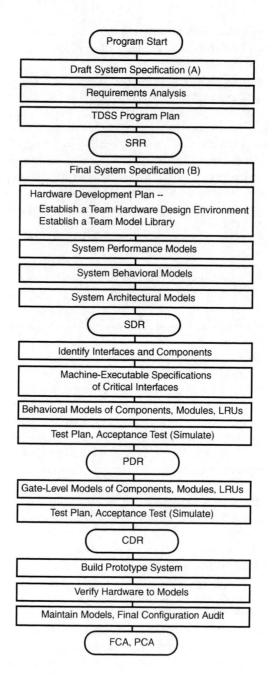

FIGURE 20.2 TDSS hardware development flow.

If everything has been done properly, the lowest-level models (and the hardware) will produce results that are traceable back through the design chain all the way to the system specification.

Experience on many programs has shown that the vast majority of system defects can be traced back to misunderstandings due to ambiguity in the specifications. These misunderstandings arise because different hardware designers interpret specifications in completely logical but, unfortunately different, perhaps erroneous, ways. The errors are often not discovered until hardware is actually built and tested. By this time, schedule and budgetary pressures usually preclude leisurely analysis and re-design. So a "workaround" is sought that will overcome the problem, but always at a compromise to system performance.

If a workaround cannot be found, then the program must suffer adverse budget or schedule impacts, often both.

One novel aspect of TDSS eliminates the ambiguous specifications and their associated dangers right from the start. Since a specification is meant to convey the behavior of something, the best way to do that is to provide a readable description that can also be executed on a computer to provide a demonstrable and verifiable example of the behavior being described. A simulation model (written correctly) can be an unambiguous medium in which to embody a specification. A machine-executable specification (MES) is unambiguous because its behavior can be observed under a variety of conditions. The hardware designer does not have to interpret the specification because the behavior can be observed instead. So, if all hardware developers use the same (unambiguous) executable model, they are likely to have the same fundamental understanding of the specification.

20.3.5 Software Modeling in TDSS

A complete system depends on software as well as hardware, and system software design can make or break any system—in performance, schedule, and cost. Good software can bring out the best in mediocre hardware, but poor software can bring excellent hardware to its knees.

Hardware and software are the same at all but the lowest implementation level. Indeed, system specifications and requirements are (theoretically) drawn up with no thought as to hardware/software partitioning. Before software-programmable computers were invented, everything was done in hardware. As software execution speeds increased, more time-critical functions could be handled in software, but from a system standpoint, whether a function is performed by pure hardware or is embedded in code is immaterial as long as the requirements are met and the missions can be achieved. Any complete system design effort must take into account the adequacy or inadequacy of software performance. To verify the adequacy of the "other half" of the system, software performance modeling and analysis should:

1. Account for processing delays on the system or subsystem architecture.
2. Assess the compatibility and portability of software to COTS platforms.
3. Assess the impact of Open Systems Architecture (OSA) requirements on software performance and subsystem interoperability.

Software architecture and performance can be modeled just as hardware can. In both cases, the focus is on input and output rates and amounts, latency (response time), senescence (data age), efficiency, and correct results.

20.4 Performance Modeling for System Partitioning (A Case Study)

The following case study describes a practical application of modeling and simulation used to both articulate and answer system-level questions regarding data latency such that system functional partitioning may be accomplished. Figure 20.3 represents a modern integrated avionics architecture. The system is intended to be fabricated with COTS components and is generalized to include connectivity to existing legacy subsystems.

20.4.1 System Description

This architecture includes the following functional components: Sensor Front Ends representing multiple sensors that receive signals and accomplish analog signal processing; Embedded Computing Assets consisting of computing assets, closely coupled to the sensor front ends, that perform low-latency control processing, digital signal processing, and prepare data for transmission; an ATM Dataflow Network serving as the data transfer medium upon which data are transferred between Embedded Computing Assets, Core Computing Assets, Operator Workstations, and the LAN Gateway; Core Computing Assets

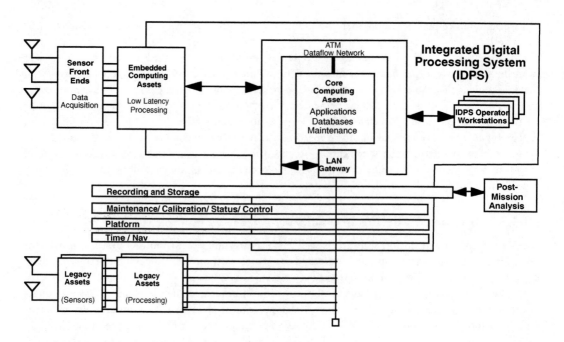

FIGURE 20.3 A generalized integrated avionics architecture.

which include computing and memory devices used to perform application programs and services such as database and maintenance functions; and Operator Workstations servings as the man-machine interface. Additionally, there are recording and storage assets, maintenance, calibration, and control assets, as well as platform-specific and time/navigation assets. Finally, there is the LAN Gateway which serves as the interface to Legacy Systems and networks.

Given the above architecture, the system designer must determine where to allocate software processes that operate on incoming sensor data. The designer must determine which processes are mission critical, which are low-latency processes and which processes are non-low-latency processes that could be allocated to application software. Can the designer allocate all processes to the Core Computing Assets and use the core as a server? What processes, if any, should remain in computing assets local to the incoming sensor data? Is there any incoming sensor data that have to be processed within a time constraint for a successful mission? None of those questions can be answered until a fundamental question is first addressed. Given that a data packet is stored in the RAM of the Embedded Computing Assets, what is the latency associated with that packet's transfer from Embedded Computing Asset RAM, across the ATM Dataflow Network, to Core Computing Asset RAM, and back to Embedded Asset RAM traversing the same path?

The example Embedded and Core Computing Assets each consist of a dual VME 64 backplane configuration, real-time Concurrent Maxion board set (including four 200-MHz R4400 CPU cards [XPU] with 128 MB of RAM per CPU, an I/O card [XIO] conforming to VME 64 standards, and a crosspoint switching architecture capable of 1.2 GB/s peak), and two Fore Systems ATM to VME adapter cards supporting OC-3 155 MB/s SONET fiber (one of these cards is used for redundant purposes only). The Embedded Computing Assets also included MIL-STD-1553, IRIG B, and MXI cards. The Core Assets also include MIL-STD-1553, IRIG B, and SCSI-compatible VME cards. The ATM Dataflow Network (DFN) consists of a Fore Systems ASX-200BX 16-port switch capable of switching up to 1.2 GB/s data across a nonblocking switch fabric. This switch is used to connect the ATM adapter cards using OC-3 155 MB/s SONET fiber. The system is decomposed into the component block diagram shown in Figure 20.4. This is the component block diagram that was used as the basis for a performance model.

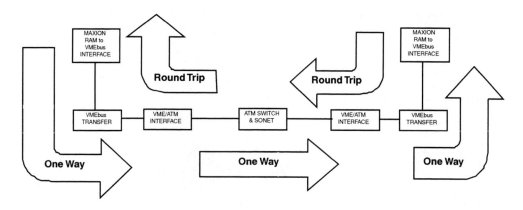

FIGURE 20.4 Component block diagram.

20.4.2 Model Development

Modeling and simulation of the system was accomplished using a commercial computer-aided engineering (CAE) tool (the OPNET modeling tool from Mil-3). This tool is an event-driven simulation tool particularly well suited for computer and network modeling. The tool utilizes hierarchical domains defined as Network, Process, and State domains. A Network domain model for the system was built using the Component Block Diagram shown in Figure 20.4. The Network domain of OPNET allows the specification of nodes and selection and modeling of data transfer pipes (OC-3 SONET, Ethernet, etc.). The Process domain allows the modeler to break down each network node into a configuration of queues, processors, transmitters, receivers, and packet generators. The State domain of the tool allows each processor and queue to be further decomposed into states, with definable transitions between states. Each state of a particular process is defined using C-based programming and simulation kernel calls to simulate the behavior of the process when it arrives at that state. To this end, any protocol behavior can be modeled with the proper combination of states and state behavior definitions. The tool also allows the modeler to probe the simulation at any point in the model to take measurements (throughput, queue capacities, delays, etc.) and analyze the results of these probes while the simulation is running. The system was broken down into modeled components as shown in Figure 20.5 below.

The model for the system was set up to transmit packets (1024 bytes) from RAM located in the Embedded Computing Assets, across the ATM DFN, to the RAM located in the Core Computing Assets, and then back to the RAM of the Embedded Computing Assets via the same path. During the simulation, the computing assets and ATM DFN were loaded at various levels with other data (a known characteristic of the real system was that other data with specific size and frequency rates would also be utilizing the ATM DFN and computer resources at various points throughout the system).

Insertion of these data at various points throughout the model was achieved using the modeling tool's generator modules. This allowed the model to be data loaded at various points in a manner consistent with how the real system would be data loaded during normal operation. For this system, the known system characteristics and protocols explicitly modeled in this example include but are not limited to: I/O read and write delays to/from the Core Computing Assets, CPU wait cycles due to caching operations, crosspoint architecture arbitration, TCP/IP protocol overhead and buffering operations, VMEbus protocol including service interrupt and acknowledgment and delays associated with the various addressing modes of VME operation, ATM segmentation and reassembly delays, ATM call setup delays, ATM fabric switching, and output port buffer delays.

Also included in the model were known data sources that were used to simulate other traffic on the ATM Dataflow Network. A file of System Loading Parameters was also developed made up of a collection of all the modeling parameters available. These include, but were not strictly limited to VMEbus addressing mode, TCP socket buffer thresholds, ATM switching delay parameters, SONET data transport overhead,

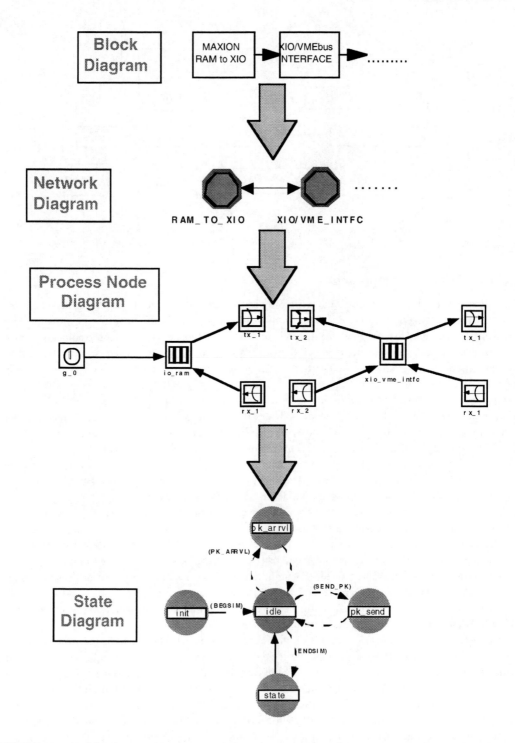

FIGURE 20.5   Modeling decomposition process includes component to network to process to state diagram deveopment.

proprietary communications layer overhead, and many others. The simulation was run 10 times with the System Loading Parameters held constant. The average latency over the 10 simulation runs was plotted as a function of those parameters. The parameters were changed, and another 10 simulations were run at that particular level of system loading. System Loading Parameters were varied from 0.5 to 1.0 in increments of 0.5.

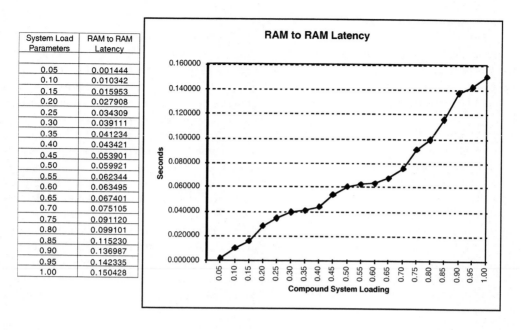

FIGURE 20.6 Case study simulation results.

20.4.3 Modeling Results

The modeling simulation results are shown in Figure 20.6. Each point on the plot represents 10 simulation runs. During each simulation run, the average round trip latency for the target data packets was measured. In a lightly loaded system, according to the simulation results, the target data could be expected to traverse the round trip path presented earlier in an average time of time of 0.001444 s. In a heavily loaded system, again using the simulation results, the data packets experienced a latency of up to 0.150140 s. This was the maximum result just prior to 1.0 loading.

These results were provided to the system designer as performance-based partitioning data and were used to allocate system functions between the Embedded Computing Assets and the Core Computing Assets.

20.4.4 Summary

The actual modeling and simulation effort that formed the basis for this case study was performed as part of a functional allocation process for an airborne system. The results allowed for system partitioning around a 150-ms boundary. By determining the latency early in the design phase, the design could be optimized to make best use of the available processing assets and avoid costly reallocations later in the system development.

This example further demonstrated several advantages to modeling and simulation of a system vs. "vendor data sheet" calculations. First, vendors often times report "best case" or "burst mode" performance characteristics. Modeling and simulation allows parameters of the system to be varied such that "heavily loaded" or "degraded performance" modes of the system may be investigated. Another distinct advantage is that the model is able to generate synchronous data, asynchronous data, and data based on probability distributions. Data latencies in the model are based on modeled event-triggered protocol disciplines rather than deriving answers on a calculator or spreadsheet. Not every aspect of a real system can be modeled completely in every tool, but the known characteristics of the system that relate to performance form a quality model for performance estimation in a tool as was used here.

20.5   Research Issues and Summary

As mentioned above, not every aspect of a real system can be modeled completely in every tool, but the known characteristics of the system that relate to performance form a quality model for performance estimation. Clearly, market forces will continue to drive the quantity, quality, completeness, and rate of change of system engineering environments. As a practical matter, better integration and traceability between requirements automation systems and system/software engineering environments are called for. A powerful result of this integration will be the improved traceability between proposed system concepts, their driving requirements, the resultant technical configuration(s), and their cost and schedule impacts.

The requirement to integrate the trade space across technical, program, financial, and market boundaries is likely to continue and remain incomplete. To paraphrase Brooks, there is more common ground than the technical community will admit, and there are differences that most managers do not understand [Brooks, 1995]. Automation alone will not reduce the complexities of this effort until there is a common, multidisciplined, quantitative definition of cost vs. price in system performance trades. As to the automation issue, the size, content, and format issues of current and legacy technical, financial, programmatic data bases will continue to grow and diverge until the stewards of college curricula produce graduates that solve more problems than they create [Parnas, 1989]. These new practitioners will develop future information systems and engineering environments which encompass the disciplines, language(s), and methods critical to the practice.

Defining Terms

ATM: Asynchronous Transfer Mode
CAD: Computer Automated Design
CAE: Computer-Aided Engineering
CDR: Critical Design Review
COTS: Commercial-off-the-shelf (products)
FCA: Functional Configuration Audit
MES: Machine Executable Specifications
MIPS: Millions of instructions per second, or, misleading indicator of processor speed
MITL: Man-in-the-loop
OC"n":   Optical Carrier Level n (i.e., OC-3, a SONET specification)
OSA: Open Systems Architecture
PCA: Physical Configuration Audit
PDR: Preliminary Design Review
RAM: Random Access Memory
SDR: System Design Review
SMITL: Simulated-man-in-the-loop
SONET: Synchronous Optical Network
SRR: System Requirements Review
TDSS: Top Down System Simulation

References

Allen, Arnold O., 1994. *Computer Performance Analysis with Mathematica.* Academic Press, New York.

Bell, C. G. and Newell, A., 1973. *Computer Structures: Readings and Examples.* McGraw-Hill, New York.

Bratley, P., Fox, B., and Schrage, L., 1987. *A Guide to Simulation,* 2nd ed., Springer-Verlag, New York.

Brooks J.R. and Fredrick P., 1995. *The Mythical Man Month.* Addison-Wesley, Reading, MA.

Donnelly, C. F., 1992. Evaluating the IOBIDS Specification Using Gate-Level System Simulation, in *Proc. IEEE Natl. Aerosp. Electron. Conf.*, p. 748.

Hennessy, J. L. and Patterson, D. A., 1990. *Computer Architecture: A Quantitative Approach.* Morgan Kaufmann, San Francisco, CA.

Kleinrock, L., 1975. Theory, *Queueing Systems,* Vol. 1, John Wiley & Sons, New York.

Liu, Hsi-Ho, 1992. Software Issues in Hardware Development, in *Computer Engineering Handbook.* McGraw-Hill, New York.

Parnas, D. L., 1989. Education for Computing Profressionals, in Tech. Rep. 89-247 ISSN-0836-0227.

Portelli, W., Oseth, T., and Strauss, J. L., 1989. Demonstration of Avionics Module Exchangeability via Simulation (DAMES) Program Overview, in *Proc. IEEE Natl. Aerosp. Electron. Conf.,* pp. 660.

Strauss, J. L., 1994. The Third Possibility, in *Modeling and Simulation of Embedded Systems,* Proc. Embedded Computing Inst., pp. 160.

Swangim, J., Strauss, J. L., et al., 1989. Challenges of Tomorrow—The Future of Secure Avioncs, in *Proc. IEEE Natl. Aerosp. Electron. Conf.,* pp. 580.

Walker, R. A. and Thomas, D. E., 1985. A Model of Design Representation and Synthesis, in *Proc. 22nd Design Automation Conf.,* pp. 453–459.

Further Information

Arnold Allen's 1994 text, *Computer Performance Analysis with Mathematica,* Academic Press, New York, is an excellent introduction to computing systems performance modeling.

For Performance Modeling and Capacity Planning the following organizations provide information through periodicals and specialized publications:

- The Computer Measurement Group (CMG); 414 Plaza Drive, Suite 209, Westmont, IL 60559, (708) 655-1812
- Institute for Capacity Management; P.O. Box 82847, Phoenix, AZ 85071, (602) 997-7374
- ACM Sigmetrics; 11 West 42nd Street, New York, NY 10036, (212) 869-7440

For Computer-Aided Engineering the following list of World Wide Web sites provide information on vendors of major tools and environments:

- www.zycad.com
- www.mil-3.com
- www.mentorgraphics.com
- www.synopsis.com
- www.rational.com

21

Formal Methods

Sally C. Johnson
NASA Langley Research Center

Ricky W. Butler
NASA Langley Research Center

21.1 Introduction

With each new generation of aircraft, the requirements for digital avionics systems become increasingly complex, and their development and validation consumes an ever-increasing percentage of the total development cost of an aircraft. The introduction of life-critical avionics, where failure of the computer hardware or software can lead to loss of life, brings new challenges to avionics validation. The FAA recommends that catastrophic failures of the aircraft be "so unlikely that they are not anticipated to occur during the entire operational life of all airplanes of one type" and suggests probabilities of failure on the order of 10^{-9} per flight hour [FAA, 1988].

There are two major reliability factors to be addressed in the design of ultrareliable avionics: hardware component failures and design errors. Physical component failures can be handled by using redundancy and voting. This chapter addresses the problem of design errors. Design errors are errors introduced in the development phase rather than the operational phase. These may include errors in the specification of the system, discrepancies between the specification and the design, and errors made in implementing the design in hardware or software. The discussion in this chapter centers mainly around software design errors; however, the increasing use of complex, custom-designed hardware instead of off-the-shelf components that have stood the test of time makes this discussion equally relevant for hardware.

The problem with software is that the complexity exceeds our ability to have intellectual control over it. Our intuition and experience is with continuous systems, but software exhibits discontinuous behavior. We are forced to separately reason about or test millions of sequences of discrete state transitions. Testing of software sufficient to assure ultrahigh reliability has been shown to be infeasible for systems of realistic complexity [Butler and Finelli, 1993], and fault tolerance strategies cannot be relied upon because of the demonstrated lack of independence between design errors in multiple versions of software [Knight and Leveson, 1986]. Since the reliability of ultrareliable software cannot be quantified, life-critical avionics software must be developed in a manner that concentrates on producing a correct design and implementation rather than on quantifying reliability after a product is built. This chapter describes how rigorous analysis employing formal methods can be applied to the software development process. While not yet standard practice in industry,* formal methods is included as an alternate means of compliance in the DO178B standard for avionics software development.

21.2 Fundamentals of Formal Methods

Formal methods is the use of formal mathematical reasoning or logic in the design and analysis of computer hardware and software. Mathematical logic serves the computer system designer in the same way that calculus serves the designer of continuous systems—as a notation for describing systems and as an analytical tool for calculating and predicting the behavior of systems.

Formal logic provides rules for constructing arguments that are sound because of their form and independent of their meaning, and consequently can be manipulated with a computer program. Formal logic provides rules for manipulating formulas in such a manner that, given valid premises, only valid conclusions are deducible. The manipulations are called a proof. If the premises are true statements about the world, then the soundness theorems of logic guarantee that the conclusion is also a true statement about the world. Furthermore, assumptions about the world are made explicit and are separated from the rules of deduction.

Formal methods can be roughly divided into two basic components: specification and verification. Each of these is discussed below.

21.2.1 Formal Specification

Formal specification is the use of notations derived from formal logic to define (1) the requirements that the system is to achieve, (2) a design to accomplish those requirements, and (3) the assumptions about the world in which a system will operate. The requirements explicitly define the functionality required from the system as well as enumerating any specific behaviors that the system must meet, such as safety properties.

A design specification is a description of the system itself. In practice, a set of hierarchical specifications is often produced during the design process ranging from a high-level, abstract representation of the system to a detailed implementation specification, as shown in Figure 21.1. The highest-level specification might

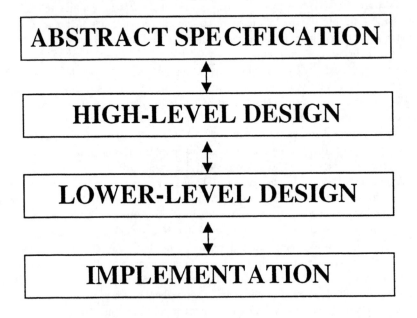

FIGURE 21.1 Hierarchy of formal specifications.

---

*However, Intel, Motorola, and AMD are using formal methods in the development of microchips.

describe only the basic requirements or functionality of the system in a state-transition format. The lowest-level specification would detail the algorithms used to implement the functionality. The hierarchical specification process guides the design and analysis of a system in an orderly manner, facilitating better understanding of the basic abstract functionality of the system as well as unambiguously clarifying the implementation decisions. Additionally, the resulting specifications serve as useful system documentation for later analysis and modification.

An avionics system can be thought of as a complex interface between the pilot and the aircraft. Thus, the specification of the avionics system relies on a set of assumptions about the behaviors of the pilot and the response of the aircraft. Likewise, the specification of a modular subsystem of an integrated avionics system will rely on a set of assumptions about the behaviors and interfaces of the other subsystems. The unambiguous specification of interfaces between subsystems also prevents the problem of developers of different subsystems interpreting the requirements differently and arriving at the system integration phase of software development with incompatible subsystems.

21.2.2 Formal Verification

Formal verification is the use of proof methods from formal logic to (1) analyze a specification for certain forms of consistency and completeness, (2) prove that the design will satisfy the requirements, given the assumptions, and (3) prove that a more detailed design implements a more abstract one. The formal verifications may simply be paper-and-pencil proofs of correctness; however, the use of a mechanical theorem prover to ensure that all of the proofs are valid lends significantly more assurance and credibility to the process. The use of a mechanical prover forces the development of an argument for an ultimate skeptic who must be shown every detail.

In principle, formal methods can accomplish the equivalent of exhaustive testing, if applied to the complete specification hierarchy from requirements to implementation. However, such a complete verification is rarely done in practice because it is difficult and time-consuming. A more pragmatic strategy is to concentrate the application of formal methods on the most critical portions of a system. Parts of the system design that are particularly complex or difficult to comprehend can also be good candidates for more rigorous analysis. Although a complete formal verification of a large, complex system is impractical at this time, a great increase in confidence in the system can be obtained by the use of formal methods at key locations in the system.

There are several publicly available theorem-proving toolkits. These tools automate some of the tedious steps associated with formal methods, such as *typechecking* of specifications and conducting the simplest of proofs automatically. However, these tools must be used by persons skilled in mathematical logic to perform rigorous proofs of complex systems.

21.2.3 Limitations

For many reasons, formal methods do not provide an absolute guarantee of perfection, even if checked by an automated theorem prover. First, formal methods cannot guarantee that the top-level specification is what was intended. Second, formal methods cannot guarantee that the mathematical model of the physical world, such as aircraft control characteristics, is accurate. The mathematical model is merely a simplified representation of the physical world. Third, the formal verification process often is only applied to part of the system, usually the critical components. Finally, there may be errors in the formal verification tools themselves.* Nevertheless, formal methods provide a significant capability for discovering/removing errors in large portions of the design space.

21.3 Example Application

The techniques of formal specification and verification of an avionics subsystem will be demonstrated on a very simplified example of a mode-control panel. An informal, English-language specification of

---

*We do not believe this to be a significant source of undetected errors in formal specifications or verifications.

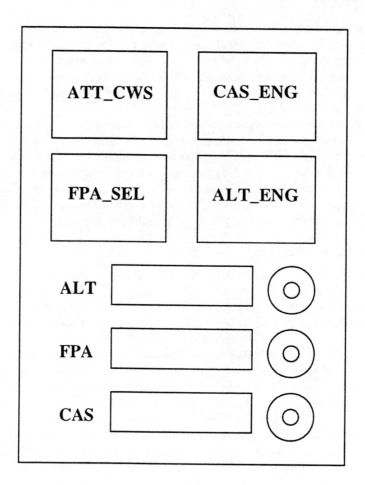

FIGURE 21.2 Mode-control panel.

the mode-control panel, representative of what software developers typically encounter in practice, will be presented. The process of clarifying and formalizing the English specification into a formal specification, often referred to as requirements capture, will then be illustrated.

21.3.1 English Specification of the Example System

This section presents the informal, English-language specification of the example system. The English specification is annotated with section numbers to facilitate references back to the specification in later sections.

1. *The mode-control panel contains four buttons for selecting modes and three displays for dialing in or displaying values, as shown in Figure 21.2. The system supports the following four modes:*

    attitude control wheel steering (att_cws)*,*
    flight-path angle selected (fpa_sel)*,*
    altitude engage (alt_eng)*, and*
    calibrated air speed (cas_eng)*.*

    Only one of the first three modes can be engaged at any time. However, the cas_eng *mode can be engaged at the same time as any of the other modes. The pilot engages a mode by pressing the corresponding button on the panel. One of the three modes,* att_cws*,* fpa_sel*, or* alt_eng*, should* be engaged at all times. Engaging any of the first three modes will automatically cause the other two to be disengaged since only one of these three modes can be engaged at a time.

2. *There are three displays on the panel: air speed, flight-path angle, and altitude. The displays usually show the current values for the air speed, flight-path angle, and altitude of the aircraft. However, the pilot can enter a new value into a display by dialing in the value using the knob next to the display. This is the target or "preselected" value that the pilot wishes the aircraft to attain. For example, if the pilot wishes to climb to 25,000 feet, he will dial 250 into the altitude display window* and then press the* `alt_eng` *button to engage the altitude mode. Once the target value is achieved or the mode is disengaged, the display reverts to showing the "current" value.***

3. *If the pilot dials in an altitude that is more than 1200 feet above the current altitude and then presses the* `alt_eng` *button, the altitude mode will not directly engage. Instead, the altitude engage mode will change to "armed" and the flight-path angle select mode is engaged. The pilot must then dial in a flight-path angle for the flight-control system to follow until the aircraft attains the desired altitude. The flight-path angle select mode will remain engaged until the aircraft is within 1200 feet of the desired altitude, then the altitude engage mode is automatically engaged.*

4. *The calibrated air speed and the flight-path angle values need not be preselected before the corresponding modes are engaged — the current values displayed will be used. The pilot can dial in a different value after the mode is engaged. However, the altitude must be preselected before the altitude engage button is pressed. Otherwise the command is ignored.*

5. *The calibrated air speed and flight-path angle buttons toggle on and off every time they are pressed. For example, if the calibrated air speed button is pressed while the system is already in calibrated air speed mode, that mode will be disengaged. However, if the attitude control wheel steering button is pressed while the attitude control wheel steering mode is already engaged, the button is ignored. Likewise, pressing the altitude engage button while the system is already in altitude engage mode has no effect.*

Because of space limitations, only the mode-control panel interface itself will be modeled in this example. The specification will only include a simple set of commands the pilot can enter plus the functionality needed to support mode switching and displays. The actual commands that would be transmitted to the flight-control computer to maintain modes, etc. are not modeled.

21.3.2 Formally Specifying the Example System

In this section, we will describe the process of formally specifying the mode-control panel described in English in the previous section. Quotes cited from the English specification will be annotated with the corresponding section number in parentheses. The goal is to completely describe the system requirements in a mathematical notation, yet not overly constrain the implementation.

This system collects inputs from the pilot and maintains the set of modes that are currently active. Thus, it is appropriate to model the system as a state machine. The state of the machine will include the set of modes that are active, and the pilot inputs will be modeled as events that transition the system from the current state (S_c) to a new state (S_n):

$$S_c \longrightarrow S_n$$

The arrow represents a transition function, `nextstate`, which is a function of the current state and an event, say `ev`:

$$S_n = \texttt{nextstate}(S_c, \texttt{ev}).$$

---

*Altitude is expressed in terms of "flight level," which is measured in increments of 100 ft.

**In a real mode-control panel, the buttons would be lit with various colored lights to indicate which modes are currently selected and whether "preselected" or "current" values are being displayed. Because of space limitations, the display lights will not be included in the example specification.

The goal of the formal specification process is to provide an unambiguous elaboration of the nextstate function. This definition must be complete; i.e., it must provide a next state for all possible events and all possible states. Thus, the first step is to elaborate all possible events and the content of the state of the machine. The system will be specified using the PVS specification language [Owre et al., 1993].

21.3.2.1 Events

The pilot interacts with the mode-control panel by pressing the mode buttons and by dialing preselected values into the display. The pilot actions of pressing one of the four buttons will be named as follows: press_att_cws, press_cas_eng, press_alt_eng, and press_fpa_sel. The actions of dialing a value into a display will be named as follows: input_alt, input_fpa, and input_cas. The behavior of the mode-control panel also depends upon the following inputs it receives from sensors: alt_reached, fpa_reached, and alt_gets_near. In PVS, the set of events are specified as follows:

```
events: TYPE = {press_att_cws, press_cas_eng, press_fpa_sel,
                press_alt_eng, input_alt, input_fpa, input_cas,
                alt_gets_near, alt_reached, fpa_reached}
```

21.3.2.2 State Description

The *state* of a system is a collection of attributes that represents the system's operation. In the example system, the set of active modes would certainly be a part of the system state. Also, the values in the displays that show altitude, flight-path angle, and air speed, and the readings from the airplane sensors would be included in the state. Formalization of the system state description entails determining which attributes are necessary to fully describe the system's operation, then choosing a suitable formalism for representing those attributes.

One possible approach to describing the system state for the example is to use a set to delineate which modes are active. For example, {att_cws, cas_eng} would represent the state of the system where both att_cws and cas_eng are engaged but alt_eng and fpa_sel are not engaged. Also, the alt_eng mode has the additional capability of being armed. Thus, a better approach to describing the example system state is to associate with each mode one of the following values: off, armed, or engaged. In PVS, a type or domain can be defined with these values:

```
mode_status: TYPE = {off, armed, engaged}
```

The state descriptor is as follows:*

```
[# % RECORD
att_cws: mode_status,
cas_eng: mode_status,
fpa_sel: mode_status,
alt_eng: mode_status,
#] % END
```

For example, the record [att_cws=engaged, cas_eng=engaged, fpa_sel=off, alt_eng=off] would indicate a system where both att_cws and cas_eng are engaged, fpa_sel is off and alt_eng is off. However, there is still a problem with the state descriptor; in the example system only alt_eng can be armed. Thus, more restrictive domains are needed for the modes other than alt_eng. This can be accomplished by defining the following sub-types of mode_status:

```
off_eng: TYPE={mode: mode_status | mode = off OR
                                    mode = engaged}
```

---

*In the PVS language, a record definition begins with [# and ends with #]. The % denotes that the remainder of the line contains a comment and is ignored.

The type off_eng has two values: off and engaged. The state descriptor is thus corrected to become:

```
[# % RECORD
att_cws: off_eng,
cas_eng: off_eng,
fpa_sel: off_eng,
alt_eng: mode_status
#] % END
```

The mode panel also maintains the state of the displays. To simplify the example, the actual values in the displays will not be represented. Instead, the state descriptor will only keep track of whether the value is a preselected value or the actual value read from a sensor. Thus, the following type is added to the formalism:

```
disp_status: TYPE = {pre_selected, current}
```

and three additional fields are added to the state descriptor:

```
alt_disp: disp_status,
fpa_disp: disp_status, and
cas_disp: disp_status.
```

The behavior of the mode-control panel does not depend upon the actual value of "altitude" but rather on the relationship between the actual value and the preselected value in the display. The following "values" of altitude are sufficient to model this behavior:

```
altitude_vals: TYPE = {away, near_pre_selected,
                          at_pre_selected},
```

and the following is added to the state descriptor:

```
altitude: altitude_vals.
```

The final state descriptor is

```
states: TYPE = [# % RECORD
      att_cws: off_eng,
      cas_eng: off_eng,
      fpa_sel: off_eng,
      alt_eng: mode_status,
      alt_disp: disp_status,
      fpa_disp: disp_status,
      cas_disp: disp_status,
      altitude: altitude_vals
      #] END
```

21.3.2.3 Formal Specification of Nextstate Function

Once the state descriptor is defined, the next step is to define a function to describe the system's operation in terms of state transitions. The nextstate function can be defined in terms of ten subfunctions, one for each event, as follows:

```
event: VAR events
st: VAR states =
nextstate(st,event): states
```

```
CASES event OF
  press_att_cws: tran_att_cws(st),
  press_cas_eng: tran_cas_eng(st),
  press_fpa_sel: tran_fpa_sel(st),
  press_alt_eng: tran_alt_eng(st),
  input_alt    : tran_input_alt(st),
  input_fpa    : tran_input_fpa(st),
  input_cas    : tran_input_cas(st),
  alt_gets_near: tran_alt_gets_near(st),
  alt_reached  : tran_alt_reached(st),
  fpa_reached  : tran_fpa_reached(st)
ENDCASES
```

The CASES statement is equivalent to an IF-THEN-ELSIF-ELSE construct. For example, if the event is press_fpa_sel then nextstate(st,event) = tran_fpa_sel(st). The step is to define each of these subfunctions.

21.3.2.4 Specifying the att_cws Mode

The tran_att_cws function describes what happens to the system when the pilot presses the att_cws button. This must be specified in a manner that covers all possible states of the system. According to the English specification, the action of pressing this button attempts to engage this mode if it is off. Changing the att_cws field to engaged is specified as follows:

```
st WITH [att_cws := engaged].
```

The WITH statement is used to alter a record in PVS. This expression produces a new record that is identical to the original record st in every field except att_cws. Of course, this is not all that happens in the system. The English specification also states that, "*Only one of the* [att_cws, fpa_sel, *or* cas_eng] *modes can be engaged at any time*" (1). Thus, the other modes must become something other than engaged. It is assumed that this means they are turned off. This would be indicated as:

```
st WITH [att_cws:= engaged,  fpa_sel:= off,  alt_eng:= off].
```

The English specification also states that when a mode is disengaged, "*…the display reverts to showing the "current value:*" (2):

```
st WITH [att_cws := engaged, fpa_sel := off, alt_eng := off,
         alt_disp := current, fpa_disp := current].
```

The English specification also says that, "*…if the attitude control wheel steering button is pressed while the attitude control wheel steering mode is already engaged, the button is ignored*" (5). Thus, the full definition is

```
tran_att_cws(st): states =
  IF att_cws(st) = off THEN
      st WITH [att_cws := engaged, fpa_set := off,
               alt_eng := off, alt_disp := current,
               fpa_disp := current]
  ELSE st %% IGNORE: state is not altered at all
  ENDIF
```

The formal specification has elaborated exactly what the new state will be. The English specification does not address what happens to a preselected alt_eng display or preselected fpa_sel display when the att_cws button is pressed. However, the formal specification does explicitly indicate what will happen: these modes are turned off. The process of developing a complete mathematical specification has uncovered this ambiguity in the English specification. If the displays should not be changed when the corresponding mode is off, then the following specification would be appropriate:

```
tran_att_cws(st): states =
  IF att_cws(st) = off THEN
    st WITH [att_cws := engaged,
             fpa_sel := off,
             alt_eng := off,
       alt_disp := IF alt_eng(st) = off THEN alt_disp(st)
                ELSE current ENDIF,
       fpa_disp := IF fpa_sel(st) = off THEN fpa_disp(st)
                ELSE current ENDIF]
  ELSE st %% IGNORE: state is not altered at all
  ENDIF
```

We realize that this situation will arise in several other cases as well. In particular, whenever a mode becomes engaged, should any other preselected displays be changed to current or should they remain preselected? We decide to consult the system designers. They agree that the displays should be returned to current and suggest that the following be added to the English specification:

6. *Whenever a mode other than* cas_eng *is engaged, all other preselected displays should be returned to* current.

21.3.2.5 Specifying the `cas_eng` Mode

The tran_cas_eng function describes what happens to the system when the pilot presses the cas_eng button. This is the easiest mode to specify because its behavior is completely independent of the other modes. Pressing the cas_eng button merely toggles this mode on and off. The complete function is

```
tran_cas_eng(st): states =
     IF cas_eng(st) = off THEN
       st WITH [cas_eng := engaged]
     ELSE st WITH [cas_eng := off, cas_disp := current]
     ENDIF
```

This specification states that if the cas_eng mode is currently off, pressing the button will engage the mode. If the mode is engaged, pressing the button will turn the mode off. Thus, the button acts like a switch.

21.3.2.6 Specifying the `fpa_sel` Mode

The tran_fpa_sel function describes the behavior of the system when the fpa_sel button is pressed. The English specification states that this mode "*need not be preselected*" (4). Thus, whether the mode is off or preselected, the outcome is the same:

```
IF fpa_sel(st) = off THEN
    st WITH [fpa_sel := engaged, att_cws := off,
             alt_eng := off, alt_disp := current]
```

Note that not only is the fpa_sel mode engaged, but att_cws and alt_eng are turned off as well. This was included because the English specification states that, "*Engaging any of the first three modes will automatically cause the other two to be disengaged*" (1). Also note that this specification indicates that alt_disp is set to "current." The English specification states that, "*Once the target value is achieved or the mode is disengaged, the display reverts to showing the 'current' value*" (2). Thus, the altitude display must be specified to return to "current" status. If the alt_eng mode was not currently active, the WITH update does not actually change the value, but merely updates that attribute to the value it already holds.

Since PVS requires that functions be completely defined, we must also cover the case where fpa_sel is already engaged. We consult the English specification and find, "*The calibrated air speed and flight-*

path angle buttons toggle on and off every time they are pressed." We interpret this to mean that if the `fpa_sel` button is pressed while the mode is engaged, the mode will be turned off. This is specified as follows:

```
st WITH [fpa_sel := off, fpa_disp := current]
```

Because the mode is disengaged, the corresponding display is returned to current. We realize that we also must cover the situation where the `alt_eng` mode is armed and the `fpa_sel` is engaged. In fact, Section (3) of the English specification indicates that this will occur when one presses the `alt_eng` button and the airplane is far away from the preselected altitude. However, Section (3) does not tell us whether the disengagement of `fpa_sel` will also disengage the armed `alt_eng` mode. We decide to consult the system designers. They inform us that pressing the `fpa_sel` button should turn off both the `fpa_sel` and `alt_eng` mode in this situation. Thus, we modify the state update statement as follows:

```
st WITH [fpa_sel := off,  alt_eng := off,
         fpa_disp := current, alt_disp := current]
```

The complete specification is thus:

```
tran_fpa_sel(st): states =
   IF fpa_sel(st) = off THEN
    st WITH [fpa_sel := engaged, att_cws  := off,
             alt_eng := off, alt_disp := current]
    ELSE st WITH [fpa_sel := off,  alt_eng := off,
             fpa_disp := current, alt_disp := current]
   ENDIF
```

The perspicacious reader may have noticed that there is a mistake in this formal specification. The rest of us will discover it when a proof is attempted using a theorem prover in the later section entitled, "Formal Verification of the Example System."

21.3.2.7 Specifying the `alt_eng` Mode

The `alt_eng` mode is used to capture a specified altitude and hold it. This is clearly the most difficult of the four to specify since it has a complicated interaction with the `fpa_sel` mode.

The English specification states that, "*The altitude must be preselected before the altitude engage button is pressed*" (4). This is interpreted to mean that the command is simply ignored if it has not been preselected. Consequently, the specification of `tran_alt_eng` begins:

```
tran_alt_eng(st): states =
   IF alt_disp(st) = pre_selected THEN
      ...
   ELSE st % IGNORE
   ENDIF
```

This specifies that the system state will change as a result of pressing the `alt_eng` button only if the `alt_disp` is preselected.

We must now proceed to specify the behavior when the `IF` expression is true. The English specification indicates that if the aircraft is more than 1200 ft from the target altitude, this request will be put on hold (the mode is said to be armed) and the `fpa_sel` mode will be engaged instead. The English specification also says that, "*The pilot must then dial in a flight-path angle at this point*" (3). The question arises whether the `fpa_sel` engagement should be delayed until this is done. Another part of the English specification offers a clue, "*The calibrated air speed and flight-path angle values need not be preselected before the corresponding modes are engaged"* (4). Although this specifically addresses the case of pressing the fpa_sel button and not the situation where the alt_eng button indirectly turns this mode on, we suspect that the behavior is the same. Nevertheless, we decide to check with the system designers to make sure. The system designers explain that this is the correct interpretation and that this is the reason the mode is called "flight-path angle select" rather than "flight-path angle engage."

The behavior must be specified for the two situations: when the airplane is near the target and when it is not. There are several ways to specify this behavior. One way is for the state specification to contain the current altitude in addition to the target altitude. This could be included in the state vector as two numbers:

```
target_altitude:  number
actual_altitude:  number
```

The first number contains the value dialed in and the second the value last measured by a sensor. The specification would then contain:

```
IF  abs(target_altitude  -  actual_altitude)  >  1200  THEN
```

where abs is the absolute value function. If the behavior of the mode-control panel were dependent upon the target and actual altitudes in a multitude of ways, this would probably be the proper approach. However, in the example system the behavior is only dependent upon the relation of the two values to each other. Therefore, another way to specify this behavior is by abstracting away the details of the particular values and only storing information about their relative values in the state descriptor record. In particular, the altitude field of the state record can take on one of the following three values:

away	the preselected value is $>$ 1200 feet away
near_pre_selected	the preselected value is $<=$ 1200 feet away
at_pre_selected	the preselected value is $=$ the actual altitude

The two different situations can then be distinguished as follows:

```
IF altitude(st) = away THEN
```

When the value is not away, the alt_eng mode is immediately engaged. This is specified as follows:

```
st WITH [alt_eng := engaged, att_cws := off,
         fpa_sel := off,  fpa_disp := current
```

Note that not only is the alt_eng mode engaged, but this specification indicates that several other modes are affected just as in the tran_fpa_sel subfunction.

Now the behavior of the system must be specified for the other case, when the aircraft is away from the target altitude. In this case fpa_sel is engaged and alt_eng is armed:

```
ELSE
  st WITH [fpa_sel := engaged, att_cws := off,
           alt_eng := armed]
```

As before, the att_cws mode is also turned off.

So far we have not considered whether the behavior of the system should be different if the alt_eng mode is already armed or engaged. The English specification states that, *"Pressing the altitude engage button while the system is already in altitude engage mode has no effect"* (5). However, there is no information about what will happen if the mode is armed. Once again, the system designers are consulted, and we are told that the mode-control panel should ignore the request in this case as well. The complete specification of `tran_alt_eng` becomes:

```
tran_alt_eng(st): states =
   IF alt_eng(st) = off AND alt_disp(st) = pre_selected THEN
        IF altitude(st) /= away THEN %% ENGAGED
            st WITH [att_cws := off, fpa_sel := off,
                    alt_eng := engaged, fpa_disp := current]
        ELSE st WITH [att_cws := off, fpa_sel := engaged,
                    alt_eng := armed] %% ARMED
        ENDIF
   ELSE st %% IGNORE request
   ENDIF
```

Note that the last `ELSE` takes care of both the armed and engaged cases.

21.3.2.8 Input to Displays

The next three events that can occur in the system are `input_alt`, `input_fpa`, and `input_cas`. These occur when the pilot dials a value into one of the displays. The `input_alt` event corresponds to the subfunction of `nextstate` named `tran_input_alt`. The obvious thing to do is to set the appropriate field as follows:

```
                st WITH [alt_disp := pre_selected]
```

This is certainly appropriate when `alt_eng` is off. However, we must carefully consider the two cases: (1) when the `alt_eng` mode is armed, and (2) when it is engaged. In this case, the pilot is changing the target value after the `alt_eng` button has been pressed. The English specification required that the `alt_eng` mode be preselected before it could become engaged, but did not rule out the possibility that the pilot could change the target value once it was armed or engaged. We consult the system designers once again. They inform us that entering a new target altitude value should return the `alt_eng` mode to off and the pilot must press the `alt_eng` button again to reengage the mode. We add the following to the English specification:

7. *If the pilot dials in a new altitude while the* `alt_eng` *button is already engaged or armed, then the* `alt_eng` *mode is disengaged and the* `att_cws` *mode is engaged.*

The reason given by the system designers was that they didn't want the altitude dial to be able to automatically trigger a new active engagement altitude. They believed it was safer to force the pilot to press the `alt_eng` button again to change the target altitude.

Thus, the specification of `tran_input_alt` is

```
    tran_input_alt(st): states =
      IF alt_eng(st) = off THEN
        st WITH [alt_disp := pre_selected]
      ELSE % alt_eng(st) = armed OR alt_eng(st) = engaged THEN
         st WITH [alt_eng := off, alt_disp := pre_selected,
                  att_cws := engaged,
                  fpa_sel := off, fpa_disp := current]
      ELSE st
      ENDIF
```

The other input event functions are similar:

```
    tran_input_fpa(st): states =
      IF fpa_sel(st) = off THEN st WITH [fpa_disp :  = pre_selected]
      ELSE st ENDIF
```

```
tran_input_cas(st): states =
    IF cas_eng(st) = off THEN st WITH [cas_disp := pre_selected]
    ELSE st ENDIF
```

21.3.2.9 Other Actions

There are other events that are not initiated by the pilot but that still affect the mode-control panel; in particular, changes in the sensor input values. As described previously, rather than including the specific values of the altitude sensor, the state descriptor only records which of the following is true of the preselected altitude value: away, near_pre_selected or at_pre_selected. Events must be defined that correspond to significant changes in the altitude so as to affect the value of this field in the state. Three such events affect the behavior of the panel:

alt_gets_near	the altitude is now near the preselected value
alt_reached	the altitude reaches the preselected value
alt_gets_away	the altitude is no longer near the preselected value

The transition subfunction associated with the first event must consider the case where the *alt_eng* mode is armed because the English specification states that, *"The flight-path angle select mode will remain engaged until the aircraft is within 1200 feet of the desired altitude, then the altitude engage mode is automatically engaged"* (3). Thus we have:

```
tran_alt_gets_near(st): states =
    IF alt_eng(st) = armed THEN
        st WITH [altitude := near_pre_selected,
                 alt_eng := engaged, fpa_sel := off]
    ELSE st WITH [altitude := near_pre_selected]
    ENDIF
```

The subfunction associated with the second event is similar because we cannot rule out the possibility that the event alt_reached may occur without alt_gets_near occurring first:

```
tran_alt_reached(st): states =
    IF alt_eng(st) = armed THEN
        st WITH [altitude := at_pre_selected,
                 alt_disp := current, alt_eng := engaged,
                 fpa_sel := off]
    ELSE st WITH [altitude := at_pre_selected,
                  alt_disp := current]
    ENDIF
```

Note that in this case, the alt_disp field is returned to current because the English specification states, *"Once the target value is achieved or the mode is disengaged, the display reverts to showing the 'current' value"* (2).

However, the third event is problematic in some situations. If the alt_eng mode is engaged, is it even possible for this event to occur? The flight-control system is actively holding the altitude of the airplane at the preselected value. Thus, unless there is some major external event such as a windshear phenomenon, this should never occur. Of course, a real system should be able to accommodate such unexpected events. However, to shorten this example, it will be assumed that such an event is impossible. The situation where the alt_eng mode is not engaged or armed would not be difficult to model, but also would not be particularly interesting to demonstrate.

There are three possible events corresponding to each of the other displays. These are all straightforward and have no impact on the behavior of the panel other than changing the status of the corresponding display. For example, the event of the airplane reaching the preselected flight-path angle is fpa_reached.

The specification of the corresponding subfunction `tran_fpa_reached` is

```
tran_fpa_reached(st): states =
        st WITH [fpa_disp   := current]
```

21.3.2.10 Initial State

The formal specification must include a description of the state of the system when the mode-control panel is first powered on. One way to do this would be to define a particular constant, say `st0`, that represents the initial state:

```
st0: states = (# att_cws   := engaged, cas_eng   := off,
                  fpa_sel := off, alt_eng   := off,
                  alt_disp  := current, fpa_disp   := current,
                  cas_disp  := current, altitude   := away #)
```

Alternatively, one could define a predicate (i.e., a function that returns true or false) that indicates when a state is equivalent to the initial state:

```
is_initial(st) : bool =
    att_cws(st) = engaged AND cas_eng(st)   = off AND
    fpa_sel(st) = off AND alt_eng(st)   = off AND
    alt_disp(st) = current AND fpa_disp(st)   = current AND
    cas_disp(st)  = current
```

Note that this predicate does not specify that the altitude field must have the value "away." Thus, this predicate defines an equivalence class of states, not all identical, in which the system could be initially. This is the more realistic way to specify the initial state since it does not designate any particular altitude value.

21.3.3 Formal Verification of the Example System

The formal specification of the mode-control panel is complete. But how does the system developer know that the specification is correct? Unlike the English specification, the formal specification is known to be detailed and precise. But it could be unambiguously wrong. Since this is a requirements specification, there is no higher-level specification against which to prove this one. Therefore, ultimately the developer must rely on human inspection to insure that the formal specification is "what was intended." Nevertheless, the specification can be analyzed in a formal way. In particular, the developer can postulate properties that he believes should be true about the system and attempt to prove that the formal specification satisfies these properties. This process serves to reinforce the belief that the specification is what was intended. If the specification cannot be proven to meet the desired properties, the problem in the specification must be found or the property must be modified until the proof can be completed. In either case, the developer's understanding of and confidence in the system is increased.

In the English specification of the mode-control panel, there were several statements made that characterize the overall behavior of the system. For example, *"One of the three modes [att_cws, fpa_sel, or alt_eng] should be engaged at all times"* (1). This statement can be formalized, and it can be proven that no matter what sequence of events occurs, this will remain true of the system. Properties such as this are often called system *invariants*. This particular property is formalized as follows:

```
att_cws(st) = engaged
OR fpa_sel(st) = engaged
OR alt_eng(st) = engaged
```

Another system invariant can be derived from the English specification: *"Only one of the first three modes [att_cws, fpa_sel, alt_eng] can be engaged at any time"* (1). This can be specified in several ways. One possible way is as follows:

```
(alt_eng(st)/= engaged OR fpa_sel(st)/= engaged)  AND
(att_cws(st) =  engaged IMPLIES
       alt_eng(st)/= engaged AND fpa_sel(st)/= engaged)
```

Finally, it would be prudent to insure that whenever `alt_eng` is armed that `fpa_sel` is engaged:

```
(alt_eng(st) = armed  IMPLIES  fpa_sel(st) = engaged).
```

All three of these properties can be captured in one *predicate* (i.e., a function that is true or false) as follows:

```
valid_state(st):  bool =
     (att_cws(st) =  engaged OR fpa_sel(st) =  engaged
                               OR alt_eng(st) =  engaged)
   AND (alt_eng(st)/= engaged OR fpa_sel(st)/= engaged)
   AND (att_cws(st)  = engaged IMPLIES
        alt_eng(st)/= engaged AND fpa_sel(st)/= engaged)
   AND  (alt_eng(st) = armed IMPLIES
           fpa_sel(st) = engaged)
```

The next step is to prove that this is always true of the system. One way to do this is to prove that the initial state of the system is valid and that if the system is in a valid state before an event then it is in a valid state after an event, no matter what event occurs. In other words, we must prove the following two theorems:

```
initial_valid: THEOREM is_initial(st)  IMPLIES  valid_state(st)
nextstate_valid: THEOREM valid_state(st) IMPLIES
                         valid_state(nextstate(st,event))
```

These two theorems effectively prove by induction that the system can never enter a state that is not valid.* Both of these theorems are proved by the single PVS command, GRIND. The PVS system replays the proofs in 17.8 s on a 450 MHz PC (Linux) with 384 MB of memory.

As mentioned earlier, the specification of `fpa_sel` contains an error. On the attempt to prove the nextstate_valid theorem on the erroneous version of `fpa_sel` described earlier, the prover stops with the following sequent:

```
   nextstate_valid :
 [21]  fpa_sel(st!1) = engaged
 [22]  press_fpa_sel?(event!1)
   u-------
 [1]  att_cws(st!1) = engaged
 [2]  alt_eng(st!1) = engaged
 [3]  press_att_cws?(event!1)
 [4]  press_alt_eng?(event!1)
```

---

*In order for this strategy to work, the invariant property (i.e., `valid_state`) must be sufficiently strong for the induction to go through. If it is too weak, the property may not be provable in this manner even though it is true. This problem can be overcome by either strengthening the invariant (i.e., adding some other terms) or by decomposing the problem using the concept of reachable states. Using the latter approach, one first establishes that a predicate "reachable (st)" delineates all of the reachable states. Then one proves that all reachable states are valid, i.e., reachable (st) $\Rightarrow$ valid_state(st).

The basic idea of a sequent is that one must prove that one of the statements after the |------- is provable from the statements before it. In other words, one must prove:

$$[-1] \ \text{AND} \ [-2] \ === > \ [1] \ \text{OR} \ [2] \ \text{OR} \ [3] \ \text{OR} \ [4]$$

We see that formulas [3] and [4] are impossible, because `press_fpa_sel?(event!1)` tells us that `event!1 = press_fpa_sel` and not `press_att_cws` or `press_alt_eng`. Thus, we must establish [1] or [2]. However, this is impossible. There is nothing in this sequent to require that `att_cws(st!1) = engaged` or that `alt_eng(st!1) = engaged`. Thus, it is obvious at this point that something is wrong with the specification or the proof. It is clear that the difficulty surrounds the case when the event `press_fpa_sel` occurs, so we examine `tran_fpa_sel` more closely. We realize that the specification should have set `att_cws` to engaged as well as turning off the `fpa_sel` mode and `alt_eng` mode:

```
tran_fpa_sel(st): states =
  IF fpa_sel(st) = off THEN
     st WITH [fpa_sel := engaged, att_cws := off,
           alt_eng := off, alt_disp := current]
  ELSE st WITH [fpa_sel := off,  alt_eng := off,
                att_cws := engaged,
                fpa_disp := current, alt_disp := current]
  ENDIF
```

This modification is necessary because otherwise the system could end up in a state where no mode was currently active. After making the correction, the proof succeeds.

Thus we see that formal verification can be used to establish global properties of a system and to detect errors in the specifications.

21.3.4 Alternative Methods of Specifying Requirements

Many systems can be specified using the state-machine method illustrated in this chapter. However, as the state-machine becomes complex, the specification of the state transition functions can become exceedingly complex. Therefore, many different approaches have been developed to elaborate this machine. Some of the more widely known are decision tables, decision trees, state-transition diagrams, state-transition matrices, Statecharts, Superstate, and R-nets [Davis, 1988].

Although these methods effectively accomplish the same thing—the delineation of the state machine—they vary greatly in their input format. Some use graphical notation, some use tables, and others use language constructs. Aerospace industries have typically used the table-oriented methods because they are considered the most readable when the specification requires large numbers of pages. Although there is insufficient space to discuss any particular method in this chapter, expression of a part of the mode-control panel using a table-oriented notation will be illustrated briefly.

Consider the following table describing the `att_cws` mode:

`att_cws`	Event	New Mode
`off`	`press_att_cws`	`engaged`
	`press_fpa_sel WHEN fpa_sel /= off`	`engaged`
	`input_alt WHEN alt_eng /= off`	`engaged`
	`all others`	`off`
`engaged`	`press_att_cws`	`engaged`
	`press_alt_eng WHEN alt_eng = off AND alt_disp = pre_selected`	`off`
	`press_fpa_sel WHEN fps_sel = off`	`off`
	`All others`	`engaged`

The first column lists all possible states of `att_cws` prior to the occurrence of an event. The second column lists all possible events, and the third column lists the new state of the `att_cws` mode after the occurrence of the event. Note also that some events may include other conditions. This table could be specified in PVS as follows:

```
att_cws_off: AXIOM att_cws(st) = off IMPLIES
att_cws(nextstate(event, st)) =
        IF (event = press_att_cws) OR
           (event = press_fpa_sel AND fpa_sel(st) /= off) OR
           (event = input_alt AND alt_eng(st) /= off) THEN engaged
        ELSE off
        ENDIF
att_cws_eng: AXIOM att_cws(st) = engaged IMPLIES
      att_cws(nextstate(event,st)) =
        IF (event = press_alt_eng AND alt_eng(st) = off
                        AND alt_disp(st) = pre_selected) OR
           (event = press_fpa_sel AND fps_sel(st) = off) THEN off
        ELSE engaged
        ENDIF
```

This approach requires that nextstate be defined in pieces (axiomatically) rather than definitionally. If this approach is used, it is necessary to analyze the formal specification to make sure that nextstate is completely defined. In particular, it must be shown that the function's behavior is defined for all possible values of its arguments and that there are no duplications. This must be performed manually if the tables are not formalized. The formalization can be done using a general-purpose theorem prover such as PVS or using a special-purpose analysis tool such as Tablewise* [Hoover and Chen, 1994].

When the specification can be elaborated in a finite-state machine, there are additional analysis methods available that are quite powerful and fully automatic. These are usually referred to as model-checking techniques. Some of the more widely known tools are SMV [Burch et al., 1992] and Murphi [Burch and Dill, 1994]. These require that the state-space of the machine be finite. Our example specification has a finite state space. However, if the values of chosen and measured altitude had not been abstracted away, the state-space would have been infinite.

21.4 Some Additional Observations

The preceding discussion illustrates the process that one goes through in translating an English specification into a formal one. Although the example system was contrived to demonstrate this feature, the process demonstrated is typical for realistic systems, and the English specification for the example is actually more complete than most because the example system is small and simple.

The formal specification process forces one to clearly elaborate the behavior of a system in detail. Whereas the English specification must be examined in multiple places and interpreted to make a judgment about the desired system's behavior, the formal specification completely defines the behavior. Thus, the requirements capture process includes making choices about how to interpret the informal specification. Traditional software development practices force the developer to make these interpretation choices (consciously or unconsciously) during the process of creating the design or implementation. Many of the choices are hidden implicitly in the implementation without being carefully thought out or verified by the system designers, and the interpretations and clarifications are seldom faithfully recorded in the requirements document. On the other hand, the formal methods process exposes these ambiguities early in the design process and forces early and clear decisions, which are fully documented in the formal specification.

---

*The Tablewise tool was previously named Tbell.

A more detailed version of this paper:

R. W. Butler, An Introduction to Requirements Capture Using PVS: Specification of a Simple Autopilot, NASA TM-110255, May 1996, pp. 33.

is available at `http://techreports.larc.nasa.gov/ltrs/ltrs.html`. Recent work has looked at using formal methods to detect and eliminate mode confusion in flight guidance systems [Miller and Potts, 1999; Butler et al., 1998].

Defining Terms

Invariant: A property of a system that always remains true throughout all operational modes.
Mode: A particular operational condition of a system. The mode-control panel controls switching between operational conditions of the flight-control system.
Predicate: A function that returns true or false.
State: A particular operational condition of a system. A common method of representing the operational functioning of a system is by enumerating all of the possible system states and transitions between them. This is referred to as a state-transition diagram or finite-state machine representation.
Typechecking: Verification of consistency of data types in a specification. The detailed use of data types to differentiate between various kinds of objects, when supported by automated typechecking, can make a specification more readable, maintainable, and reliable.

References

Burch, J.R., Clarke, E.M., McMillan, K.L., Dill, D.L., and Hwang, L.J., 1992. Symbolic Model Checking: 10^{20} States and Beyond, *Inf. Comput.* 98(2):142–170.

Burch, J.R. and Dill, David L., 1994. Automatic Verification of Pipelined Microprocessor Control, *Computer-Aided Verification, CAV'94*, Stanford, CA, pp. 68–80, June.

Butler, R.W. and Finelli, G.B. 1993. The Infeasibility of Quantifying the Reliability of Life-Critical Real-Time Software, *IEEE Trans. Software Eng,* 19(1):3–12.

Butler, R. W., Miller, S. P., Potts, J. N., and Carreno, V. A, A Formal Methods Approach to the Analysis of Mode Confusion, *17th Digital Avionics Syst. Conf.,* Bellevue, WA, October 31–November 6, 1998.

Davis, A.M., 1988. A Comparison of Techniques for the Specification of External System Behavior., *CACM,* 31(9):1098–1115.

FAA, 1988. System Design and Analysis, Advisory Circular AC 25.1309-1A, U.S. Department of Transportation, Washington, D.C., June.

Hoover, D.N. and Chen, Z., 1994. Tbell: A Mathematical Tool for Analyzing Decision Tables, NASA Contractor Rep. 195027, November.

Knight, J.C. and Leveson, N.G., 1991. An Experimental Comparison of Software Fault-Tolerance and Fault Elimination, *IEEE Trans. Software Eng.,* SE-12(1):96–109.

Miller, S.P. and Potts, J. N., 1999. Detecting Mode Confusion Through Formal Modeling and Analysis, NASA/CR-1999-208971, January.

Owre, S., Shankar, N., and Rushby, J.M., 1993. The PVS Specification Language (Beta Release), Computer Science Laboratory, SRI International, Menlo Park, CA, 1993.

Further Information

A good introduction to the fundamentals of mathematical logic is presented in *Mathematical Logic* by Joseph R. Schoenfield.

The application of formal methods to digital avionics systems is discussed in detail in *Formal Methods and Digital Systems Validation for Airborne Systems*, NASA Contractor Report 4551, by John Rushby. Rushby's report was written for certifiers of avionics systems and gives useful insights into how formal methods can be effectively applied to the development of ultra-reliable avionics systems.

Two NASA Guidebooks on formal methods: "Formal Methods Specification and Verification Guidebook for Software and Computer Systems, Volume I: Planning and Technology Insertion" [NASA/TP-98-208193], 1998, and "Formal Methods Specification and Analysis Guidebook for the Verification of Software and Computer Systems, Volume II: A Practitioner's Companion" [NASA-GB-001-97], 1997 are available on the Web at `http://eis.jpl.nasa.gov/quality/Formal_Methods/`.

The complete specification and proofs for the mode-control panel example described in this chapter can be obtained at `http://shemesh.larc.nasa.gov/fm/ftp/larc/mode-control-ex/`. Several other formal methods application examples and papers can also be obtained from that site.

22

Electronic Hardware Reliability

Arun Ramakrishnan
University of Maryland

Toby Syrus
University of Maryland

Michael Pecht
University of Maryland

22.1 Introduction

Reliability is the ability of a product to perform as intended (i.e., without failure and within specified performance limits) for a specified time, in its life cycle application environment. To achieve product reliability over time demands an approach that consists of a set of tasks, each requiring total engineering and management commitment and enforcement. These tasks impact electronic hardware reliability through the selection of materials, structural geometries and design tolerances, manufacturing processes and tolerances, assembly techniques, shipping and handling methods, operational conditions, and maintenance and maintainability guidelines.[1] The tasks are as follows:

1. Define realistic product requirements and constraints determined by the life cycle application profile, required operating and storage life, performance expectations, size, weight, and cost.

The manufacturer and the customer must jointly define the product requirements in the light of both the customer's needs and the manufacturer's capability to meet those needs.

2. Define the product life cycle environment by specifying all relevant assembly storage, handling, shipping, and operating conditions for the fielded product. This includes all stress and loading conditions.

3. Characterize the materials and the manufacturing and assembly processes. Variabilities in material properties and manufacturing processes can induce failures. A knowledge of the variability is required to assess design margins and possible trade-offs with weight, size, and cost.

4. Select the parts required for the product, using a well-defined assessment procedure that ensures that the parts selected have sufficient quality and integrity, are capable of delivering the expected performance and reliability in the application, and will be available to sustain the product throughout its life cycle.

5. Identify the potential failure sites and failure mechanisms by which the product can be expected to fail. Critical parts, part details, and potential failure modes and mechanisms must be identified early in the design, and appropriate measures must be implemented to assure design control. Potential architectural and stress interactions must also be defined and assessed.

6. Design to the usage and process capability of the product (i.e., the quality level that can be controlled in manufacturing and assembly), considering the potential failure sites and failure mechanisms. The design stress spectra, the part test spectra, and the full-scale test spectra must be based on the anticipated life cycle usage conditions. The proposed product must survive the life cycle environment, be optimized for manufacturability, quality, reliability, and cost-effectiveness, and be available to the market in a timely manner.

7. Qualify the product manufacturing and assembly processes. Key process characteristics in all the manufacturing and assembly processes required to make the part must be identified, measured, and optimized. Tests should be conducted to verify the results for complex products. The goal of this step is to provide a physics-of-failure basis for design decisions, with an assessment of all possible failure mechanisms for the anticipated product. If all the processes are in control and the design is valid, then product testing is not warranted and is therefore not cost-effective. This represents a transition from product test, analysis, and screening to process test, analysis, and screening.

8. Monitor and control the manufacturing and assembly processes addressed in the design, so that process shifts do not arise. Each process may involve screens and tests to assess statistical process control.

9. Manage the life cycle usage of the product using closed loop management procedures. This includes realistic inspection and maintenance procedures.

22.2   Product Requirements and Constraints

A product's requirements and constraints are defined in terms of customer demands and the company's core competencies, culture, and goals. If the product is for direct sale to end users, marketing usually takes the lead in defining the product's requirements and constraints through interaction with the customer's marketplace, examination of the current product sales figures, and analysis of the competition. Alternatively, if the product is a subsystem that fits within a larger product, the requirements and constraints are determined by the product into which the subsystem fits. The results of capturing product requirements and constraints allow the design team to choose product parts that conform to product-specific and company objectives.

The definition process begins with the identification of an initial set of requirements and constraints defined by either the marketing activity (or in some cases by a specific customer), or by the product into which the subsystem fits. The initial requirements are formulated into a requirements document, where they are prioritized. The requirements document needs to be approved by several groups of people, ranging from engineers to management to customers (the specific people involved in the approval will vary with the organization and the product). Once the requirements are approved, the engineering team prepares a preliminary specification indicating the exact set of requirements that are practical to implement. Disconnects between the requirements document and the preliminary specification become the topic of trade-off analyses (usually cost/performance trade-offs), and if, after analyses and negotiation, all the requirements cannot be implemented, the requirements document may be modified. When the requirements document and the preliminary specifications are agreed upon, a final specification is prepared and the design begins.

22.3 The Product Life Cycle Environment

The product life cycle environment goes hand in hand with the product requirements. The life cycle environment affects product design and development decisions, qualification and specification processes, parts selection and management, quality assurance, product safety, warranty and support commitments, and regulatory conformance.

The product life cycle environment describes the assembly, storage, handling, and scenario for the use of the product, as well as the expected severity and duration of these environments, and thus contains the necessary load input information for failure assessment and the development of design guidelines, assembly guidelines, screens, and tests. Specific load conditions may include steady-state temperatures, temperature ranges, temperature cycles, temperature gradients, humidity levels, pressure levels, pressure gradients, vibrational or shock loads and transfer functions, chemically aggressive or inert environments, acoustic levels, sand, dust, and electromagnetic radiation levels. In electrical systems, stresses caused by power, current, and voltage should also be considered. These conditions may influence the reliability of the product either individually or in combination with each other. Since the performance of a product over time is often highly dependent on the magnitude of the stress cycle, the rate of change of the stress, and the variation of the stress with time and space, the interaction between the application profile and the internal conditions must be specified in the design.

The product life cycle environment can be divided into three parts: the application and life profile conditions, the external conditions under which the product must operate, and the internal product-generated stress conditions. The application and life profile conditions include the application length, the number of applications in the expected life of the product, the product use or non-use profile (storage, testing, transportation), the deployment operations, and the maintenance concept or plan. This information is used to group usage platforms (whether the product will be installed in a car, boat, airplane, satellite, or underground), to develop duty cycles (on-off cycles, storage cycles, transportation cycles, modes of operation, and repair cycles), to determine design criteria, to develop screens and test guidelines, and to develop support requirements to sustain attainment of reliability and maintainability objectives.

The external operational conditions include the anticipated environment(s) and the associated stresses that the product will be required to survive. These conditions are usually determined through experimentation and through the use of numerical simulation techniques. Experiments are performed by creating environmental parameter monitoring systems consisting of sensors placed near and within the product that are capable of monitoring the loads that the product experiences. A sensor's function is to convert a physical variable input into, in most cases, an electrical output that is directly related to the physical variable. Signals can be transmitted to either local or remote output devices, enabling data to be collected in a safe and secure manner. Numerical simulation techniques combine material properties, geometry, and product architecture information with environmental data to determine the life cycle environment based on external stresses. Whenever credible data are not available, the worst-case design load must be estimated. A common cause of failure is the use of design factors related to average loads, without adequate consideration being given to the extreme conditions that may occur during the product's life cycle.[2]

The internal operational conditions are associated with product-generated stresses, such as power consumption and dissipation, internal radiation, and release or outgassing of potential contaminants.

If the product is connected to other products or subsystems in a system, the stresses associated with the interfaces (i.e., external power consumption, voltage transients, voltage spikes, electronic noise, and heat dissipation) must also be included.

Life cycle stresses can cause strength degradation in materials, for example, combined stresses can accelerate damage and reduce the fatigue limit. In such cases, protective measures must be taken to mitigate the life cycle environment by the use of packaging, provision of warning labels and instructions, and protective treatment of surfaces. The measures to be taken must be identified as appropriate to assembly, storage, transportation, handling, operation, and maintenance. Protection against extreme loads may not always be possible, but should be considered whenever practicable. When overload protection is provided, a reliability analysis should be performed on the basis of the maximum anticipated load, keeping the tolerances of the protection system in mind.[2] If complete protection is not possible, the design team must specify appropriate maintenance procedures for inspection, cleaning, and replacement.

An example of the scenario for use of a product is a flight application, which can involve engine warm-up, taxi, climb, cruising, maneuvers, rapid descent, and emergency landing. Each part of the application will be associated with a set of load conditions, such as time, cycles, acceleration, velocity, vibration, shocks, temperature, humidity, and electrical power cycles. Together, these loads comprise a load history of the product.

22.4 Characterization of Materials, Parts, and Manufacturing Processes

Design is intrinsically linked to the materials, parts, interfaces, and manufacturing processes used to establish and maintain the functional and structural integrity of the product. It is unrealistic and potentially dangerous to assume defect-free and perfect-tolerance materials, parts, and structures. Materials often have naturally occurring defects, and manufacturing processes can introduce additional defects in the materials, parts, and structures. The design team must also recognize that the production lots or vendor sources for parts that comprise the design are subject to change, and variability in parts characteristics is likely to occur during the fielded life of a product.

Design decisions involve the selection of parts, materials, and controllable process techniques using processes appropriate to the scheduled production quantity. Any new parts, materials, and processes must be assessed and tested before being put into practice, so that training for production personnel can be planned, quality control safeguards can be set up, and alternative second sources can be located. Often, the goal is to maximize part and configuration standardization, to increase package modularity for ease in fabrication, assembly, and modification, to increase flexibility of design adaptation to alternate uses, and to utilize common fabrication processes. Design decisions also involve choosing the best material interfaces and the best geometric configurations, given the product requirements and constraints.

22.5 Parts Selection and Management

Product differentiation, which determines market share gain and loss, often motivates a company to adopt new technologies and insert them into their mainstream products. However, while technological advances continue to fuel product development, two factors, management decisions regarding when and how a new technology will be used, and accurately assessing risks associated with a technology, differentiate the winners from the losers. Few companies have failed because the right technology was not available; far more have failed when a technology was not effectively managed.

The methodology, shown in Figure 22.1, provides an "eyes-on, hands-off" approach to parts selection and management, which enables organizations to:

- Employ risk assessment and mitigation techniques to address technology insertion;
- Organize and conduct fact-finding processes to select parts with improved quality, integrity, application-specific reliability, and cost-effectiveness;

Assessments performed for each part

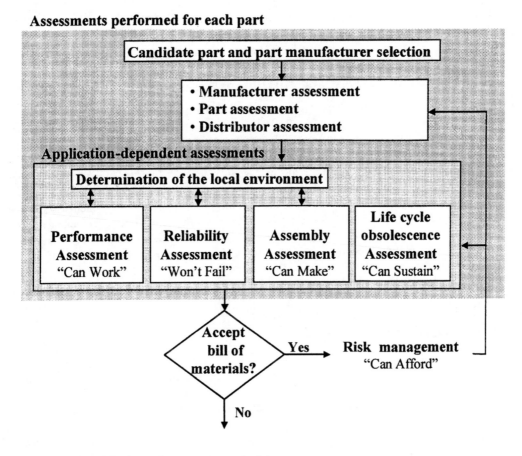

FIGURE 22.1   Parts selection and management methodology.

- Make an informed organization-wide decision about parts selection and management, based upon organization resources, policies, culture, goals, and customer demands;
- Understand and evaluate the local environment the part sees within a product's life cycle, and thereby choose the most appropriate technique to fit the part to its intended environmental requirements;
- Maximize product supportability by preparing for and meeting the challenge of parts becoming obsolete during product life; and
- Improve supply-chain interactions and communications with regulatory agencies to minimize time to profit.

22.5.1   Candidate Part and Part Manufacturer Selection

A candidate part is one that conforms to the functional, electrical, and mechanical requirements of the product, considering product requirements, technology direction, and development. In addition, a candidate part must conform to availability and cost constraints. Availability of an electronic part is a measure of the ease with which the part can be procured. Availability is assessed by determining the amount of inventory at hand, the number of parts required for units in production and forecasted, the economic order quantity for the part(s), the lead time(s) between placing an order for the part(s) and receiving the part(s), production schedules and deadlines, and part discontinuation plans. The cost of the part is assessed relative to the product's budget during candidate part selection. In many cases, a part similar to the required one will have already been designed and tested. This "preferred part" is typically mature, in the sense that the variabilities in manufacturing, assembly, and field operation that could cause problems will have already been identified and corrected. Many design groups maintain a list of preferred parts of proven performance, cost, availability, and reliability.

22.5.2 Manufacturer, Part, and Distributor Assessment

In the manufacturer assessment, the part manufacturer's ability to produce parts with consistent quality is evaluated, and in the part assessment, the candidate part's quality and integrity is gauged. The distributor assessment evaluates the distributor's ability to provide parts without affecting the initial quality and integrity, and to provide certain specific services, such as part problem and change notifications. The equipment supplier's parts selection and management team defines the minimum acceptability criteria for this assessment, based on the equipment supplier's requirements. If the part satisfies the minimum acceptability criteria, the candidate part then moves to "application-dependent assessments."

If the part is found unacceptable due to nonconformance with the minimum acceptability criteria, some form of equipment supplier intervention may be considered.[3,4] If equipment supplier intervention is not feasible due to economic or schedule considerations, the candidate part may be rejected. If, however, equipment supplier intervention is considered necessary, then the intervention action items should be identified, and their cost and schedule implications should be analyzed through the "risk management" process step.

22.5.3 Performance Assessment

The goal of performance assessment is to evaluate the ability of the part to meet the functional, mechanical, and electrical performance requirements. In order to increase performance, products often incorporate features that tend to make them less reliable than proven, lower-performance products. Increasing the number of parts, although improving performance, also increases product complexity, and may lead to lower reliability unless compensating measures are taken.[5] In such situations, product reliability can be maintained only if part reliability is increased or part redundancy is built into the product. Each of these alternatives, in turn, must be assessed against the incurred cost. The trade-off between performance, reliability, and cost is a subtle issue, involving loads, functionality, system complexity, and the use of new materials and concepts.

In general, there are no distinct stress boundaries for parameters such as voltage, current, temperature, and power dissipation, above which immediate failure will occur and below which a part will operate indefinitely.[6] However, there is often a minimum and a maximum stress limit beyond which the part will not function properly, or at which the increased complexity required will not offer an advantage in cost-effectiveness. Part manufacturers' ratings or users' procurement ratings are generally used to determine these limiting values. Equipment manufacturers who integrate such parts into their products need to adapt their design so that the parts do not experience conditions beyond their absolute maximum ratings, even under the worst possible operating conditions (e.g., supply voltage variations, load variations, and signal variations).[7] It is the responsibility of the parts selection and management team to establish that the electrical, mechanical, and functional performance of the part is suitable for the operating conditions of the particular product. If a product must be operated outside the manufacturer-specified operating conditions, then uprating* may have to be considered.

Part manufacturers need to assess the capability of a part over its entire intended life cycle environment, based on the local environment that is determined. If the parametric and functional requirements of the system cannot be met within the required local environment, then the local environment may have to be modified, or a different part may have to be used.

---

*The term *uprating* was coined by Michael Pecht to distinguish it from *upscreening*, which is a term used to describe the practice of attempting to create a part equivalent to a higher quality by additional screening of a part (e.g., screening a JANTXV part to JANS requirements).

22.5.4 Reliability Assessment

Reliability assessment results provide information about the ability of a part to meet the required performance specifications in its life cycle application environment for a specified period of time. Reliability assessment is conducted through the use of integrity test data, virtual qualification results, or accelerated test results. The reliability assessment process is shown in Figure 22.2.

Integrity is a measure of the appropriateness of the tests conducted by the manufacturer and of the part's ability to survive those tests. Integrity monitoring tests are conducted by the part manufacturer to monitor part/process changes and the ongoing material or process changes specific to the part. Integrity test data (often available from the part manufacturer) is examined in light of the application life cycle stresses and the applicable failure modes and mechanisms. If the magnitude and duration of the application life cycle loads are less severe than those of the integrity tests, and if the test sample size and results are acceptable, then the part reliability is acceptable. However, if the magnitude and duration of the application life cycle loads are more severe than those encountered during the integrity tests, then integrity test data cannot be used to validate part reliability in the application, and virtual qualification should be considered.

Virtual qualification is a simulation-based methodology used to identify the dominant failure mechanisms associated with the part under the life cycle loads, to determine the acceleration factor for a given set of accelerated test parameters, and to determine the time-to-failures corresponding to the identified failure mechanisms. Virtual qualification allows the operator to optimize the part parameters (e.g., dimensions, materials) so that the minimum time-to-failure of any part is greater than the expected product life.

If virtual qualification proves insufficient to validate part reliability, accelerated testing should be performed. Once the appropriate test procedures, conditions, and sample sizes are determined, accelerated testing can be conducted by either the part manufacturer, the equipment supplier, or third-party test facilities. Accelerated testing results are used to predict the life of a product in its field application by computing an acceleration factor that correlates the accelerated test conditions and the actual field conditions. Whether integrity test data, virtual qualification results, accelerated test results, or a combination thereof are used, each applicable failure mechanism to which the part is susceptible must be addressed.

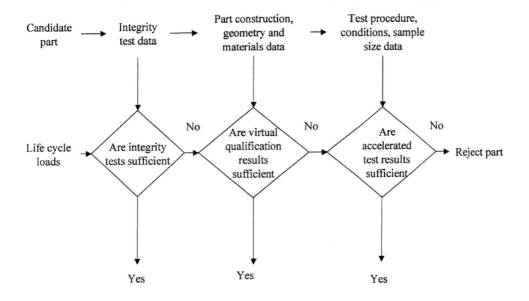

FIGURE 22.2 Reliability assessment process.

If part reliability is not ensured through the reliability assessment process, the equipment supplier must consider an alternate part or product redesign. If redesign is not considered a viable option, the part should be rejected, and an alternate part must be selected. If the part must be used in the application, redesign options may include thermal management techniques, vibration damping, and modification of assembly parameters. If product design changes are made, part reliability must be reassessed.

22.5.5 Assembly Issues

A part may be unacceptable from an assembly viewpoint if (1) it is incompatible with the assembly equipment or process; (2) it is impossible or impractical to wire the part into the product (routing compatibility), or (3) it cannot be acceptably tested or reworked. Assembly compatibility addresses whether a product that contains the part can be manufactured (assembled). Routing compatibility assesses if the candidate part can be routed within a specific application on the selected board. Test and rework acceptability assess whether the candidate part can be adequately and economically tested and reworked during assembly and maintenance.

22.5.5.1 Assembly Compatibility

Parts must conform to a range of constraints associated with their assembly into products. There are three categories of assembly constraints that must be considered when designing a product:

- Assembly process compatibility — Assembly process compatibility involves comparing the part's size, shape, and mounting method to the process that will be used to assemble the boards containing the part.
- Proximity to other structures — Proximity checking involves checking the location of the component relative to other parts assembled on the board and the edge of the board. Proximity checking includes evaluating the orientation (rotation) of the part.
- Artwork verification — Artwork verification involves checking the board layout for the correct orientation and location of fiducials (alignment marks), alignment holes, and other structures necessary to facilitate assembly.

There are three possible outcomes from assembly compatibility and proximity checking: cannot be assembled, can be assembled with a corresponding cost and yield penalty, and can be assembled with no cost or yield penalties. Artwork verification is decoupled from parts selection.

22.5.5.2 Routing Compatibility

Routing compatibility pertains to the layout and routing of an application. If the selection of a particular part causes significant layout or routing problems within the board, the part may be rejected. Rejection of a part is usually based on its use of routing resources within the board. Two routing issues must be considered:

- How much board area is required to wire the part to the rest of the product?
- How many layers of the board are required to "escape route" the part?

Escape routing is only applicable if the part has an area array format connection to the board, for example, a flip chip or ball grid array package. A component is virtually always "routable," given a sufficient number of board layers. If the rest of the parts on the board are known, routing estimation techniques can be used to determine the effective routing limited footprint of a part under the constraints posed by the board design rules (lines, spaces, via/hole capture pad diameter) and layer count. If a candidate part exceeds the fraction of board wiring resources budgeted to it based on board growth and cost constraints, it may be rejected.

A limiting requirement for some parts is escape routing. If a part's I/Os are in an area array format (as opposed to a peripheral format), the part cannot be wired into the product until all of its I/Os are routed out from under the part. The process of liberating I/Os from an array is called escape routing.

22.5.5.3 Test and Rework Acceptability

Test and rework costs are important criteria in determining whether a part is acceptable or not. The cost of testing the part (to a specified quality level) prior to assembly and the cost of replacing the part if it needs to be repaired after it is assembled must be considered.

The cost of testing a part is related to the level of testing performed by the part manufacturer, whether the part is in a package or bare, the function that the part performs, the number of gates or bits in the part, and the test equipment. If the part does not come from the manufacturer fully tested (e.g., a bare die), then test costs may need to be assessed. Test costs include the cost of creating the test patterns (or obtaining them from the manufacturer) and the cost of applying the test to the part. Predicting testing costs is of little value unless the corresponding test coverage (fraction of defects detected by the test) is also predicted.

Another key assembly-related cost is the cost of replacing a part that has been identified as defective during the assembly process. The cost of removing a defective part is a function of how the part is mounted to the board, the size of the part, and its proximity to other parts.

22.5.6 Life Cycle Mismatch Assessment

Lengthy design, qualification, and production processes inherent in electronic industries often cause parts to become obsolete before the first product is produced.[8] Furthermore, to cater to market demands and remain competitive, part manufacturers often introduce new parts and discontinue older parts. In general, electronic products go through six phases during their life cycle: design, manufacturing, growth, maturity, decline, and discontinuance. A life cycle mismatch occurs between a product and its constituent parts if the parts are not available to support the product throughout its life cycle. When factors such as lead time, risk of part obsolescence, or estimation of the product market are ignored or improperly judged during the design phase, the consequences can be costly. The obsolete part can inhibit the functioning of the product, idle the assembly line, lead to dissatisfied customers, and cause a loss of reputation to the company. The net outcome can be a financial loss for the company.

A successful life cycle mismatch assessment process is one that prevents, if possible, the selection of parts that are already obsolete or soon to be discontinued. This strategy reduces the risk associated with a life cycle mismatch between a product and its parts. The part selection depends on the degree of mismatch and the flexibility to adopt an obsolescence management strategy (e.g., redesign, lifetime buy, buy from after-market sources, part substitution). The strategy is intended to mitigate obsolescence risks associated with using the part at some future point in the life cycle of the product. If the equipment supplier finds the life cycle mismatch between part and product unacceptable, the part is unsuitable and should be rejected.

22.5.7 Risk Management

After a part is accepted, resources must be applied to managing the life cycle of the part, including supply chain management, obsolescence assessment, manufacturing and assembly feedback, manufacturer warranties management, and field failure and root-cause analysis. It is important to consider the process of managing the part and all the risks associated with the long-term use of the part throughout its life cycle during the part selection process. The risk management process is characterized using the risks identified in the parts selection process to determine the resources needed to support a part throughout its application life cycle, thus minimizing the probability of a failure. The key metric used to determine whether risks should be managed or not is resources, which include time, data, opportunity, and money.

The risks associated with including a part in the product fall into two categories:

- Managed risks: risks that the product development team chooses to proactively manage by creating a management plan and performing a prescribed regimen of monitoring the part's field performance, manufacturer, and manufacturability; and
- Unmanaged risks: risks that the product development team chooses not to proactively manage.

If risk management is considered necessary, a plan should be prepared. The plan should contain details about how the part is monitored (data collection), and how the results of the monitoring feed back into various parts selection and management processes. The feasibility, effort, and cost involved in management processes prior to the final decision to select the part must be considered.

Feedback regarding the part's assembly performance, field performance, and sales history may be essential to ascertain the validity of the predictions made during the part selection process. If the feedback calls for changes in selection criteria, they should be incorporated into the part selection process. Prospective parts should be judged based on the altered part selection criteria. Part monitoring data may also be needed to make changes in parts that are already in use. For example, part monitoring field data might indicate that a change in operating conditions is required for the part to perform satisfactorily.

22.6 Failure Modes and Mechanisms

Failure mechanisms are the physical processes by which stresses can damage the materials used to build the product. Investigation of the possible failure modes and mechanisms of the product aids in developing failure-free and reliable designs. The design team must be aware of all possible failure mechanisms if they are to design hardware capable of withstanding loads without failing. Failure mechanisms and their related models are also important for planning tests and screens to audit the nominal design and manufacturing specifications, as well as the level of defects introduced by excessive variability in manufacturing and material parameters. Numerous studies focusing on material failure mechanisms and physics-of-failure-based damage models and their role in obtaining reliable electronic products have been illustrated in a series of tutorials comprising all relevant wearout and overstress failures.[9-23]

Catastrophic failures due to a single occurrence of a stress event when the intrinsic strength of the material is exceeded are termed overstress failures. Failure mechanisms due to monotonic accumulation of incremental damage beyond the endurance of the material are termed wearout mechanisms.[24] When the damage exceeds the endurance limit of the component, failure will occur. Unanticipated large stress events can either cause an overstress (catastrophic) failure, or shorten life by causing the accumulation of wearout damage. Examples of such stresses are accidental abuse and acts of God. On the other hand, in well-designed and high-quality hardware, stresses should cause only uniform accumulation of wearout damage; the threshold of damage required to cause eventual failure should not occur within the usage life of the product.

Electrical performance failures can be caused by individual components with improper electrical parameters, such as resistance, impedance, capacitance, or dielectric properties, or by inadequate shielding from electromagnetic interference (EMI) or particle radiation. Failure modes can manifest as reversible drifts in transient and steady-state responses, such as delay time, rise time, attenuation, signal-to-noise ratio, and crosstalk. Electrical failures common in electonic hardware include overstress mechanisms due to electrical overstress (EOS) and electrostatic discharge (ESD), such as dielectric breakdown, junction breakdown, hot electron injection, surface and bulk trapping, and surface breakdown, and wearout mechanisms such as electromigration and stress-driven diffusive voiding.

Thermal performance failures can arise due to incorrect design of thermal paths in an electronic assembly. This includes incorrect conductivity and surface emissivity of individual components, as well as incorrect convective and conductive paths for heat transfer. Thermal overstress failures are a result of heating a component beyond critical temperatures such as the glass-transition temperature, melting point, fictile point, or flash point. Some examples of thermal wearout failures are aging due to depolymerization, intermetallic growth, and interdiffusion. Failures due to inadequate thermal design may be manifested as components running too hot or too cold and causing operational parameters to drift beyond specifications, although the degradation is often reversible upon cooling. Such failures can be caused either by direct thermal loads or by electrical resistive loads, which in turn generate excessive localized thermal stresses. Adequate design checks require proper analysis for thermal stress, and should include conductive, convective, and radiative heat paths.

Mechanical performance failures include those that may compromise the product performance without necessarily causing any irreversible material damage, such as abnormal elastic deformation in response to mechanical static loads, abnormal transient response (such as natural frequency or damping) to dynamic loads, and abnormal time-dependent reversible (anelastic) response, as well as failures that cause material damage, such as buckling, brittle and/or ductile fracture, interfacial separation, fatigue crack initiation and propagation, creep, and creep rupture. To take one example, excessive elastic deformations in slender structures in electronic packages can sometimes constitute functional failure due to overstress loads such as excessive flexing of interconnection wires, package lids, or flex circuits in electronic devices, causing shorting and/or excessive crosstalk. However, when the load is removed, the deformations (and consequent functional abnormalities) disappear completely without any permanent damage.

Radiation failures are principally caused by uranium and thorium contaminants, and secondary cosmic rays. Radiation can cause wearout, aging, embrittlement of materials, and overstress soft errors in electronic hardware, such as logic chips. Chemical failures occur in adverse chemical environments that result in corrosion, oxidation, or ionic surface dendritic growth. There may also be interactions between different types of stresses. For example, metal migration may be accelerated in the presence of chemical contaminants and composition gradients, and thermal loads can accelerate a failure mechanism due to a thermal expansion mismatch.

Failure modes and effects analysis (FMEA) is an evaluation process for analyzing and assessing the potential failures in a product. Its objectives are to:

1. Identify the causes and effects of each failure mode in every part in the product;
2. Ascertain the effects of each failure mode on product operation and personnel safety;
3. Assess each potential failure according to the effects on other portions of the systems; and
4. Provide a recommendation to eliminate the causes of the failure modes or compensate for their effects.

Failure effects may be considered at subsystem and at overall system levels.

There are two approaches to FMEA: functional and hardware. The functional approach, which should be used when the product definition has been identified, begins with the initial product indenture level, and proceeds downwards through lower levels. The top level shows the gross operational requirements of the product, while the lower levels represent progressive expansions of the individual functions of the preceding level. This documentation is prepared down to the level necessary to establish the hardware, software, facilities, and personnel and data requirements of the system.

The hardware approach to FMEA should be used when the design team has access to schematics, drawings, and other engineering and design data normally available once the system has matured beyond the functional design stage. This approach begins with obtaining all the information available on the design, including specifications, requirements, constraints, intended applications, drawings, stress data, test results, and so on, to the extent they are available at that time. The approach then proceeds in a part level-up fashion.

Once the approach for the analysis is selected, the product is defined in terms of a functional block diagram and a reliability block diagram. If the product operates in more than one mode in which different functional relationships or part operating modes exist, then these must be considered in the design. FMEA should involve an analysis of possible sneak circuits in the product, that is, an unexpected path or logic flow that can initiate an undesired function or inhibit a desired function. Effects of redundancy must also be considered by evaluating the effects of the failure modes assuming that the redundant system or subsystem is or is not available. The FMEA is then performed using a worksheet, and working to the part or subsystem level considered appropriate, keeping the design data available in mind. A fish-bone diagram of the product, showing all the possible ways in which the product can be expected to fail, is often used in the process. The analysis should take all the failure modes of every part into account, especially when the effects of a failure are serious (e.g., high warranty costs, reliability reputation, safety). FMEA should be started as soon as initial design information is available, and should be performed iteratively as the design evolves, so that the analysis can be used to improve the design and to provide documentation of the eventually completed design.

22.7   Design Guidelines and Techniques

Generally, products are replaced with other products, and the replaced product can be used as a baseline for comparisons with products to be introduced. Lessons learned from the baseline comparison product can be used to establish new product parameters, to identify areas of focus in new product designs, and to avoid the mistakes of the past.

Once the parts, materials, processes, and stress conditions are identified, the objective is to design a product using parts and materials that have been sufficiently characterized in terms of how they perform over time when subjected to the manufacturing and application profile conditions. Only through a methodical design approach using physics-of-failure and root-cause analysis can a reliable and cost-effective product be designed. A physics-of-failure-based reliability assessment tool must exhibit a diverse array of capabilities:

1. It should be able to predict the reliability of components under a wide range of environmental conditions;
2. It should be able to predict the time-to-failure for fundamental failure mechanisms; and
3. It should consider the effect of different manufacturing processes on reliability.

All of these can be accomplished by the use of tools such as virtual qualification and accelerated testing. Design guidelines that are based on physics-of-failure models can also be used to develop tests, screens, and derating factors. Tests based on physics-of-failure models can be designed to measure specific quantities, to detect the presence of unexpected flaws, and to detect manufacturing or maintenance problems. Screens can be designed to precipitate failures in the weak population while not cutting into the design life of the normal population. Derating or safety factors can be determined to lower the stresses for the dominant failure mechanisms.

In using design guidelines, there may not be a unique path to follow. Instead, there is a general flow in the design process. Multiple branches may exist, depending on the input design constraints. The design team should explore an adequate number of these branches to gain confidence that the final design is the best for the prescribed input information. The design team should also assess the use of guidelines for the complete design, and not those limited to specific aspects of an existing design. This does not imply that guidelines cannot be used to address only a specific aspect of an existing design, but the design team may have to trace through the implications that a given guideline suggests.

22.7.1   Protective Architectures

In designs where safety is an issue, it is generally desirable to design in some means for preventing a part, structure, or interconnection from failing, or from causing further damage when it fails. Fuses and circuit breakers are examples of elements used in electronic products to sense excessive current drain and to disconnect power from the concerned part. Fuses within circuits safeguard parts against voltage transients or excessive power dissipation, and protect power supplies from shorted parts. As another example, thermostats can be used to sense critical temperature limiting conditions, and to shut down the product or a part of the system until the temperature returns to normal. In some products, self-checking circuitry can also be incorporated to sense abnormal conditions and make adjustments to restore normal conditions, or to activate switching means to compensate for the malfunction.[6]

In some instances, it may be desirable to permit partial operation of the product after a part failure in preference to total product failure. By the same reasoning, degraded performance of a product after failure of a part is often preferable to complete stoppage. An example is the shutting down of a failed circuit whose function is to provide precise trimming adjustment within a deadband* of another control

---

*When the input in a control system changes direction, an initial change in the input has no effect on the output. This amount of side-to-side play in the system for which there is no change in the output is referred to as the deadband. The deadband is centered about the output.

product; acceptable performance may thus be achieved, perhaps under emergency conditions, with the deadband control product alone.[6]

Sometimes, the physical removal of a part from a product can harm or cause failure in another part by removing either load, drive, bias, or control. In such cases, the first part should be equipped with some form of interlock mechanism to shut down or otherwise protect the second part. The ultimate design, in addition to its ability to act after a failure, should be capable of sensing and adjusting for parametric drifts to avert failures.

In the use of protective techniques, the basic procedure is to take some form of action, after an initial failure or malfunction, to prevent additional or secondary failures. By reducing the number of failures, techniques such as enhancing product reliability can be considered, although they also affect availability and product effectiveness. Equally important considerations are the impacts of maintenance, repair, and part replacement. For example, if a fuse protecting a circuit is replaced, the following questions need to be answered: What is the impact when the product is re-energized? What protective architectures are appropriate for postrepair operations? What maintenance guidance must be documented and followed when fail-safe protective architectures have or have not been included?

22.7.2 Stress Margins

A properly designed product should be capable of operating satisfactorily with parts that drift or change with variables such as time, temperature, humidity, pressure, altitude, etc. as long as the interconnects and the other parameters of the parts are within their rated tolerances. To guard against out-of-tolerance failures, the design team must consider the combined effects of tolerances on parts to be used in manufacture, of subsequent changes due to the range of expected environmental conditions, of drifts due to aging over the period of time specified in the reliability requirement, and of tolerances in parts used in future repair or maintenance functions. Parts and structures should be designed to operate satisfactorily at the extremes of the parameter ranges, and allowable ranges must be included in the procurement or reprocurement specifications.

Statistical analysis and worst-case analysis are methods of dealing with part and structural parameter variations. In statistical analysis, a functional relationship is established between the output characteristics of the structure and the parameters of one or more of its parts. In worst-case analysis, the effect that a part has on product output is evaluated on the basis of end-of-life performance values or out-of-specification replacement parts.

22.7.3 Derating

Derating is a technique by which either the operational stresses acting on a device or structure are reduced relative to the rated strength, or the strength is increased relative to the allocated operating stress levels. Reducing the stress is achieved by specifying upper limits on the operating loads below the rated capacity of the hardware. For example, manufacturers of electronic hardware often specify limits for supply voltage, output current, power dissipation, junction temperature, and frequency. The equipment design team may decide to select an alternative component or make a design change that ensures that the operational condition for a particular parameter, such as temperature, is always below the rated level. The component is then said to have been derated for thermal stress.

The derating factor, typically defined as the ratio of the rated level of a given stress parameter to its actual operating level, is actually a margin of safety or margin of ignorance, determined by the criticality of any possible failures and by the amount of uncertainty inherent in the reliability model and its inputs. Ideally, this margin should be kept to a minimum to maintain the cost-effectiveness of the design. This puts the responsibility on the reliability engineer to identify the rated strength, the relevant operating stresses, and the reliability as unambiguously as possible.

To be effective, derating criteria must target the right stress parameter to address modeling of the relevant failure mechanisms. Field measurements may also be necessary, in conjunction with modeling simulations, to identify the actual operating stresses at the failure site. Once the failure models have been quantified, the impact of derating on the effective reliability of the component for a given load can be determined. Quantitative correlations between derating and reliability enable design teams and users to effectively tailor the margin of safety to the level of criticality of the component, leading to better and more cost-effective use of the functional capacity of the component.

22.7.4 Redundancy

Redundancy permits a product to operate even though certain parts and interconnections have failed, thus increasing its reliability and availability. Redundant configurations can be classified as either active or standby. Elements in active redundancy operate simultaneously in performing the same function. Elements in standby redundancy are designed so that an inactive one will, or can, be switched into service when an active element fails. The reliability of the associated function increases with the number of standby elements (optimistically assuming that the sensing and switching devices of the redundant configuration are working perfectly, and that the failed redundant components are replaced before their companion components fail).

A design team may often find that redundancy is

- The quickest way to improve product reliability if there is insufficient time to explore alternatives, or if the part is already designed;
- The cheapest solution, if the cost of redundancy is economical in comparison with the cost of redesign; and/or
- The only solution, if the reliability requirement is beyond the state of the art.

On the other hand, in weighing its disadvantages, the design team may find that redundancy will:

- Prove too expensive, if the parts, redundant sensors, and switching devices are costly;
- Exceed the limitations on size and weight;
- Exceed the power limitations, particularly in active redundancy;
- Attenuate the input signal, requiring additional amplifiers (which increase complexity); and/or
- Require sensing and switching circuitry so complex as to offset the reliability advantage of redundancy.

22.8 Qualification and Accelerated Testing

Qualification includes all activities that ensure that the nominal design and manufacturing specifications will meet or exceed the desired reliability targets. Qualification validates the ability of the nominal design and manufacturing specifications of the product to meet the customer's expectations, and assesses the probability of survival of the product over its complete life cycle. The purpose of qualification is to define the acceptable range of variabilities for all critical product parameters affected by design and manufacturing, such as geometric dimensions, material properties, and operating environmental limits. Product attributes that are outside the acceptable ranges are termed defects, since they have the potential to compromise product reliability.[25]

Qualification tests should be performed only during initial product development, and immediately after any design or manufacturing changes in an existing product. Once the product is qualified, routine lot-to-lot requalification is redundant and an unnecessary cost item. A well-designed qualification procedure provides economic savings and quick turnaround during development of new products or mature products subject to manufacturing and process changes.

Investigating failure mechanisms and assessing the reliability of products where longevity is required may be a challenge, since a very long test period under the actual operating conditions is necessary to obtain sufficient data to determine actual failure characteristics. One approach to the problem of obtaining meaningful qualification data for high-reliability devices in shorter time periods is using methods such as virtual qualification and accelerated testing to achieve test-time compression. However, when qualifying the reliability of a product for overstress mechanisms, a single cycle of the expected overstress load may be adequate, and acceleration of test parameters may not be necessary. This is sometimes called proof-stress testing.

22.8.1 Virtual Qualification

Virtual qualification is a process that requires significantly less time and money than accelerated testing to qualify a part for its life cycle environment. This simulation-based methodology is used to identify and rank the dominant failure mechanisms associated with the part under life cycle loads, to determine the acceleration factor for a given set of accelerated test parameters, and to determine the time-to-failure corresponding to the identified failure mechanisms. Each failure model comprises a stress analysis model and a damage assessment model. The output is a ranking of different failure mechanisms, based on the time-to-failure. The stress model captures the product architecture, while the damage model depends on a material's response to the applied stress. This process is therefore applicable to existing as well as new products. The objective of virtual qualification is to optimize the product design in such a way that the minimum time-to-failure of any part of the product is greater than its desired life. Although the data obtained from virtual qualification cannot fully replace those obtained from physical tests, it can increase the efficiency of physical tests by indicating the potential failure modes and mechanisms that the operator can expect to encounter.

Ideally, a virtual qualification process will involve identification of quality suppliers, computer-aided physics-of-failure qualification, and a risk assessment and mitigation program. The process allows qualification to be readily incorporated into the design phase of product development, since it allows design, test, and redesign to be conducted promptly and cost-effectively. It also allows consumers to qualify off-the-shelf components for use in specific environments without extensive physical tests. Since virtual qualification reduces emphasis on examining a physical sample, it is imperative that the manufacturing technology and quality assurance capability of the manufacturer be taken into account. The manufacturer's design, production, test, and measurement procedures must be evaluated and certified. If the data on which the virtual qualification is performed are inaccurate or unreliable, all results are suspect. In addition, if a reduced quantity of physical tests is performed in the interest of simply verifying virtual results, the operator needs to be confident that the group of parts selected is sufficient to represent the product. Further, it should be remembered that the accuracy of the results using virtual qualification depends on the accuracy of the inputs to the process, i.e., the accuracy of the life cycle loads, the choice of the failure models used, the choice of the analysis domain (for example, 2D, pseudo-3D, full 3D), the constants in the failure model, the material properties, and so on. Hence, to obtain a reliable prediction, the variabilities in the inputs should be specified using distribution functions, and the validity of the failure models should be tested by conducting accelerated tests.

22.8.2 Accelerated Testing

Accelerated testing involves measuring the performance of the test product at loads or stresses that are more severe than would normally be encountered, to enhance the damage accumulation rate within a reduced time period. The goal of such testing is to accelerate time-dependent failure mechanisms and the damage accumulation rate to reduce the time to failure. The failure mechanisms and modes in the accelerated environment must be the same as (or quantitatively correlated with) those observed under actual usage conditions, and it must be possible to quantitatively extrapolate from the accelerated environment to the usage environment with some reasonable degree of assurance.

Accelerated testing begins by identifying all the possible overstress and wearout failure mechanisms. The load parameter that directly causes the time-dependent failure is selected as the acceleration parameter, and is commonly called the accelerated load. Common accelerated loads include thermal loads, such as temperature, temperature cycling, and rates of temperature change; chemical loads, such as humidity, corrosives, acid, and salt; electrical loads, such as voltage, or power; and mechanical loads, such as vibration, mechanical load cycles, strain cycles, and shock/impulses. The accelerated environment may include a combination of these loads. Interpretation of results for combined loads requires a quantitative understanding of their relative interactions and the contribution of each load to the overall damage.

Failure due to a particular mechanism can be induced by several acceleration parameters. For example, corrosion can be accelerated by both temperature and humidity; and creep can be accelerated by both mechanical stress and temperature. Furthermore, a single accelerated stress can induce failure by several wearout mechanisms simultaneously. For example, temperature can accelerate wearout damage accumulation not only by electromigration, but also by corrosion, creep, and so on. Failure mechanisms that dominate under usual operating conditions may lose their dominance as the stress is elevated. Conversely, failure mechanisms that are dormant under normal use conditions may contribute to device failure under accelerated conditions. Thus, accelerated tests require careful planning if they are to represent the actual usage environments and operating conditions without introducing extraneous failure mechanisms or nonrepresentative physical or material behavior. The degree of stress acceleration is usually controlled by an acceleration factor, defined as the ratio of the life of the product under normal use conditions to that under the accelerated condition. The acceleration factor should be tailored to the hardware in question, and can be estimated from an acceleration transform (that is, a functional relationship between the accelerated stress and the life cycle stress), in terms of all the hardware parameters.

Once the failure mechanisms are identified, it is necessary to select the appropriate acceleration load; to determine the test procedures and the stress levels; to determine the test method, such as constant stress acceleration or step-stress acceleration; to perform the tests; and to interpret the test data, which includes extrapolating the accelerated test results to normal operating conditions. The test results provide failure information for improving the hardware through design and/or process changes. Accelerated testing includes:

- Accelerated test planning and development: Accelerated test planning and development is used to develop a test program that focuses on the potential failure mechanisms and modes that were identified during virtual qualification as the weak links under life cycle loads. The various issues addressed in this phase include designing the test matrix and test loads, analysis, design and preparation of the test device, setting up the test facilities (e.g., test platforms, stress monitoring schemes, failure monitoring and data acquisition schemes), fixture design, effective sensor placement, and data collection and post-processing schemes.

- Test device characterization: Test device characterization is used to identify the contribution of the environment on the test device in the accelerated life tests.

- Accelerated life testing: Accelerated life testing evaluates the vulnerability of the product to the applied life cycle due to wearout failure mechanisms. This step yields a meaningful assessment of life cycle durability only if it is preceded by the steps discussed above. Without these steps, accelerated life testing can only provide comparisons between alternate designs if the same failure mechanism is precipitated.

- Life assessment: Life assessment is used to provide a scientific and rational method to understand and extrapolate accelerated life testing failure data to estimate the life of the product in the field environment.

Detailed failure analysis of failed samples is a crucial step in the qualification and validation program. Without such analyses and feedback to the design team for corrective action, the purpose of the qualification program is defeated. In other words, it is not adequate to simply collect failure data. The key is to use the test results to provide insights into, and consequent control over, relevant failure mechanisms and to prevent them, cost-effectively.

22.9 Manufacturing Issues

Manufacturing and assembly processes can significantly impact the quality and reliability of hardware. Improper assembly and manufacturing techniques can introduce defects, flaws, and residual stresses that act as potential failure sites or stress raisers later in the life of the product. If these defects and stresses can be identified, the design analyst can proactively account for them during the design and development phase.

Auditing the merits of the manufacturing process involves two crucial steps. First, qualification procedures are required, as in design qualification, to ensure that manufacturing specifications do not compromise the long-term reliability of the hardware. Second, lot-to-lot screening is required to ensure that the variabilities of all manufacturing-related parameters are within specified tolerances.[25,26] In other words, screening ensures the quality of the product by precipitating latent defects before they reach the field.

22.9.1   Process Qualification

Like design qualification, process qualification should be conducted at the prototype development phase. The intent at this step is to ensure that the nominal manufacturing specifications and tolerances produce acceptable reliability in the product. The process needs requalification when process parameters, materials, manufacturing specifications, or human factors change.

Process qualification tests can be the same set of accelerated wearout tests used in design qualification. As in design qualification, overstress tests may be used to qualify a product for anticipated field overstress loads. Overstress tests may also be exploited to ensure that manufacturing processes do not degrade the intrinsic material strength of hardware beyond a specified limit. However, such tests should supplement, not replace, the accelerated wearout test program, unless explicit physics-based correlations are available between overstress test results and wearout field-failure data.

22.9.2   Manufacturability

The control and rectification of manufacturing defects has typically been the concern of production and process-control engineers, but not of the design team. In the spirit and context of concurrent product development, however, hardware design teams must understand material limits, available processes, and manufacturing process capabilities to select materials and construct architectures that promote producibility and reduce the occurrence of defects, increasing yield and quality. Therefore, no specification is complete without a clear discussion of manufacturing defects and acceptability limits. The reliability engineer must have clear definitions of the threshold for acceptable quality, and of what constitutes nonconformance. Nonconformance that compromises hardware performance and reliability is considered a defect. Failure mechanism models provide a convenient vehicle for developing such criteria. It is important for the reliability analyst to understand which deviations from specifications can compromise performance or reliability, and which deviations are benign and can be accepted.

A defect is any outcome of a process (manufacturing or assembly) that impairs or has the potential to impair the functionality of the product at any time. The defect may arise during a single process or may be the result of a sequence of processes. The yield of a process is the fraction of products that are acceptable for use in a subsequent manufacturing sequence or product life cycle. The cumulative yield of the process is approximately determined by multiplying the individual yields of each of the individual process steps. The source of defects is not always apparent, because defects resulting from a process can go undetected until the product reaches some downstream point in the process sequence, especially if screening is not employed.

It is often possible to simplify the manufacturing and assembly processes to reduce the probability of workmanship defects. As processes become more sophisticated, however, process monitoring and control are necessary to ensure a defect-free product. The bounds that specify whether the process is within tolerance limits, often referred to as the process window, are defined in terms of the independent variables to be controlled within the process and the effects of the process on the product or the dependent product variables. The goal is to understand the effect of each process variable on each product parameter to formulate control limits for the process, that is, the points on the variable scale where the defect rate begins to possess a potential for causing failure. In defining the process window, the upper and lower limits of each process variable beyond which it will produce defects must be determined. Manufacturing processes must be contained in the process window by defect testing, analysis of the causes of defects, and elimination of defects by process control, such as by closed-loop corrective action systems. The establishment of an effective feedback path to report process-related defect data is critical. Once this is done and the process window is determined, the process window itself becomes a feedback system for the process operator.

Several process parameters may interact to produce a different defect than would have resulted from an individual parameter acting independently. This complex case may require that the interaction of various process parameters be evaluated in a matrix of experiments. In some cases, a defect cannot be detected until late in the process sequence. Thus, a defect can cause rejection, rework, or failure of the product after considerable value has been added to it. These cost items due to defects can return on investments by adding to hidden factory costs. All critical processes require special attention for defect elimination by process control.

22.9.3 Process Verification Testing

Process verification testing is often called screening. Screening involves 100% auditing of all manufactured products to detect or precipitate defects. The aim of this step is to preempt potential quality problems before they reach the field. In principle, screening should not be required for a well-controlled process. When uncertainties are likely in process controls, however, screening is often used as a safety net.

Some products exhibit a multimodal probability density function for failures, with a secondary peak during the early period of their service life due to the use of faulty materials, poorly controlled manufacturing and assembly technologies, or mishandling. This type of early-life failure is often called infant mortality. Properly applied screening techniques can successfully detect or precipitate these failures, eliminating or reducing their occurrence in field use. Screening should only be considered for use during the early stages of production, if at all, and only when products are expected to exhibit infant mortality field failures. Screening will be ineffective and costly if there is only one main peak in the failure probability density function. Further, failures arising due to unanticipated events such as acts of God (lightning, earthquakes) may be impossible to screen cost-effectively.

Since screening is done on a 100% basis, it is important to develop screens that do not harm good components. The best screens, therefore, are nondestructive evaluation techniques, such as microscopic visual exams, X-rays, acoustic scans, nuclear magnetic resonance (NMR), electronic paramagnetic resonance (EPR), and so on. Stress screening involves the application of stresses, possibly above the rated operational limits. If stress screens are unavoidable, overstress tests are preferred to accelerated wearout tests, since the latter are more likely to consume some useful life of good components. If damage to good components is unavoidable during stress screening, then quantitative estimates of the screening damage, based on failure mechanism models must be developed to allow the design team to account for this loss of usable life. The appropriate stress levels for screening must be tailored to the specific hardware. As in qualification testing, quantitative models of failure mechanisms can aid in determining screen parameters.

A stress screen need not necessarily simulate the field environment, or even utilize the same failure mechanism as the one likely to be triggered by this defect in field conditions. Instead, a screen should exploit the most convenient and effective failure mechanism to stimulate the defects that can show up in the field as infant mortality. Obviously, this requires an awareness of the possible defects that may occur in the hardware and extensive familiarity with the associated failure mechanisms.

Unlike qualification testing, the effectiveness of screens is maximized when screens are conducted immediately after the operation believed to be responsible for introducing the defect. Qualification testing is preferably conducted on the finished product or as close to the final operation as possible; on the other hand, screening only at the final stage, when all operations have been completed, is less effective, since failure analysis, defect diagnostics, and troubleshooting are difficult and impair corrective actions. Further, if a defect is introduced early in the manufacturing process, subsequent value added through new materials and processes is wasted, which additionally burdens operating costs and reduces productivity.

Admittedly, there are also several disadvantages to such an approach. The cost of screening at every manufacturing station may be prohibitive, especially for small batch jobs. Further, components will experience repeated screening loads as they pass through several manufacturing steps, which increases the risk of accumulating wearout damage in good components due to screening stresses. To arrive at a screening matrix that addresses as many defects and failure mechanisms as feasible with each screen test, an optimum situation must be sought through analysis of cost-effectiveness, risk, and the criticality of the defects. All defects must be traced back to the root cause of the variability.

Any commitment to stress screening must include the necessary funding and staff to determine the root cause and appropriate corrective actions for all failed units. The type of stress screening chosen should be derived from the design, manufacturing, and quality teams. Although a stress screen may be necessary during the early stages of production, stress screening carries substantial penalties in capital, operating expense, and cycle time, and its benefits diminish as a product approaches maturity. If almost all of the products fail in a properly designed screen test, the design is probably incorrect. If many products fail, a revision of the manufacturing process is required. If the number of failures in a screen is small, the processes are likely to be within tolerances and the observed faults may be beyond the resources of the design and production process.

22.10 Summary

Reliability is not a matter of chance or good fortune; rather, it is a rational consequence of conscious, systematic, rigorous efforts at every stage of design, development, and manufacture. High product reliability can only be assured through robust product designs, capable processes that are known to be within tolerances, and qualified components and materials from vendors whose processes are also capable and within tolerances. Quantitative understanding and modeling of all relevant failure mechanisms provide a convenient vehicle for formulating effective design, process, and test specifications and tolerances.

The physics-of-failure approach is not only a tool to provide better and more effective designs, but it also helps develop cost-effective approaches for improving the entire approach to building electronic products. Proactive improvements can be implemented for defining more realistic performance requirements and environmental conditions, identifying and characterizing key material properties, developing new product architectures and technologies, developing more realistic and effective accelerated stress tests to audit reliability and quality, enhancing manufacturing-for-reliability through mechanistic process modeling and characterization to allow pro-active process optimization, increasing first-pass yields, and reducing hidden factory costs associated with inspection, rework, and scrap.

When utilized early in the concept stage of a product's development, reliability serves as an aid to determine feasibility and risk. In the design stage of product development, reliability analysis involves methods to enhance performance over time through the selection of materials, design of structures, choice of design tolerance, manufacturing processes and tolerances, assembly techniques, shipping and handling methods, and maintenance and maintainability guidelines. Engineering concepts such as strength, fatigue, fracture, creep, tolerances, corrosion, and aging play a role in these design analyses. The use of physics-of-failure concepts coupled with mechanistic and probabilistic techniques are often required to understand the potential problems and trade-offs, and to take corrective actions. The use of factors of safety and worst-case studies as part of the analysis is useful in determining stress screening and burn-in procedures, reliability growth, maintenance modifications, field testing procedures, and various logistics requirements.

Defining Terms

Accelerated testing: Tests conducted at stress levels that are more severe than the normal operating levels, in order to enhance the damage accumulation rate within a reduced time period.

Damage: The extent of a product's degradation or deviation from a defect-free state.

Derating: Practice of subjecting parts to lower electrical or mechanical stresses than they can withstand to increase the life expectancy of the part.

Failure mechanism: A process (such as creep, fatigue, or wear) through which a defect nucleates and grows as a function of stresses (such as thermal, mechanical, electromagnetic, or chemical loadings) ultimately resulting in the degradation or failure of a product.

Failure mode: Any physically observable change caused by a failure mechanism.

Integrity: A measure of the appropriateness of the tests conducted by the manufacturer and the part's ability to survive those tests.

Overstress failures: Catastrophic sudden failures due to a single occurrence of a stress event that exceeds the intrinsic strength of a material.

Product performance: The ability of a product to perform as required according to specifications.

Qualification: All activities that ensure that the nominal design and manufacturing specifications will meet or exceed the reliability goals.

Quality: A measure of a part's ability to meet the workmanship criteria of the manufacturer.

Reliability: The ability of a product to perform as intended (i.e., without failure and within specified performance limits) for a specified time, in its life cycle application environment.

Wearout failures: Failures due to accumulation of incremental damage, occurring when the accumulated damage exceeds the material endurance limit.

References

1. Pecht, M., *Integrated Circuit, Hybrid, and Multichip Module Package Design Guidelines—A Focus on Reliability*, John Wiley & Sons, New York, 1994.
2. O'Connor, P., *Practical Reliability Engineering*, John Wiley & Sons, New York, 1991.
3. Jackson, M., Mathur, A., Pecht, M., and Kendall, R., Part Manufacturer Assessment Process, *Qual. Reliab. Eng. Int.*, 15, 457, 1999.
4. Jackson, M., Sandborn, P., Pecht, M., Hemens-Davis, C., and Audette, P., A Risk-Informed Methodology for Parts Selection and Management, *Qual. and Reliab. Eng. Int.*, 15, 261, 1999.
5. Lewis, E.E., *Introduction to Reliability Engineering*, John Wiley & Sons, New York, 1996.
6. Sage, A.P. and Rouse, W.B., *Handbook of Systems Engineering and Management*, John Wiley & Sons, New York, 1999.
7. IEC Standard 60134, Rating systems for electronic tubes and valves and analogous semiconductor devices, (Last reviewed in July 1994 by the IEC Technical Committee 39 on Semiconductors), 1961.
8. Stogdill, R. C., Dealing with obsolete parts. *IEEE Des. Test Comput.*, 16(2), 17, 1999.
9. Dasgupta, A. and Pecht, M., Failure mechanisms and damage models, *IEEE Trans. Reliab.*, 40(5), 531, 1991.
10. Dasgupta, A. and Hu, J.M., Failure mechanism models for brittle fracture, *IEEE Trans. Reliab.*, 41(3), 328, 1992.
11. Dasgupta, A. and Hu, J.M., Failure mechanism models for ductile fracture, *IEEE Trans. Reliab.*, 41(4), 489, 1992.
12. Dasgupta, A. and Hu, J.M., Failure mechanism models for excessive elastic deformation, *IEEE Trans. Reliab.*, 41(1), 149, 1992.
13. Dasgupta, A. and Hu, J.M., Failure mechanism models for plastic deformation, *IEEE Trans. Reliab.*, 41(2), 168, 1992.
14. Dasgupta, A. and Haslach, H.W., Jr., Mechanical design failure models for buckling, *IEEE Trans. Reliab.*, 42(1), 9, 1993.
15. Engel, P.A., Failure models for mechanical wear modes and mechanisms, *IEEE Trans. Reliab.*, 42(2), 262, 1993.
16. Li, J. and Dasgupta, A., Failure mechanism models for material aging due to interdiffusion, *IEEE Trans. Reliab.*, 43(1), 2, 1994.

17. Li, J. and Dasgupta, A., Failure-mechanism models for creep and creep rupture, *IEEE Trans. Reliab.*, 42(3), 339, 1994.
18. Dasgupta, A., Failure mechanism models for cyclic fatigue, *IEEE Trans. Reliab.*, 42(4), 548, 1993.
19. Young, D. and Christou, A., Failure mechanism models for electromigration, *IEEE Trans. Reliab.*, 43(2), 186, 1994.
20. Rudra, B. and Jennings, D., Failure mechanism models for conductive-filament formation, *IEEE Trans. Reliab.*, 43(3), 354, 1994.
21. Al-Sheikhly, M. and Christou, A., How radiation affects polymeric materials, *IEEE Trans. Reliab.*, 43(4), 551, 1994.
22. Diaz, C., Kang, S.M., and Duvvury, C., Electrical overstress and electrostatic discharge, *IEEE Trans. Reliab.*, 44(1), 2, 1995.
23. Tullmin, M. and Roberge, P.R., Corrosion of metallic materials, *IEEE Trans. Reliab.*, 44(2), 271, 1995.
24. Upadhyayula, K. and Dasgupta, A., Guidelines for physics-of-failure based accelerated stress testing, Annu. Reliab. Maintainability Symp. 1998 Proc., Int. Symp. Prod. Qual. Integrity, 345, 1998.
25. Pecht, M., Dasgupta, A., Evans, J. W., and Evans, J. Y., *Quality Conformance and Qualification of Microelectronic Packages and Interconnects*, John Wiley & Sons, New York, 1994.
26. Kraus, A., Hannemann, R., Pecht, M., *Semiconductor Packaging: A Multidisciplinary Approach*, John Wiley & Sons, New York, 1994.

Further Information

Microelectronics Reliability: http://www.elsevier.com/locate/microrel
IEEE Transactions on Reliability: http://www.ewh.ieee.org/soc/rs/transactions.htm

23

Certification of Civil Avionics

Frank McCormick
Certification Services, Inc.

23.1   Introduction

Almost all aspects of the design, production, and operation of civil aircraft are subject to extensive regulation by governments. This chapter describes the most significant regulatory involvement a developer is likely to encounter in the certification of avionics.

Certification is a critical element in the safety-conscious culture on which civil aviation is based. The legal purpose of avionics certification is to document a regulatory judgment that a device meets all applicable regulatory requirements and can be manufactured properly. At another level, beneath the legal and administrative machinery of regulatory approval, certification can be regarded differently. It can be thought of as an attempt to predict the future. New equipment proposed for certification has no service history. Certification tries, in effect, to provide credible predictions of future service experience for new devices — their influences on flight crews, their safety consequences, their failure rates, and their maintenance needs. Certification is not a perfect predictor, but historically it has been quite a good one.

In this chapter, for the most part, certification activities appropriate to the U.S. Federal Aviation Administration (FAA) are discussed. However, be aware that the practices of civil air authorities elsewhere, while generally similar to those of the FAA, often differ in detail or scope. Toward the end of this chapter, some differences between the FAA and the European Joint Aviation Authorities, or JAA, headquartered in Hoofddorp, the Netherlands, will be illustrated.

Expensive misunderstandings can result from differences among regulators. Moreover, the rules and expectations of every authority, the FAA included, change over time. For current guidance, authoritative sources should be consulted.

This chapter discusses the following topics:

- The FAA regulatory basis
- The Technical Standard Order (TSO) system for equipment approval
- The Supplemental Type Certificate (STC) system for aircraft modification
- Use of FAA designees in lieu of FAA personnel
- System requirements definition
- Safety assessments
- Environmental qualification
- Software assurance
- Production approvals
- The Joint Aviation Authorities

Conceptually, the certification of avionics is straightforward, indeed almost trivial: the applicant simply defines the product, establishes its regulatory requirements, and demonstrates that those requirements have been met. The reality is, of course, more problematic.

It is a truism that for any proposed avionics system a suitable market must exist. As with any commercial pursuit, adequate numbers of avionics units must be sold at margins sufficient to recover investments made in the product. Development costs must be controlled if the project is to survive. Warranty and support costs must be predicted and managed. The choices made in each of these areas will affect and be affected by certification.

This chapter is an introduction to certification of avionics. It is not a complete treatment of the subject. Some important topics are discussed only briefly. Many situations that come up in real-life certification projects are not addressed.

Good engineering should not be confused with good certification. A new avionics device can be brilliantly conceived and flawlessly designed, yet ineligible for certification. Good engineering is a prerequisite to good certification, but the two are not synonymous.

Certification has a strong legalistic element and is more craft than science. Almost every project raises some odd regulatory-approval quirk during its development. Certification surprises are rarely pleasant, but surprises can be minimized or eliminated by maintaining open and honest communication with the cognizant regulators.

23.2   Regulatory Basis of the Federal Aviation Administration

The FAA, created in 1958, acts primarily through publication and enforcement of the Federal Aviation Regulations, or FARs. FARs are organized by sections known as Parts. The FAR Parts covering most avionics-related activity are listed below:

- Part 1 — Definitions and Abbreviations
- Part 21 — Certification Procedures for Products and Parts
- Part 23 — Airworthiness Standards: Normal, Utility, Acrobatic, and Commuter Category Airplanes
- Part 25 — Airworthiness Standards: Transport Category Airplanes
- Part 27 — Airworthiness Standards: Normal Category Rotorcraft
- Part 29 — Airworthiness Standards: Transport Category Rotorcraft
- Part 33 — Airworthiness Standards: Aircraft Engines

- Part 34 — Fuel Venting and Exhaust Emission Requirements for Turbine Engine-Powered Airplanes
- Part 39 — Airworthiness Directives
- Part 91 — General Operating and Flight Rules
- Part 121 — Operating Requirements: Domestic, Flag, and Supplemental Operations
- Part 183 — Representatives of the Administrator

Only a subset of these regulations will apply to any given project. Much of the job of managing a certification program well lies in identifying the complete but minimum set of regulations applicable to a project.

23.3   FAA Approvals of Avionics Equipment

The FARs provide several different forms of approval for electronic devices installed aboard civil aircraft. Of these, most readers will be concerned primarily with approvals under the Technical Standard Order (TSO) system, approvals under a Supplemental Type Certificate (STC), or approvals as part of a Type Certificate, Amended Type Certificate, or Service Bulletin.*

23.3.1   Technical Standard Order

An approval under the Technical Standard Order (TSO) system is common. TSOs are regulatory instruments that recognize the broad use of certain classes of products, parts, and devices. TSOs apply to more than avionics; they can apply to any product with the potential for wide use, from seat belts and fire extinguishers to tires and oxygen masks. Indeed, that is the guiding principle behind TSOs — they must be widely useful. Considerable FAA effort goes into the sponsorship and adoption of a TSO. The agency would have little interest in publishing a TSO for a device with limited application.

TSOs contain product specifications, required data submittals, marking requirements, and various instructions and limitations. Many TSOs are associated with avionics: flight-deck instruments, communications radios, ILS receivers, navigation equipment, collision avoidance systems, and flight data recorders, to name just a few.

TSO-C113, "Airborne Multipurpose Electronic Displays," is representative of avionics TSOs. Electronic display systems are used for various purposes: display of attitude, airspeed, or altitude, en route navigation display, guidance during precision approach, display of engine data or aircraft status, maintenance alerts, passenger entertainment, and so on. The same physical display device could potentially be used for any or all of these functions, and on many different aircraft types. Recognizing this broad applicability, the FAA published TSO-C113 so that developers could more easily adapt a generic display device to a variety of applications. TSO-C113 is typical, calling out requirements for the following data:

- Explanation of applicability
- Exceptions and updated wording
- References to related regulations, data, and publications
- Requirements for environmental testing
- Requirements for software design assurance
- Requirements for the marking of parts
- Operating instructions
- Equipment limitations
- Installation procedures and limitations
- Schematics and wiring diagrams
- Equipment specifications

---

*Newly developed equipment has sometimes been installed as part of a field approval under an FAA Form 337, though this has become rarer and is disallowed in most cases.

- Parts lists
- Drawing list
- Equipment calibration procedures
- Corrective maintenance procedures

When an avionics manufacturer applies for a TSO approval, and the manufacturer's facilities and data comply with the terms of the TSO, the manufacturer receives a TSO Authorization from the FAA. A TSO Authorization represents approval of both design data and manufacturing rights. That is, the proposed device is deemed to be acceptable in its design, and the applicant has demonstrated the ability to produce identical units.

In TSO-based projects, the amount of data actually submitted to the FAA varies by system type, by the FAA's experience with particular applicants, and by FAA region. In one case, an applicant might be required to submit a great deal of certification data; in another, a one-page letter from an applicant might be adequate for issuance of a TSO Authorization. On any new project, it is unwise to presume that all regulatory requirements are known. Consistency is a high priority for the FAA, but regional differences among agency offices do exist. Early discussion with the appropriate regulators will ensure that the expectations of agency and applicant are mutually understood and agreed on.

For more information on TSOs, see FAA Advisory Circular 20-110J, "Index of Aviation Technical Standard Orders;" FAA Order 8110.31, "TSO Minimum Performance Standard;" and FAA Order 8150.1, "Technical Standard Order Procedures."

Note that a TSO does not grant approval for installation in an aircraft. Although data approved under a TSO can be used to support an installation approval, the TSO Authorization itself applies only to the equipment in question. Installation approvals must be pursued through other means (see next section) and are not necessarily handled by an avionics equipment manufacturer.

23.3.2 Supplemental Type Certificate

A Supplemental Type Certificate (STC) is usually granted to someone other than the aircraft manufacturer, who wishes to modify the design of an existing aircraft. Retrofits and upgrades of avionics equipment are common motivations for seeking STC approvals from the FAA.

In an STC, the applicant is responsible for all aspects of an aircraft modification. Those aspects typically include the following:

- Formal application for a Supplemental Type Certificate (STC)
- Negotiation of the certification basis of the relevant aircraft with the FAA
- Identification of any items requiring unusual regulatory treatment
- Preparation of a certification plan
- Performance of all analyses specified in the certification plan
- Coordination with the FAA throughout the project
- Physical modification of aircraft configuration
- Performance of all conformity and compliance inspections
- Performance of all required lab, ground, and flight testing
- Preparation of flight manual supplements
- Preparation of instructions needed for continued airworthiness
- Preparation of a certification summary
- Support of all production approvals

An applicant for an STC must be "a U.S. entity," although the exact meaning of this phrase is not always clear. One common case is that of a nominally foreign firm with an office in the U.S. It is acceptable to the FAA for that U.S.-based office to apply for and hold an STC.

An applicant for an STC begins the process officially by completing and submitting FAA Form 8110-12, "Application for Type Certificate, Production Certificate, or Supplemental Type Certificate," to the cognizant FAA Aircraft Certification Office. Accompanying the application should be a description of the project and the aircraft type(s) involved, the project schedule, a list of locations where design and installation will be performed, a list of proposed Designees (discussed later in this chapter), and, if desired, a request for an initial meeting with the FAA. The FAA will assign a project number, appoint a manager for the project, schedule a meeting if one was requested, and send to the applicant an acknowledgment letter with these details.

The applicant must determine the certification basis of the aircraft to be modified. The certification basis is the sum of all applicable FAA regulations (at specified amendment levels) and any binding guidance that apply to the aircraft and project in question. Regulations tend to become more stringent over time, and complying with later rules may be more time-consuming and expensive than with earlier rules.

A certification basis is established by reference to the Type Certificate Data Sheet (TCDS) for each affected aircraft and through negotiation with the FAA. For example, an applicant might propose that a certification basis be those rules in effect at the time of original aircraft certification, whereas the FAA may require the applicant to comply with regulations in effect at the time of STC application. The differences between these two positions can be numerous and significant. Except in the simplest cases, they are a crucial topic for early discussions with the FAA.

Complex avionics systems, extensive aircraft modifications, and novel system architectures all raise the odds that something in a project will be unusual and will not fit neatly into the normal regulatory framework. For such activities, an applicant might wish to propose compliance based on other regulatory mechanisms: alternative means of compliance, findings of equivalent safety, exemptions, or special conditions. If so, generic advice is largely useless. By their nature, these activities are unusual and require close coordination with the FAA.

An STC applicant must prepare a certification plan. The plan should include the following:

- A brief description of the modification and how compliance is to be substantiated
- A summary of the Functional Hazard Assessment (see "Safety Assessment" later in this chapter)
- A list of proposed compliance documentation, including document numbers, titles, authors, and approving or recommending Designees, if applicable (the role of Designees is described in more detail later in this chapter)
- A compliance checklist, listing the applicable regulations from the certification basis, their amendment number, subject, means of compliance, substantiating documents, and relevant Designees
- A definition of Minimum Dispatch Configuration
- If used, a list of the proposed FAA Designees, including name, Designee number, appointing FAA office, classification, authorized areas, and authorized functions
- A project schedule, including dates for data submittals, test plan submittals, tests (with their locations), conformity inspections, installation completion, ground and flight testing, and project completion

Some FAA Aircraft Certification Offices require all Designated Engineering Representatives (see next section) participating in a project to sign an FAA Form 8110-3, "Statement of Compliance with the Federal Aviation Regulations," recommending approval of a certification plan.

Extensive analysis and testing are generally required to demonstrate compliance. Results of these analyses and tests must be preserved. Later in this chapter, three of the most important of these activities — safety assessments, environmental qualification, and software assurance — will be discussed along with another engineering topic, development and handling of system requirements.

The FAA's involvement in an STC is a process, not an act. Most FAA specialists support multiple projects concurrently, and matching the schedules of applicant and agency requires planning. This planning is the applicant's responsibility. Missed deadlines and last-minute surprises on the part of an applicant can result in substantial delays to a project, as key FAA personnel are forced to reschedule their time, possibly weeks or months later than originally planned.

The STC process assumes modification of at least one prototype aircraft. It is in the aircraft modification that all the engineering analysis — aircraft performance, structural and electrical loading, weight and balance, human factors, and so on — comes together. Each component used in an aircraft modification must either be manufactured under an approved production system or examined formally for conformance to its specifications. This formal examination is known as "parts conformity inspection." A completed aircraft modification is then subject to an "installation conformity inspection." In complex installations or even complex parts, progressive conformity inspections may be required. Conformity inspections are conducted by an FAA Inspector or a Designee authorized by the FAA — a Designated Manufacturing Inspection Representative (DMIR) or Designated Airworthiness Representative (DAR) (see next section).

Compliance inspections, as distinct from conformity inspections, verify through physical inspection that a modification complies with the applicable FARs. Typical of compliance inspections is an examination of modified wiring on an aircraft. A compliance inspection is conducted by an FAA engineer or authorized Designated Engineering Representative (again, see next section).

For significant projects involving ground and flight testing, the FAA will issue a Type Inspection Authorization (TIA). The TIA details all the inspections, ground tests, and flight tests necessary to complete the certification program. Prior to issuing a TIA, the FAA should have received and reviewed all of the descriptive and compliance data for the project. The FAA has recently added an item to its TIA procedures: the flight test risk assessment. The risk assessment seeks to identify and mitigate any perceived risks in flight tests that include FAA personnel, based on data supplied by the applicant.

New avionics equipment installed as part of an STC will usually impose new and different procedures on flight crews. An applicant will, in most cases, document new procedures in a supplement to an approved flight manual. In complex cases, it may also be necessary to provide a supplement to an operations manual.

An applicant must provide instructions for the continued airworthiness of a modified airplane. Penetrations of the pressure vessel by, say, wiring or tubing may require periodic inspection. Actuators associated with a new subsystem may need scheduled maintenance. Instructions for continued airworthiness are usually a supplement to a maintenance manual but may also include supplements to an illustrated parts catalog, a structural repair manual, structural inspection procedures, or component maintenance manuals.

Much of this discussion has been more applicable to transport aircraft than to smaller aircraft. Regulatory requirements for the smaller (FAR Part 23) aircraft are, in some respects, less stringent than for transport aircraft. Yet even for transports, not everything described above is required in every circumstance. Early discussion between applicant and regulator is the quickest way to determine what actually needs to be done.

Some avionics developers may find it desirable to pursue an STC through an organization called a Designated Alteration Station (DAS). A DAS can, if properly authorized by the FAA, perform all the work associated with a given aircraft modification and issue an STC. In this approach, the developer might not deal with FAA personnel at all. Key issues are ownership of the STC rights and handling of production approvals.

For more information on STCs, see FAA Advisory Circular 21-40, "Application Guide for Obtaining a Supplemental Type Certificate." For more information on DASs, see FAA Advisory Circular 21.431-1A, "Designated Alteration Station Authorization Procedures."

23.3.3　Type Certificate, Amended Type Certificate, and Service Bulletin

Approvals as part of a Type Certificate, Amended Type Certificate, or Service Bulletin are tied to the certification activities of airframers or engine manufacturers. For development programs involving these kinds of approvals, an avionics supplier's obligations are roughly similar to those imposed by an STC project, though detailed requirements can vary greatly. Avionics suppliers participating in an aircraft- or engine-development program can and should expect to receive certification guidance from the manufacturer of the aircraft or engine. Hence, these cases will not be considered further here.

23.4 FAA Designees

In the U.S., any applicant may deal directly with the FAA. Unlike many other civil air authorities, the FAA does not collect fees for its services from applicants. However (and also unlike other agencies), the FAA can at its discretion appoint individuals who meet certain qualifications to act on its behalf. These appointees, called Designees, receive authorizations under FAR Part 183 and act in a variety of roles. Some are physicians authorized to issue medical certificates to pilots. Others are examiners authorized to issue licenses to new pilots. Still others are inspectors authorized to approve maintenance work.

Avionics developers are most likely to encounter FAA Designated Engineering Representatives (DERs) and either Designated Manufacturing Inspection Representatives (DMIRs) or Designated Airworthiness Representatives (DARs).

All Designees must possess authorizations from the FAA appropriate to their activities. DERs can approve engineering data just as the FAA would. Flight Test Pilot DERs can conduct and approve the results of flight tests in new or modified aircraft. DMIRs and DARs can perform conformity inspections of products and installations, and DARs can issue Airworthiness Certificates. When acting in an authorized capacity, a Designee is legally a representative of the FAA; in most respects, he or she is the FAA for an applicant's purposes. Nevertheless, there are practical differences in conduct between the FAA and its Designees.

The most obvious difference is that an applicant actually hires and pays a Designee, and thus has more flexibility in managing his or her time on the project. The resulting benefits in project scheduling can more than offset the costs of the Designee. In addition, experienced Designees can be sources of valuable guidance and recommendations. The FAA, by contrast, restricts itself to findings of compliance. That is, the agency will simply tell an applicant whether or not submitted data complies with the regulations. If data are judged noncompliant, the FAA will not, in most cases, tell an applicant how to bring it into compliance. A Designee, however, can assist an applicant with recovery strategies or, better yet, steer an applicant toward compliant approaches in the first place.

The FAA often encourages the use of Designees by applicants. An applicant must define and propose the use of Designees, by name, to the FAA Aircraft Certification Office (ACO) for each project. If the proposed Designees are acceptable to the ACO, the ACO will coordinate with its manufacturing counterpart and delegate certain functions to the specified Designees. Those Designees are then obliged to act as surrogates for the relevant FAA personnel on that project, providing oversight and ultimately approving or recommending approval of compliant data.

Although an applicant's use of Designees is discretionary, the realities of the FAA workload and scheduling may make the use of Designees a pragmatic necessity. Whenever Designees are considered for inclusion in a project, their costs and benefits should be evaluated with the same care devoted to any other engineering resource. For more information, see FAA Order 8100.8, "Designee Management Handbook;" FAA Order 8110.37C, "Designated Engineering Representatives (DER) Guidance Handbook;" and FAA Order 8130.28A, "Airworthiness Designee Management Program."

This chapter has so far dealt mainly with the definitions and practices of FAA regulation. There is, of course, a great deal of engineering work to be done in any avionics development. Four engineering topics of great interest to the FAA are the handling of system requirements, performance of a safety assessment, environmental qualification, and software assurance.

23.5 System Requirements

Avionics developers must document the requirements of their proposed systems, ideally in ways that are easily controlled and manipulated. Many experienced practitioners regard the skillful capture of requirements as the single most important technical activity on any project. A system specification is the basis for descriptions of normal and abnormal operation, functional testing, training and maintenance procedures, and much else. A brief treatment of the topic here does not imply that it can be approached superficially. On the contrary, system specification is so important that a large body of literature exists for it elsewhere (see Chapter 21 for a starting point). Requirements definition is supported by many acceptable methods. Each company evolves its own practices in this area.

Over the years, many types of avionics systems have come to be described by *de facto* standardized requirements, easing the burden of both engineering and certification. New systems, though, are free to differ from tradition in arbitrary ways. Applicants should expect such differences to be scrutinized closely by regulators and customers, who may demand additional justification and substantiation for the changes.

Proper requirements are the foundation for well-designed avionics. Whatever the sources of requirements, and whatever the methods used for their capture and refinement, an applicant must be able to demonstrate that a new system's requirements — performance, safety, maintenance, continued airworthiness, and so on — have been addressed comprehensively. Some projects simply tabulate requirements manually, along with the means of compliance for each requirement. Others implement large, sophisticated databases to control requirements and compliance information. Compliance is generally shown through analysis, test, inspection, demonstration, or some combination thereof.

23.6   Safety Assessment

Early in a project — the earlier the better — developers should consider the aircraft-level hazards associated with their proposed equipment. This is the first of possibly several steps in a safety assessment of a new system.

There is an explicit correlation between the severity of a system's hazards and the scrutiny to which that system is subjected. With a few notable exceptions,* systems that are inconsequential from a safety standpoint receive little attention. Systems whose improper operation can result in aircraft damage or loss of life receive a great deal of attention and require correspondingly greater engineering care and substantiation.

Unsurprisingly, there is an inverse relationship between the severity of a system's hazards and the frequency with which those hazards can be tolerated. Minor annoyances might be tolerable every thousand or so flight hours. Catastrophic hazards, by contrast, must occur less frequently than once in every billion flight hours. In addition, the regulations for transport aircraft require that no single failure, regardless of probability, result in a catastrophic hazard, implying that any such hazard must arise from two or more independent failures occurring together.

Initial considerations of hazards should be formalized in a Functional Hazard Assessment (FHA) for the proposed system. An FHA should address hazards only at levels associated directly with operation of the system in question. For example, an autopilot FHA would consider the hazards of an uncommanded hardover or oscillation of a control surface. A display-system FHA would consider the hazards of blank, frozen, and active-but-misleading displays during various phases of flight.

In general, if an FHA concludes that misbehavior of a system has little or no effect on continued safe flight and landing, no further work is needed for the safety assessment. On the other hand, if the FHA confirms that a system can pose nontrivial risk to the aircraft or its occupants, then investigation and analysis must continue. The additional work, if needed, will likely involve preparation of a Preliminary System Safety Assessment, Fault Tree Analysis, Failure Modes and Effects Analysis, Common Cause Analysis, and a final System Safety Assessment.

In the absence of a specific aircraft installation, assumptions must be made regarding avionics usage to make progress on a safety assessment. This is true in TSO approvals, for example, if design assurance levels are not specified in the TSO or if developers contemplate hazards or usage different from those assumed

---

*For example, failures of flight data recorders, cockpit voice recorders, and emergency locator transmitters have no effect on continued safe flight and landing. Conventional safety-assessment reasoning would dismiss these devices from failure-effect considerations. However, the systems obviously perform important functions, and the FAA defines them as worthy of more attention than suggested by a safety assessment. For more discussion of this topic, refer to Software Assurance in this chapter for a description of software levels assigned to flight data recorders.

in the TSO. There are pitfalls* in unthinking acceptance and use of generic hazard classifications and software levels (see Software Assurance later in this chapter and in Chapter 27), even for standard products. Technologies can change quickly; regulations cannot. The gap between what is technically possible and what can be approved sometimes leads to conflicting requirements, bewildering difficulties, and delays in bringing to market devices that offer improvements to safety, operating economics, or both. The solution is early agreement with the appropriate regulators concerning the requirements applicable to a new device.

The details of safety assessments are outside the scope of this chapter. For an introduction to safety-related analysis, refer to the following:

- Chapters 21 and 22 of this book
- ARP4754 — Systems Integration Requirements Guidelines; Society of Automotive Engineers Inc., 1994
- ARP4761** — Guidelines and Tools for Conducting the Safety Assessment Process on Civil Airborne Systems and Equipment; Society of Automotive Engineers Inc., 1994
- NUREG-0492 — Fault Tree Handbook; U.S. Nuclear Regulatory Commission, 1981
- FAA Advisory Circular 25.1309-1A — System Design Analysis, 1988
- FAA Advisory Circular 23.1309-1C*** — Equipment, Systems, and Installations in Part 23 Airplanes, 1999
- Safeware: System Safety and Computers — Nancy G. Leveson, Addison-Wesley Publishing Company, 1995
- Systematic Safety: Safety Assessment of Aircraft Systems — Civil Aviation Authority (UK), 1982

Customers routinely demand that some failures, even those associated with minor hazards, be less frequent than required by regulation — that is, the customer's requirement is more stringent than the FAA's. Economic issues such as dispatch reliability and maintenance costs are the usual motivation, and meeting the customer's specification automatically satisfies the regulatory requirement.

Some TSOs refer to third-party guidance material, usually in the form of equipment-performance specifications from organizations such as RTCA**** and the Society of Automotive Engineers. TSOs, Advisory Circulars, and these third-party specifications can explicitly call out hazard levels and software assurance levels. If such prescriptions apply to a given project, developers may simply adopt the prescriptions given for use in their own safety assessments. Developers, of course, must still substantiate their claims to the prescribed levels.

In addition to a safety assessment, an analysis of equipment reliability may be required to predict average times between failures of the equipment. Although this analysis is often performed by safety analysts, the focus is different. Whereas a safety assessment is concerned with the operational consequences and

---

*A given TSO might specify a software level (see Software Assurance section in this chapter), and a TSOA could certainly be granted on that basis. However, actual installation of such a device on an aircraft might require a higher software level. For example, an airspeed sensor containing Level C software could be approved under TSO-C2d, but the sensor could not then be used to supply a transport aircraft with primary air data, because that function requires Level A software.

**ARP 4754 and ARP 4761 are expected to be recognized by a new FAR/JAR advisory circular, AC/ACJ 25.1309-1B. At this writing, the advisory circular has not been adopted.

***An applicant developing avionics exclusively for general aviation airplanes should pay special attention to Advisory Circular 23.1309-1C. The Advisory Circular offers regulatory relief from many requirements that would otherwise apply. In particular, for some functions on several classes of small airplanes it allows software assurance at lower levels than would be the case for transport aircraft.

****RTCA Inc., formerly known as Radio Technical Corporation of America, is a nonprofit association of U.S.-based aeronautical organisations from both government and industry. RTCA seeks sound technical solutions to problems involving the application of electronics and telecommunications to aeronautical operations. RTCA tries to resolve such problems by mutual agreement of its members (cf. EUROCAE).

probabilities of system failures, a reliability analysis is concerned with the frequency of failures of particular components in a system.

23.7   Environmental Qualification

Environmental qualification is invariably required of avionics. The standard in this area is RTCA/DO-160D, "Environmental Conditions and Test Procedures for Airborne Equipment" (RTCA, 1997). DO-160D specifies testing for temperature range, humidity, crashworthiness, vibration, susceptibility to radiated and conducted radio frequencies, lightning tolerance, and other environmental factors.

It is the responsibility of applicants to identify environmental tests appropriate to their systems. Whenever choices for environmental testing are unclear, guidance from FAA personnel or DERs is in order.

To receive certification credit, environmental testing must be performed on test units whose configurations are controlled and acceptable for the tests in question. Conformity inspection may be necessary for test articles not manufactured in accordance with a production approval. An approved test plan, test setup conformity inspection, and formal witnessing of tests by FAA specialists or Designees are often required. In all cases, an applicant must document and retain evidence of equipment configurations, test setups, test procedures, and test results.

For further information on environmental testing, see Chapter 25.

23.8   Software Assurance

Software has become increasingly important in avionics development and has assumed a correspondingly higher profile in certification. It is frequently the dominant consideration in certification planning.

Regulatory compliance for software can be shown by conforming to the guidelines described in RTCA/DO-178B, "Software Considerations in Airborne Systems and Equipment Certification" (RTCA, 1992). DO-178B was developed jointly by RTCA and the European Organisation for Civil Aviation Equipment (EUROCAE).*

DO-178B is not a development standard for software. It is an assurance standard. DO-178B is neutral with respect to development methods. Developers are free to choose their own methods, provided the results satisfy the assurance criteria of DO-178B in the areas of planning, requirements definition, design and coding, integration, verification, configuration management, and quality assurance.

DO-178B defines five software levels, A through E, corresponding to hazard classifications derived from the safety assessment discussed earlier. At one extreme, Level A software is associated with functions whose anomalous behavior could cause or contribute to a catastrophic failure condition for the aircraft. Obvious examples of Level A software include fly-by-wire primary control systems and full-authority digital engine controllers. At the other extreme, passenger entertainment software is almost all Level E, because its failure has no safety-related effects.

A sliding scale of effort exists within DO-178B: the more critical the software, the more scrutiny that must be applied to it. Level A software generates more certification data than does Level B software, Level B generates more than does Level C, and so on.

Avionics customers sometimes insist on software assurance levels higher than those indicated by a safety assessment. This is purely a contractual matter. Confusion can be avoided by separating a customer's contractual wishes from regulatory compliance data submitted to the FAA or to DERs. Certification submittals should be based on the safety assessment rather than on the contract. If a safety assessment concludes that a given collection of software should be Level C, but that software's customer wants it to be Level B, then the applicant should submit to the FAA plans and substantiating data for Level C software.

---

*RTCA/DO-178B is equivalent to EUROCAE/ED-12B, "Considerations sur le Logiciel en Vue de la Certification des Systemes et Equipments de Bord." (EUROCAE, 1992.)

Any additional evidence needed to demonstrate contractual compliance to Level B should be an issue between supplier and customer. That evidence is not required for certification and should become a regulatory matter only in unusual circumstances.*

FAA guidance itself sometimes requires that software be assured to a level higher than indicated by the safety assessment. This is not uncommon in equipment required for dispatch but whose failures do not threaten continued safe flight and landing. For example, a flight data recorder must be installed and operating in most scheduled-flight aircraft, but failure of a recorder during a flight would have no effect on the ability of a crew to carry on normally. Thus, from a safety assessment viewpoint, a flight data recorder has no safety-related failure conditions. Based on that, the recorder's software would be classified as Level E, implying that the software need not receive any FAA scrutiny. This, of course, violates common sense — the FAA plainly has a regulatory interest in the proper operation of flight data recorders. To resolve this mismatch, the FAA requires at least Level D compliance for any software associated with a dispatch-required function.

Digital technology predates DO-178B. Many software-based products were developed and approved before DO-178B became available. If an applicant is making minor modifications to equipment approved under an older standard, it may be possible to preserve that older standard as the governing criteria for the update. More frequently, the FAA will require new or changed software to meet the guidelines of DO-178B, with unchanged software "grandfathered" in the new approval. When transport airplanes are involved in such cases, an Issue Paper dealing with use of "legacy" software is likely to be included in the certification basis of the airplane by the FAA's Transport Airplane Directorate. In a few cases, the FAA may require a wholesale rework of a product to meet current standards.

The question of how much software data to submit to the FAA arises routinely. It is impractical to consider submitting all software data to the FAA. An applicant can realistically submit only a fraction of the data produced during software development. Applicants should propose and negotiate that data subset with the FAA. Whether submitted formally or not, an applicant should retain and preserve all relevant data (see DO-178B, Section 9.4, as a starting point). The FAA can examine applicants' facilities and data at any time. It is the applicant's responsibility to ensure that all relevant data are controlled, archived, and retrievable.

For more information on software-assurance guidelines, see Chapter 27 of this book, the FAA software home page on the World Wide Web at ⟨ www.faa.gov/avr/air/air100/sware/sware.htm⟩, and the supplemental information to DO-178B published by RTCA.

In recent years, the FAA has paid growing attention to programmable logic devices: application-specific integrated circuits, field-programmable gate arrays, and so on. Findings of compliance for these devices are often handled by FAA software specialists or delegated to DERs with software authorizations. The agency's increased scrutiny is intended to ensure that acceptable processes are being followed during development of such devices. The FAA has a generic issue paper addressing compliance for the devices. If proposed electronic equipment contains programmable logic devices, an applicant should expect the FAA to tailor its generic issue paper to the project and to include the tailored issue paper in the certification basis of that project.

Though little guidance is available officially at this writing, applicants should also note that FAA concern has increased with respect to assurance of all avionics hardware design processes. A great deal of effort in industry and government has been spent to specify acceptable practices in this area, primarily through the joint efforts of RTCA Special Committee 180 and EUROCAE Working Group 46 ("Design Assurance Guidance for Airborne Electronic Hardware"). See RTCA document DO-254 (2000) for further information.

---

*It is usually prudent to avoid setting precedents of additional work beyond that required by regulations. Of course, applicants are always free to do additional work — developers often do, for their own reasons — and if the regulations seem inappropriate or inadequate, applicants should seek to improve the regulations. Precedents are powerful things, for both good and ill, in any regulatory regime. New precedents often have unintended and surprising consequences.

23.9   Manufacturing Approvals

It is not enough to obtain design approval for avionics equipment. Approval to manufacture and mark production units must be obtained as well. Parts manufactured in accordance with an approved production system do not require parts conformity inspections.

With a TSO, as explained earlier, the approvals of design and manufacturing actually go together. In order to receive a TSO Authorization, the applicant must demonstrate not just an acceptable prototype but also an ability to manufacture the article.

An STC holder must demonstrate production capabilities separately. After obtaining an STC approval the holder may apply for Parts Manufacturer Approval (PMA) authority to produce the parts necessary to support the STC. PMA approvals are issued by the FAA Manufacturing Inspection District Office responsible for the applicant. An STC applicant who will need subsequent PMA authority should plan and prepare for PMA from the beginning of a project. Alternatively, an STC holder may assign production rights to others, who would then hold PMA authority for the parts in question.

23.10   The Joint Aviation Authorities

The European Joint Aviation Authorities (JAA) is an influential aviation body internationally. The JAA represents European states (32 at this writing) that have agreed to cooperate in developing and implementing common safety standards and procedures for civil aviation. These standards and procedures are codified in the Joint Aviation Requirements (JARs).

Although the JAA develops and adopts JARs in the areas of aircraft operations, aircraft maintenance, and the licensing of aviation personnel, this chapter is mainly concerned with the JARs affecting aircraft design and certification. These include rules for the certification of airplanes (JAR-23, JAR-25), sailplanes and powered sailplanes (JAR-22), helicopters (JAR-27, JAR-29), engines (JAR-E), auxiliary power units (JAR-APU), and equipment (JAR-TSO).

There is a great deal of similarity between the JARs and the FARs, as well as between the JAA's and FAA's advisory material. Indeed, the JAA and the FAA made commitments in 1992 to harmonize "where appropriate, to the maximum extent possible" the JARs and FARs. The harmonization effort for airworthiness rules is expected to be completed in the year 2000. After that, the JAA and the FAA intend to engage in joint rulemaking to encourage the uniformity of new regulatory material. On at least four new aircraft programs, the JAA and the FAA have agreed to work together in a process dubbed "Cooperative and Concurrent Certification."

Still, there are differences. The following list illustrates a few of the differences:

- The JAA is not a regulatory body. Whereas the FAA defines and enforces its rules under its own authority, JAA actions are carried out through its member states and their national authorities. For example, on a development program for a new aircraft or engine, the JAA member states will assign specialists to a Joint Certification Team that acts on behalf of all the JAA members. At the successful completion of the team's evaluations, Type Certificates for the new aircraft or engine are issued not by the JAA itself, but by the member state. Thus, each Type Certificate remains a national artifact, subject to regulation by the national authority of the issuing state.

- Although JAA member countries have various forms of delegation to organizations, the JAA has no individual delegation mechanism equivalent to the FAA Designee system. Certification work is performed by JAA specialists directly or in concert with their counterparts at other non-JAA civil air authorities.

- Fees are charged to each applicant for JAA certification work.

- On some certification programs, the JAA and FAA have disagreed over the intensity of disruptive electromagnetic fields to which aircraft should be subjected during tests.

- The JAA requirements in some areas, such as operation with two engines failed on a three-engine airplane, and operation at negative load factors, differ from those of the FAA.

These examples are chosen largely at random. They serve only to illustrate that JAA/FAA harmonization is not complete. Any U.S. applicant whose certification project has a European or, for that matter any other international component, should investigate the implications thoroughly.

23.11 Summary

Certification can be straightforward, but like any other developmental activity, it must be managed. At the beginning of a project, applicants should work with their regulators to define expectations on both sides. During development, open communication should be maintained among suppliers, customers, and regulators. In a well-run project, evidence of compliance with regulatory requirements will be produced with little incremental effort, almost as a side-effect of good engineering during the normal course of work. The cumulative result will, in the end, be a complete demonstration of compliance, soon followed by certification.

Regulatory officials, whether FAA employees or Designees, work best and are most effective when they are regarded as part of an applicant's development team.

An applicant is obliged to demonstrate compliance with the applicable regulations, nothing more. However, partial information from an applicant can lead to misunderstandings and delays, and attempts to resolve technical disagreements with regulators through nontechnical means rarely have the desired effect.

In the past, regrettably, large investments have been made in systems that could not be approved by the FAA. In order to avoid such outcomes, applicants are well advised to hold early discussions with appropriate FAA personnel or Designees.

Defining Terms

Certification: Legal recognition, through issuance of a certificate by a civil aviation authority, that a product, service, organization, or person complies with that authority's requirements.

Certification basis: The sum of all current regulations applicable to a given project at the time application is made to a civil aviation authority to begin a certification process.

Designee: An individual authorized by the FAA under FAR Part 183 to act on behalf of the agency in one or more specified areas.

Issue Paper: Instrument administered by an FAA Directorate to define and control a substantial understanding between an applicant and the FAA, such as formal definition of a certification basis or a finding of equivalent safety, or to provide guidance on a specific topic, such as approval methods for programmable logic devices.

PMA: Parts Manufacturer Approval, by which the FAA authorizes the production of parts for replacement and modification, based on approved designs.

Special Condition: A modification to a certification basis, necessary if an applicant's proposed design features or circumstances are not addressed adequately by existing FAA rules; in effect, a new regulation, administered by an FAA Directorate, following public notice and a public comment period of the proposed new rule.

STC: Supplemental Type Certificate, by which the FAA approves the design of parts and procedures developed to perform major modifications to the design of existing aircraft.

TSOA: Technical Standard Order Authorization, the mechanism by which the FAA approves design data and manufacturing authority for products defined by a Technical Standard Order (see also ⟨http://www.faa.gov/avr/air/AIR100/tsohome.htm⟩).

Further Information

Certification Services, Inc.: www.certification.com
European Organisation for Civil Aviation Equipment (EUROCAE): www.eurocae.org
Federal Aviation Administration (FAA): www.faa.gov
Joint Aviation Authorities (JAA): www.jaa.nl
RTCA: www.rtca.org
Society of Automotive Engineers (SAE): www.sae.org

24

Processes for Engineering a System

James N. Martin
The Aerospace Corporation

24.1 Introduction

In April 1995, the G47 Systems Engineering Committee of the Electronic Industries Alliance (EIA) chartered a working group to convert the interim standard EIA/IS 632 into a full standard. This full standard was developed and released in December 1998 as ANSI/EIA-632-1998.

The interim standard (IS), EIA/IS 632, was titled "Systems Engineering." The full standard was expanded in scope to include *all the technical processes for engineering a system*. It is intended to be a higher-level abstraction of the activities and tasks found in the IS version plus those other technical activities and tasks deemed to be essential to the engineering of a system.

This chapter describes the elements of these processes and related key concepts. The intended purpose is to give the reader of the standard some background in its development and to help other standards activities in developing their own standard. There is a paper that describes the evolution from an interim standard to the full standard [Martin, 1998].

This standard is intended to be a "top tier" standard for the *processes essential to engineering a system*. It is expected that there will be second- and third-tier standards that define specific practices related to certain disciplines (e.g., systems engineering, electrical engineering, software engineering) and industry domains (e.g., aircraft, automotive, pharmaceutical, building, and highway construction).

It is important to understand several things that are *not* covered by this standard:

1. It does *not* define what "systems engineering" is;
2. It does *not* define what a "systems engineer" is supposed to do; and
3. It does *not* define what a "systems engineering organization" is supposed to do.

24.2 Structure of the Standard

The standard is organized as shown below:

Clause 1	*Scope*
Clause 2	*Normative references*
Clause 3	*Definitions and acronyms*
Clause 4	*Requirements*
Clause 5	*Application context*
Clause 6	*Application key concepts*
Annex A	*Glossary*
Annex B	*Enterprise-based life cycle*
Annex C	*Process task outcomes*
Annex D	*Planning documents*
Annex E	*System technical reviews*
Annex F	*Unprecedented and precedented development*
Annex G	*Requirement relationships*

24.3 Role of the EIA 632 Standard

Implementation of the requirements of EIA 632 are intended to be through establishment of enterprise policies and procedures that define the requirements for application and improvement of the adopted processes from the standard. This is illustrated in Figure 24.1.

24.4 Heritage of EIA 632

Figure 24.2 shows the relationship between EIA 632 and other standards on systems engineering. Some of the key software engineering standards are shown for comparison since there has been an intimate relationship between the development of both types of standards. There has been much activity recently in unifying the processes contained in each.

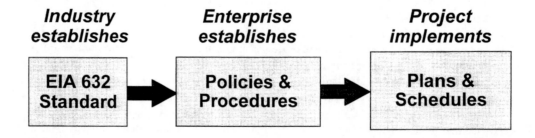

FIGURE 24.1   Role of the standard in relation to development projects. (Adapted from ANSI/EIA-632-1998. With permission.)

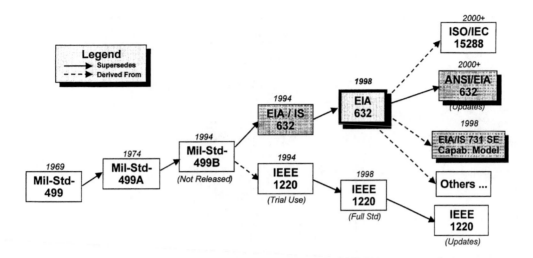

FIGURE 24.2 Heritage of systems engineering standards.

24.5 The Processes

Figure 24.3 shows the processes described in EIA 632 and their relationship to one another. Each enterprise will determine which of these processes are implemented by systems engineering personnel, and how they are allocated to the organizational elements of the enterprise and its functional disciplines.

24.5.1 Process Hierarchy

The processes for engineering a system are grouped into the five categories as shown in Figure 24.4. This grouping was made for ease of organizing the standard and is not a required structure for process implementation. Traditional systems engineering is most often considered to include two of these processes: Requirements Definition and Systems Analysis. Often, Planning and Assessment are included in what is called systems engineering management.

24.5.2 Technical Management Processes

Technical Management provides oversight and control of the technical activities within a development project. The processes necessary to accomplish this are shown in Figure 24.5.

24.5.3 Acquisition and Supply Processes

Acquisition and Supply provides the mechanism for a project to supply its own products to a customer or higher-level project and to acquire the necessary products for its own product development activities. The processes necessary to accomplish this are shown in Figure 24.6.

24.5.4 System Design Processes

System Design provides the activities for a project to define the relevant requirements for its product development effort and to design solutions that meet these requirements. The processes necessary to accomplish this are shown in Figure 24.7.

24.5.5 Product Realization Processes

Product Realization provides the activities for a project to implement the product designs and to transition these products to their place of use. The processes necessary to accomplish this are shown in Figure 24.8.

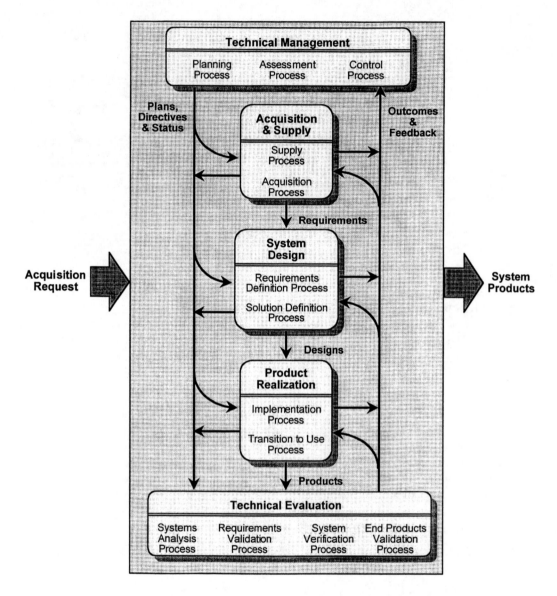

FIGURE 24.3   Top-level view of the processes for engineering a system. (From ANSI/EIA-632-1998. With permission.)

24.5.6 Technical Evaluation Processes

Technical Evaluation provides activities for a project to analyze the effectiveness of its proposed designs, validate the requirements and end products, and to verify that the system and its product meet the specified requirements. The processes necessary to accomplish this are shown in Figure 24.9.

24.6 Project Context

These "technical" processes fit into a larger context of a project (see Figure 24.10), and the project resides in some sort of enterprise, which in turn resides in an environment external to the enterprise. There are processes in the project and within the enterprise (but outside the project) that significantly affect the successful implementation of the technical processes.

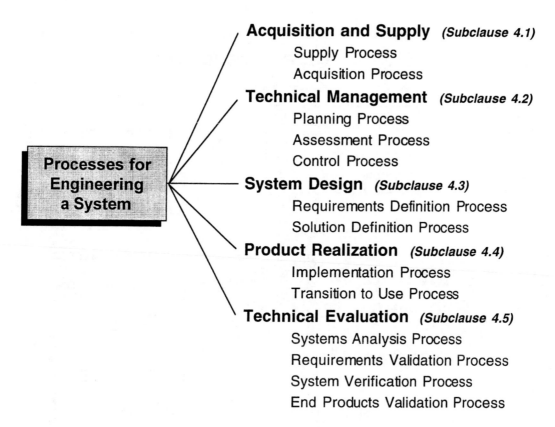

Processes for Engineering a System

Acquisition and Supply *(Subclause 4.1)*
  Supply Process
  Acquisition Process

Technical Management *(Subclause 4.2)*
  Planning Process
  Assessment Process
  Control Process

System Design *(Subclause 4.3)*
  Requirements Definition Process
  Solution Definition Process

Product Realization *(Subclause 4.4)*
  Implementation Process
  Transition to Use Process

Technical Evaluation *(Subclause 4.5)*
  Systems Analysis Process
  Requirements Validation Process
  System Verification Process
  End Products Validation Process

FIGURE 24.4 Hierarchical view of the processes for engineering a system. (From ANSI/EIA-632-1998. With permission.)

24.7 Key Concepts

To understand the processes as described in this standard, it is essential to understand the distinct use of certain terms and the conceptual models that underlie each process. Some of the key terms are system, product, verification, and validation. Some of the key concepts are building block, end products, associated processes, and development layers.

24.7.1 The System and Its Products

What is a system? The term "system" is commonly used to mean the set of hardware and software components that are developed and delivered to a customer. This standard uses this term in a broader sense in two aspects.

First, the system that needs to be developed consists of not only the "operations product" (that which is delivered to the customer and used by a user), but also the enabling products associated with that operations product. The operations product consists of one or more end products (so-called since these are the elements of the system that "end up" in the hands of the ultimate user). The associated processes are performed using enabling products that "enable" the end products to be put into service, kept in service, and retired from service.

Second, the end products that need to be developed often go beyond merely the hardware and software involved. There are also people, facilities, data, materials, services, and techniques. This is illustrated in Figure 24.12.

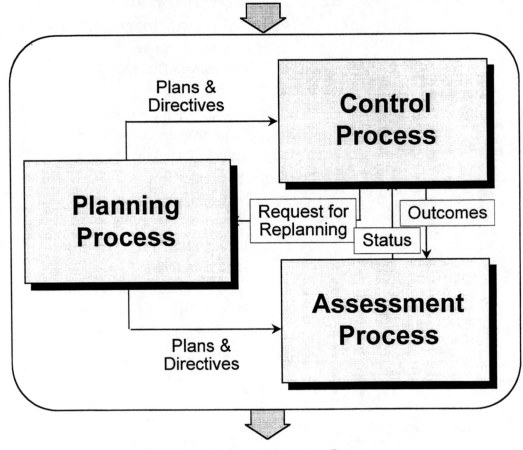

FIGURE 24.5   Technical Management Processes. (From ANSI/EIA-632-1998. With permission.)

This is not intended to be an exhaustive list of the "basic" product types since these will vary depending on the particular business or technology domain. For example, in the television industry, "media" is certainly one of the system elements that constitute the overall system that is developed. "CBS News" might be considered the system, for example, with end products like:

Airwaves, worldwide web (media)
Cameras, monitors (hardware)
Schedule management tools, video compression algorithms (software)
Camera operators, news anchor (personnel)
Studio, broadcasting tower (facilities)
Script, program guide (data)
Pictures, stories (materials)
Airplane transportation, telephone (services)
Presentation method, editing procedures (techniques)

Note that any or all of these end products could be "off-the-shelf." But some of them may need to be conceived, designed, and implemented. Even if one of these items is truly off-the-shelf, it may still need

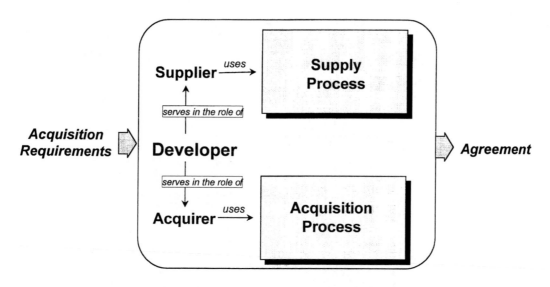

FIGURE 24.6 Acquisition and Supply Processes. (From ANSI/EIA-632-1998. With permission.)

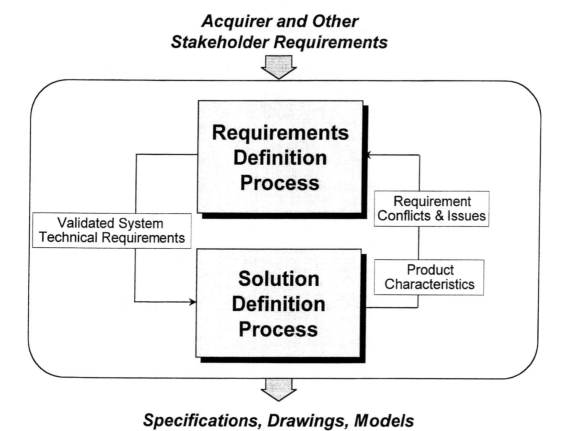

FIGURE 24.7 System Design Processes. (From ANSI/EIA-632-1998. With permission.)

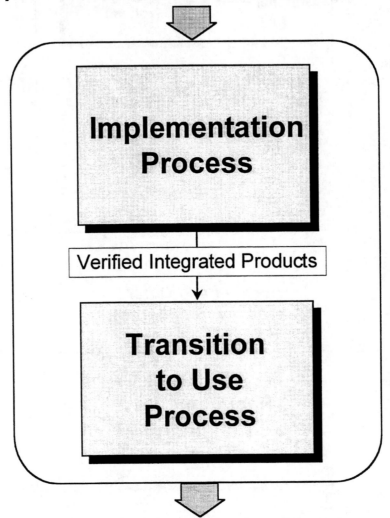

Specified Requirements
Supplier- or Acquirer-Provided Products

Implementation Process

Verified Integrated Products

Transition to Use Process

Agreement Satisfaction
Other Stakeholder Satisfaction

FIGURE 24.8   Product Realization Processes. (From ANSI/EIA-632-1998. With permission.)

some enabling product to allow effective use of that item. For example, even if you can use existing editing procedures, you may need to develop a training program to train the editors.

24.7.2   Building Block Framework

As we can see from the description above of the "CBS News" system, the nonhardware/software items may be crucial to successful realization of the whole system. You may also need to develop the associated processes along with the relevant enabling products. If we tie all these elements together, we can illustrate this using the so-called "building block" shown in Figure 24.13.

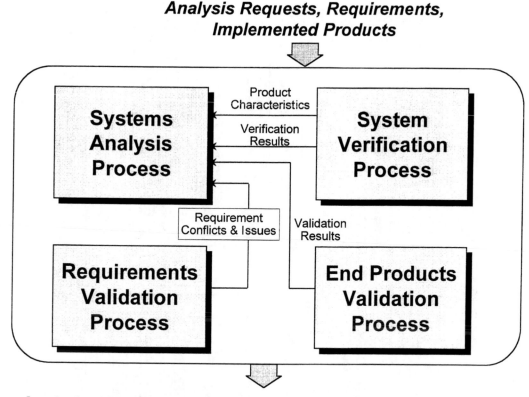

Analysis Requests, Requirements, Implemented Products

Analytical Models & Assessments, Validated Requirements, Verified Products, Validated Products

FIGURE 24.9 Technical Evaluation Processes. (From ANSI/EIA-632-1998. With permission.)

There are seven associated processes related to these sets of enabling products. These processes are used at various points in the life of a product (sometimes called "system life cycle elements"). Hence, the use of the building block concept is intended to help the developer in ensuring that the full life cycle of the end product is properly considered.

Note that each end product can consist of subsystems. Each of these subsystems may need its own development. The building block can be used at the various "development layers" of the total system. These development layers are illustrated in Figure 24.14.

24.7.3 Development of Enabling Products

As mentioned above, the enabling products may need to be developed also. For each associated process, there could be enabling products that either exist already or that need some degree of development. Figure 24.15 shows how enabling products related to the deployment process have their own building blocks.

24.7.4 Relationship Between the Building Blocks and the Processes

The building blocks relevant to a particular system development can be "stacked" into a System Breakdown Structure (SBS). The System Design Process and Verification and Validation Processes have a special relationship to the SBS, as shown in Figure 24.16.

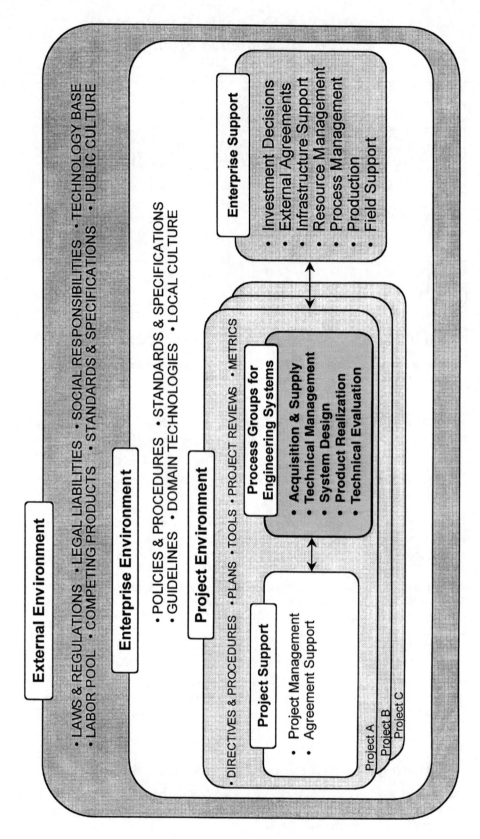

FIGURE 24.10   The technical processes in the context of a project and an enterprise. (From ANSI/EIA-632-1998. With permission.)

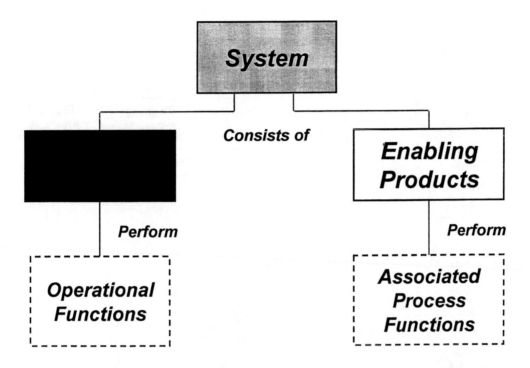

FIGURE 24.11 Constituents of the system. (From ANSI/EIA-632-1998. With permission.)

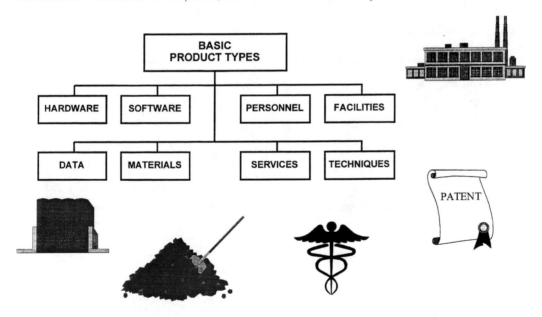

FIGURE 24.12 Different product types that make up a system.

24.7.5 Hierarchy of Building Blocks

Typically, a project is developing a system of such a size that it will require more than one building block to complete the development. Figure 24.17 shows an example of how several building blocks would be related to each other within a project. Notice also how an "adjacent" Project A has building blocks that might have interfaces to the systems under development by Project B. There also may be building blocks above and below a project.

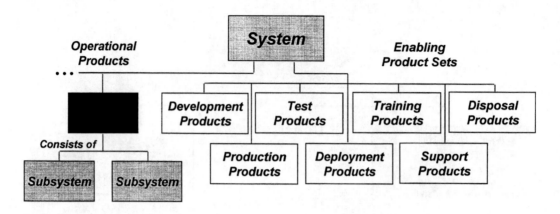

FIGURE 24.13   The building block concept. (From ANSI/EIA-632-1998. With permission.)

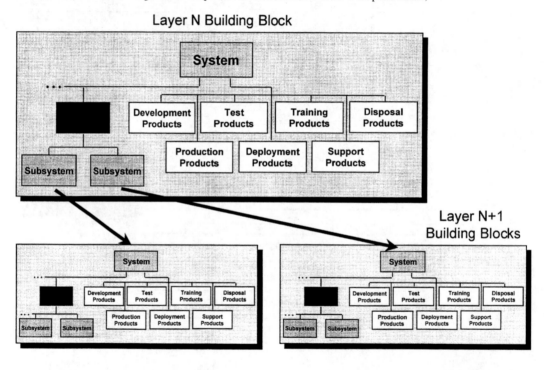

FIGURE 24.14   Development layers concept. (From ANSI/EIA-632-1998. With permission.)

24.7.6   Requirements

Requirements are especially useful for engineering a system, especially when the system consists of several building blocks. Requirements act as a kind of "decoupling mechanism" between the building blocks. They are the "terms and conditions" between the layers of development. Requirements start out as vague, usually nontechnical stakeholder requirements and evolve into more technically precise and verifiable technical requirements. This is illustrated in Figure 24.18.

Lower-level requirements must be derived from higher-level requirements. This is usually accomplished through logical and physical analyses and design activities to determine the requirements that apply to the design solution. This is illustrated in Figure 24.19 which shows the different types of requirements dealt with by the processes described in EIA 632.

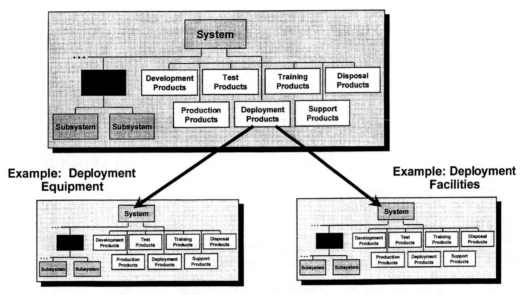

Other Deployment Enabling Products Possibly Needing Development:

v **Deployment Procedures (e.g. installation manual)**

v **Deployment Personnel (e.g. field technicians, installers)**

v **Deployment Services (e.g. shipping & receiving, transportation)**

FIGURE 24.15 Development of enabling products. (Adapted from ANSI/EIA-632-1998. With permission.)

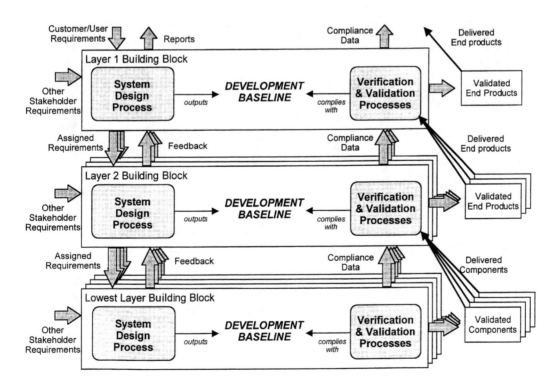

FIGURE 24.16 Development layers in a project context. (Adapted from ANSI/EIA-632-1998. With permission.)

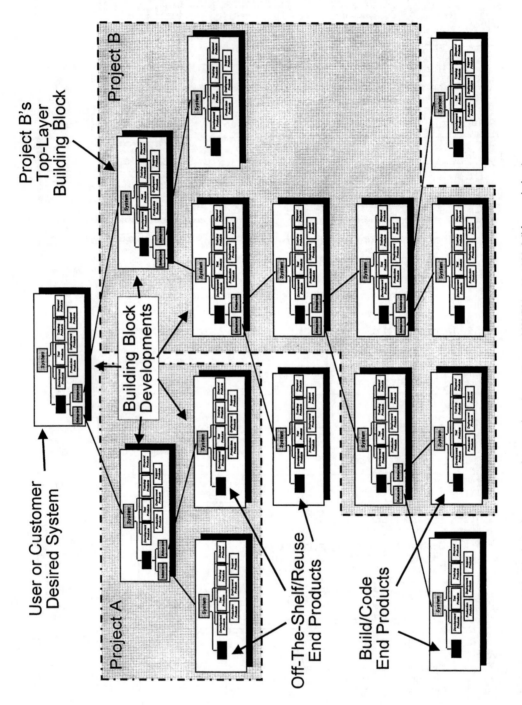

FIGURE 24.17 Building blocks in the context of a several projects. (From ANSI/EIA-632-1998. With permission.)

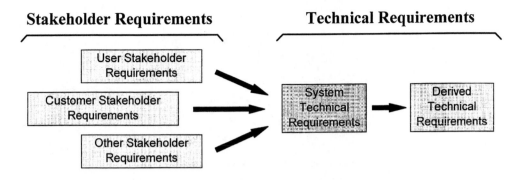

FIGURE 24.18 Evolution of requirements. (*Source*: ANSI/EIA-632-1998. With permission.)

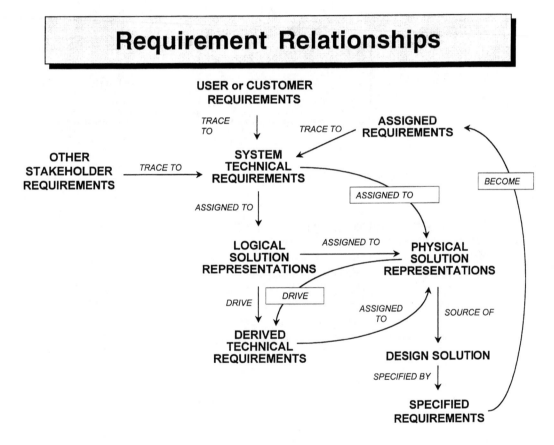

FIGURE 24.19 Types of requirements and their interrelationships. (From ANSI/EIA-632-1998. With permission.)

Requirements generally do not come from a particular customer. And even when they do, as is often the case in government-sponsored developments, they do not represent the full spectrum of the requirements that will eventually need to be defined for development of the products of a system.

User Requirements. User requirements are often nontechnical in nature and usually conflict with each other. These need to be translated into "technical requirements" that are in the vernacular of the engineers on a project and are specific to the domain of the technology to be used in the system.

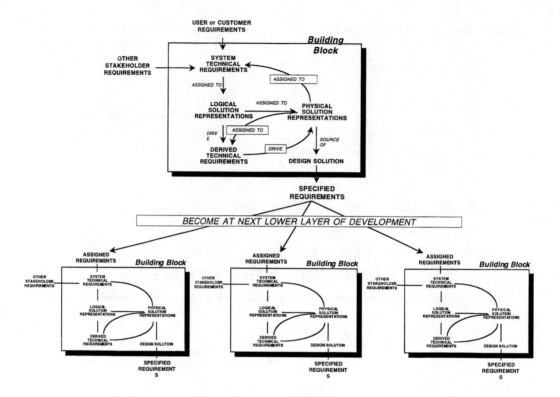

FIGURE 24.20 Layers of requirements as they relate to the layers of development. (Adapted from ANSI/EIA-632-1998. With permission.)

Customer Requirements. Customers will often translate their perceived user requirements into a set of requirements to be given to a development project. These also, like the user requirements, are often not technical enough and usually conflict with each other. Moreover, a particular project may have more than one customer for its products.

Stakeholder Requirements. In defining the technical requirements for the system, one must consider that there are other stakeholders for that system. For example, the manufacturing organization within your company may desire that you use their existing processes, tools, facilities, etc. There are constraints associated with these "enabling products." Hence, you must identify the stakeholder requirements that drive the development.

Requirements Decomposition. Once the system technical requirements are defined, further requirements are often "derived" by analyzing both the logical and physical aspects of the system. Some of these derived requirements, plus whatever system technical requirements are relevant, are "specified" for the system, its end products, and possibly subsystems of those end products.

Given these specified requirements for an item, the process starts all over again for items needing further development. In other words, another layer of building blocks may be required to develop an item far enough along such that the bottommost item can be procured from a supplier. This is illustrated in Figure 24.20.

24.7.7 Functional, Performance, and Interface Requirements

Another way to look at requirements is to categorize them as functional (what the product must accomplish), performance (how well the product must accomplish each function), and interface (under what conditions must the functions be accomplished). This is illustrated in Figure 24.21.

- # **Functional Requirements**
  - » *What* an item is to accomplish
    - Behavior of an item
    - An effect produced
    - Action or service to be performed

- # **Performance Requirements**
  - » *How well* an item is to accomplish a function
    - ... like how much, how often, how many, how few, . . .

- # **Interface Requirements**
  - » *Conditions* of interaction between items
    - ... could be functional, physical, logical, . . .

FIGURE 24.21 Types of requirements.

24.7.8 Verification and Validation

Verification and validation are both very important in engineering a system. Verification is the act of determining if a product meets its specified requirements. Validation is the act of determining if a product satisfies its stakeholders. This is illustrated in Figure 24.22

Defining Terms

The following definitions are extracted from the standard.

Building block: A representation of the conceptual framework of a system that is used for organizing the requirements, work, and other information associated with the engineering of a system. An element in the structured decomposition of the system.

Enabling product: Item that provides the means for (a) getting an end product into service, (b) keeping it in service, or (c) ending its service.

End product: The portion of a system that performs the operational functions and is delivered to an acquirer.

Process: A set of interrelated tasks that, together, transform inputs into outputs.

Product: (1) An item that consists of one or more of the following: hardware, software, firmware, facilities, data, materials, personnel, services, techniques, and processes; (2) a constituent part of a system.

Requirement: Something that governs what, how well, and under what conditions a product will achieve a given purpose.

Stakeholder: An enterprise, organization, or individual having an interest or a stake in the outcome of the engineering of a system.

Subsystem: A grouping of items that perform a set of functions within a particular end product.

System: An aggregation of end products and enabling products to achieve a given purpose.

Validation: (1) Confirmation by examination and provision of objective evidence that the specific intended use of an end product (developed or purchased), or an aggregation of end products, is

- **Verification**
  - » Check compliance against specified requirements
  - » "Have you done the *job right*?"
  - » Two Types
    - Product & Process Qualification
      - Full compliance to specification
      - Product requalification may be necessary when Product is "redesigned"
      - Process requalification may be necessary when Process is "restarted"
    - Product Acceptance
      - Full compliance to key criteria
      - Done on every unit or on sample basis
      - Can be done prior to shipment or after installation

- **Validation**
  - » Check satisfaction of stakeholders
  - » "Have you done the *right job*?"
  - » Two Types
    - Requirements Validation
      - Check for "traceability"
      - "Have we missed any requirements?"
      - "Do we have extraneous requirements?"
    - Product Validation
      - Check if meeting "needs and expectations" of stakeholders

FIGURE 24.22   Distinction between verification and validation.

accomplished in an intended usage environment; (2) confirmation by examination that requirements (individually and as a set) are well formulated and ar usable for intended use.
Verification: Confirmation by examination and provision of objective evidence that the specified
requirements to which an end product is built, coded, or assembled have been fulfilled.

References

ANSI/EIA 632, *Processes for Engineering a System*. 1998.
EIA/IS 632, *Systems Engineering*. 1994.
Martin, J. N., "Evolution of EIA 632 from an Interim Standard to a Full Standard." Proc. INCOSE 1998
    Int. Symp., 1998.

Further Information

The purpose of this chapter is to give a high-level overview of the processes for engineering a system
described in ANSI/EIA 632. Further details can be found in the full publication of the standard or by
contacting the EIA directly at 2500 Wilson Boulevard, Arlington, Virginia or at their Website
(www.eia.org).

25

Electromagnetic Environment (EME)

Richard Hess
Honeywell

25.1   Introduction

The advent of digital electronic technology in electrical/electronic systems has enabled unprecedented expansion of aircraft system functionality and evolution of aircraft function automation. As a result, systems incorporating such technology are used more and more to implement aircraft functions, including Level A systems that affect the safe operation of the aircraft; however, such capability does not come free. The EME (electromagnetic environment) is a form of energy, which is the same type of energy (electrical) that is used by electrical/electronic equipment to process and transfer information. As such, this environment represents a fundamental threat to the proper operation of systems that depend on such equipment. It is a common mode threat that can defeat fault-tolerant strategies reliant upon redundant electrical/electronic systems.

Electrical/electronic systems, characterized as Level A, provide functions that can affect the safe operation of an aircraft and depend upon information (i.e., guidance, control, etc.) processed by electronic equipment. Thus, the EME threat to such systems may translate to a threat to the airplane itself. The computers associated with modern aircraft guidance and control systems are susceptible to upset from lightning and sources that radiate RF at frequencies predominantly between 1 and 500 MHz and produce aircraft internal field strengths of 5 to 200 V/m or greater. Internal field strengths greater than 200 V/m are usually periodic pulses with pulsewidths less than 10 μs. Internal lightning-induced voltages and currents can range from approximately 50 V and 20 A to over 3000 V and 5000 A.

Electrical/electronic system susceptibility to such an environment has been suspect as the cause of "nuisance disconnects," "hardovers," and "upsets." Generally, this form of system upset occurs at significantly lower levels of EM field strength than that which could cause component failure, leaves no trace, and is usually nonrepeatable.

25.2 EME Energy Susceptibility

It is clear that the sources of electromagnetic (EM) threats to the electrical/electronic system, either digital or analog, are numerous. Although both respond to the same threats, there are factors that can make the threat response to a momentary transient (especially intense transients like those that can be produced by lightning) far more serious in digital processing systems than in analog systems. For example, the information bandwidth and, therefore, the upper noise response cutoff frequency in analog devices is limited to, at most, a few megahertz. In digital systems it is often in excess of 100 MHz and continues to increase. This bandwidth difference, which is at least 10 times more severe in digital systems, allows substantially more energy and types of energy to be coupled into the digital system. Moreover, the bandwidths of analog circuits associated with autopilot and flight management systems are on the order of 50 Hz for servo loops and much less for other control loops (less than 1 Hz for outer loops). Thus, if the disturbance is short relative to significant system time constants, even though an analog circuit device possessing a large gain and a broad bandwidth may be momentarily upset by an electromagnetic transient, the circuit will recover to the proper state. It should be recognized that, to operate at high speeds, proper circuit card layout control and application of high-density devices is a must. When appropriate design tools (signal integrity, etc.) are applied, effective antenna loop areas of circuit card tracks become extremely small, and the interfaces to a circuit card track (transmission line) are matched. Older (1970s–1980s) technology with wirewrap backplanes and processors built with discrete logic devices spread over several circuit cards were orders of magnitude more susceptible. Unlike analog circuits, digital circuits and corresponding computational units, once upset, may not recover to the proper state and may require external intervention to resume normal operation. It should be recognized that (for a variety of reasons) large-gain bandwidth devices are and have been used in the digital computing platforms of aircraft systems. A typical discrete transistor can be upset with 10^{-5} J, 2000 V at 0.1 mA for 50 μs. A typical integrated circuit can be upset with only 10^{-9} J, 20 V at 1 μA for 50 μs. As time goes on and processor semiconductor junction feature sizes get smaller and smaller, this problem becomes worse.

It should be noted that in addition to upset, lightning-induced transients appearing at equipment interfaces can, because of the energy they possess, produce hard faults (i.e., damage circuit components) in interface circuits of either analog or digital equipment. Mechanical, electromechanical, electrohydraulic, etc. elements associated with conventional (not electronic primary flight controls with associate "smart" actuators) servo loops and control surface movement are either inherently immune or vastly more robust to EME energy effects than the electronic components in an electrical/electronic system.

Immunity of electronic components to damage is a consideration that occurs as part of the circuit design process. The circuit characteristic (immunity to damage) is influenced by a variety of factors:

1. Circuit impedances (resistance, inductance, capacitance), which may be distributed as well as lumped;
2. The impedances around system component interconnecting loops along with the characteristic (surge) impedance of wiring interfacing with circuit components;
3. Properties of the materials used in the construction of a component (e.g., thick-film/thin-film resistors);
4. Threat level (open circuit voltage/short circuit current), resulting in a corresponding stress on insulation, integrated circuit leads, PC board trace spacing, etc.; and
5. Semiconductor device nonlinearities (e.g., forward biased junctions, channel impedance, junction/gate breakdown).

Immunity to upset for analog processors is achieved through circuit design measures, and for digital processors it is achieved through architectural as well as circuit design measures.

25.2.1 Soft Faults

Digital circuit upset, which has also been known by the digital computer/information processing community as a "soft fault," is a condition known to occur even in relatively benign operating environments. Soft faults occur despite the substantial design measures (timing margins, transmission line interconnects, ground and power planes, clock enablers of digital circuits) to achieve a relatively high degree of integrity in digital processor operation.

In a normal operating environment, the occurrence of soft faults within digital processing systems is relatively infrequent and random. Such occasional upset events should be treated as probabilistic in nature and can be the result of:

- Coincidence of EME energy with clocked logic clock edges, etc.
- Occasional violation of a device's operational margin (resulting margin from the design, processing, and manufacturing elements of the production cycle).

From this perspective, the projected effect of a substantial increase in the severity of the electromagnetic environment will be an increased probability of a soft fault occurrence. That is, in reality a soft fault may or may not occur at any particular point in time but, on the average, soft faults will occur more frequently with the new environmental level.

Once developed, software is "burned into nonvolatile" memory (becomes "firmware"); the result will be a special purpose real-time digital electronic technology data processing machine with the inherent potential for "soft faults." Because it is a hardware characteristic, this potential exists even when a substantial amount of attention is devoted to developing "error-free" operating system and application programs(s) (software) for the general purpose digital machine (computing platform, digital engine, etc.).

25.2.2 MTBUR/MTBF

In the past, service experience with digital systems installed on aircraft has indicated that the confirmed failure rates equal or exceed predicted values that were significantly better than previous generation analog equipment. However, the unscheduled removal rate remains about the same. In general, the disparity in mean time between unscheduled removal (MTBUR) and the mean time between failure (MTBF) continues to be significant. The impact of this disparity on airline direct operating costs is illustrated in Figure 25.1.

To the extent that soft faults contribute to the MTBUR/MTBF disparity, any reduction in soft fault occurrence and propagation could translate into reduction of this disparity.

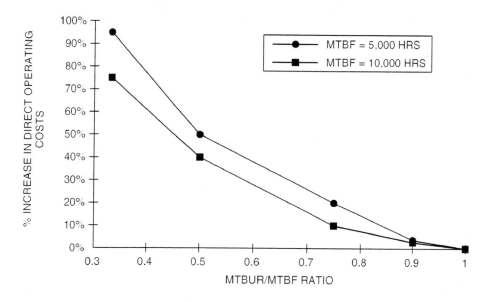

FIGURE 25.1 MTBUR/MTBF ratio impact of operating costs.

25.3 Civil Airworthiness Authority Concerns

The Federal Aviation Administration (FAA) and European Joint Aviation Authorities (more commonly known as JAA) have identified the lightning and High Intensity Radiated Field (HIRF) elements of the EME as a safety issue for aircraft functions provided by electrical/electronic systems.

The following factors, identified by the FAA and JAA, have led to this concern about lightning and HIRF effects:

- Increased reliance on electrical and electronic systems to perform functions that may be necessary for the continued safe flight and landing of the aircraft.
- Reduction of the operating power level of electronic devices that may be used in electrical and electronic systems, which may cause circuits to be more reactive to induced lightning and RF voltages and currents leading to malfunction or failure.
- Increased percentage of composite materials in aircraft construction. Because of their decreased conductivity, composite materials may result in less inherent shielding by the aircraft structure.
- Since current flowing in the lightning channel will be forced (during lightning attachment) into and through the aircraft structure without attenuation, decreased conductivity for aircraft structure materials can be particularly troubling for lightning.

The direct effects of lightning (dielectric puncture, blasting, melting, fuel ignition, etc.) have been recognized as flight hazards for decades, and in 1972 the SAE formed the AE4 Special Task F (which later became AE4L) to address this issue. In the early 1980s, the FAA began developing policy relative to the effects of lightning on electrical/electronic systems (indirect effects) and AE4L supported the FAA and JAA by providing the technical basis for international standards (rules/regulations) and guidance material that, for aircraft type certification, would provide acceptable means for demonstrating compliance to those rules/regulations. AE4L also supported RTCA Special Committee 135 (SC-135) to integrate lightning environment conditions and test procedures into airborne equipment standards (DO-160) and the EUROCAE standards counterpart (ED-14). In 1987, EUROCAE formed Working Group 31 to be the European counterpart of AE4L.

In 1986, the FAA and JAA identified High Energy Radio Frequency (HERF) electromagnetic fields as an issue for aircraft electrical/electronic systems. Some time later the term HERF was changed to its present designation, which is High Intensity Radiated Field (HIRF). Subsequent to the FAA identifying HIRF as a safety issue, SAE and EUROCAE formed Committee AE4R and Working Group 33, respectively, to support the FAA and JAA in much the same way as was the case with AE4L and lightning. In addition, unlike the case for lightning, RTCA SC-135 formed a HIRF working group (the corresponding European group was already part of EUROCAE/WG33) to integrate HIRF requirements into DO-160/ED-14.

In the interim between the absence and existence of a rule for lightning and HIRF, special conditions have been or are issued to applicants for aircraft type certification (TC, STC, ATC). The rationale for the special condition is given in words to the effect:

These series of aircraft will have novel or unusual design features associated with the installation of new technology electrical and electronic systems, which perform critical or essential functions. The applicable airworthiness regulation does not contain adequate or appropriate safety standards for the protection of these systems from the effects of lightning and radio frequency (RF) energy. This notice contains the additional safety standards that the Administrator considers necessary to ensure that critical and essential functions the new technology electrical and electronic systems perform are maintained when the airplane is exposed to lightning and RF energy.

Presently, the FAA's Federal Aviation Regulations (FARs) have been updated to include the "indirect effects" of lightning, but not HIRF. In the time period between the absence and existence of a rule for HIRF, special conditions for HIRF are being issued to applicants for aircraft certification. However, the FAA has established the Aviation Rule-Making Advisory Committee, which in turn established the

Electromagnetic Effects Harmonization Working Group (EEHWG) to develop the rule-making package for HIRF and for amendments to the lightning rules.

Portable electronic devices (PEDs) have not been identified for regulatory action, but in 1992 the FAA requested the RTCA to study the EME produced by PEDs. In response to an FAA request relative to PEDs, RTCA formed Special Committee 177 (SC-177) in 1992. In 1996, SC-177 issued a report titled "Portable Electronic Devices Carried Onboard Aircraft" (DO-233). Currently, control of PEDs and their associated electromagnetic (EM) emissions are handled by integrating some of the RTCA recommendations into airline policy regarding instructions (prohibition of personal cellular phone use, turn-off of PEDs during taxi, take-off, and landing, etc.) given to passengers.

25.3.1 EME Compliance Demonstration for Electrical/Electronic Systems

FAA/JAA FAR(s)/JAR(s) require compliance demonstration either explicitly or implicitly for the following EME elements:

- Lightning
- HIRF (FAA)
- HIRF (JAA)
- EMC

At the aircraft level, the emphasis should be on lightning and HIRF because most of the energy and system hazards arise from these threats. Their interaction with aircraft systems is global and also the most complex, requiring more effort to understand. Intrasystem electromagnetic emissions fall under the broad discipline of EMC. PEDs are a source of EM emissions that fall outside of the categories of equipment normally included in the EMC discipline. Like lightning and HIRF, the interaction of PED emissions with aircraft electrical/electronic systems is complex and could be global.

The electrical and/or electronic systems that perform functions "critical" to flight must be identified by the applicant with the concurrence of the cognizant FAA ACO. This may be accomplished by conducting a functional hazard assessment and, if necessary, preliminary system safety assessments (see SAE ARP 4761). The term "critical" means those functions whose failure would contribute to, or cause, a catastrophic failure condition (loss of aircraft). Table 25.1 provides the relationship between function failure effects and development assurance levels associated with those systems that implement functions that can affect safe aircraft operation.

TABLE 25.1 Nomenclature Cross Reference Between AC25.1309 and SAE-ARP 4754

Failure Condition Classification	Development Assurance Level
Catastrophic	Level A
Severe Major/Hazardous	Level B
Major	Level C
Minor	Level D
No Effect	Level E

The terms "Level A," etc. designate particular system development assurance levels. System development assurance levels refer to the rigor and discipline of processes used during system development (design, implementation, verification/certification, production, etc.). It was deemed necessary to focus on the development processes for systems based upon "highly integrated" or "complex" (whose safety cannot be shown solely by test and whose logic is difficult to comprehend without the aid of analytical tools) elements, i.e., primarily digital electronic elements.

Development assurance activities are ingredients of the system development processes. As has been noted, systems and appropriate associated components are assigned "development assurance levels" based on failure condition classifications associated with aircraft-level functions implemented by systems and components.

The rigor and discipline needed in performing the supporting processes will vary, depending on the assigned development assurance level.

There is no development process for aircraft functions. Basically, they should be regarded as intrinsic to the aircraft and are categorized by the role they play for the aircraft (control, navigation, communication, etc.). Relative to safety, they are also categorized (from FAA advisory material) by the effect of their failures, i.e., catastrophic, severe major/hazardous, major, etc.

EMC has been included in FAA regulations since the introduction of radio and electrical/electronic systems into aircraft. Electrical equipment, controls, and wiring must be installed so that operation of any one unit, or system of units, will not adversely affect the simultaneous operation of any other electrical unit or system essential to aircraft safe operation. Cables must be grouped, routed, and spaced so that damage to essential circuits will be minimized if there are faults in heavy current-carrying cables. In showing compliance with aircraft electrical/electronic system safety requirements with respect to radio and electronic equipment and their installations, critical environmental conditions must be considered. Radio and electronic equipment, controls, and wiring must be installed so that operation of any one component or system of components will not adversely affect the simultaneous operation of any other radio or electronic unit, or system of units, required by aircraft functions.

Relative to safety and electrical/electronic systems, the systems, installations, and equipment whose functioning is required for safe aircraft operation must be designed to ensure that they perform their intended functions under all foreseeable operating conditions. Aircraft systems and associated components, considered separately and in relation to other systems, must be designed so that:

- The occurrence of any failure condition that would prevent the continued safe flight and landing of the airplane is extremely improbable.

- The occurrence of any other failure condition that would reduce the capability of the airplane or the ability of the crew to cope with adverse operating conditions is improbable.

25.3.2   EME Energy Propagation

As has been noted in the introductory paragraph and illustrated in Figure 25.2, lightning and HIRF are threats to the overall aircraft. Since they are external EME elements, of the two, lightning produces the most intense environment, particularly by direct attachment.

Both lightning and HIRF interactions produce internal fields. Lightning can also produce substantial voltage drops across the aircraft structure. Such structural voltages provide another mechanism (in addition to internal fields) for energy to propagate into electrical/electronic systems. Also, the poorer the conductivity of structural materials, the greater the possibility that there are

- Voltage differences across the structure
- Significant lightning diffusion magnetic fields
- Propagation of external environment energy

Figure 25.3 gives the HIRF spectrum and associated aircraft/installations features of interest.

In general, the propagation of external EME energy into the aircraft interior and electrical/electronic systems is a result of complex interactions of the EME with the aircraft exterior structures, interior structures, and system installations (see Figures 25.3 through 25.7). Figure 25.8 gives representative transfer functions, in the frequency domain, of energy propagation into electrical/electronic systems, and Figure 25.9 provides time domain responses to a lightning pulse resulting from transfer functions having the low-frequency characteristic $V_o(f) = kf[Hi(f)]$ and a high frequency "moding" (resonant) characteristic (e.g., open loop voltage of cabling excited by a magnetic field; see Figure 25.8).

Paths of electromagnetic wave entry from the exterior to the interior equipment regions are sometimes referred to as points of entry. Examples of points of entry may be seams, cable entries, windows, etc. As noted, points of entry are driven by the local environment, not the incident environment. The internal field levels are dependent on both the details of the point of entry and the internal cavity.

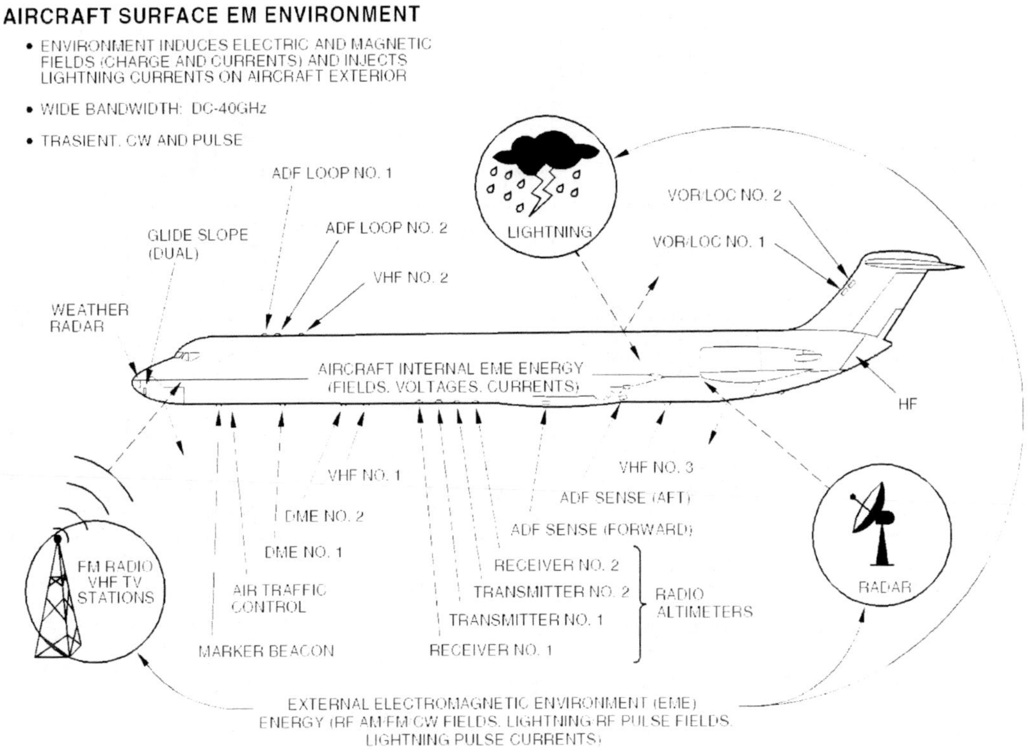

AIRCRAFT SURFACE EM ENVIRONMENT

- ENVIRONMENT INDUCES ELECTRIC AND MAGNETIC FIELDS (CHARGE AND CURRENTS) AND INJECTS LIGHTNING CURRENTS ON AIRCRAFT EXTERIOR

- WIDE BANDWIDTH: DC-40GHz

- TRASIENT, CW AND PULSE

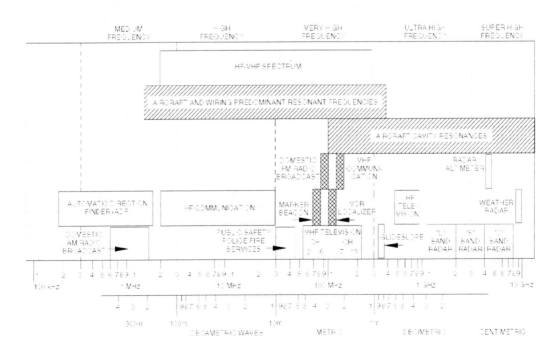

FIGURE 25.2 External EME (HIRF, lightning) interaction.

FIGURE 25.3 RF spectrum and associated installation dimensions of interest.

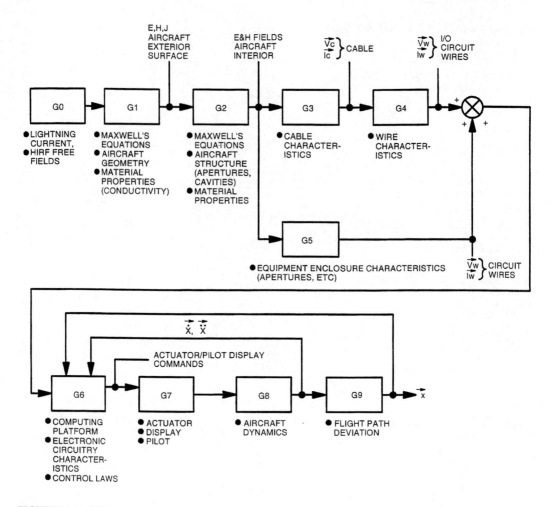

FIGURE 25.4   EME propagation process transfer function perspective.

Resulting internal fields can vary over a wide range of intensity, wave shape, and wave impedance. (Below 10 MHz within a metal aircraft, the magnetic fields due to lightning predominate because of the electric field shielding properties of metal skins. For HIRF "high-frequency" bands in some internal regions, internal field levels may exceed the incident field levels.)

The EME local to the equipment or system within the installation (the EME energy coupled to installation wiring which appears at equipment interface circuits) and the degree of attenuation or enhancement achieved for any region are the product of many factors such as external EME characteristics, materials, bonding of structure, dimensions and geometric form of the region, and the location and size of any apertures allowing penetration into the aircraft (G0 through G5 of Figure 25.4 which could have any of the characteristics of Figure 25.8.)

In HIRF high-frequency bands (frequencies on the order of 100 MHz and higher) the internal field resulting from such influences, as noted above, will in most cases produce a nonuniform field within the region or location of the system or equipment. The field cannot be considered as uniform and homogeneous. The field will not necessarily allow the adoption of single-point measurement techniques for the accurate determination of the equivalent internal field for to be used as the test level for systems. Several hot spots typically exist within any subsection of the aircraft. This is particularly true at cavity resonant conditions. Intense local effects are experienced at all frequencies in the immediate vicinity of any apertures for a few wavelengths away from the aperture itself. For apertures small with respect to wavelength, measurements of the fields within the aperture would yield fields much larger than those

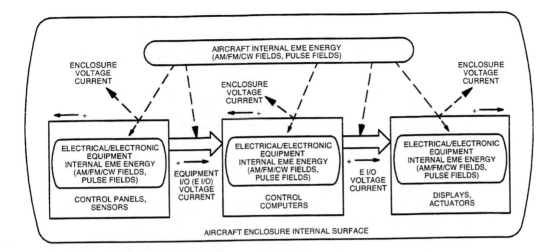

- External energy penetrates to interior via apertures, composites, seams, joints, and antennas
- Voltages and currents induced on flight control system components and cables
  - RF energy below 1 megahertz - induced coupling at these frequencies is inefficient and thus will probably be of lesser concern
  - RF energy between 1 and 300 megahertz is of major concern as aircraft wiring, when their lengths are on the order of a wavelength divided by two ($\lambda/2$) or longer at these frequencies, acts as a highly efficient antenna
  - RF energy coupling to aircraft wiring drops off at frequencies above 300 megahertz (at these higher frequencies, the EM energy tends to couple through box apertures rather than through aircraft wiring)

FIGURE 25.5   Aircraft internal EME energy electrical/electronic system.

further inside the aircraft because the fields fall off inversely proportional to radius cubed. For apertures on the order of a wavelength in size or larger, the fields may penetrate unattenuated.

The HIRF spectrum of RF energy that couples into aircraft wiring and electrical/electronic systems can be summarized into three basic ranges:

- HIRF energy below 1 MHz — induced coupling at these frequencies is inefficient and thus will be of lesser concern.
- HIRF energy between 1 and 400 MHz — induced coupling is of major concern since aircraft wiring acts as a highly efficient antenna at these frequencies.
- HIRF energy above 400 MHz — coupling to aircraft wiring drops off at frequencies above 400 MHz. At these higher frequencies the EM energy tends to couple through equipment apertures and seams and to the quarter wavelength of wire attached to the line replaceable unit (LRU). In this frequency range, aspects of equipment enclosure construction become important.

The extension of electrical/electronic systems throughout the aircraft ranges from highly distributed (e.g., flight controls) to relatively compact. Wiring associated with distributed systems penetrates several aircraft regions. Some of these regions may be more open to the electromagnetic environment than others, and wiring passing through the more open regions is exposed to a higher environment. Thus, at frequencies below 400 MHz, the wiring of a highly distributed system could have a relatively wide range of induced voltages and currents that would appear at equipment interface circuits.

The flight deck of the aircraft is an example of an open region. The windscreen "glass" presents approximately zero attenuation to an incoming field at and above the frequency for which its perimeter is one wavelength. Some enhancement above the incident field level generally exists in and around the aperture at this resonance condition.

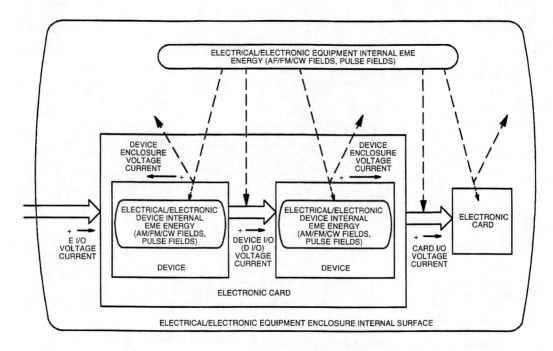

- Voltage, fields, currents and charge on system components penetrate into equipment enclosure interiors via holes, seams, and airplane wiring (cables)
- Energy (voltage and current) picked up by wires and printed conductors on cards and carried to electronic devices

FIGURE 25.6   Electrical/electronic equipment internal EME interaction electrical/electronic circuitry.

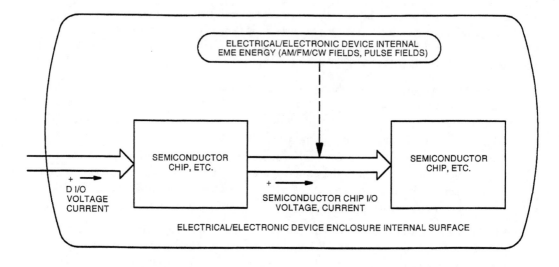

- Card and device conductors carry energy to the semiconductor chips, etc.
- Possible effects
    - Damage
    - Upset

FIGURE 25.7   Electrical/electronic device internal EME interaction electrical/electronic circuitry.

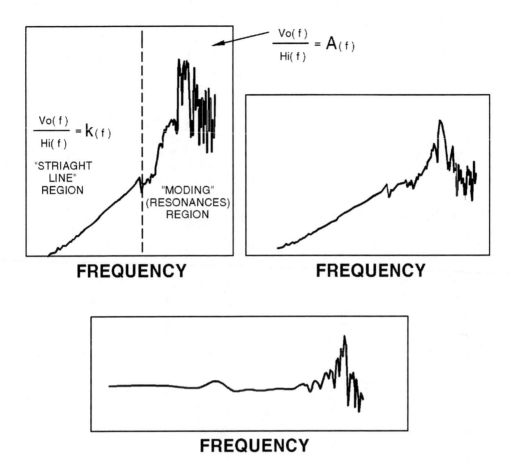

FIGURE 25.8 Frequency domain representation of EME energy attenuation/coupling transfer functions.

Lightning is a transient electromagnetic event, as is the resulting internal environment. Relative to a spectral representation, lightning energy would be concentrated in the zero to 50 MHz range (most energy is below 3 MHz). However, since lightning is such an intense transient, significant energy can be present up to and sometimes above 10 MHz.

Relative to the higher frequency range (above 100 MHz) strong resonances of aircraft interior volumes (cavities) such as the flight deck, equipment bay, etc, could occur. At the very high frequencies the EME can be in the form of both very intense and very short duration. From a cavity resonance issue, since the time constant of a relatively good cavity resonator is on the order of 1 μs, the pulse can be gone before significant field energy is developed within the cavity.

25.4 Architecture Options for Fault Mitigation

New system architecture measures have been evolving which could complement/augment traditional schemes to provide protection against EME energy effects. Architecture options can be applied at the overall system level or within the digital computing platform for the system. These options include the following:

- Distributed bus architecture
- Error Detection and Corrective (EDC) schemes
- Fiber optic data transfer
- Computation recovery

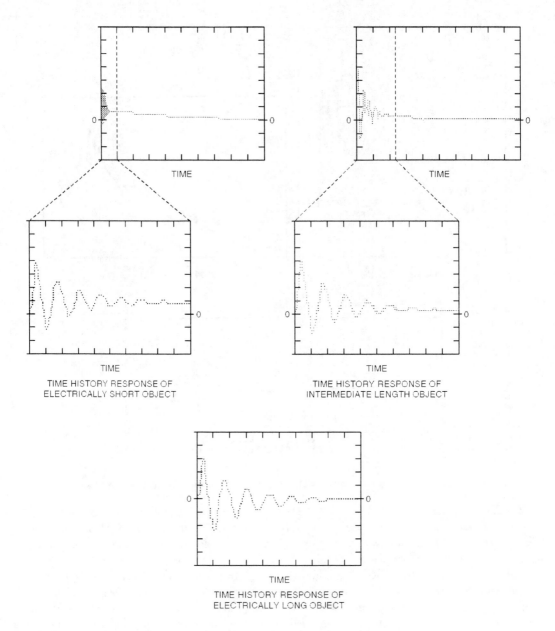

FIGURE 25.9 Responses for lightning EM pulse field interaction with objects of different "electrical lengths."

25.4.1 Electrical/Electronic System

In the past, soft faults in digital avionics were physically corrected by manual intervention, recycle power, etc. More recently, system-level measures for the automatic correction of soft faults have begun to be developed. It is perceived that significant benefits can be gained through soft fault protection measures designed into the basic system mechanization. System-level soft fault protection methodologies provide the ability to tolerate disruption of either input/output data or internal computation. Accordingly, there are two distinct classes of disruption:

- Disruption at the system equipment interface boundary causing corruption of data flowing to or from the affected subsystem.

- Disruption that reaches within system equipment to corrupt internal data and computation. As a worst case scenario, it must be presumed that any memory elements within the computational machine (registers, memory, etc.) may be affected at the time of disruption.

The short-term disruption of input/output data at an equipment boundary can be managed via a variety of existing methodologies. Data errors must be detected and the associated data suppressed until the error status is cleared. The data processing algorithm should tolerate data loss without signaling a hard fault. The length of time that can be tolerated between valid refreshes depends on the data item and the associated time constants (response) of the system and corresponding function being implemented.

The ability to tolerate disruption that reaches computation and memory elements internal to system equipment without propagation of the associated fault effect is a more difficult problem. For systems with redundant channels, this means tolerance of the disruption without loss of any of the redundant channels. Fault clearing and computation recovery must be rapid enough to be "transparent" relative to functional operation and flight deck effects.

Such computational recovery requires that the disruption be detected and then the state of the affected system be restored. Safety-critical systems are almost always mechanized with redundant channels. Outputs of channels are compared in real time, and an errant channel is blocked from propagating a fault effect. One means available for safety-critical systems to detect disruption is the same cross-channel monitor. If a miscompare between channels occurs, a recovery is attempted. For a hard fault, the miscompare condition will not have been remedied by the recovery attempt.

A basic approach to "rapid" computational recovery would be to transmit function state variable data from valid channels to the channel that has been determined faulted and for which a recovery is to be attempted (Figure 25.10). However, the cross-channel mechanization is ineffective against a disruption that has the potential to affect all channels.

25.4.2 Digital Computing Platform

The platform for the Airplane Information Management System (AIMS) used on Boeing 777 aircraft and Versatile Integrated Avionics (VIA) technology is an example of an architectural philosophy in the design of computing platforms. Essentially, VIA is a repackaged version of the AIMS technology. As mentioned, first-generation digital avionics have been plagued with high MTBUR (no-fault-found) rates.

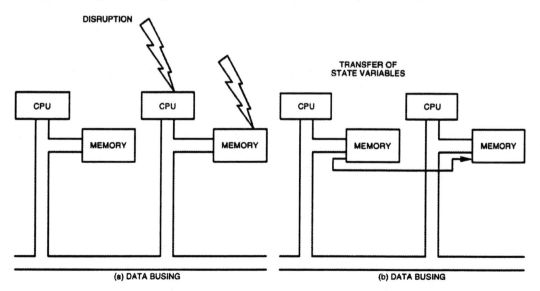

FIGURE 25.10 Redundant CPUs cross-lane recovery (can accomplish some degree of "rapid" recovery).

One primary goal of the Boeing 777 program was to greatly improve operational readiness and associated life-cycle cost performance for the airlines. The AIMS functionally redundant, self-checking pairs architecture was specifically selected to attack these problems. The high integration supported by AIMS required a very comprehensive monitoring environment that is ideal for in-channel "graceful" recovery.

In AIMS, the more dramatic step of making hardware monitoring active on every CPU clock cycle was taken. All computing and I/O management resources are lockstep compared on a processor cycle-by-cycle basis. All feasible hardware soft or hard faults are detected. In this approach, if a soft or hard fault event occurs, the processor module is immediately trapped to service handlers and no data can be exported. In past systems, the latency between such an event and eventual detection (or washout) was the real culprit. The corrupted data would propagate through computations and eventually affect some output. To recover, drastic actions (reboots or rearms) were often necessary. In AIMS, critical functions such as displays (because the flight crew could "see" hiccups) have a "shadowing" standby computational resource. The shadow sees the same input set at the same time as the master self-checking pair. If the master detects an event, within nanoseconds the faulty unit is blocked from generating outputs. The Honeywell SAFEbus® system detects the loss of output by the master and immediately passes the shadow's correct data for display.

In the faulted processor module, the core system has two copies of processor "state data" fundamental in the self-checking pair. Unlike past systems where the single thread processor may be so defective it cannot record any data, at least one half of the AIMS self-checking pair should be successful. Thus, the process of diagnosing hardware errors involves comparing what each half of the pair thought was going on. Errors, down to processor address, control, or data bits can be easily isolated. If the event was a soft fault, the core system allows a graceful recovery before the processor module is again allowed to export data. On the surface it appears to be a more sensitive system. However, even with the comprehensive monitoring (potentially a brittle operation), from the standpoint of a self-checking (dual-lockstep) pairs processor data comparison, in these platforms the automatic recovery capabilities should provide a compensating, more robust operation. In other words, from a macro time perspective, system functions will continue to be performed even though, on a micro time basis, a soft fault occurred.

In addition to the isolation of hardware faults (hard or soft), effective temporal and physical partitioning for execution of application software programs involving a variety of software levels has been achieved by the monitoring associated with the self-checking pairs processor and a SAFEbus® communication technology approach.

Defining Terms

DO-160: RTCA Document 160, Environmental Conditions and Test Procedures for Airborne Equipment, produced by RTCA Special Committee 135. Harmonized with ED-14.

ED-14: EUROCAE Document 14, Counterpart to DO-160, produced by EUROCAE Working Groups 14, 31, and 33. Harmonized with DO-160.

EMC: Electromagnetic Compatibility is a broad discipline dealing with EM emissions from and susceptibility to electrical/electronic systems and equipment.

EME: Electromagnetic Environment, which for commercial aircraft, consists of lightning, HIRF, and the electrical/electronic system and equipment emissions (intra and inter) portion (susceptibility not included) of EMC.

MTBF: Mean Time Between Failures (World Airlines Technical Operations Glossary).

MTBUR: Mean Time Between Unscheduled Removals (World Airlines Technical Operations Glossary).

EUROCAE: European Organization for Civil Aviation Equipment; for the European aerospace community, serving a role comparable to that of the RTCA and SAE.

PED: Portable Electronic Device, an emerging source of EM emissions not included in the EMC discipline.

References

1. AC25.1309, "System Design and Analysis."
2. SAE ARP-4761, "Guidelines and Tools for Conducting the Safety Assessment Process on Civil Airborne Systems and Equipment," issued December, 1996.
3. SAE ARP-4754, "Certification Consideration for Highly Integrated or Complex Aircraft Systems," issued November, 1996.
4. R.F. Hess, "Options for Aircraft Function Preservation in the Presence of Lightning," *Int. Conf. Lightning Static Electr.*, Toulouse, France, June 1999.
5. Aviation Regulations
   XX.581   Lightning Protection
   XX.1316 System Lightning Protection
6. AC/AMJ 20-136, "Protection of Aircraft Electrical/Electronic Systems Against the Indirect Effects of Lightning."
7. N8110.67 (FAA Notice), "Guidance for the Certification of Aircraft Operating in High Intensity Radiated Field (HIRF) Environments."
8. SAE ARP5413, "Certification of Aircraft Electrical/Electronic Systems for the Indirect Effects of Lightning," issued August, 1999.
9. SAE AE4L Report: AE4L-87-3 ("Orange Book"), "Certification of Aircraft Electrical/Electronic Systems for the Indirect Effects of Lightning," September 1996 (original publication February 1987).
10. SAE ARP5412, "Aircraft Lightning Environment and Related Test Waveforms," 1999.
11. SAE Report: AE4L-97-4, "Aircraft Lightning Environment and Related Test Waveforms Standard," July 1997.
12. EUROCAE ED-14D/RTCA DO-160D, "Environmental Conditions and Test Procedures for Airborne Equipment."
13. MIL-STD-464, "Electromagnetic Environmental Effects Requirements for Systems."
14. Clarke, Clifton A. and Larsen, William E., FAA Report DOT/FAA/CT 86/40, "Aircraft Electromagnetic Compatibility," June 1987.
15. Hess, R.F. "Implications Associated with the Operation of Digital Data Processing in the Relatively Harsh EMP Environments Produced by Lightning," *Int. Aerosp. and Ground Conf. Lightning and Static Elect.*, Paris, France, June 1985.

26

Ada

John G. P. Barnes

John Barnes Informatics

26.1 Introduction

Ada 95 is a comprehensive high-level programming language especially suited for the professional development of large or critical programs for which correctness and robustness are major considerations. This chapter gives a brief account of the background to the development of Ada 95 (and its predecessor Ada 83), its place in the overall language scene, and then outlines some of the major features of the language.

Ada 95 is a direct descendant of, and highly compatible with, Ada 83, which was originally sponsored by the U.S. Department of Defense for use in the embedded system application area. Ada 83 became an ANSI standard in 1983 and an ISO standard in 1987.

Time and technology do not stand still and accordingly, after several years' use, it was decided that the Ada 83 language standard should be revised in the light of experience and changing requirements. One major change was that Ada was now being used for many areas of application other than the embedded systems for which it was originally designed. Much had also been learned about new programming paradigms such as object oriented programming.

A set of requirements for the revision was published in 1990. An interesting aspect of the requirements is that they cover a number of specialized application areas. It seemed likely that it would be too costly to implement a language meeting all these requirements in their entirety on every architecture. On the other hand, one of the strengths of Ada is its portability, and the last thing anyone wanted was the anarchy of uncontrolled subsets. As a consequence, Ada 95 comprises a core language plus a small number of specialized annexes. All compilers have to implement the core language and vendors can choose to implement zero, one, or more annexes according to the needs of their markets.

Having established the requirements, the language design was contracted to Intermetrics Inc. under the technical leadership of S. Tucker Taft with continued strong interaction with the user community. The new ISO standard was published on February 15, 1995 and so the revised language is called Ada 95.

26.1.1 Software Engineering

It should not be thought that Ada is just another programming language. Ada is about software engineering, and by analogy with other branches of engineering it can be seen that there are two main problems with the development of software: the need to reuse software components as much as possible and the need to establish disciplined ways of working.

As a language, Ada (and hereafter by Ada we mean Ada 95) largely solves the problem of writing reusable software components (or at least, because of its excellent ability to prescribe interfaces it provides an enabling technology in which reusable software can be written).

Many years' use of Ada 83 has shown that Ada is living up to its promise of providing a language which can reduce the cost of both the initial development of software and its later maintenance. The main advantage of Ada is simply that it is reliable. The strong typing and related features ensure that programs contain few surprises; most errors are detected at compile time and of those that remain many are detected by run-time constraints. This aspect of Ada considerably reduces the costs and risks of program development compared, for example, with C and its derivatives such as C++. Moreover an Ada compilation system includes the facilities found in separate tools such as lint and make for C. Even if Ada is seen as just another programming language, it reaches parts of the software development process that other languages do not reach.

The essence of Ada 95 is that it adds extra flexibility to the inherent reliability of Ada 83, thereby producing an outstanding language suitable for the development needs of applications well into the new millenium.

Two kinds of applications stand out where Ada is particularly relevant. The very large and the very critical. Very large applications, which inevitably have a long lifetime, require the cooperative effort of large teams. The information hiding properties of Ada, and especially the way in which integrity is maintained across compilation unit boundaries, are invaluable in enabling such developments to progress smoothly. Furthermore, if and when the requirements change and the program has to be modified, the structure, and especially the readability of Ada, enables rapid understanding of the original program even if it has been modified by a different team.

Very critical applications are those that just have to be correct otherwise people or the environment may be damaged. Obvious examples occur in avionics, railway signaling, process control, and medical applications. Such programs may not be large, but they have to be very well understood and often mathematically proven to be correct. The full flexibility of Ada is not appropriate in this case, but the intrinsic reliability of the strongly typed kernel of the language is exactly what is required. Indeed many certification agencies dictate the properties of acceptable languages and while they do not always explicitly demand a subset of Ada, nevertheless the same properties are not provided by any other practically available language.

Ada is thus very appropriate for avionics applications which embrace both the large and the critical.

26.1.2 Abstraction and Freedom

The evolution of programming languages essentially concerns the use of abstraction to hide unnecessary and harmful details of the program.

Thus "expression abstraction" in any language (such as Fortran or Pascal) hides the use of the machine registers to evaluate expressions, and "control abstraction" in Algol and Pascal hides the goto's and labels which had to be explicitly used in early versions of languages such as Fortran. A more recent advance is "data abstraction". This means separating the details of the representation of data from the abstract operations defined upon the data.

Older languages take a very simple view of data types. In all cases the data are directly described in numerical terms. Thus if the data to be manipulated are not really numerical (they could be traffic light colors) then some mapping of the abstract type must be made by the programmer into a numerical type (usually integer). This mapping is purely in the mind of the programmer and does not appear in the written program except perhaps as a comment.

Pascal introduced a certain amount of data abstraction as instanced by the enumeration type. Enumeration types allow us to talk about the traffic light colors in their own terms without having to know how they are represented in the computer.

Another form of data abstraction concerns visibility. It has long been recognized that the traditional block structure of Algol and Pascal is not adequate. For example, it is not possible in Pascal to write two procedures to operate on some common data and make the procedures accessible without also making the data directly accessible. Many languages have provided control of visibility through separate compilation; this technique is adequate for medium-sized systems, but since the separate compilation facility usually depends upon some external system, total control of visibility is not gained. Ada 83 was probably the first practical language to bring together these various forms of data abstraction.

Another language which made an important contribution to the development of data abstraction is Simula 67 with its concept of class. This leads us into the paradigm now known as Object Oriented Programming (OOP) which is currently in vogue. There seems to be no precise definition of OOP, but its essence is a flexible form of data abstraction providing the ability to define new data abstractions in terms of old ones and allowing dynamic selection of types.

All types in Ada 83 are static and thus Ada 83 is not classed as a truly Object Oriented language but as an Object Based language. However, Ada 95 includes all the essential functionality associated with OOP such as polymorphism and type extension.

We are probably too close to the current scene to achieve a proper perspective. It remains to be seen just how useful "object abstraction" actually is. Indeed it might well be that inheritance and other aspects of OOP turn out to be unsatisfactory by obscuring the details of types although not hiding them completely; this could be argued to be an abstraction leak making the problems of program maintenance harder if OOP is overused.

Another concept relevant to the design of languages is freedom. Freedom takes two forms. Freedom to do whatever we want on the one hand, and freedom from dangers and difficulties on the other. In general terms, modern society seems obsessed with freedom to do one's own thing and is less concerned with freedom from unpleasant consequences.

In terms of a programming language we need both freedoms at appropriate points. For large parts of a program we need freedom from inadvertent errors; this is best provided by a controlled framework in which the details of the machine and other parts of the system are hidden through various forms of abstraction. However, there are areas where freedom to get down to the raw hardware is vital; this especially applies to embedded applications where access to interrupt mechanisms and autonomous transfer of data are vital for proper responses to external events.

The merit of Ada is that it provides both kinds of freedom: freedom from errors by the use of abstraction and yet freedom to get at the underlying machine and other systems when necessary.

A brief survey of how Ada relates to other languages would not be complete without mention of C and C++. These have a completely different evolutionary trail than the classic Algol-Pascal-Ada route.

The origin of C can be traced back to the CPL language devised in the early 1960s. From it emerged the simple system programming language BCPL and from that B and then C. The essence of BCPL was the array and pointer model which abandoned any hope of strong typing and (with hindsight) a proper mathematical model of the mapping of the program onto a computing engine. Even the use of : = for assignment was lost in this evolution which reverted to the confusing use of = as in Fortran. About the only feature of the elegant CPL code remaining in C is the unfortunate braces {} and the associated compound statement structure which has now been abandoned by all other languages in favor of the more reliable bracketed form originally proposed by Algol 68.

C is an example of a language which almost has all the freedom to do things, but with little freedom from difficulties. Of course there is a need for a low-level systems language with functionality like C. It is, however, unfortunate that the interesting structural ideas in C++ have been grafted onto the fragile C foundation. As a consequence, although C++ has many important capabilities for data abstraction, including inheritance and polymorphism, it is all too easy to break these abstractions and create programs that violently misbehave or are exceedingly hard to understand and maintain.

The designers of Ada 95 have striven to incorporate the positive dynamic facilities of the kind found in C++ onto the firm foundation provided by Ada 83. Ada 95 is thus an important advance along the evolution of abstraction. It incorporates full object abstraction in a way that is highly reliable without incurring excessive run-time costs.

26.1.3 From Ada 83 to Ada 95

In this section (and especially for the benefit of those familiar with Ada 83), we briefly survey the main changes from Ada 83 to Ada 95. As we said above, one of the great strengths of Ada 83 is its reliability. The strong typing ensures that most errors are detected at compile time while many of those remaining are detected by various checks at run time. Moreover, the compile-time checking extends across compilation unit boundaries. This reliability aspect of Ada considerably reduces the costs and risks of program development (especially for large programs) compared with weaker languages which do not have such a rigorous model.

However, after a number of years' experience it became clear that some improvements were necessary to completely satisfy the present and the future needs of users from a whole variety of application areas. Four main areas were perceived as needing attention.

- Object oriented programming. Recent experience with other languages has shown the benefits of the object oriented paradigm. This provides much flexibility and, in particular, it enables a program to be extended without editing or recompiling existing and tested parts of it.
- Program libraries. The library mechanism is one of Ada's great strengths. Nevertheless the flat structure in Ada 83 is a hindrance to fine visibility control and to program extension without recompilation.
- Interfacing. Although Ada 83 does have facilities to enable it to interface to external systems written in other languages, these have not proved to be as flexible as they might. For example, it has been particularly awkward to program call-back mechanisms which are very useful, especially when using graphical user interfaces.
- Tasking. The Ada 83 rendezvous model provides an advanced description of many paradigms. However, it has not turned out to be entirely appropriate for shared data situations where a static monitor-like approach brings performance benefits. Furthermore, Ada 83 has a rather rigid approach to priorities and it is not easy to take advantage of recent deeper understanding of scheduling theory which has emerged since Ada was first designed.

The first three topics are really all about flexibility, and so a prime goal of the design of Ada 95 has been to give the language a more open and extensible feel without losing its inherent integrity and efficiency. In other words to get a better balance of the two forms of freedom.

The additions to Ada 95 which contribute to this more flexible feel are the extended or tagged types, the hierarchical library, and the greater ability to manipulate pointers or references. The tagged types and hierarchical library together provide very powerful tools for programming by extension.

In the case of the tasking model, the introduction of protected types allows a more efficient implementation of standard paradigms of shared data access. This brings with it the benefits of speed provided by low-level primitives such as semaphores without the risks incurred by the use of such unstructured primitives. Moreover, the clearly data oriented view brought by the protected types fits in naturally with the general spirit of the object oriented paradigm. Other improvements to the tasking model allow a more flexible response to interrupts and other changes of state.

The remainder of this chapter is a brief survey of most of the key features of Ada 95. Some will be familiar to those who know Ada 83, but much will be new or appear in a new light.

26.2 Key Concepts

Ada is a large language since it addresses many important issues relevant to the programming of practical systems in the real world. It is, for instance, much larger than Pascal which, unless it extended in some way, is really only suitable for training purposes (for which it was designed) and for small personal programs. Similarly, Ada is much larger than C although perhaps about the same size as C++. But a big difference is the stress which Ada places on integrity and readability. Some of the key issues in Ada are

- Readability—it is recognized that professional programs are read much more often than they are written. It is important therefore to avoid an overly terse notation which, although allowing a program to be written down quickly, makes it almost impossible to be read except perhaps by the original author soon after it was written.
- Strong typing—this ensures that each object has a clearly defined set of values and prevents confusion between logically distinct concepts. As a consequence, many errors are detected by the compiler which in other languages would have led to an executable but incorrect program.
- Programming in the large—mechanisms for encapsulation, separate compilation, and library management are necessary for writing portable and maintainable programs of any size.
- Exception handling—it is a fact of life that programs of consequence are rarely perfect. It is necessary to provide a means whereby a program can be constructed in a layered and partitioned way so that the consequences of unanticipated events in one part can be contained.
- Data abstraction—as mentioned earlier, extra portability and maintainability can be obtained if the details of the representation of data can be kept separate from the specifications of the logical operations on the data.
- Object oriented programming—in order to promote the reuse of tested code, the type of flexibility associated with OOP is important. Type extension (inheritance), polymorphism, and late binding are all desirable especially if achieved without loss of type integrity.
- Tasking—for many applications it is important that the program be conceived as a series of parallel activities rather than just as a single sequence of actions. Building appropriate facilities into a language rather than adding them later via calls to an operating system gives better portability and reliability.
- Generic units—in many cases the logic of part of a program is independent of the types of the values being manipulated. A mechanism is therefore necessary for the creation of related pieces of program from a single template. This is particularly useful for the creation of libraries.
- Interfacing—programs do not live in isolation and it is important to be able to communicate with systems written in other languages.

An overall theme in designing Ada was concern for the programming process. Programming is a human activity and a language should be designed to be helpful. An important aspect of this is enabling errors to be detected early in the overall process. For example, care was taken that wherever possible a single typographical error would result in a program that does not compile rather than in a program that still compiles but does the wrong thing.

26.2.1 Overall Structure

One of the most important objectives of software engineering is to be able to reuse existing pieces of a program so that the effort of writing new coding is kept to a minimum. The concept of a library of program components naturally emerges, and an important aspect of a programming language is therefore its ability to express how to use the items in a library.

Ada recognizes this situation and introduces the concept of library units. A complete Ada program is assembled as a main subprogram (itself a library unit) which calls upon the services of other library units.

The main subprogram takes the form of a procedure of an appropriate name. The service library units can be subprograms (procedures or functions) but they are more likely to be packages. A package is a group of related items such as subprograms but may contain other entities as well.

Suppose we wish to write a program to print out the square root of some number. We can expect various library units to be available to provide us with a means of computing square roots and doing input and output. Our job is merely to write a main subprogram to incorporate these services as we wish. We will suppose that the square root can be obtained by calling a function in our library whose name is Sqrt. We will also suppose that our library includes a package called Simple_IO containing various simple input-output facilities. These facilities might include procedures for reading numbers, printing numbers, printing strings of characters, and so on.

Our program might look like

```
with Sqrt, Simple_IO;
procedure Print_Root is
  use Simple_IO;
  X: Float;
begin
  Get(X);
  Put(Sqrt(X));
end Print_Root;
```

The program is written as a procedure called Print_Root preceded by a *with* clause giving the names of the library units which it wishes to use. Between **is** and **begin** we can write declarations, and between **begin** and **end** we write statements. Broadly speaking, declarations introduce the entities we wish to manipulate and statements indicate the sequential actions to be performed.

We have introduced a variable X of type Float which is a predefined language type. Values of this type are a set of certain floating point numbers and the declaration of X indicates that X can have values only from this set. In our example a value is assigned to X by calling the procedure Get which is in our package Simple_IO.

Writing

use Simple_IO ;

gives us immediate access to the facilities in the package Simple_IO. If we had omitted this use clause we would have had to write

Simple_IO.Get(X);

in order to indicate where Get was to be found.

The procedure then contains the statement

Put(Sqrt(X));

which calls the procedure Put in the package Simple_IO with a parameter which in turn is the result of calling the function Sqrt with the parameter X.

Some small-scale details should be noted. The various statements and declarations all terminate with a semicolon; this is unlike some other languages such as Pascal where semicolons are separators rather than terminators. The program contains various identifiers such as **procedure**, Put and X. These fall into two categories. A few identifiers (69 in fact) such as **procedure** and **is** are used to indicate the structure of the program; they are reserved and can be used for no other purpose. All others, such as Put and X, can be used for whatever purpose we desire. Some of these, notably Float in our example, have a predefined meaning but we can nevertheless reuse them if we so wish although it might be confusing to do so. For clarity we write the reserved words in lower-case bold and capitalize the others. This is purely a notational convenience; the language rules do not distinguish the two cases except when we consider the manipulation of characters themselves. Note also how the underline character is used to break up long identifiers into meaningful parts.

Finally, observe that the name of the procedure, Print_Root, is repeated between the final **end** and the terminating semicolon. This is optional but is recommended so as to clarify the overall structure, although this is obvious in a small example such as this.

Our program is very simple; it might be more useful to enable it to cater for a whole series of numbers and print out each answer on a separate line. We could stop the program somewhat arbitrarily by giving it a value of zero.

```
with Sqrt, Simple_IO;
procedure Print_Roots is
  use Simple_IO;
  X: Float;
begin
  Put("Roots of various numbers");
  New_Line(2);
  loop
    Get(X);
    exit when X = 0.0;
    Put(" Root of ");
    Put(X);
    Put(" is ");
    if X < 0.0 then
      Put(" not calculable ");
    else
      Put(Sqrt(X));
    end if;
    New_Line;
  end loop;
  New_Line;
  Put("Program finished");
  New_Line;
end Print_Roots;
```

The output has been enhanced by the calls of further procedures New_Line and Put in the package Simple_IO. A call of New_Line will output the number of new lines specified by the parameter (which is of the predefined type Integer); the procedure New_Line has been written in such a way that if no parameter is supplied then a default value of 1 is assumed. There are also calls of Put with a string as argument. This is in fact a different procedure from the one that prints the number X. The compiler knows which is which because of the different types of parameters. Having more than one procedure with the same name is known as overloading. Note also the form of the string; this is a situation where the case of the letters does matter.

Various new control structures are also introduced. The statements between **loop** and **end loop** are repeated until the condition X = 0.0 in the **exit** statement is found to be true; when this is so the loop is finished and we immediately carry on after **end loop**. We also check that X is not negative; if it is we output the message "not calculable" rather than attempting to call Sqrt . This is done by the **if** statement; if the condition between **if** and **then** is true, then the statements between **then** and **else** are executed, otherwise those between **else** and **end if** are executed.

The general bracketing structure should be observed: **loop** is matched by **end loop** and **if** by **end if**. All the control structures of Ada have this closed form, rather than the open form of Pascal and C that can lead to poorly structured and incorrect programs.

We will now consider in outline the possible general form of the function Sqrt and the package Simple_IO that we have been using. The function Sqrt will have a structure similar to that of our main subprogram; the major difference will be the existence of parameters.

```
function Sqrt(F: Float) return Float is
  R: Float;
begin
  -- compute value of Sqrt(F) in R
  return R;
end Sqrt;
```

We see here the description of the formal parameters (in this case only one) and the type of the result. The details of the calculation are represented by the comment which starts with a double hyphen. The return statement is the means by which the result of the function is indicated. Note the distinction between a function which returns a result and is called as part of an expression, and a procedure which does not have a result and is called as a single statement.

The package Simple_IO will be in two parts: the specification which describes its interface to the outside world, and the body which contains the details of how it is implemented. If it just contained the procedures that we have used, its specification might be

```
package Simple_IO is
  procedure Get(F: out Float);
  procedure Put(F: in Float);
  procedure Put(S: in String);
  procedure New_Line(N: in Integer := 1);
end Simple_IO;
```

The parameter of Get is an **out** parameter because the effect of calling Get as in Get(X); is to transmit a value *out* from the procedure to the actual parameter X. The other parameters are all **in** parameters because the value goes *in* to the procedures.

Only a part of the procedures occurs in the package specification; this part is known as the procedure specification and gives just enough information to enable the procedures to be called. We see also the two overloaded specifications of Put, one with a parameter of type Float and the other with a parameter of type String. Finally, note how the default value of 1 for the parameter of New_Line is indicated.

The package body for Simple_IO will contain the full procedure bodies plus any other supporting material needed for their implementation and is naturally hidden from the outside user. In vague outline it might look like

```
with Ada.Text_IO;
package body Simple_IO is
    ...
  procedure Get(F: out Float) is
    ...
  begin
    ...
  end Get;
  -- other procedures similarly
end Simple_IO;
```

The with clause shows that the implementation of the procedures in Simple_IO uses the more general package Ada.Text_IO. The notation indicates that Text_IO is a child package of the package Ada. It should also be noticed how the full body of Get repeats the procedure specification which was given in the corresponding package specification. Note that the package Text_IO really exists whereas Simple_IO is a figment of our imagination made up for the purpose of our example. We will say more about Text_IO in a moment.

The example in this section has briefly revealed some of the overall structure and control statements of Ada. One purpose of this section has been to stress that the idea of packages is one of the most important concepts in Ada. A program should be conceived as a number of components which provide services to and receive services from each other.

Perhaps this is an appropriate point to mention the special package Standard which exists in every implementation and contains the declarations of all the predefined identifiers such as Float and Integer. We can assume access to Standard automatically and do not have to give its name in a with clause.

26.2.2 Errors and Exceptions

We introduce this topic by considering what would have happened in the example in the previous section if we had not tested for a negative value of X and consequently called Sqrt with a negative argument. Assuming that Sqrt has itself been written in an appropriate manner, then it clearly cannot deliver a value to be used as the parameter of Put. Instead an exception will be raised. The raising of an exception indicates that something unusual has happened and the normal sequence of execution is broken. In our case the exception might be Constraint_Error which is a predefined exception declared in the package Standard. If we did nothing to cope with this possibility then our program would be terminated and no doubt the Ada Run Time System will give us a message saying that our program has failed and why. We can, however, look out for an exception and take remedial action if it occurs. In fact we could replace the conditional statement

```
if X < 0.0 then
  Put ("not calculable");
else
  Put (Sqrt (X) );
end if;
```

by

```
begin
  Put (Sqrt (X) );
exception
  when Constraint_Error =>
    Put ("not calculable");
end;
```

This fragment of a program is an example of a block. If an exception is raised by the sequence of statements between **begin** and **exception**, then control immediately passes to the one or more statements following the handler for that exception, and these are obeyed instead. If there were no handler for the exception (it might be another exception such as Storage_Error) then control passes up the flow hierarchy until we come to an appropriate handler or fall out of the main subprogram, which then becomes terminated as we mentioned, with a message from the Run Time System.

The above example is not a good illustration of the use of exceptions since the event we are guarding against can easily be tested for directly. Nevertheless it does show the general idea of how we can look out for unexpected events and leads us into a brief consideration of errors in general.

From the linguistic viewpoint, an Ada program may be incorrect for various reasons. There are two main error categories, according to how they are detected.

- Many errors are detected by the compiler—these include simple punctuation mistakes such as leaving out a semicolon or attempting to violate the type rules such as mixing up colors and fish. In these cases the program is said to be illegal and will not be executed.

- Other errors are detected when the program is executed. An attempt to find the square root of a negative number or divide by zero are examples of such errors. In these cases an exception is raised as we have just seen, and we have an opportunity to recover from the situation.

One of the main goals in the design of Ada was to ensure that human errors fall into the first category most of the time so that their detection and correction is a straightforward matter.

26.2.3 Scalar Type Model

We have said that one of the key benefits of Ada is its strong typing. This is well illustrated by the enumeration type. Consider

```
declare
   type Color is (Red, Amber, Green);
   type Fish is (Cod, Hake, Plaice);
   X, Y: Color;
   A, B: Fish;
begin
   X := Red;            -- ok
   A := Hake;           -- ok
   B := X;              -- illegal
   ...
end;
```

Here we have a block which declares two enumeration types, Color and Fish, and two variables of each type, and then performs various assignments. The declarations of the types gives the allowed values of the types. Thus the variable X can only take one of the three values Red, Amber, or Green. The fundamental rule of strong typing is that we cannot assign a value of one type to a variable of a different type. So we cannot mix up colors and fish and thus our (presumably accidental) attempt to assign the value of X to B is illegal and will be detected during compilation.

There are three enumeration types predefined in the package Standard. One is

```
type Boolean is (False, True);
```

which plays a fundamental role in control flow. Thus the predefined relational operators such as $<$ produce a result of this type and such a value follows **if** as we saw in the construction

```
if X < 0.0 then
```

in the example above. The other predefined enumeration types are Character and Wide_Character. The values of these types are the 8-bit ISO Latin-1 characters and the 16-bit ISO Basic Multilingual Plane characters; these types naturally play an important role in input-output. The literal values of these types include the printable characters and these are represented by placing them in single quotes thus 'X' or 'a' or indeed '''.

The other fundamental types are the numeric types. One way or another, all other data types are built out of enumeration types and numeric types. The two major classes of numeric types are the integer types and floating point types (there are also fixed-point types which are rather obscure and deserve no further mention in this brief overview). The integer types in fact are subdivided into signed integer types (such as Integer) and unsigned or modular types. All implementations will have the types Integer and Float. In addition, if the architecture is appropriate, an implementation may have other predefined numeric types, Long_Integer, Long_Float, Short_Float, and so on. There will also be specific integer types for an implementation depending upon the supported word lengths such as Integer_16 and corresponding unsigned types such as Unsigned_16.

One of the problems of numeric types is how to obtain both portability and efficiency in the face of variations in machine architecture. In order to explain how this is done in Ada it is convenient to introduce the concept of a derived type. (We will deal with derived types in more detail when we come to object oriented programming).

The simplest form of derived type introduces a new type which is almost identical to an existing type except that it is logically distinct. If we write

```
type Light is new Color;
```

then `Light` will, like `Color`, be an enumeration type with literals `Red`, `Amber` and `Green`. However, values of the two types cannot be arbitrarily mixed since they are logically distinct. Nevertheless, in recognition of the close relationship, a value of one type can be converted to the other by explicitly using the destination type name. So we can write

```
declare
  type Light is new Color;
  C: Color;
  L: Light;
begin
  L := Amber; -- the light amber, not the color
  C := Color(L); -- explicit conversion
  ...
end;
```

whereas a direct assignment

$$C := L; \qquad \text{-- illegal}$$

would violate the strong typing rule and this violation would be detected during compilation.

Returning now to our numeric types, if we write

```
type My_Float  is new Float;
```

then `My_Float` will have all the operations ($+$, $-$, etc.) of `Float` and in general can be considered as equivalent. Now suppose we transfer the program to a different computer on which the predefined type `Float` is not so accurate and that `Long_Float` is necessary. Assuming that the program has been written using `My_Float` rather than `Float`, then replacing the declaration of `My_Float` by

```
type My_Float is new Long_Float;
```

is the only change necessary. We can actually do better than this by directly stating the precision that we require, thus

```
type My_Float is digits 7;
```

will cause `My_Float` to be based on the smallest predefined type with at least seven decimal digits of accuracy.

A similar approach is possible with integer types so that rather than using the predefined types `Integer` or `Long_Integer`, we can give the range of values required thusly:

```
type My_Integer is range 21000_000 ... + 1000_000;
```

The point is that it is not good practice to use the predefined numeric types directly when writing professional programs which may need to be portable.

26.2.4 Arrays and Records

Ada naturally enables the creation of composite array and record types. Arrays may actually be declared without giving a name to the underlying type (the type is then said to be anonymous), but records always have a type name.

```
                                    1

                                  1   1

                                1   2   1

                              1   3   3   1

                            1   4   6   4   1

FIGURE 26.1   Pascal's triangle.        1   5   10  10   5   1
```

As an example of the use of arrays suppose we wish to compute the successive rows of Pascal's triangle. This is usually represented as shown in Figure 26.1. The reader will recall that the rows are the coefficients in the expansion of $(1 + x)^n$ and that a neat way of computing the values is to note that each one is the sum of the two diagonal neighbors in the row above.

Suppose that we are interested in the first 10 rows. We could declare an array to hold such a row by

```
Pascal: array (0 .. 10) of Integer;
```

and now assuming that the current values of the array Pascal correspond to row $n-1$, with the component Pascal(0) being 1, then the next row could be computed in a similar array Next by

```
Next(0) := 1;
for I in 1 .. N-1 loop
  Next(I) := Pascal(I-1) + Pascal(I);
end loop;
Next(N) := 1;
```

and then the array Next could be copied into the array Pascal.

This illustrates another form of loop statement where a controlled variable I takes successive values from a range; the variable is automatically declared to be of the type of the range which in this case is Integer. Note that the intermediate array Next could be avoided by iterating backwards over the array; we indicate this by writing **reverse** in front of the range, thus

```
Pascal(N) := 1;
  for I in reverse 1 .. N-1 loop
    Pascal(I) := Pascal(I-1) + Pascal(I);
  end loop;
```

We can also declare arrays of several dimensions. So if we wanted to keep all the rows of the triangle we might declare

```
Pascal2: array (0 .. 10, 0 .. 10) of Integer;
```

and then the loop for computing row n would be

```
Pascal2(N, 0) := 1;
for I in 1 .. N - 1 loop
  Pascal2(N, I) := Pascal2(N-1, I-1) + Pascal2(N-1, I);
end loop;
Pascal2(N, N) := 1;
```

We have declared our arrays without giving a name to their type. We could alternatively have written

```
type Row is array (0 .. Size) of Integer;
Pascal, Next: Row;
```

where we have given the name Row to the type and then declared the two arrays Pascal and Next. There are advantages to this approach as we will see later. Incidentally, the bounds of an array do not have to be constant, they could be any computed values such as the value of some variable Size.

We conclude this brief discussion of arrays by observing that the type String is in fact an array whose components are of the enumeration type Character. Its declaration (in the package Standard) is

type String **is array** (Positive **range** <>) **of** Character;

and this illustrates a form of type declaration which is said to be indefinite because it does not give the bounds of the array; these have to be supplied when an object is declared:

Buffer: String(1 .. 80);

Incidentally, the identifier Positive in the declaration of the type String denotes what is known as a subtype of Integer; values of the subtype Positive are the positive integers and so the bounds of all arrays of type String must also be positive—the lower bound is of course typically 1, but need not be.

A record is an object comprising a number of named components typically of different types. We always have to give a name to a record type. If we were manipulating a number of buffers then it would be convenient to declare a record type containing the buffer and an indication of the start and finish of that part of the buffer actually containing useful data.

```
type Buffer is
  record
    Data: String(1 .. 80);
    Start, Finish: Integer;
  end record;
```

An individual buffer could then be declared by

My_Buffer: Buffer;

and the components of the buffer can then be manipulated using a dotted notation to select the individual components

```
My_Buffer.Start  := 1;
My_Buffer.Finish := 3;
My_Buffer.Data(1 .. 3)  :== "XYZ";
```

Note that this assigns values to the first three components of the array using a so-called slice.

Whole array and record values can be created using aggregates, which are simply a set of values in parentheses separated by commas. Thus we could assign appropriate values to Pascal and My_Buffer by

```
Pascal(0 .. 4)  :=  (1, 4, 6, 4, 1);
My_Buffer  :=  (('X', 'Y', 'Z', others => ' '), 1, 3);
```

where in the latter case we have in fact assigned all 80 values to the array Data and used **others** to indicate that after the three useful characters the remainder of the array is padded with spaces. Note also the nesting of parentheses.

This concludes our brief discussion on simple arrays and records. We will show how record types can be extended in a moment.

26.2.5 Access Types

The last section showed how the scalar types (numeric and enumeration types) may be composed into arrays and records. The other vital means for creating structures is through the use of access types (the Ada name for pointer types); access types allow list processing and are typically used with record types.

The explicit manipulation of pointers or references has been an important feature of most languages since Algol 68. References rather dominated Algol 68 and caused problems, and the corresponding pointer facility in Pascal is rather austere. The pointer facility in C, on the other hand, provides raw flexibility which is open to abuse and quite insecure and thus the cause of many wrong programs.

Ada provides both a high degree of reliability and considerable flexibility through access types. Ada access types must explicitly indicate the type of data to which they refer. The most general form of access types can refer to any data of the type concerned but we will restrict ourselves in this overview to those which just refer to data declared in a storage pool (the Ada term for a heap).

For example, suppose we wanted to declare various buffers of the type in the previous section. We might write

```
type Buffer_Ptr is access Buffer;
Handle: Buffer_Ptr;
...
Handle := new Buffer;
```

This allocates a buffer in the storage pool and sets a reference to it into the variable `Handle`. We can then refer to the various components of the buffer indirectly using the variable `Handle`

```
Handle.Start  := 1;
Handle.Finish := 3;
```

and we can refer to the complete record as `Handle.**all**`. Note that `Handle.Start` is strictly an abbreviation for `Handle.**all**.Start`.

Access types are of particular value for list processing where one record structure contains an access value to another record structure. The classic example is typified by

```
type Cell;
type Cell_Ptr is access Cell;

type Cell is
  record
    Value: Data;
    Next: Cell_Ptr;
end record;
```

The type `Cell` is a record containing a component of some type `Data` plus a component `Next` which can refer to another similar record. Note the partial declaration of the type `Cell`. This is required in the declaration of the type `Cell_Ptr` because of the inherent nature of the circularity of the declarations.

Access types can be used to refer to any type, although records are common. Access types may also be used to refer to subprograms and this is particularly important when communicating with programs in other languages.

26.2.6 Error Detection

We mentioned earlier that an overall theme in the design of Ada was concern for correctness and that errors should be detected early in the programming process. As a simple example consider a fragment of a program controlling the crossing gates on a railroad. First we have an enumeration type describing the state of a signal:

```
type Signal is (Clear, Caution, Danger);
```

and then perhaps

```
if The_Signal = Clear then
    Open_Gates;
    Start_Train;
end if;
```

It is instructive to consider how this might be written in C and then to consider the consequences of various simple programming errors. C does not have enumeration types as such, so the signal values have to be held as an integer (**int**) with code values such as 0, 1, and 2 representing the three states. This already has potential for errors because there is nothing in the C language that can prevent us from assigning a silly value such as 4 to a signal whereas it is not even possible to attempt such a thing when using the Ada enumeration type.

The corresponding text in C is

```
if (The_Signal == 0)
    {Open_Gates();
    Start_Train();
    }
```

It is interesting to consider what would happen in the two languages if we make various typographical errors. Suppose first that we accidentally type an extra semicolon at the end of the first line. The Ada program then fails to compile and the error is immediately drawn to our attention; the C program, however, still compiles and the condition is ignored (since it then controls no statements). The C program consequently always opens the gates and starts the train irrespective of the state of the signal!

Another possibility is that one of the = signs might be omitted in C. The equality then becomes an assignment and also returns the result as the argument for the test. So the program still compiles, the signal is always set clear no matter what its previous state, and of course the gates are then opened and the train is started on its perilous journey. The corresponding error in Ada might be to write : = instead of = and of course the program will then not compile.

However, many errors cannot be detected at compile time. For example using My_Buffer we might write

```
Index: Integer;
...
Index := 81;
...
My_Buffer.Data(Index) := 'x';
```

which attempts to write to the 81st component of the array, which does not exist. Such assignments are checked in Ada at run time and Constraint_Error would be raised so that the integrity of the program is not violated.

The corresponding instructions in C would undoubtedly overwrite an adjacent piece of storage and probably corrupt the value in My_Buffer.Start.

It often happens that variables such as Index can only sensibly have a certain range of values; this can be indicated by introducing a subtype

```
subtype Buffer_Index is Integer range 1 ..80;
Index: Buffer_Index;
```

or by indicating the constraint directly

```
          Index: Integer range 1 .. 80;
```

This has the advantage that the attempt to assign 81 to Index is itself checked and prevented, so the error is detected even earlier.

The reader may feel that such checks will make the program slower. Studies have shown that provided the ranges are appropriate then the overhead should be minimal. For example, having ensured that `Index` cannot have a value out of range there is no need to apply checks to `My_Buffer.Data(Index)` as well. In any event, if we are really confident that the program is correct, the checks can be switched off for production use.

26.3 Abstraction

As mentioned earlier, abstraction in various forms seems to be the key to the development of programming languages.

We saw above how we can declare a type for the manipulation of a buffer

```
type Buffer is
  record
    Data: String(1 .. 80);
    Start: Integer;
    Finish: Integer;
  end record;
```

in which the component `Data` actually holds the characters in the buffer and `Start` and `Finish` index the ends of the part of the buffer containing useful information. We also saw how the various components might be updated and read using normal assignment.

However, such direct assignment is often unwise since the user could inadvertently set inconsistent values into the components or read nonsense components of the array. A much better approach is to create an Abstract Data Type (ADT) so that the user cannot see the internal details of the type but can only access it through various subprogram calls which define an appropriate protocol.

This can be done using a package containing a private type. Let us suppose that the protocol allows us to reload the buffer (possibly not completely full) and to read one character at a time. Consider the following

```
package Buffer_System is    -- visible part

  type Buffer is private;

  procedure Load(B: out Buffer; S: in String);
  procedure Get(B: in out Buffer; C: out Character);

private                     -- private part
  Max: constant Integer := 80;
  type Buffer is
    record
      Data: String(1 .. Max);
      Start: Integer := 1;
      Finish: Integer := 0;
    end record;

end Buffer_System;

package body Buffer_System is

  procedure Load(B: out Buffer; S: in String) is
  begin
    B.Start := 1;
    B.Finish := S'Length;
    B.Data(B.Start .. B.Finish) := S;
```

```
    end Load;

  procedure Get(B: in out Buffer; C: out Character) is
  begin
    C := B.Data(B.Start);
    B.Start := B.Start + 1;
    end Get;
  end Buffer_System;
```

With this formulation the client can only access the information in the visible part of the specification which is the bit before the word **private**. In this visible part the declaration of the type Buffer merely says that it is private and the full declaration then occurs in the private part. There are thus two views of the type Buffer; the external client just sees the partial view whereas within the package the code of the server subprograms can see the full view. The specifications of the server subprograms are naturally also declared in the visible part.

The net effect is that the user can declare and manipulate a buffer by simply writing

```
My_Buffer: Buffer;
...
Load(My_Buffer, Some_String);
...

Get(My_Buffer, A_Character);
```

but the internal structure is quite hidden. There are two advantages: one is that the user cannot inadvertently misuse the buffer and the second is that the internal structure of the private type could be rearranged if necessary; provided that the protocol is maintained, the user program will not need to be changed.

This hiding of information and consequent separation of concerns is very important and illustrates the benefit of data abstraction. The design of appropriate interface protocols is the key to the development and subsequent maintenance of large programs.

The astute reader will note that we have not bothered to ensure that the buffer is not loaded when there is still unread data in it or the string is too long to fit, nor read from when it is empty. We could rectify this by declaring our own exception called perhaps Error in the visible part of the specification thus

```
                      Error: exception;
```

and then check within Load by for example

```
  if S'Length > Max or B.Start <= B.Finish then
    raise Error;
  end if;
```

This causes our own exception to be raised if the buffer is overloaded.

As a minor point note the use of the constant Max so that the literal 80 only appears in one place. Note also the attribute Length which applies to any array and gives the number of its components. The upper and lower bounds of an array S are incidentally given by S'First and S'Last.

Another point is that the parameter Buffer of Get is marked as **in out** because the procedure both reads the initial value of Buffer and updates it.

Finally, note that the components Start and Finish of the record have initial values in the declaration of the record type; these ensure that when a buffer is declared these components are assigned sensible values and thereby indicate that the buffer is empty. An alternative, of course, would be to provide a procedure Reset but the user might forget to call it.

26.3.1   Objects and Inheritance

As its name suggests, object oriented programming (OOP) concerns the idea of programming around objects. A good example of an object in this sense is the variable My_Buffer of the type Buffer in the previous section. We conceive of the object such as a buffer as a coordinated whole and not just as the sum of its components. Indeed, the external user cannot see the components at all but can only manipulate the buffer through the various subprograms associated with the type.

Certain operations upon a type are called the primitive operations of the type. In the case of the type Buffer they are the subprograms declared in the package specification along with the type itself and which have parameters or a result of the type. In the case of a type such as Integer, the primitive operations are those such as + and − which are predefined for the type (and indeed they are declared in Standard along with the type Integer itself and so fit the same model).

Other important ideas in OOP are

- The ability to define one type in terms of another and especially as an extension of another; this is type extension,
- The ability for such a derived type to inherit the primitive operations of its parent and also to add to and replace such operations; this is inheritance,
- The ability to distinguish the specific type of an object at run time from among several related types and in particular to select an operation according to the specific type; this is (dynamic) polymorphism.

In an earlier section we showed how the type Light was derived from Color and also showed how numeric portability could be aided by deriving a numeric type such as My_Float from one of the predefined types. These were very simple forms of inheritance; the new types inherited the primitive operations of the parent; however, the types were not extended in any way and underneath were really the same type. The main benefit of such derivation is simply to provide a different name and thereby to distinguish the different uses of the same underlying type in order to prevent us from inadvertently using a Light when we meant to use a Color.

The more general case is where we wish to extend a type in some way and also to distinguish objects of different types at run time. The most natural form of type for the purposes of extension, of course, is a record where we can consider extension as simply the addition of further components. The other point is that if we need to distinguish the type at run time then the object must contain an indication of its type. This is provided by a hidden component called the tag. Type extension in Ada is thus naturally carried out using tagged record types.

As a simple example suppose we wish to manipulate various kinds of geometrical objects. We can imagine that the kinds of objects form a hierarchy as shown in Figure 26.2.

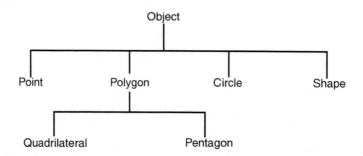

FIGURE 26.2   A hierarchy of geometrical objects.

All objects will have a position given by their *x*- and *y*-coordinates. So we declare the root of the hierarchy as

```ada
type Object is tagged
  record
    X_Coord: Float;
    Y_Coord: Float;
  end record;
```

Note carefully the introduction of the reserved word **tagged**. This indicates that values of the type carry a tag at run time and that the type can be extended. The other types of geometrical objects will be derived (directly or indirectly) from this type. For example we could have

```ada
type Circle is new Object with
  record
      Radius: Float;
  end record;
```

and the type `Circle` then has the three components `X_Coord`, `Y_Coord`, and `Radius`. It inherits the two coordinates from the type `Object` and the component `Radius` is added explicitly.

Sometimes it is convenient to derive a new type without adding any further components. For example:

```ada
type Point is new Object with null record;
```

In this last case we have derived `Point` from `Object` but naturally not added any new components. However, since we are dealing with tagged types we have to explicitly add **with null record**; to indicate that we did not want any new components. This has the advantage that it is always clear from a declaration whether a type is tagged or not.

A private type can also be marked as tagged

```ada
type Shape is tagged private ;
```

and the full type declaration must then (ultimately) be a tagged record

```ada
type Shape is tagged
  record …
```

or derived from a tagged record such as `Object`. On the other hand we might wish to make visible the fact that the type `Shape` is derived from `Object` and yet keep the additional components hidden. In this case we would write

```ada
package Hidden_Shape is
  type Shape is new Object with private;    -- client view
  …
private
  type Shape is new Object with             -- server view
    record
      -- the private components
    end record;
end Hidden_Shape;
```

In this last case it is not necessary for the full declaration of `Shape` to be derived directly from the type `Object`. There might be a chain of intermediate derived types (it could be derived from `Circle`); all that matters is that `Shape` is ultimately derived from `Object`.

The primitive operations of a type are those declared in the same package specification as the type and that have parameters or result of the type. On derivation these operations are inherited by the new type. They can be overridden by new versions and new operations can be added and these then become primitive operations of the new type and are themselves naturally inherited by any further derived type.

Thus we might have declared a function giving the distance from the origin

```
function Distance(O: in Object) return Float is
begin
  return Sqrt(O.X_Coord**2 + O.Y_Coord**2);
end Distance;
```

The type Circle would then sensibly inherit this function. If however, we were concerned with the area of an object then we might start with

```
function Area(O: in Object) return Float is
begin
  return 0.0;
end Area;
```

which returns zero since a raw object has no area. The abstract concept of an area applies also to a circle and so it is appropriate that a function Area be defined for the type Circle. However, to inherit the function from the type Object is clearly inappropriate and so we explicitly declare

```
function Area(C: in Circle) return Float is
begin
  return Pi*C.Radius**2;
end Area;
```

which then overrides the inherited operation. We can perhaps summarize these ideas by saying that the specification is always inherited whereas the implementation may be inherited but can be replaced.

It is possible to convert a value from the type Circle to Object and vice versa. From circle to object is straightforward, we simply write

```
O: Object := (1.0,  0.5);
C: Circle := (0.0,  0.0,  34.7);
...

O := Object(C);
```

which effectively ignores the third component. However, conversion in the other direction requires the provision of a value for the extra component and this is done by an extension aggregate, thus:

$$C := (O \text{ \textbf{with} } 41.2);$$

where the expression O is extended after **with** by the values of the extra components written just as in a normal aggregate. In this case we only had to give a value for the radius.

It is important to remember that the primitive operations are those declared in the same package as the type. Thus the type Object and the functions Distance and Area might be in a package Objects. And then a package Shapes might contain the types Circle and Point and a type Triangle with sides A, B, and C and appropriate functions returning the area. The overall structure would then be

```
package Objects is
  type Object is tagged
    record
      X_Coord: Float;
      Y_Coord: Float;
    end record;
```

```ada
    function Distance(O: Object) return Float;
    function Area(O: Object) return Float;
  end Objects;

package body Objects is
    function Distance(O: Object) return Float is
    begin
      return Sqrt(O.X_Coord**2 + O.Y_Coord**2);
    end Distance;

    function Area(O: Object) return Float is
    begin
      return 0.0;
    end Area;
end Objects;

with Objects; use Objects;
package Shapes is
    type Point is new Object with null record;

    type Circle is new Object with
      record
        Radius: Float;
      end record;

    function Area(C: Circle) return Float;

    type Triangle is new Object with
      record
        A, B, C: Float;
      end record;

    function Area(T: Triangle) return Float;
end Shapes;

package body Shapes is
    function Area(C: Circle) return Float is
    begin
      return Pi * C.Radius**2;
    end Area;

    function Area(T: Triangle) return Float is
      S: constant Float:= 0.5 * (T.A + T.B + T.C);
    begin
      return Sqrt(S * (S - T.A) * (S - T.B) * (S - T.C));
    end Area;
end Shapes;
```

Note that we can put the use clause for Objects immediately after the **with** clause.

26.3.2 Classes and Polymorphism

In the last section we showed how to declare a hierarchy of types derived from the type Object. We saw how on derivation further components and operations could be added and that operations could be replaced.

However, it is very important to note that an operation cannot be taken away nor can a component be removed. As a consequence we are guaranteed that all the types derived from a common ancestor will have all the components and operations of that ancestor.

So in the case of the type Object, all types in the hierarchy derived from Object will have the common components such as their coordinates and the common operations such as Distance and Area. Since they have these common properties it is natural that we should be able to manipulate a value of any type in the hierarchy without knowing exactly which type it is, provided that we only use the common properties. Such general manipulation is done through the concept of a class.

Ada carefully distinguishes between the set of types such as Object plus all its derivatives on the one hand, and an individual type such as Object itself on the other hand. A set of such types is known as a class. Associated with each class is a type called the class-wide type which for the set rooted at Object is denoted by Object'Class. The type Object is referred to as a specific type when we need to distinguish it from a class-wide type.

We can of course have subclasses. For example, Polygon'Class represents the set of all types derived from and including Polygon. This is a subset of the class Object'Class. All the properties of Object'Class will also apply to Polygon'Class but not vice versa. For example, although we have not shown it, the type Polygon will presumably contain a component giving the length of the sides. Such a component will belong to all types of the class Polygon'Class but not to Object'Class.

As a simple example of the use of a class-wide type consider the following function

```
function Moment(OC: Object'Class) return Float is
begin
   return OC.X_Coord * Area(OC);
end Moment;
```

Those who recall their school mechanics will remember that the moment of a force about a fulcrum is the product of the weight multiplied by the distance from the fulcrum. So in our example, taking the *x*-axis as being horizontal, the moment of a geometrical object about the origin is the *x*-coordinate multiplied by the weight which we can take as being proportional to the area (and for simplicity we have assumed is just the area).

This function has a formal parameter of the class-wide type Object'Class. This means it can be called with an actual parameter whose type is any specific type in the class comprising the set of all types derived from Object. Thus we could write

```
C:  Circle ...
M:  Float;
...
M  := Moment(C);
```

Within the function Moment we can naturally refer to the specific object as OC. Since we know that the object must be of a specific type in the Object class, we are guaranteed that it will have a component OC.X_Coord. Similarly we are guaranteed that the function Area will exist for the type of the object since it is a primitive operation of the type Object and will have been inherited by (and possibly overridden for) every type derived from Object. So the appropriate function Area is called and the result multiplied by the *x*-coordinate and returned as the result of the function Moment.

Note carefully that the particular function Area to be called is not known until the program executes. The choice depends upon the specific type of the parameter and this is determined by the tag of the object passed as actual parameter; remember that the tag is a sort of hidden component of the tagged type. This selection of the particular subprogram according to the tag is known as *dispatching* and is a vital aspect of the dynamic behavior provided by polymorphism.

Dispatching only occurs when the actual parameter is of a class-wide type; if we call Area with an object of a specific type such as C of type Circle, then the choice is made at compile time. Dispatching is often called late binding because the call is only bound to the called subprogram late in the compile-link-execute process. The binding to a call of Area with the parameter C of type Circle is called static binding because the subprogram to be called is determined at compile time.

Observe that the function Moment is not a primitive operation of any type; it is just an operation of Object'Class and it happens that a value of any specific type derived from Object can be implicitly converted to the class-wide type. Class wide types do not have primitive operations and so no inheritance is involved.

It is interesting to consider what would have happened if we had written

```
function Moment(O: Object) return Float is
begin
  return O.X_Coord * Area(O);
end Moment;
```

where the formal parameter is of the specific type Object. This always returns zero because the function Area for an Object always returns zero. If this function Moment were declared in the same package as Object then it would be a primitive operation of Object and thus inherited by the type Circle. However, the internal call would still be to the function Area for the type Object and not to the type Circle and so the answer would still be zero. This is because the binding is static and inheritance simply passes on the same code. The code mechanically works on a Circle because it only uses the Object part of the circle (we say it sees the Object view of the Circle); but unfortunately it is not what we want. Of course, we could override the inherited operation by writing

```
function Moment(C: Circle) return Float is
begin
  return C.X_Coord * Area(C);
end Moment;
```

but this is tedious and causes unnecessary duplication of similar code. The proper approach for such general situations is to use the original class-wide version with its internal dispatching; this can be shared by all types without duplication and always calls the appropriate function Area.

A major advantage of using a class-wide operation such as Moment is that a system using it can be written, compiled, and tested without knowing all the specific types to which it is to be applied. Moreover, we can then add further types to the system without recompilation of the existing tested system.

For example we could add a further type

```
type Pentagon is new Object with...
function Area(P: Pentagon) return Float;
...
Star: Pentagon := ...
...
Put("Moment of star is");
Put(Moment(Star));
```

and then the old existing tried and tested Moment will call the new Area for the Pentagon without being recompiled. (It will of course have to be relinked.)

This works because of the mechanism used for dynamic binding; the essence of the idea is that the class-wide code has dynamic links into the new code and this is accessed via the tag of the type. This creates a very flexible and extensible interface ideal for building up a system from reusable components.

One problem with class-wide types is that we cannot know how much space might be occupied by an arbitrary object of the type because the type might be extended. So although we can declare an object of a class-wide type it has to be initialized and thereafter that object can only be of the specific type of that initial value. Note that a formal parameter of a class-wide type such as in Moment is allowed because the space is provided by the actual parameter.

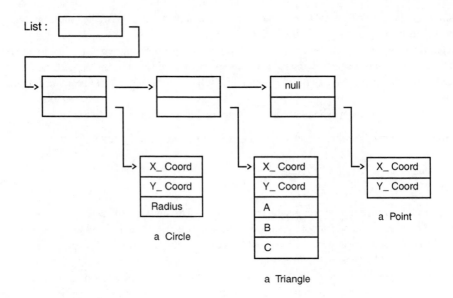

FIGURE 26.3 A chain of objects.

Another similar restriction is that we cannot have an array of class-wide components (even if initialized) because the components might be of different specific types of different sizes and thus impossible to index efficiently.

One consequence of these necessary restrictions is that it is very natural to use access types with object oriented programming since there is no problem with pointing to objects of different sizes at different times. Thus, suppose we wanted to manipulate a series of geometrical objects; it is very natural to declare these in free storage as required. They can then be chained together on a list for processing. Consider

```
type Pointer is access Object'Class;
type Cell;
type Cell_Ptr is access Cell;

type Cell is
  record
    Next: Cell_Ptr;
    Element: Pointer;
  end record;
```

which enables us to create cells which can be linked together; each cell has an element which is a pointer to any geometrical object.

We can now imagine that a number of objects have been created and linked together to form a list as in Figure 26.3. We assume that this chain is accessed through a variable called List of the type Cell_Ptr.

We can now easily process the objects on the list and might, for example, compute the total moment of the set of objects by calling the following function

```
function Total_Moment(The_List: Cell_Ptr) return Float is
  Local: Cell_Ptr := The_List;
  Result: Float := 0.0;
begin
  loop
    if Local = null then        -- end of list
      return Result;
```

```
      end if;
      Result := Result + Moment(Local.Element.all);
      Local := Local.Next;
   end loop;
 end Total_Moment;
```

We conclude this brief survey of the OOP facilities in Ada by considering abstract types. It is sometimes the case that we would like to declare a type as the foundation for a class of types with certain common properties, but without allowing objects of the original type to be declared. For example, we probably would not want to declare an object of the raw type Object. If we wanted an object without any area then it would be appropriate to declare a Point. Moreover, the function Area for the type Object is dubious since it usually has to be overridden anyway. But it is important to be able to ensure that all types derived from Object do have an Area so that the dispatching in the function Moment always works. We can achieve this by writing

```
package Objects is
  type Object is abstract tagged
    record
      X_Coord: Float;
      Y_Coord: Float;
    end record;

  function Distance(O: in Object) return Float;
  function Area(O: in Object) return Float is abstract;
end Objects;

package body Objects is
  function Distance(O: Object) return Float is
  begin
    return Sqrt(O.X_Coord ** 2 + O.Y_Coord ** 2);
  end Distance;
end Objects;
```

In this formulation the type Object and the function Area are marked as abstract. It is illegal to declare an object of an abstract type, and an abstract subprogram has no body and so cannot be called. On deriving a concrete (that is, nonabstract) type from an abstract type any abstract inherited operations must be overridden by concrete operations. Note that we have declared the function Distance as not abstract; this is largely because we know that it will be appropriate anyway.

This approach has a number of advantages; we cannot declare a raw Object by mistake, we cannot inherit the silly function Area, and we cannot make the mistake of declaring the function Moment for the specific type Object (because it would then contain a call of the abstract function Area).

But despite the type being abstract, we can declare the function Moment for the class-wide type Object'Class. This always works because of the rule that we cannot declare an object of an abstract type; any actual object passed as parameter must be of a concrete type and will have an appropriate function Area to which it can dispatch.

26.3.3 Genericity

We have seen how class-wide types provide us with dynamic polymorphism. This means that we can manipulate several types with a single construction and that the specific type is determined dynamically, that is at run time. In this section we introduce the complementary concept of static polymorphism, where again we have a single construction for the manipulation of several types but in this case the choice of type is made statically at compile time.

An important objective of Software Engineering is to reuse existing software components. However, the strong typing model of Ada (even with class-wide types) sometimes gets in the way unless we have a method of writing software components which can be used for various different types. For example, the program to do a sort is largely independent of what it is sorting — all it needs is a rule for comparing the values to be sorted.

So we need a means of writing pieces of software which can be parameterized as required for different types. In Ada this is done by the generic mechanism. We can make a package or subprogram generic with respect to one or more parameters which can include types. Such a generic unit provides a template from which we can create genuine packages and subprograms by so-called instantiation. The full details of Ada generics are quite extensive but the following brief introduction will illustrate the ideas.

The standard package for the input and output of floating point values in text form is generic with respect to the actual floating type. This is because we want a single package to cope with all the possible floating types, such as the underlying machine types `Float` and `Long_Float` as well as the portable type `My_Float`. Its specification is

```
generic
  type Num is digits <>;
package Float_IO is
  ...
  procedure Get(Item: out Num; ...);
  procedure Put(Item: in Num; ...);
  ...
end Float_IO;
```

where we have omitted various details relating to the format. The one generic parameter is Num and the notation **digits** <> indicates that it must be a floating point type and echoes the declaration of `My_Float` using **digits** 7.

In order to create an actual package to manipulate values of the type `My_Float`, we write

```
package My_Float_IO is new Float_IO(My_Float);
```

which creates a package with the name `My_Float_IO` where the formal type `Num` has been replaced throughout with our actual type `My_Float`. As a consequence, procedures `Get` and `Put` taking parameters of the type `My_Float` are created and we can then call these as required.

The kind of parameterization provided by genericity is similar but rather different to that provided through class-wide types. In both cases the parameterization is over a related set of types with a set of common properties. Such a related set of types is termed a class.

A common form of class is a derivation class where all the types are derived from a common ancestor such as the type `Object`. Tagged derivation classes form the basis of dynamic polymorphism as we have seen.

But there are also broader forms of class such as the set of all floating-point types which are allowed as actual parameters for the generic package `Float_IO`. The parameters of generic units use these broader classes. For example, a very broad class is the set of all types having assignment. Such a class would be a suitable basis for writing a generic sort routine.

The two forms of polymorphism work together; a common form of generic package is one which takes as a parameter a type from a tagged derivation class. This may be used to provide important capabilities such as multiple inheritance.

26.3.4 Object Oriented Terminology

It is perhaps convenient at this point to compare the Ada terminology with that used by other object oriented (OO) languages such as Smalltalk and C++.

About the only term in common is inheritance. Ada 83 has always had inheritance although not type extension, which only occurs with tagged record types. Untagged record types are called structs in some languages. Ada actually uses the term object to denote variables and constants in general, whereas in the OO sense an object is an instance of an Abstract Data Type (ADT).

Many languages use *class* to denote what Ada calls a specific tagged type (or more strictly, an ADT consisting of a tagged type plus its primitive operations). The reason for this difference is that Ada uses the word class exclusively to refer to a group of related types. Ada classes are not just those groups of types related by derivation, but also groups with broader correspondence as used for generic parameter matching.

The Ada approach clarifies the distinction between the group of types and a single type and strengthens that clarification by introducing class-wide types as such. Some languages use the term class for both specific types and the group of types, with much resulting confusion in terms of description but also in understanding the behavior of the program and keeping track of the real nature of an object.

Primitive operations of Ada tagged types are often called methods or virtual functions. The call of a primitive operation in Ada is bound statically or dynamically according to whether the parameter is of a specific type or class wide. The rules in C++ are more complex and depend upon whether the parameter is a pointer and also whether the call is prefixed by its class name. In Ada, an operation with class-wide formal parameters is always bound statically although it applies to all types of the class.

Dispatching is the Ada term for calling a primitive operation with dynamic binding, and indeed subprogram calls through access types are also a form of dynamic binding.

Abstract types correspond to abstract class in many languages. Abstract subprograms are pure virtual member functions in C++. Note that Eiffel uses the term deferred rather than abstract.

Ancestor or parent type and descendant or derived type become superclass and subclass. The Ada concept of subtype (a type with a constrained set of values) has no correspondence in other languages which do not have range checks, and has no relationship to subclass. A subtype can never have more values than its base type, whereas a descendant type (subclass to other languages) can never have fewer values than its parent type.

Generic units are templates; in Ada they carry type checking with them, whereas many languages treat templates as raw macros.

Another important point is that many languages have no encapsulation mechanism other than so-called classes, whereas Ada has the package largely unrelated to type extension and inheritance and the private type. The effect of private and protected operations in C++ is provided in Ada by a combination of private types and child packages; the latter are kinds of friends.

26.3.5 Tasking

No survey of abstraction in Ada would be complete without a brief mention of tasking. It is often necessary to write a program as a set of parallel activities rather than just as one sequential program.

Most programming languages do not address this issue at all. Some argue that the underlying operating system provides the necessary mechanisms and that they are therefore unnecessary in a programming language. Such arguments do not stand up to careful examination for two main reasons:

- Built-in syntactic constructions provide a degree of reliability which cannot be obtained through a series of individual operating system calls.
- General operating systems do not provide the degree of control and timing required by many applications.

An Ada program can thus be written as a series of interacting tasks. There are two main ways in which tasks can communicate: directly by sending messages to each other and indirectly by accessing shared data. Direct communication between Ada tasks is achieved by one task calling an entry in another task. The calling (client) task waits while the called (server) task executes an accept statement in response to the call; the two tasks are closely coupled during this interaction, which is called a rendezvous. Controlled access to shared data is vital in tasking applications if interference is to be avoided. For example, returning to the character buffer example, it would be a disaster if a task started to read the buffer while another task was updating it with further information since the component B.Start could be changed and a component of the Data array read by Get before the buffer had been correctly updated by Load. Ada prevents such interference by a construction known as a protected object.

The general syntactic form of both tasks and protected objects is similar to that of a package. They have a specification part prescribing the interface, a private part containing hidden details of the interface, and a body stating what they actually do. The general client-server model can thus be expressed as

```ada
task Server is
  entry Some_Service(Formal: Data);
end;

task body Server is
begin
  ...
  accept Some_Service(Formal: Data) do
    -- statements providing the service
    ...
  end;
  ...
end Server;

task Client;

task body Client is
begin
  ...
  Server.Some_Service(Actual);
  ...
end Client;
```

A good example of the form of a protected object is given by the buffer example, which could be rewritten as follows

```ada
protected type Buffer(Max: Integer) is      -- visible part
  procedure Load(S: in String);
  procedure Get(C: out Character);
private                                      -- private part
  Data: String(1 .. Max);
  Start: Integer := 1;
  Finish: Integer := 0;
end Buffer;

protected body Buffer is

  procedure Load(S: in String) is
  begin
    Start := 1;
    Finish := S'Length;
    Data(Start .. Finish) := S;
  end Load;
```

```
  procedure Get(C: out Character) is
  begin
    C := Data(Start);
    Start := Start + 1;
  end Get;

end Buffer;
```

This construction uses a slightly different style to the package. It is a type in its own right whereas the package exported the type. As a consequence, the calls of the procedures do not need to pass explicitly the parameter referring to the buffer and, moreover, within their bodies the references to the private data are naturally taken to refer to the current instance. Note also that the type is parameterized by the discriminant Max and so we can supply the actual size of a particular buffer when it is declared. Statements using the protected type might thus look like

```
B: Buffer(80);
...
B.Load(Some_String);
...
B.Get(A_Character);
```

Although this formulation prevents disastrous interference between several clients, nevertheless it does not prevent a call of Load from overwriting unread data. As before we could insert tests and raise an exception. But the proper approach is to cause the tasks to wait if circumstances are not appropriate. This is done through the use of entries and barriers. The protected type might then be

```
protected type Buffer(Max: Integer) is
  entry Load(S: in String);
  entry Get(C: out Character);
private
  Data: String(1 .. Max);
  Start: Integer := 1;
  Finish: Integer := 0;
end Buffer;

protected body Buffer is

  entry Load(S: in String) when Start > Finish is
  begin
    Start := 1;
    Finish := S'Length;
    Data(Start .. Finish) := S;
  end Load;

  entry Get(C: out Character) when Start <= Finish is
  begin
    C := Data(Start);
    Start := Start + 1;
  end Get;

end Buffer;
```

In this formulation, the procedures are replaced by entries and each entry body has a barrier condition. A task calling an entry is queued if the barrier is false and is only allowed to proceed when the barrier becomes true. This construction is very efficient because task switching is minimized.

26.4 Programs and Libraries

A complete program is put together out of various separately compiled units. In developing a very large program it is inevitable that it will be conceived as a number of subsystems, themselves each composed from a number of separately compiled units.

We see at once the risk of name clashes between the various parts of the total. It would be all too easy for the designers of different parts of the system to reuse popular package names such as `Error_Messages` or `Debug_Info`, and so on. In order to overcome this and other related problems, Ada has a hierarchical naming scheme at the library level. Thus, a package `Parent` may have a child package with the name `Parent.Child`.

We immediately see that if our total system breaks down into a number of major parts such as acquisition, analysis, and report, then name clashes will be avoided if it is mapped into three corresponding library packages plus appropriate child units. There is then no risk of a clash between `Analysis.Debug_Info` and `Report.Debug_Info` because the names are quite distinct.

The naming is hierarchical and can continue to any depth. A good example of the use of this hierarchical naming scheme is found in the standard libraries which are provided by every implementation of Ada.

We have already mentioned the package `Standard` which is an intrinsic part of the language. All library units can be considered to be children of `Standard` and it should never be necessary to explicitly mention `Standard` at all.

In order to reduce the risk of clashes with the user's own names, the predefined library comprises just three packages, each of which has a number of children. The three packages are `System`, which is concerned with the control of storage and similar implementation matters; `Interfaces`, which is concerned with interfaces to other languages and the intrinsic hardware types; and finally `Ada` which contains the bulk of the predefined library.

In this overview we will briefly survey the main package `Ada`. The package `Ada` itself is simply

```
package Ada is
  pragma Pure(Ada);
end Ada;
```

and the various predefined units are children of `Ada`. The pragma indicates that `Ada` has no variable state (this concept is important for sharing in distributed systems, a topic well outside the scope of this discussion).

Important child packages of `Ada` are

`Numerics`; this contains the mathematical library providing the various elementary functions, random number generators, and facilities for complex numbers.

`Characters`; this contains various packages for classifying and manipulating characters as well as the names of all the characters in the Latin-1 set.

`Strings`; this contains packages for the manipulation of strings of various kinds: fixed length, bounded, and unbounded.

`Text_IO`, `Sequential_IO`, and `Direct_IO`; these and other packages provide a variety of input-output facilities.

Those familiar with Ada 83 will note that the predefined library units of Ada 83 have now become child units of `Ada`. Compatibility is achieved because of the predefined renamings of these child units as library units such as

```
with Ada.Text_IO;
package Text_IO renames Ada.Text_IO;
```

Most library units are packages and it is easy to think that all library units must be packages; indeed only a package can have child units. Of course, the main subprogram is a library subprogram and any library package can have child subprograms. A library unit can also be a generic package or subprogram and even an instantiation of a generic package or subprogram; this latter fact is often overlooked.

We have introduced the hierarchical library as simply a naming mechanism. It also has important information hiding and sharing properties. An example is that a child unit can access the information in the private part of its parent; but of course other units cannot see into the private part of a package. This and related facilities enable a group of units to share private information while keeping the information hidden from external clients.

It should also be noted that a child package does not need to have a *with* clause or *use* clause for its parent; this emphasizes the close relationship of the hierarchical structure and parallels the fact that we never need a *with* clause for Standard because all units are children of Standard.

26.4.1 Input-Output

The Ada language is defined in such a way that all input and output is performed in terms of other language features. There are no special intrinsic features just for input and output. In fact input-output is just a service required by a program and so is provided by one or more Ada packages. This approach runs the attendant risk that different implementations will provide different packages and program portability will be compromised. In order to avoid this, the language defines certain standard packages that will be available in all implementations. Other, more elaborate, packages may be appropriate to special circumstances and the language does not prevent this. Indeed very simple packages such as our purely illustrative Simple_IO may also be appropriate. We will now briefly describe how to use some of the features for the input and output of simple text.

Text input-output is performed through the use of the standard package Ada.Text_IO. Unless we specify otherwise, all communication will be through two standard files, — one for input and one for output — and we will assume that (as is likely the case for most implementations) these are such that input is from the keyboard and output is to the screen. The full details of Text_IO cannot be described here but if we restrict ourselves to just a few useful facilities it looks a bit like

```
with Ada.IO_Exceptions;
package Ada.Text_IO is
  type Count is …          -- an integer type
  …
  procedure New_Line(Spacing: in Count := 1);
  …
  procedure Get(Item: out Character);
  procedure Put(Item: in Character);
  procedure Put(Item: in String);
  …
  -- the package Float_IO outlined above
  -- plus a similar package Integer_IO
  …
end Ada.Text_IO;
```

Note first that this package commences with a *with* clause for the package Ada.IO_Exceptions. This further package contains the declaration of a number of exceptions relating to a variety of things which can go wrong with input-output. For most programs the most likely problem to arise is probably Data_Error, which would occur, for example, if we tried to read in a number from the keyboard but then accidentally typed in something which was not a number at all or was in the wrong format.

The next thing to note is the outline declaration of the type Count. This is an integer type having similar properties to the type Integer and almost inevitably with the same implementation (just as the type My_Integer might be based on Integer). The parameter of New_Line is of the type Count rather than plain Integer, although since the parameter will typically be a literal such as 2 (or be omitted so that the default of 1 applies) this will not be particularly evident.

A single character can be output by, for example:

```
Put ('A');
```

and a string of characters by

```
Put ("This Is a string of characters");
```

A value of the type My_Float can be output in various formats. But first we have to instantiate the package Float_IO mentioned above, and which is declared inside Ada.Text_IO. Having done that, we can call Put with a single parameter, the value of type My_Float to be output, in which case a standard default format is used, or we can add further parameters controlling the format.

If we do not supply any format parameters then an exponent notation is used with 7 significant digits — 1 before the point and 6 after (the 7 matches the precision given in the declaration of My_Float). There is also a leading space or minus sign. The exponent consists of the letter E followed by the exponent sign ($+$ or $-$) and then a two digit decimal exponent.

We can override the default by providing three further parameters which give, respectively, the number of characters before the point, the number of characters after the point, and the number of characters after E. However, there is still only one digit before the point. If we do not want exponent notation then we simply specify the last parameter as zero and then get normal decimal notation.

The effect is shown by the following statements with the output given as a comment. For clarity, the output is surrounded by quotes and s designates a space; in reality there are no quotes and spaces are spaces.

```
Put (12.34);              -- "s1.234000E+01"
Put (12.34, 3, 4, 2);     -- "ss1.2340E+1"
Put (12.34, 3, 4, 0);     -- "s12.3400"
```

The output of values of integer types follows a similar pattern. In this case we similarly instantiate the generic package Integer_IO inside Ada.Text_IO which applies to all integer types with the particular type such as My_Integer.

We can then call Put with a single parameter, the value of type My_Integer, in which case a standard default field is used, or we can add a further parameter specifying the field. The default field is the smallest that will accommodate all values of the type My_Integer allowing for a leading minus sign. Thus, for the range of My_Integer, the default field is 8. It should be noticed that if we specify a field which is too small then it is expanded as necessary. So

```
Put (123);           -- "sssss123"
Put (123, 4);        -- "s123"
Put (123, 0);        -- "123"
```

Simple text input is similarly performed by a call of Get with a parameter that must be a variable of the appropriate type.

A call of Get with a floating or integer parameter will expect us to type in an appropriate number at the keyboard; this must have a decimal point if the parameter is of a floating type. It should also be noted that leading blanks (spaces) and newlines are skipped. A call of Get with a parameter of type Character will read the very next character, and this can be neatly used for controlling the flow of an interactive program, thus

```
C: Character;
...
Put("Do you want to stop?  Answer Y if so.");
Get(C);
if C ='Y' then
    ...
```

For simple programs that do not have to be portable, the effort of doing the instantiations is not necesssary if we just use the predefined types Integer and Float, since the predefined library contains nongeneric versions with the names Ada.Integer_Text_IO and Ada.Float_Text_IO, respectively. So all we need to write is

```
use Ada.Integer_Text_IO, Ada.Float_Text_IO;
```

and then we can call Put and Get without more ado.

26.4.2 Numeric Library

The numeric library comprises the package Ada.Numerics plus a number of child packages. The package Ada.Numerics is as follows:

```
package Ada.Numerics is
  pragma Pure(Numerics);
  Argument_Error: exception;
  Pi: constant := 3.14159_26535_ ...;
  e : constant := 2.71828_18284_ ...;
end Ada.Numerics;
```

This contains the exception Argument_Error which is raised if something is wrong with the argument of a numeric function (such as attempting to take the square root of a negative number) and the two useful constants Pi and e.

One child package of Ada.Numerics provides the familiar elementary functions such as Sqrt and is another illustration of the use of the generic mechanism. Its specification is

```
generic
  type Float_Type is digits <>;
package Ada.Numerics.Generic_Elementary_Functions is
  function Sqrt(X: Float_Type'Base) return Float_Type'Base;
  ...    -- and so on
end;
```

Again, there is a single generic parameter giving the floating type. In order to call the function Sqrt we must first instantiate the generic package much as we did for Float_IO, thus (assuming appropriate *with* and *use* clauses)

```
package My_Elementary_Functions is
  new Generic_Elementary_Functions(My_Float);
use My_Elementary_Functions;
```

and we can then write a call of Sqrt directly.

It should be noted that there is a nongeneric version for the type Float with the name Elementary_Functions for the convenience of those using the predefined type Float.

A little point to note is that the parameter and result of Sqrt is written as Float_Type'Base; the reason for this is to avoid unnecessary range checks.

We emphasize that the exception Ada.Numerics.Argument_Error is raised if the parameter of a function such as Sqrt is unacceptable. This contrasts with our hypothetical function Sqrt introduced earlier, which we assumed raised the predefined exception Constraint_Error when given a negative parameter. It is generally better to declare and raise our own exceptions rather than use the predefined ones.

Two other important child packages are those for the generation of random numbers. One returns a value of the type Float within the range 0 to 1 and the other returns a random value of a discrete type (a discrete type is an integer type or an enumeration type). We will look superficially at the latter; its specification is as follows

```
generic
   type Result_Subtype is (<>);
package Ada.Numerics.Discrete_Random is
   type Generator is limited private;
   function Random(Gen: Generator) return Result_Subtype;
   ... -- plus other facilities
end Ada.Numerics.Discrete_Random;
```

This introduces a number of new points. The most important is the form of the generic formal parameter which indicates that the actual type must be a discrete type. The pattern echoes that of an enumeration type in much the same way as that for the floating generic parameter in Generic_Elementary_Functions echoed the declaration of a floating type. Thus we see that the discrete types are another example of a class of types.

A small point is that the type Generator is declared as limited. This simply means that assignment is not available for the type (or at least not for the partial view as seen by the client).

The random number generator is used as in the following fragment which illustrates the simulation of tosses of a coin

```
use Ada.Numerics;
type Coin is (Heads, Tails);
package Random_Coin is new Discrete_Random(Coin);
use Random_Coin;
G: Generator;
C: Coin;
loop
   C := Random(G);
   ...
end loop;
...
```

Having declared the type Coin we then instantiate the generic package. We then declare a generator and use it as the parameter of successive calls of Random. The generator technique enables us to declare several generators and thus run several independent random sequences at the same time.

26.4.3 Running a Program

We are now in a position to put together a complete program using the proper input-output facilities. As an example, we will rewrite the procedure Print_Roots and also use the standard mathematical library. For simplicity we will first use the predefined type Float. The program becomes

```
with Ada.Text_IO;
with Ada.Float_Text_IO;
with Ada.Numerics.Elementary_Functions;
procedure Print_Roots is
   use Ada.Text_IO;
```

```ada
  use Ada.Float_Text_IO;
  use Ada.Numerics.Elementary_Functions;

  X: Float;
begin
  Put("Roots of various numbers");

  ... -- and so on as before

end Print_Roots;
```

Note that we can put the *use* clauses inside the procedure Print_Roots as shown or we can place them immediately after the *with* clauses.

If we want to write a portable version, then the general approach would be

```ada
with Ada.Text_IO;
with Ada.Numerics.Generic_Elementary_Functions;
procedure Print_Roots is
  type My_Float is digits 7;
  package My_Float_IO is new Ada.Text_IO.Float_IO(My_Float);
  use My_Float_IO;
  package My_Elementary_Functions is
      new Ada.Numerics.Generic_Elementary_Functions(My_Float);
  use My_Elementary_Functions;

  X: My_Float;
begin
    Put("Roots of various numbers");

  ... -- and so on as before

end Print_Roots;
```

To have to write all that introductory stuff each time is rather a burden, so we will put it in a standard package of our own and then compile it once so that it is permanently in our program library and can then be accessed without more ado. We include the type My_Integer as well, and write

```ada
with Ada.Text_IO;
with Ada.Numerics.Generic_Elementary_Functions;
package Etc is
  type My_Float is digits 7;
  type My_Integer is range -1000_000 .. +1000_000;
  package My_Float_IO is new Ada.Text_IO.Float_IO(My_Float);
  package My_Integer_IO is new Ada.Text_IO.Integer_IO(My_Integer);

  package My_Elementary_Functions is
      new Ada.Numerics.Generic_Elementary_Functions(My_Float);
end Etc;
```

and having compiled Etc our typical program can look like

```ada
with Ada.Text_IO, Etc;
use Ada.Text_IO, Etc;
procedure Program is
  use My_Float_IO, My_Integer_IO, My_Elementary_Functions;
  ...
  ...
end Program;
```

An alternative approach, rather than declaring everything in the one package Etc, is to first compile a tiny package just containing the types My_Float and My_Integer and then to compile the various instantiations as individual library packages (remember that a library unit can be just an instantiation).

```ada
package My_Numerics is
  type My_Float is digits 7;
  type My_Integer is range -1000_000 .. +1000_000;
end My_Numerics;

with My_Numerics; use My_Numerics;
with Ada.Text_IO;
package My_Float_IO is new Ada.Text_IO.Float_IO(My_Float);

with My_Numerics; use My_Numerics;
with Ada.Text_IO;
package My_Integer_IO is new Ada.Text_IO.Integer_IO(My_Integer);

with My_Numerics; use My_Numerics;
with Ada.Numerics.Generic_Elementary_Functions;
package My_Elementary_Functions is
  new Ada.Numerics.Generic_Elementary_Functions(My_Float);
```

With this approach we only need to include (via with clauses) the particular packages as required and our program is thus likely to be smaller if we do not need them all. We could even arrange the packages as an appropriate hierarchy, thus:

```ada
with Ada.Text_IO;
package My_Numerics.My_Float_IO is
    new Ada.Text_IO.Float_IO(My_Float);

with Ada.Text_IO;
package My_Numerics.My_Integer_IO is
    new Ada.Text_IO.Integer_IO(My_Integer);

with Ada.Numerics.Generic_Elementary_Functions;
package My_Numerics.My_Elementary_Functions is
    new Ada.Numerics.Generic_Elementary_Functions(My_Float);
```

One important matter remains to be addressed, and that is how to build a complete program. A major benefit of Ada is that consistency is maintained between separately compiled units so that the integrity of strong typing is preserved across compilation unit boundaries. It is therefore illegal to build a program out of inconsistent units. The exact means whereby this is achieved will depend upon the implementation.

A related issue is the order in which units are compiled. This is dictated by the idea of dependency. There are three main causes of dependency:

- A body depends upon the corresponding specification
- A child depends upon the specification of its parent
- A unit depends upon the specifications of those it mentions in a *with* clause

The key rule is that a unit can only be compiled if all those units on which it depends are present in the library environment. An important consequence is that although a specification and body can be compiled separately, the specification must always be present before the body can be compiled. Similarly if a unit is modified, then typically all those units that depend on it will also have to be recompiled in order to preserve consistency of the total program.

This brings us to the end of our brief survey of the main features of Ada. We have in fact encountered most of the main concepts, although very skimpily in some cases.

Important topics not discussed are

- Type parameterization through the use of discriminants. This is particularly important in OOP for enabling one type to be parameterized by another and provides the capabilities of multiple inheritance.
- More general access types. These provide the flexibility to reference objects on the stack as well as the heap. However, the rules are carefully designed to avoid "dangling references" without undue loss of flexibility. Access to subprogram types are important for interaction with software written in other languages and especially for programming call-back.
- Limited types. These are types which cannot be assigned. They are particularly helpful for modeling objects in the real world which by their nature are never copied. They are also useful for controlling resources.
- Controlled types. These are forms of tagged types which have automatic initialization and finalization. They enable the reliable programming of servers that keep track of resources used by clients in a foolproof manner. Controlled types also provide the capability for user-defined assignment.

It should also be noted that we have not discussed the specialized annexes at all. These provide important capabilities in a variety of areas. Very briefly, their contents are as follows:

Systems Programming. This covers a number of low-level features such as in-line machine instructions, interrupt handling, shared variable access, task identification, and per-task attributes. This annex is a prior requirement for the Real-Time Systems annex.

Real-Time Systems. This annex addresses various scheduling and priority issues including setting priorities dynamically, scheduling algorithms, and entry queue protocols. It also includes detailed requirements on a monotonic time package `Ada.Real_Time` (as distinct from the core package `Calendar` which might go backwards because of time zone or daylight-saving changes). There are also suggested tasking restrictions which might be appropriate for the development of very efficient run time systems for specialized applications.

Distributed Systems. The core language introduces the idea of a partition whereby one coherent "program" is distributed over a number of partitions each with its own main task. This annex defines active and passive partitions and interpartition communication using statically and dynamically bound remote subprogram calls.

Information Systems. The core language includes basic support for decimal types. This annex defines additional attributes and a number of packages providing detailed facilities for manipulating decimal values and conversion to and from external formats using picture strings.

Numerics. This annex addresses the special needs of the numeric community. It defines generic packages for the definition and manipulation of complex numbers. It also defines the accuracy requirements for numerics.

Safety and Security. This annex addresses restrictions on the use of the language and requirements of compilation systems for programs to be used in safety-critical and related applications where program security is vital.

As the reader will appreciate, many of these topics are important for avionics applications. The provision of these annexes ensures that these specialized capabilites are provided by all relevant implementations in a standard manner.

References

Barnes, J. (Ed.), 1997, *Ada 95 Rationale*, LNCS 1247, Springer Verlag.

Barnes, J.G.P., 1998. *Programming in Ada 95*, 2nd editiion, Addison-Wesley, Reading, MA.

Intermetrics Inc., 1995. *Annotated Ada 95 Reference Manual*, Intermetrics Inc., Burlington, MA.

International Organization for Standardization, 1995. *Information Technology — Programming Languages — Ada. Ada Reference Manual.* ISO/IEC 8652:1995(E).

Taft, S. T. and Duff, R.A. (Eds.), 1997, *Ada 95 Reference Manual*, LNCS 1246, Springer Verlag.

Further Information

The official definition of Ada 95 is the *Reference Manual for the Ada Programming Language* [ISO, 1995]. This has been reprinted [Taft and Duff, 1997]. An annotated version containing detailed explanatory information is also available [Intermetrics, 1995]. There is also an accompanying *Rationale* document giving a broad overview of the language and especially the reasons for the changes from Ada 83 [Barnes, 1997].

This chapter is based on a somewhat condensed form of Chapters 1 to 4 of *Programming in Ada 95* by the author and published by Addison-Wesley [Barnes, 1998]. The author is grateful to Addison-Wesley for their permission to use the material presented here. *Programming in Ada 95* is a comprehensive description of the core of Ada 95 with many illustrative examples plus exercises and answers. It extends to 23 chapters and includes full coverage of the predefined library, an overview of the annexes, and a CD containing an Ada compiler and a number of complete example programs.

27

RTCA
DO-178B/EUROCAE
ED-12B

Thomas K. Ferrell
Ferrell and Associates Consulting

Uma D. Ferrell
Ferrell and Associates Consulting

27.1 Introduction

This chapter provides a summary of the document RTCA DO-178B, *Software Considerations in Airborne Systems and Equipment Certification*,[1] with commentary on the most common mistakes made in understanding and applying DO-178B. The joint committee of RTCA Special Committee 167 and EUROCAE* Working Group 12 prepared RTCA DO-178B** (also known as EUROCAE ED-12B), and it was subsequently published by RTCA and by EUROCAE in December 1992. DO-178B provides guidance for the production of software for airborne systems and equipment such that there is a level of confidence in the correct functioning of that software in compliance with airworthiness requirements. DO-178B represents industry consensus opinion on the best way to ensure safe software. It should also be noted that although DO-178B does not discuss specific development methodologies or management activities, there is clear evidence that by following rigorous processes, cost and schedule benefits may be realized. The verification activities specified in DO-178B are particularly effective in identifying software problems early in the development process.

---

*European Organization for Civil Aviation Equipment.

**DO-178B and ED-12B are copyrighted documents of RTCA and EUROCAE, respectively. In this chapter, DO-178B shall be used to refer to both the English version and the European equivalent. This convention was adopted solely as a means for brevity.

27.1.1 Comparison with Other Software Standards

DO-178B is a mature document, having evolved over the last 20 years through two previous revisions, DO-178 and DO-178A. It is a consensus document that represents the collective wisdom of both the industry practitioners and the certification authorities. DO-178B is self-contained, meaning that no other software standards are referenced except for those that the applicant produces to meet DO-178B objectives. Comparisons have been made between DO-178B and other software standards such as MIL-STD-498, MIL-STD-2167A, IEEE/EIA-12207, IEC 61508, and U.K. Defense Standard 0-55. All of these standards deal with certain aspects of software development covered by DO-178B. None of them has been found to provide complete coverage of DO-178B objectives. In addition, these other standards lack objective criteria and links to safety analyses at the system level. However, organizations with experience applying these other standards often have an easier path to adopting DO-178B.

Advisory Circular AC 20-115B specifies DO-178B as an acceptable means, but not the only means, for receiving regulatory approval for software in systems or equipment being certified under a Technical Standard Order (TSO) Authorization, Type Certificate (TC), or Supplemental Type Certificate (STC). Most applicants use DO-178B to avoid the work involved in showing that other means are equivalent to DO-178B. Even though DO-178B was written as a guideline, it has become the standard practice within the industry. DO-178B is officially recognized as a *de facto* international standard by the International Organization for Standardization (ISO).

27.1.2 Document Overview

DO-178B consists of 12 sections, 2 annexes, and 3 appendices as shown in Figure 27.1.

Section 2 and Section 10 are designed to illustrate how the processes and products discussed in DO-178B relate to, take direction from, and provide feedback to the overall certification process. Integral Processes detailed in Sections 6, 7, 8, and 9, support the software life cycle processes noted in Sections 3, 4, and 5.

Section 11 provides details on the life cycle data and Section 12 gives guidance to any additional considerations. Annex A, discussed in more detail below, provides a leveling of objectives. Annex B provides the document's glossary. The glossary deserves careful consideration since much effort and care was given to precise definition of the terms. Appendices A, B, C, and D provide additional material

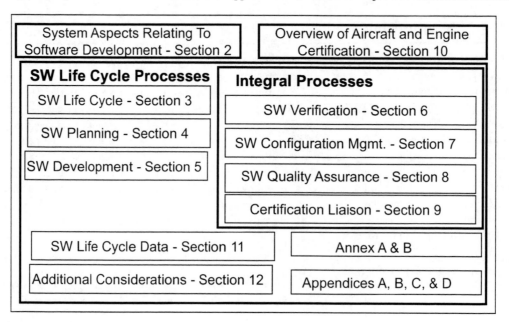

FIGURE 27.1   Document structure.

	Objective		Applicability by SW Level				Output		Control Category by SW Level			
	Description	Ref.	A	B	C	D	Description	Ref.	A	B	C	D
1	Low-level Requirements comply with High-level Requirements.	6.3.2a	●	●	○		Software Verification Results	11.14	②	②	②	②

FIGURE 27.2   Example objective from Annex A.

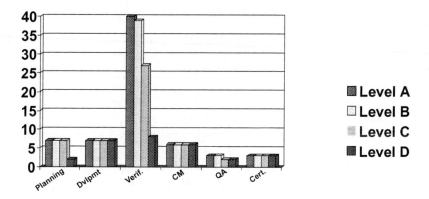

FIGURE 27.3   Objectives over the software development life cycle.

including a brief history of the document, the index, a list of contributors, and a process improvement form, respectively. It is important to note that with the exception of the appendices and some examples embedded within the text, the main sections and the annexes are considered normative, i.e., required to apply DO-178B.

The 12 sections of DO-178B describe processes and activities for the most stringent level of software.* Annex A provides a level by level tabulation of the objectives for lower levels of software.** This leveling is illustrated in Figure 27.2 extracted from Annex A Table A-4, Verification of Outputs of Software Design Process.

In addition to the leveling of objectives, Annex A tables serve as an index into the supporting text by way of reference, illustrate where independence is required in achieving the objective, which data items should include the objective evidence, and how that evidence must be controlled. More will be said about the contents of the various Annex A tables in the corresponding process section of this text. If an applicant adopts DO-178B for certification purposes, Annex A may be used as a checklist to achieve these objectives. The FAA's position is that if an applicant provides evidence to satisfy the objectives, then the software is DO-178B compliant. Accordingly, the FAA's checklists for performing audits of DO-178B developments are based on Annex A tables.

Before discussing the individual sections, it is useful to look at a breakout of objectives as contained in Annex A. While DO-178B contains objectives for the entire software development life cycle, there is a clear focus on verification as illustrated by Figure 27.3. Although at first glance it appears that there is only one objective difference between levels A and B, additional separation between the two is accomplished through

---

*Levels are described in Section 27.1.3, "Software as part of system."

**Software that is determined to be at level E is outside the scope of DO-178B.

the relaxation of independence requirements. Independence is achieved by having the verification or quality assurance of an activity performed by a person other than the one who initially conducted the activity. Tools may also be used to achieve independence.

27.1.3 Software as Part of the System

Application of DO-178B fits into a larger system of established or developing industry practices for systems development and hardware. The system level standard is SAE ARP4754, *Certification Considerations for Highly-Integrated or Complex Aircraft Systems.*[2] The relationship between system, software, and hardware processes is illustrated in Figure 27.4.

The interfaces to the system development process were not well defined at the time DO-178B was written. This gap was filled when ARP4754 was published in 1996. DO-178B specifies the information flow between system processes and software processes. The focus of the information flow from the system process to the software process is to keep track of requirements allocated to software, particularly those requirements that contribute to the system safety. The focus of information flow from the software process to the system process is to ensure that changes in the software requirements, including the introduction of derived requirements (those not directly traceable to a parent requirement), do not adversely affect system safety.

The idea of system safety, although outside the scope of DO-178B, is crucial to understanding how to apply DO-178B. The regulatory materials governing the certification of airborne systems and equipment define five levels of failure conditions. The most severe of these is *catastrophic*, meaning failures that result in the loss of ability to continue safe flight and landing. The least severe is *no effect*, where the failure results in no loss of operational capabilities and no increase in crew workload. The intervening levels define various levels of loss of functionality resulting in corresponding levels of workload and potential for loss of life. These five levels map directly to the five levels of software defined in DO-178B. This mapping is shown in Figure 27.5.

It is important to note that software is never certified as a standalone entity. A parallel exists for the hardware development process and flow of information between hardware processes and system process. Design trade-offs between software processes and hardware processes are also taken into consideration

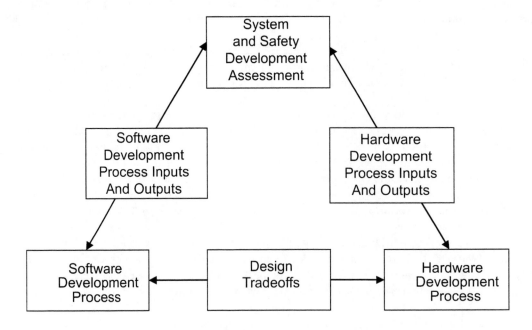

FIGURE 27.4   Relationship between system development process and the software development process.

Failure Condition	DO-178B Software Level
Catastrophic	A
Hazardous	B
Major	C
Minor	D
No effect	E

FIGURE 27.5 Software levels.

at the system level. Software levels may be lowered by using protective software or hardware mechanisms elsewhere in the system. Such architectural methods include partitioning, use of hardware or software monitors, and architectures with built-in redundancy.

27.2 Software Life-Cycle Processes

The definition of how data are exchanged between the software and systems development processes is part of the software life-cycle processes discussed in DO-178B. Life-cycle processes include the planning process, the software development processes (requirements, design, code, and integration), and the integral processes (verification, configuration management, software quality assurance, and certification liaison). DO-178B defines objectives for each of these processes as well as outlining a set of activities for meeting the objectives.

DO-178B discusses the software life-cycle processes and transition criteria between life-cycle processes in a generic sense without specifying any particular life-cycle model. Transition criteria are defined as "the minimum conditions, as defined by the software planning process, to be satisfied to enter a process." Transition criteria may be thought of as the interface points between all of the processes discussed in DO-178B. Transition criteria are used to determine if a process may be entered or reentered. They are a mechanism for knowing when all of the tasks within a process are complete and may be used to allow processes to execute in parallel. Since different development models require different criteria to be satisfied for moving from one step to the next, specific transition criteria are not defined in DO-178B. However, it is possible to describe a set of characteristics that all well-defined transition criteria should meet. For transition criteria to successfully assist in entry from one life-cycle process to another, they should be quantifiable, flexible, well documented, and present for every process. It is also crucial that the process owners agree upon the transition criteria between their various processes.

27.2.1 Software Planning Process

DO-178B defines five types of planning data* for a software development. They are

- Plan for Software Aspects of Certification (PSAC)
- Software Development Plan
- Software Verification Plan

---

*The authors of DO-178B took great pains to avoid the use of the term "document" when referring to objective evidence that needed to be produced to satisfy DO-178B objectives. This was done to allow for alternative data representations and packaging as agreed upon between the applicant and the regulatory authority. For example, the four software plans-outlining development, verification, QA, and CM may all be packaged in a single plan, just as the PSAC may be combined with the System Certification Plan.

- Software Configuration Management Plan
- Software Quality Assurance Plan

These plans should include consideration of methods, languages, standards, and tools to be used during the development. A review of the planning process should have enough details to assure that the plans, proposed development environment, and development standards (requirements, design, and code) comply with DO-178B.

Although DO-178B does not discuss the certification liaison process until Section 9, the intent is that the certification liaison process should begin during the projects' planning phase. The applicant outlines the development process and identifies the data to be used for substantiating the means of compliance for the certification basis. It is especially important that the applicant outline specific features of software or architecture that may affect the certification process.

27.2.2 Software Development Process

Software development processes include requirements, design, coding, and integration. DO-178B allows for requirements to be developed that detail the system's functionality at various levels. DO-178B refers to these levels as high- and low-level requirements. System complexity and the design methodology applied to the system's development drive the requirements' decomposition process. The key to understanding DO-178B's approach to requirement's definition can be summed up as, "one person's requirements are another person's design." Exactly where and to what degree the requirements are defined is less important than ensuring that all requirements are accounted for in the resulting design and code, and that traceability is maintained to facilitate verification.

Some requirements may be derived from the design, architecture, or the implementation nuances of the software and hardware. It is recognized that such requirements will not have a traceability to the high-level requirements. However, these requirements must be verified and must also be considered for safety effects in the system safety assessment process.

DO-178B provides only a brief description of the design, coding, and integration processes since these tend to vary substantially between various development methodologies. The one exception to this is in the description to the outputs of each of the processes. The design process yields low-level requirements and software architecture. The coding process produces the source code, typically either in a high-order language or assembly code. The result of the integration effort is executable code resident on the target computer along with the various build files used to compile and link the executable. Each of these outputs is verified, assured, and configured as part of the integral processes.

27.3 Integral Processes

DO-178B defines four processes as integral, meaning that they overlay and extend throughout the software life cycle. These are the software verification process, software configuration management, software quality assurance, and certification liaison process.

27.3.1 Software Verification*

As noted earlier, verification objectives outnumber all others in DO-178B, accounting for over two thirds of the total. DO-178B defines verification as a combination of reviews, analyses, and testing. Verification is a technical assessment of the results of both the software development processes and the software verification process. There are specific verification objectives that address the requirements, design, coding, integration, as well as the verification process itself. Emphasis is placed at all stages to assure that there is traceability from high-level requirements to the final "as-built" configuration.

---

*Software Verification is a complex topic, which deserves in-depth treatment. The reader is directed to References 4, 5, and 6 for detailed discussion on verification approaches and explanation of terms.

Reviews provide qualitative assessment of a process or product. The most common types of reviews are requirements reviews, design reviews, and test procedure reviews. DO-178B does not prescribe how these reviews are to be conducted, or what means are to be employed for effective reviews. Best practices in software engineering process states that for reviews to be effective and consistent, checklists should be developed and used for each type of review. Checklists provide:

- Objective evidence of the review activity
- A focused review of those areas most prone to error
- A mechanism for applying "lessons learned"
- A practical traceable means for ensuring that corrective action is taken for unsatisfactory items

Review checklists can be common across projects, but they should themselves be reviewed for appropriateness and content for a particular project.

Analyses provide repeatable evidence of correctness and are often algorithmic or procedural in nature. Common types of analyses used include timing, stack, data flow, and control flow analyses. Race conditions and memory leakage should be checked as part of the timing and stack analysis. Data and control coupling analysis should include, a minimum, basic checks for set/use and may extend to a full model of the system's behavior. Many types of analyses may be performed using third-party tools. If tools are used for this purpose, DO-178B rules for tool qualification must be followed.

The third means of verification, testing, is performed to demonstrate that

- The software product performs its intended function
- The software does not demonstrate any unintended actions

The key to accomplishing testing correctly to meet DO-178B objectives in a cost-effective manner is to maintain a constant focus on requirements. This requirements-based test approach represents one of the most fundamental shifts from earlier versions of the document. As test cases are designed and conducted, requirements coverage analysis is performed to assess that all requirements are tested. A structural coverage analysis is performed to determine the extent to which the requirements-based test exercised the code. In this manner, structural coverage is used as a means of assessing overall test completion. The possible reasons for lack of structural coverage are shortcomings in requirements-based test cases or procedures, inadequacies in software requirements, compiler generated code, unreachable, or inactive code.

As part of the test generation process, tests should be written for both normal range and abnormal inputs (robustness). Tests should also be conducted using the target environment whenever possible.

Structural coverage and how much testing is required for compliance at the various levels are misunderstood topics. Level D software verification requires test coverage of high-level requirements only. No structural coverage is required.

Low-level requirements testing is required at level C. In addition, testing of the software structure to show proper data and control coupling is introduced. This coverage involves coverage of dependencies of one software component on other software component via data and control. Decision coverage is required for level B, while level A code requires Modified Condition Decision Coverage (MCDC).

For level A, structural coverage analysis may be performed on source code only to the extent that the source code can be shown to map directly to object code. The reason for this rule is that some compilers may introduce code or structure that is different from source code.

MCDC coverage criteria were introduced to retain the benefits of multiple-condition coverage while containing the exponential growth in the required number of test cases required. MCDC requires that each condition must be shown to independently affect the outcome of the decision and that the outcome of a decision changes when one condition is changed at a time. Many tools are available to determine the minimum test case set needed for DO-178B compliance. There is usually more than one set of test cases that satisfy MCDC coverage.[3] There is no firm policy on which set should be used for compliance. It is best to get an agreement with the certification authorities concerning the algorithms and tools used to determine compliance criteria.

27.3.2 Software Configuration Management

Verification of the various outputs discussed in DO-178B are only credible when there is clear definition of what has been verified. This definition or configuration is the intent of the DO-178B objectives for configuration management. The six objectives in this area are unique, in that they must be met for all software levels. This includes identification of what is to be configured, how baselines and traceability are established, how problem reports are dealt with, how the software is archived and loaded, and how the development environment is controlled.

While configuration management is a fairly well-understood concept within the software engineering community (as well as the aviation industry as a whole), DO-178B does introduce some unique terminology that has proven to be problematic. The concept of control categories is often misunderstood in a way that overall development costs are increased, sometimes dramatically. DO-178B defines two control categories (CC1 and CC2) for data items produced throughout the development.

The authors of DO-178B intended the two levels as a way of controlling the overhead costs of creating and maintaining the various data items. Items controlled as CC2 have less requirements to meet in the areas of problem reporting, baselining, change control, and storage. The easiest way to understand this is to provide an example. Problem reports are treated as a CC2 item. If problem reports were a CC1 item and a problem was found with one of the entries on the problem report itself, a second problem report would need to be written to correct the first one.

A second nuance of control categories is that the user of DO-178B may define what CC1 and CC2 are within their own CM system as long as the DO-178B objectives are met. One example of how this might be beneficial is in defining different retention periods for the two levels of data. Given the long life of airborne systems, these costs can be quite sizeable. Another consideration for archival systems selected for data retention is technology obsolescence of the archival medium as well as means of retrieval.

27.3.3 Software Quality Assurance

Software quality assurance (SQA) objectives provide oversight of the entire DO-178B process and require independence at all levels. It is recognized that it is prudent to have an independent assessment of quality. SQA is active from the beginning of the development process. SQA assures that any deviations during the development process from plans and standards are detected, recorded, evaluated, tracked, and resolved. For levels A and B, SQA is required to assure transition criteria are adhered to throughout the development process.

SQA works with the CM process to assure that proper controls are in place and applied to life cycle data. This last task culminates in the conduct of a software conformity review. SQA is responsible for assuring that the as-delivered products matches the as-built and as-verified product. The common term used for this conformity review in commercial aviation industry is "First Article Inspection."

27.3.4 Certification Liaison Process

As stated earlier, the certification liaison process is designed to streamline the certification process by ensuring that issues are identified early in the process. While DO-178B outlines twenty distinct data items to be produced in a compliant process, three of these are specific to this process and must be provided to the certifying authority. They are

- Plan for Software Aspects of Certification (PSAC)
- Software Configuration Index
- Software Accomplishment Summary

Other data items may be requested by the certification authority, if deemed necessary. As mentioned earlier, applicants are encouraged to start a dialogue with certification authorities as early in the process as possible to reach a common understanding of a means of achieving compliance with DO-178B. This is especially important as new technology is applied to avionics and as new personnel enter the field.

Good planning up front, captured in the PSAC, should minimize surprises later in the development process, thus minimizing cost. Just as the PSAC states *what you intend to do*, the accomplishment summary captures *what you did*. It is used to gauge the overall completeness of the development process and to ensure that all objectives of DO-178B have been satisfied.

Finally, the configuration index serves as an overall accounting of the content of the final product as well as the environment needed to recreate it.

27.4   Additional Considerations

During the creation of DO-178B, it was recognized that new development methods and approaches existed for developing avionics. These included incorporation of previously developed software, use of tools to accomplish one or more of the objectives required by DO-178B, and application of alternate means in meeting an objective such as formal methods. In addition, there are a small class of unique issues such as field-loadable and user-modifiable software. Section 12 collects these items together under the umbrella title of Additional Considerations. Two areas, Previously Developed Software (PDS) and Tool Qualification, are common sources of misunderstanding in applying DO-178B.

27.4.1   Previously Developed Software

PDS is software that falls in one of the following categories:

- Commercial off-the-shelf software (e.g., shrink-wrap)
- Airborne software developed to other standards (e.g., MIL-STD-498)
- Airborne software that predates DO-178B (e.g., developed to the original DO-178 or DO-178A)
- Airborne software previously developed at a lower software level

The use of one or more of these types of software should be planned for and discussed in the PSAC. In every case, some form of gap analysis must be performed to determine where specific objectives of DO-178B have not been met. It is the applicant's responsibility to perform this gap analysis and propose to the regulatory authority a means for closing any gaps. Alternate sources of development data, service history, additional testing, reverse engineering, and wrappers* are all ways of ensuring the use of PDS is safe in the new application.

In all cases, usage of PDS must be considered in the safety assessment process and may require that the process be repeated if the decision to use a PDS component occurs after the approval of PSAC. A special instance of PDS usage occurs when software is used in a system to be installed on an aircraft other than the one for which it was originally designed. Although the function may be the same, interfaces with other aircraft systems may behave differently. As before, the system safety assessment process must be repeated to assure that the new installation operates and behaves as intended.

If service history is employed in making the argument that a PDS component is safe for use, the relevance and sufficiency of the service history must be assessed. Two tests must be satisfied for the service history approach to work. First, the application for which history exists must be shown to be similar to the intended new use of the PDS. Second, there should be data, typically problem reports, showing how the software has performed over the period for which credit is sought. The authors of DO-178B intended that any use of PDS be shown to meet the same objectives required of newly developed code.

Prior to identifying PDS as part of a new system, it is prudent to investigate and truly understand the costs of proving that the PDS satisfies the DO-178B objectives. Sometimes, it is easier and cheaper to develop the code again!

---

*Wrappers is a generic term used to refer to hardward or software components that isolate and filter inputs to and from the PDS for the purposes of protecting the system from erroneous PDS behavior.

27.4.2 Tool Qualification

DO-178B requires qualification of tools when the processes noted by DO-178B are eliminated, reduced, or automated by a tool without its output being verified according to DO-178B. If the output of a tool is demonstrated to be restricted to a particular part of the life cycle, the qualification can also be limited to that part of the life cycle. Only deterministic tools can be qualified.

Tools are classified as development tools and verification tools. Development tools produce output that becomes a part of the airborne system and thus can introduce errors. Rules for qualifying development tools are fashioned after the rules of assurance for generating code. Once the need for development tool qualification is established, a tool qualification plan must be written. The rigor of the plan is determined by the nature of the tool and the level of code upon which it is being used. A tool accomplishment summary is used to show compliance with the tool qualification plan. The tool is required to satisfy the objectives at the same level as the software it produces, unless the applicant can justify a reduction in the level to the certification authority.

Verification tools cannot introduce errors but may fail to detect them or mask their presence. Qualification criterion for verification tools is the demonstration of its requirements under normal operational conditions. Compliance is established by noting tool qualification within PSAC and Software Accomplishment Summary. A tool qualification plan and a tool accomplishment summary are not required for verification tools by DO-178B although an applicant may find them useful for documenting the qualification effort.

27.5 Additional Guidance

RTCA SC-190/EUROCAE WG-52 was formed in 1997 to address issues that were raised by the industry and certification authorities in the course of applying DO-178B since its release in 1992. The membership in this committee includes over 200 industry and regulatory representatives from the U.S. and Europe. The outputs of the SC-190 consensus process are available to industry from the RTCA or EUROCAE in the form of errata, frequently asked questions, and discussion papers. These outputs have been collated and published in DO-248/ED-94.[7]

27.6 Synopsis

DO-178B provides objectives for software life-cycle processes, activities to achieve these objectives, and outlines objective evidence for demonstrating that these objectives were accomplished. The purpose of software compliance to DO-178B is to provide considerable confidence that the software is suitable for use in airborne systems. DO-178B should not be viewed as a documentation guide.

Compliance data are intended to be a consequence of the process. Complexity and extent of the required compliance data depend upon the characteristics of the system/software, associated development practices, and the interpretation of DO-178B, especially when it is applied to new technology and no precedent is available.

Finally, it has to be emphasized that DO-178B objectives do not directly deal with safety. Safety is dealt with at the system level via the system safety assessment. DO-178B objectives help verify the correct implementation of safety-related requirements that flow from the system safety assessment. Like any standard, DO-178B has good points and bad points (and even a few errors). However, careful consideration of its contents, taken together with solid engineering judgment, should result in better and safer airborne software.

References

1. RTCA DO-178B, *Software Considerations in Airborne Systems and Equipment Certification*, RTCA Inc., Washington, D.C, 1992. Copies of DO-178B may be obtained from RTCA, Inc., 1140 Connecticut Avenue, NW, Suite 1020, Washington, D.C. 20036-4001 U.S. (202) 833-9339. This document is also known as ED 12B, *Software Considerations in Airborne Systems and Equipment*

Certification, EUROCAE, Paris, 1992. Copies of ED-12B may be obtained from EUROCAE, 17, rue Hamelin, 75783 PARIS CEDEX France, (331) 4505-7188.

2. SAE ARP4754, *Certification Considerations for Highly-Integrated or Complex Aircraft Systems*, SAE, Warrendale, PA, 1996.

3. Chilenski, J.J. and Miller, P.S., Applicability of modified condition/decision coverage to software testing, *Software Eng. J.*, 193, September 1994.

4. Myers, G. J., *The Art of Software Testing*, John Wiley & Sons, New York, 1979.

5. Beizer, B., *Software Testing Techniques*, 2nd ed., Coriolis Group, Scottsdale, AZ, 1990.

6. McCracken, D. and Passafiume, M., *Software Testing and Evaluation*, Benjamin/Cummings, Menlo Park, CA, 1987.

7. RTCA DO-248, Annual Report for Clarification of DO-178B "Software Considerations in Airbone Systems and Equipment Certification; EOROCAE ED-94, Annual Report for Clarification of ED-12B "Software Considerations in Airbone Systems and Equipment Certification."

Further Information

1. The Federal Aviation Administration Web Page: www.faa.gov.

2. The RTCA Web Page: www.rtca.org.

3. Spitzer, C.R., *Digital Avionics Systems Principles and Practice*, 2nd ed., McGraw-Hill, New York, 1993.

4. Wichmann, B.A., A Review of a Safety-Critical Software Standard, National Physical Laboratory, Teddington, Middlesex, U.K. (report is not dated).

5. Herrman, D.S., Software Safety and Reliability, IEEE Computer Society Press, Washington, D.C., 1999.

28

Fault-Tolerant Avionics

Ellis F. Hitt
Battelle

Dennis Mulcare
Science Applications
International Co.

28.1 Introduction

Fault-tolerant designs are required to ensure safe operation of digital avionics systems performing flight-critical functions. This chapter discusses the motivation for fault-tolerant designs, and the many different design practices that are evolving to implement a fault-tolerant system. The designer needs to make sure the fault tolerance requirements are fully defined to select the design concept to be implemented from the alternatives available. The requirements for a fault-tolerant system include performance, dependability, and the methods of assuring that the design, when implemented, meets all requirements. The requirements must be documented in a specification of the intended behavior of a system, specifying the tolerances imposed on the various outputs from the system [Anderson and Lee, 1981].

Development of the design proceeds in parallel with the development of the methods of assurance to validate that the design meets all requirements including the fault tolerance. The chapter concludes with references to further reading in this developing field.

A fault-tolerant system provides continuous, safe operation in the presence of faults. A fault-tolerant avionics system is a critical element of flight-critical architectures which include the fault-tolerant computing system (hardware, software, and timing), sensors and their interfaces, actuators, elements, and data communication among the distributed elements. The fault-tolerant avionics system ensures integrity of output data used to control the flight of the aircraft, whether operated by the pilot or autopilot. A fault-tolerant system must detect errors caused by faults, assess the damage caused by the fault, recover from the error, and isolate the fault. It is generally not economical to design and build a system that is capable of tolerating all possible faults in the universe. The faults the system is to be designed to tolerate must be defined based on analysis of requirements including the probability of each fault occurring, and the impact of not tolerating the fault.

A user of a system may observe an error in its operation which is the result of a fault being triggered by an event. Stated another way, *a fault is the cause of an error, and an error is the cause of a failure.* A mistake made in designing or constructing a system can introduce a fault into the design of the system, either because of an inappropriate selection of components or because of inappropriate (or missing) interactions between components. On the other hand, if the design of a system is considered to be correct, then an erroneous transition can occur only because of a failure of one of the components of the system. Design faults require more powerful fault-tolerance techniques than those needed to cope with component faults. Design faults are unpredictable; their manifestation is unexpected and they generate unanticipated errors. In contrast, component faults can often be predicted, their manifestation is expected, and they produce errors which can be anticipated [Anderson and Lee, 1981].

In a non-fault-tolerant system, diagnosis is required to determine the cause of the fault that was observed as an error. Faults in avionics systems are of many types. They generally can be classified as hardware, software, or timing related. Faults can be introduced into a system during any phase of its life cycle including requirements definition, design, production, or operation.

In the 1960s, designers strived to achieve highly reliable safe systems by avoiding faults, or masking faults. The Apollo guidance and control system employed proven, highly reliable components and triple modular redundancy (TMR) with voting to select the correct output. Improvements in hardware reliability, and our greater knowledge of faults and events which trigger them, has led to improved design methods for fault-tolerant systems which are affordable.

In any fault-tolerant system, the range of potential fault conditions that must be accommodated is quite large; enumerating all such possibilities is a vital yet formidable task in validating the system's airworthiness, or its readiness for deployment. The resultant need to handle each such fault condition prompts attention to the various assurance-oriented activities that contribute to certification system airworthiness.

28.1.1   Motivation

Safety is of primary importance to the economic success of the aviation system. The designer of avionics systems must assure that the system provides the required levels of safety to passengers, aircrew, and maintenance personnel. Fault-tolerant systems are essential with the trend to increasingly complex digital systems.

Many factors necessitate fault tolerance in systems that perform functions that must be sustained without significant interruption. In avionics systems, such functions are often critical to continued safe flight or to the satisfactory conduct of a mission; hence the terms flight-critical and mission-critical. The first compelling reality is that physical components are non-ideal, i.e., they are inescapably disposed to physical deterioration or failure. Clearly then, components inherently possess a finite useful life, which varies with individual instances of the same component type. At some stage, then, any physical component will exhibit an abrupt failure or excessive deterioration such that a fault may be detected at some level of system operation.

The second contributing factor to physical faults is the non-ideal environment in which an avionics system operates. Local vibrations, humidity, temperature cycling, electrical power transients, electromagnetic interference, etc. tend to induce stress on the physical component which may cause abrupt failure or gradual deterioration. The result may be a transient or a permanent variation in output, depending on the nature and severity of the stress. The degree of induced deterioration encountered may profoundly influence the useful life of a component. Fortunately, design measures can be taken to reduce susceptibility to the various environmental effects. Accordingly, a rather comprehensive development approach is needed for system dependability, albeit fault tolerance is the most visible aspect because it drives the organization and logic of the system architecture.

The major factor necessitating fault tolerance is design faults. Tolerance of design faults in hardware and software and the overall data flow is required to achieve the integrity needed for flight-critical systems. Reliance on the hardware chip producing correct output when there is no physical failure has been shown to be risky, as demonstrated by the design error discovered in the floating point unit of a high-performance microprocessor in wide use. Because of the difficulty in eliminating all design faults, dissimilar redundancy is used to produce outputs which should be identical even though computed by dissimilar computers. Use of dissimilar redundancy is one approach to tolerating common-mode failures. A common-mode failure (CMF) occurs when copies of a redundant system suffer faults nearly simultaneously, generally due to a single cause [Lala, 1994].

28.1.2 Definitional Framework

A digital avionics system is a "hard real-time" system producing time-critical outputs that are used to control the flight of an aircraft. These critical outputs must be dependable. A dependable system is both reliable and safe. Reliability has many definitions, and is often expressed as the probability of not failing. Another definition of reliability is the probability of producing a "correct" output [Vaidya and Pradhan, 1993]. Safety has been defined as the probability that the system output is either correct, or the error in the output is detectable [Vaidya and Pradhan, 1993]. Correctness has been defined as the requirement that the output of all channels agree bit-for-bit under no-fault conditions [Lala, 1994]. Another design approach, approximate consensus, considers a system to be correct if the outputs agree within some threshold. Both approaches are in use.

Hardware component faults are often classified by extent, value, and duration [Avizienis, 1976]. Extent applies to whether the errors generated by the fault are localized or distributed; value indicates whether the fault generates fixed or varying erroneous values; duration refers to whether the fault is transient or permanent. Several studies have shown that permanent faults cause only a small fraction of all detected errors, as compared with transient faults [Sosnowski, 1994]. A recurring transient fault is often referred to as intermittent [Anderson and Lee, 1981]. Figure 28.1 depicts these classifications in the tree of faults.

Origin faults may result from a physical failure within a hardware component of a system, or may result from human-made faults. System boundary internal faults are those parts of the system's state which, when invoked by the computation activity, will produce an error, while external faults result from system interference caused by its physical environment, or from system interaction with its human environment. Origin faults classified by the time phase of creation include design faults resulting from

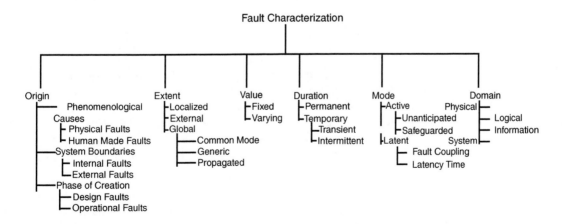

FIGURE 28.1 Fault classification.

imperfections that arise during the development of the system (from requirements specification to implementation), subsequent modifications, the establishment of procedures for operating or maintaining the system, or operational faults which appear during the system operation [Lala and Harper, 1994].

A fault is in the active mode if it yields an **erroneous state**, either in hardware or software, i.e., a state that differs from normal expectations under extant circumstances. Alternatively, a fault is described as latent when it is not yielding an erroneous state. Measures for error detection that can be used in a fault-tolerant system fall into the following broad classification [Anderson and Lee, 1981]:

1. Replications checks
2. Timing checks
3. Reversal checks
4. Coding checks
5. Reasonableness checks
6. Structural checks
7. Diagnostic checks.

These checks are discussed in Section 28.4.

During the development process, it is constructive to maintain a perspective regarding the fault attributes of Domain and Value in Figure 28.1. Basically, Domain refers to the universe of layering of fault abstractions that permit design issues to be addressed with a minimum of distraction. Value simply refers to whether the erroneous state remains fixed, or whether it indeterminately fluctuates. While proficient designers tend to select the proper abstractions to facilitate particular development activities, these associated fault domains should be explicitly noted:

- Physical: elemental PHYSICAL FAILURES of hardware components — *underlying short, open, ground faults*
- Logical: manifested LOGICAL FAULTS per device behavior — *stuck-at-one, stuck-at-zero, inverted*
- Informational: exhibited ERROR STATES in interpreted results — *incorrect value, sign change, parity error*
- System: resultant SYSTEM FAILURE provides unacceptable service — *system crash, deadlock, hardover.*

These fault domains constitute levels of design responsibilities and commitments as well as loci of fault tolerance actions per se. Thus, fault treatment and, in part, fault containment, are most appropriately addressed in the physical fault domain. Similarly, hardware fault detection and assessment are most readily managed in the logical fault domain, where the fixed or fluctuating nature of the erroneous value(s) refines the associated fault identification mechanism(s). Lastly, error recovery and perhaps some fault containment are necessarily addressed in the informational fault domain, and service continuation in the system fault domain.

For safety-critical applications, physical hardware faults no longer pose the major threat to dependability. The dominant threat is now common-mode failures. Common-mode failures (CMFs) result from faults that affect more than one fault containment region at the same time, generally due to a common cause. Fault avoidance, fault removal through test and evaluation or via fault insertion, and fault tolerance implemented using exception handlers, program checkpointing, and restart are approaches used in tolerating CMFs [Lala, 1994]. Table 28.1 presents a classification of common-mode faults with the X indicating the possible combinations of faults that must be considered which are not intentional faults.

Physical, internal, and operational faults can be tolerated by using hardware redundancy. All other faults can affect multiple fault-containment regions simultaneously. Four sources of common-mode failures need to be considered:

1. Transient (External) Faults which are the result of temporary interference to the system from its physical environment such as lightning, High Intensity Radio Frequencies (HIRF), heat, etc.;
2. Permanent (External) Faults which are the result of permanent system interference caused by its operational environment such as heat, sand, salt water, dust, vibration, shock, etc.;

TABLE 28.1   Classification of Common-Mode Faults

Phenomenological Cause		System Boundary		Phase of Creation		Duration		Common Mode Fault Label
Physical	Human Made	Internal	External	Design	Operational	Permanent	Temporary	
X			X		X		X	Transient (External) CMF
X			X		X	X		Permanent (External) CMF
	X	X		X			X	Intermittent (Design) CMF
	X	X		X		X		Permanent (Design) CMF
	X		X		X		X	Interaction CMF

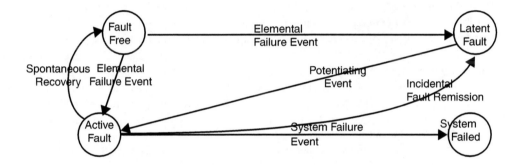

FIGURE 28.2   Hardware states (no corrective action).

3. Intermittent (Design) Faults which are introduced due to imperfections in the requirements specifications, detailed design, implementation of design, and other phases leading up to the operation of the system;
4. Permanent (Design) Faults are introduced during the same phases as intermittent faults, but manifest themselves permanently [Lala, 1994].

An **elemental physical failure**, which is an event resulting in component malfunction, produces a physical fault. These definitions are reflected in Figure 28.2, a state transition diagram that portrays four fault status conditions and associated events in the absence of fault tolerance. Here, for example, a Latent Fault Condition transitions to an Active Fault Condition due to a *Potentiating* Event. Such an event might be a functional mode change that caused the fault region to be exercised in a revealing way. Following the incidence of a sufficiently severe active fault from which spontaneous recovery is not forthcoming, a **system failure** *event* occurs wherein expected functionality can no longer be sustained. If the effects of a particular active fault are not too debilitating, a system may continue to function with some degradation in services. Fault tolerance can of course forestall both the onset of system failure and the expectation of degraded services.

The Spontaneous Recovery Event in Figure 28.2 indicates that faults can sometimes be *transient* in nature when a fault vanishes without purposeful intervention. This phenomenon can occur after an external disturbance subsides or an intermittent physical aberration ceases. A somewhat similar occurrence is provided through the Incidental Fault Remission Event in Figure 28.2. Here, the fault does not

TABLE 28.2 Delineation of Fault Conditions

Recovered Mode	Latent Mode	Active Mode
Spontaneous Recovery following Disruption	—	Erroneous State Induced by Transient Disturbance
or		or
Marginal Physical Fault Recovery	—	Passing Manifestation of Marginal Fault
—	Hard Physical Fault Remission	Passing Manifestation of Hard Fault
	or	or
—	Hard Physical Fault Latency	Persistent Manifestation of Hard Fault

vanish, but rather spontaneously reverts from an active to a latent mode due to the cessation or removal of fault excitation circumstances. Table 28.2 complements Figure 28.2 in clarifying these fault categories. Although these transient fault modes are thought to account for a large proportion of faults occurring in deployed systems, such faults may nonetheless persist long enough to appear as permanent faults to the system. In many cases then, explicit features must be incorporated into a system to ensure the timely recovery from faults that may induce improper or unsafe system operation.

Three classes of faults are of particular concern because their effects tend to be global regarding extent, where global implies impact on redundant components present in fault-tolerant systems. A **common mode-fault** is one in which the occurrence of a single physical fault at a one particular point in a system can cause coincident debilitation of all similar redundant components. This phenomenon is possible where there is a lack of protected redundancy, and the consequence would be a massive failure event. A **generic fault** is a development fault that is replicated across similar redundant components such that its activation yields a massive failure event like that due to a common mode fault. A **propagated fault** is one wherein the effects of a single fault spread out to yield a compound erroneous state. Such fault symptoms are possible when there is a lack of fault containment features. During system development, particular care must be exercised to safeguard against these classes of global faults, for they can defeat fault-tolerance provisions in a single event. Considerable design assessment is therefore needed, along with redundancy provisions, to ensure system dependability.

28.1.3 Dependability

Dependability is an encompassing property that enables and justifies reliance upon the services of a system. Hence, dependability is a broad, qualitative term that embodies the aggregate nonfunctional attributes, or "ilities," sought in an ideal system, especially one whose continued safe performance is critical. Thus, attributes like safety, reliability, availability, and maintainability, which are quantified using conditional probability formulas, can be conveniently grouped as elements of dependability. As a practical matter, however, dependability usually demands the incorporation of fault tolerance into a system to realize quantitative reliability or availability levels. Fault tolerance, moreover, may be need to achieve maintainability requirements, as in the case of on-line maintenance provisions.

For completeness, sake, it should be noted as depicted in Figure 28.3 that the attainment of dependability relies on fault avoidance and fault alleviation, as well as on fault tolerance.

This figure is complemented by Table 28.3, which emphasizes the development activities, such as analyzing fault possibilities during design to minimize the number and extent of potential fault cases. This activity is related to the criteria of containment in Figure 28.3 in that the analysis should ensure that both the number and propagation of fault cases are contained. The overall notion here is to minimize the number of fault possibilities, to reduce the prospects of their occurrences, and to ensure the safe handling of those that do happen in deployed systems.

TABLE 28.3   Ensuring Dependability

	Physical Faults	Development Faults
Fault Avoidance	Minimize by *Analysis*	Prevent by *Rigor* *Development Error*s
Fault Alleviation	*Selectivity the Incidence of Faults*	Remove by *Verification*
Fault Tolerance	*Ensure by Redundancy*	*Testing*

Note: Both physical and development fault handling may be present but any deficiency revealed is a development defect.

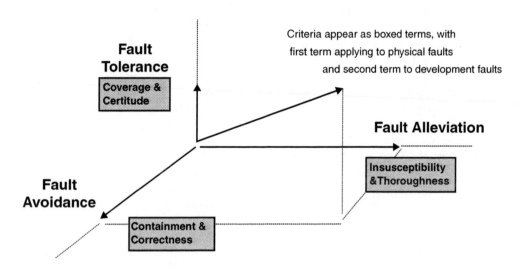

FIGURE 28.3   Dependability.

28.1.4   Fault Tolerance Options

System reliability requirements derive from the function criticality level and maximum exposure time. A **flight-critical** function is one whose loss might result in the loss of the aircraft itself, and possibly the persons on-board as well. In the latter case, the system is termed **safety-critical**. Here, a distinction can be made between a civil transport, where flight-critical implies safety-critical, and a combat aircraft. The latter admits to the possibility of the crew ejecting from an unflyable aircraft, so its system reliability requirements may be lower. A **mission-critical** function is one whose loss would result in the compromising or aborting of an associated mission. For avionics systems, a higher cost usually associates with the loss of an aircraft than with the abort of a mission (an antimissile mission to repel a nuclear weapon could be an exception). Thus, a full-time flight-critical system would normally pose much more demanding reliability requirements than a *flight-phase* mission-critical system. Next, the system reliability requirements coupled with the marginal reliabilities of system components determine the level of redundancy to ensure fault survivability, i.e., the minimum number of faults that must survive. To ensure meeting reliability requirements then, the evolution of a fault-tolerant design must be based on an interplay between design configuration commitments and substantiating analyses.

For civil transport aircraft, the level of redundancy for a flight-phase critical function like automatic all-weather landing is typically **single fail-operational**, meaning that the system should remain operable after any potential single elemental failure. Alternatively, a full-time critical function like fly-by-wire primary flight controls is typically **double fail-operational**. It should be noted that a function that is not flight-critical itself can have failure modes that threaten safety of flight. In the case of active controls to alleviate structural loads due to gusts or maneuvering, the function would not be critical where the purpose is merely to reduce structural fatigue effects. If the associated flight control surfaces have the authority during a hardover or runaway fault case to cause structural damage, then such a failure mode is safety-critical. In such cases, the failure modes must be designed to be **fail-passive**, which precludes any active mode failure effect like a hardover control surface. A failure mode that exhibits active behavior can still be **failsafe**, however, if the rate and severity of the active effects are well within the flight crews capability to manage safely.

28.1.5   Flight Systems Evolution

Beginning in the 1970s, NASA's F-8 digital fly-by-wire (DFBW) flight research program investigated the replacement of the mechanical primary flight control systems with electronic computers and electrical signal paths. The goal was to explore the implementation technology and establish the practicality of replacing mechanical linkages to the control surfaces with electrical links, thereby yielding significant weight and maintenance benefits. The F-8 DFBW architecture relied on bit-wise exact consensus of the outputs of redundant computers for fault detection and isolation [Lala et al., 1994].

The Boeing 747, Lockheed L-1011, and Douglas DC-10 utilized various implementations to provide the autoland functions which required a probability of failure of $<10^{-9}$ during the landing. The 747 used triply redundant analog computers. The L-1011 used digital computers in a dual-dual architecture, and the DC-10 used two identical channels, each consisting of dual-redundant fail-disconnect analog computers for each axis.

Since that time, the Airbus A-320 uses a full-time DFBW flight control system. It uses software design diversity to protect against common-mode failures. The Boeing 777 flight control computer architecture uses a 3 by 3 matrix of 9 processors of 3 different types. Multiversion software is also used.

28.1.6   Design Approach

It is virtually impossible to design a complex avionics system that will tolerate all possible faults. Faults can include both permanent and transient faults, hardware and software faults, and they can occur singularly or concurrently. Timing faults directly trace to the requirement of real-time response within a few milliseconds, and may be attributable to both hardware and software contributions to data latency, as well as incorrect data.

The implementation of fault tolerance entails increased system overhead, complexity, and validation challenges. The overhead, which is essentially increased system resources and associated management activities, lies in added hardware, communications, and computational demands. The expanded complexity derives from more intricate connectivity and dependencies among both hardware and software elements; the greater number of system states that may be assumed; and appreciably more involved logic to manage the system. Validation challenges are posed by the need to identify, assess, and confirm the capability to sustain system functionality under a broad range of potential fault cases. Hence, the design approach taken for fault-tolerant avionics must attain a balance between the costs incurred in implementing fault tolerance and the degree of dependability realized.

Design approach encompasses both system concepts, development methodology, and fault tolerance elements. The system concepts derive largely from the basic fault-tolerance options introduced in Section 28.1.4, with emphasis on judicious combinations of features that are adapted to given application attributes. The development methodology reduces to mutually supportive assurance-driven methods that propagate consistency, enforce accountability, and exact high levels of certitude as to system dependability. Fault tolerance design elements tend to unify the design concepts and methods in the sense of providing an orderly pattern of system organization and evolution. These fault tolerance elements, which in general should appear in some form in any fault-tolerant system, are

- Error Detection — recognition of the incidence of a fault
- Damage Assessment — diagnosis of the locus of a fault
- Fault Containment — restriction of the scope of effects of a fault

- Error Recovery — restoration of a restartable error-free state
- Service Continuation — sustained delivery of system services
- Fault Treatment — repair of fault.

A fundamental design parameter that spans these elements and constrains their mechanization is that of the granularity of fault handling. Basically, the detection, isolation, and recovery from a fault should occur at the same level of modularity to achieve a balanced and coherent design. In general, it is not beneficial or justified to discriminate or contain a fault at a level lower than that of the associated fault-handling boundary. There may be exceptions, however, especially in the case of fault detection, where a finer degree of granularity may be employed to take advantage of built-in test features or to reduce fault latency.

Depending on the basis for its instigation, fault containment may involve the inhibition of damage propagation of a physical fault and/or the suppression of an erroneous computation. Physical fault containment has to be designed into the hardware, and software error-state containment has to be designed into the applications software. In most cases, an erroneous software state must be corrected because of applications program discrepancies introduced during the delay in detecting a fault. This error recovery may entail resetting certain data object values and backtracking a control flow path in an operable processor. At this point, the readiness of the underlying architecture, including the coordination of operable components, must be ensured by the infrastructure. Typically, this activity relies heavily on system management software for fault tolerance. Service continuation, then, begins with the establishment of a suitable applications state for program restart. In an avionics system, this sequence of fault tolerance activities must take place rather quickly because of real-time functional demands. Accordingly, an absolute time budget must be defined, with tolerances for worst-case performance, for responsive service continuation.

28.2 System Level Fault Tolerance

28.2.1 General Mechanization

As discussed in Section 28.1.2, system failure is the loss of system services or expected functionality. In the absence of fault tolerance, a system may fail after just a single crucial fault. This kind of system, which in effect is zero fail-operational, would be permissible for non-critical functions. Figure 28.2, moreover, characterizes this kind of system in that continued service depends on spontaneous remission of an active fault or a fault whose consequences are not serious enough to yield a system failure.

Where the likelihood of continued service must be high, redundancy can be incorporated to ensure system operability in the presence of any permanent fault(s). Such fault-tolerant systems incorporate an additional fault status state, namely that of recovery, as shown in Figure 28.4. Here, system failure occurs only after the exhaustion of spares or an unhandled severe fault. The aforementioned level of redundancy can render it extremely unlikely that the spares will be exhausted as a result of hardware faults alone. An unhandled fault could occur only as a consequence of a design error, like the commission of a generic error wherein the presence of a fault would not even be detected.

This section assumes a system-level perspective, and undertakes to examine fault-tolerant system architectures and examples thereof. Still, these examples embody and illuminate general system-level principles of fault tolerance. Particular prominence is directed toward flight control systems, for they have motivated and pioneered much of the fault tolerance technology as applied to digital flight systems. In the past, such systems have been functionally dedicated, thereby providing a considerable safeguard against malfunction due to extraneous causes. With the increasing prevalence of integrated avionics, however, the degree of function separation is irretrievably reduced. This is actually not altogether detrimental, more avionics functions than ever are critical, and a number of benefits accrue from integrated processing. Furthermore, the system developer can exercise prerogatives that afford safeguards against extraneous faults.

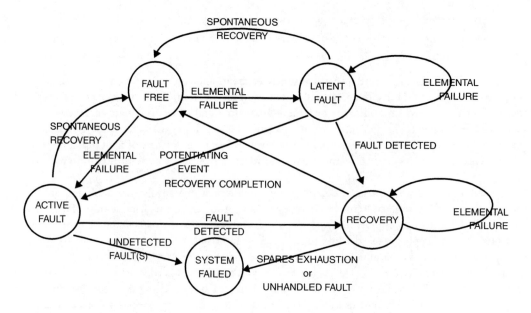

FIGURE 28.4   Hardware states (with corrective action).

28.2.2   Redundancy Options

Fault tolerance is usually based on some form of redundancy to extend system reliability through the invocation of alternative resources. The redundancy may be in hardware, software, time, or combinations thereof. There are three basic types of redundancy in hardware and software: static, dynamic, and hybrid. Static redundancy masks faults by taking a majority of the results from replicated tasks. Dynamic redundancy takes a two-step procedure for detection of, and recovery from faults. Hybrid redundancy is a combination of static and dynamic redundancy [Shin and Hagbae, 1994].

In general, much of this redundancy resides in additional hardware components. The addition of components reduces the mean-time-between-maintenance actions, because there are more electronics that can, and at some point will, fail. Since multiple, distinct faults can occur in fault-tolerant systems, there are many additional failure modes that have to be evaluated in establishing the airworthiness of the total system. Weight, power consumption, cooling, etc. are other penalties for component redundancy. Other forms of redundancy also present system management overhead demands, like computational capacity to perform software-implemented fault tolerance tasks. Like all design undertakings, the realization of fault tolerance presents trade-offs and the necessity for design optimization. Ultimately, a balanced, minimal, and validatable design must be sought that demonstrably provides the safeguards and fault survival margins appropriate to the subject application.

A broad range of redundancy implementation options exist to mechanize desired levels and types of fault tolerance. Figure 28.5 presents a taxonomy of redundancy options that may be invoked in appropriate combinations to yield an encompassing fault tolerance architecture.

This taxonomy indicates the broad range of redundancy possibilities that may be invoked in system design. Although most of these options are exemplified or described in later paragraphs, it may be noted briefly that redundancy is characterized by its category, mode, coordination, and composition aspects. Architectural commitments have to be made in each aspect. Thus, a classical system might, for example, employ fault masking using replicated hardware modules that operate synchronously. At a lower level, the associated data buses might use redundancy encoding for error detection and correction. To safeguard against a generic design error, a backup capability might be added using a dissimilar redundancy scheme.

Before considering system architectures per se, however, key elements of Figure 28.5 need to be described and exemplified. Certain of these redundancy implementation options are basic to formulating a fault-tolerant

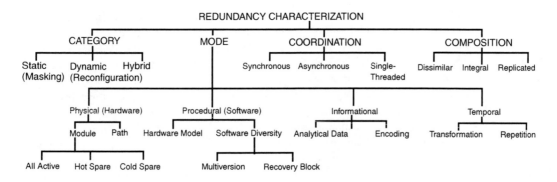

FIGURE 28.5 Redundancy classification.

TABLE 28.4 Fault Avoidance Techniques and Tools

Technique
Use of mature and formally verified components
Conformance to standards
Formal methods
Design automation
Integrated formal methods and VHDL design methodology
Simplifying abstractions
Performance common-mode failure avoidance
Software and hardware engineering practice
Design diversity

architecture, so their essential trade-offs need to be defined and explored. In particular, the elements of masking vs. reconfiguration, active vs. standby spares, and replicated vs. dissimilar redundancy are reviewed here.

No unifying theory has been developed that can treat CMFs the same way the Byzantine resilience (BR) treats random hardware or physical operational faults. Three techniques, fault-avoidance, fault-removal, and fault-tolerance are the tools available to design a system tolerant of CMFs. The most cost effective phase of the total design and development process for reducing the likelihood of CMFs is the earliest part of the program. Table 28.4 presents fault avoidance techniques and tools that are being used [Lala and Harper, 1994].

Common-mode fault removal techniques and tools include design reviews, simulation, testing, fault injection, and a rigorous quality control program. Common-mode fault tolerance requires error detection and recovery. It is necessary to corroborate the error information across redundant channels to ascertain which recovery mechanism (i.e., physical fault recovery, or common-mode failure recovery) to use. Recovery from CMF in real time requires that the state of the system be restored to a previously known correct point from which the computational activity can resume [Lala and Harper, 1994].

28.2.3 Architectural Categories

As indicated in Figure 28.5, the three categories of fault-tolerant architectures are masking, reconfiguration, and hybrid.

28.2.3.1 Fault Masking

The masking approach is classical per von Neumann's triple modular redundancy (TMR) concept, which has been generalized for arbitrary levels of redundancy. The TMR concept centers on a voter that, within a spares exhaustion constraint, precludes a faulted signal from proceeding along a signal path. The approach is passive in that no reconfiguration is required to prevent the propagation of an erroneous state or to isolate a fault.

Modular avionics systems consisting of multiple identical modules and a voter require a trade-off of reliability and safety. A "module" is not constrained to be a hardware module; a module represents an entity capable of producing an output. When safety is considered along with reliability, the module design affects both safety and reliability. It is usually expected that reliability and safety should improve with added redundancy. If a module has built-in error detection capability, it is possible to increase both reliability and safety with the addition of one module providing input to a voter. If no error detection capability exists at the module level, at least two additional modules are required to improve both reliability and safety. An error control arbitration strategy is the function implemented by the voter to decide what is the correct output, and when the errors in the module outputs are excessive so that the correct output cannot be determined, the voter may put out an unsafe signal. Reliability and safety of an n-module safe modular redundant (nSMR) depend on the individual module reliability and on the particular arbitration strategy used. No single arbitration strategy is optimal for improving both reliability and safety. Reliability is defined as the probability the voter's data output is correct and the voter does not assert the unsafe signal. Safety = Reliability plus the probability the voter asserts the unsafe signal. As system reliability and safety are interrelated, increasing system reliability may result in a decrease in system safety, and vice versa [Vaidya and Pradhan, 1993].

Voters that use bit-for-bit comparison have been employed when faults consist of arbitrary behavior on the part of failed components, even to the extreme of displaying seemingly intelligent malicious behavior [Lala and Harper, 1994]. Such faults have been called Byzantine faults. Requirements levied on an architecture tolerant of Byzantine faults (referred to as Byzantine-resilient [BR]) comprise a lower bound on the number of fault containment regions, their connectivity, their synchrony, and the utilization of certain simple information exchange protocols. No *a priori* assumptions about component behavior are required when using bit-for-bit comparison. The dominant contributor to failure of correctly designed BR system architecture is the common-mode failure.

Fault effects must be masked until recovery measures can be taken. A redundant system must be managed to continue correct operation in the presence of a fault. One approach is to partition the redundant elements into individual fault containment regions (FCRs). An FCR is a collection of components that operates correctly regardless of any arbitrary logical or electrical fault outside the region. A fault containment boundary requires the hardware components be provided with independent power and clock sources. Interfaces between FCRs must be electrically isolated. Tolerance to such physical damage as a weapons hit necessitates a physical separation of FCRs such as different avionics bays. In flight control systems, a channel may be a natural FCR. Fault effects manifested as erroneous data can propagate across FCR boundaries. Therefore, the system must provide error containment as well. This is done using voters at various points in the processing including voting on redundant inputs, voting the result of control law computations, and voting at the input to the actuator. Masking faults and errors provides correct operation of the system with the need for immediate damage assessment, fault isolation, and system reconfiguration [Lala and Harper, 1994].

28.2.3.2 Reconfiguration

Hardware interlocks provide the first level of defense prior to reconfiguration or the use of the remaining non-faulty channels. In a triplex or higher redundancy system, the majority of channels can disable the output of a failed channel. Prior to taking this action, the system will determine whether the failure is permanent or transient.

Once the system determines a fault is permanent or persistent, the next step is to ascertain what functions are required for the remainder of the mission and whether the system needs to invoke damage assessment, fault isolation, and reconfiguration of the remaining system assets. The designer of a system required for long-duration missions may undertake to implement a design having reconfiguration capability.

28.2.3.3 Hybrid Fault Tolerance

Hybrid fault tolerance uses hybrid redundancy, which is a combination of static and dynamic redundancy, i.e., masking, detection, and recovery that may involve reconfiguration. A system using hybrid redundancy will have N-active redundant modules, as well as spare (S) modules. A disagreement detector detects if the output of any of the active modules is different from the voter output. If a module output disagrees with the voter, the switching circuit replaces the failed module with a spare. A hybrid (N,S) system cannot have more than (N-1)/2 failed modules at a time in the core, or the system will incorrectly switch out the good module when two out of three have failed.

28.2.3.4 Hybrid Fault Tolerance

Hybrid fault tolerance employs a combination of masking and reconfiguration, as noted in Section 28.2.3. The intent is to draw on strengths of both approaches to achieve superior fault tolerance. Masking precludes an erroneous state from affecting system operation and thus obviating the need for error recovery. Reconfiguration removes faulted inputs to the voter so that multiple faults cannot defeat the voter. Masking and reconfiguration actions are typically implemented in a voter-comparator mechanism, which is discussed in Section 28.3.1.

Figure 28.6 depicts a hybrid TMR arrangement with a standby spare channel to yield double fail-operational capability. Upon the first active channel failure, it is switched out of the voter-input configuration, and the standby channel is switched in. Upon a second channel failure, the discrepant input to the voter is switched out. Only two inputs remain then, so a succeeding (third) channel failure can be detected but not properly be identified by the voter per se. With a voter that selects the lower of two remaining signals, and hence precludes a hardover output, a persistent miscomparison results in a fail-passive loss of system function.

An alternative double-fail operational configuration would forego the standby channel switching and simply employ a quadruplex voter. This architecture is actually rather prevalent in dedicated flight-critical systems like fly-by-wire (FBW) flight control systems. This architecture still employs reconfiguration to remove faulty inputs to the voter.

The fault tolerance design elements described in Section 28.1.6 are reflected in the fault-tolerant architecture in Figure 28.6 by way of annotations. For example, error detection is provided by the comparators; damage assessment is then accomplished by the reconfiguration logic using the various comparator states. fault containment and service continuation are both realized through the voter, which also obviates the need for error recovery. Last, fault treatment is accomplished by the faulty path switching prompted be the reconfiguration logic. Thus, this simple example illustrates at a high level how the various aspects of fault tolerance can be incorporated into an integrated design.

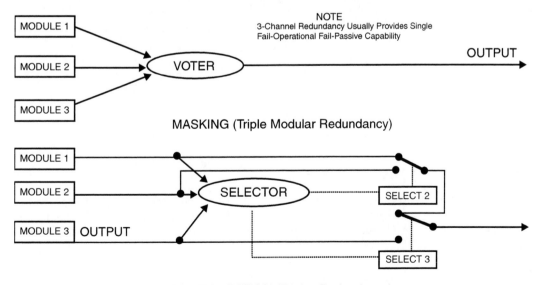

FIGURE 28.6 Masking vs. Reconfiguration.

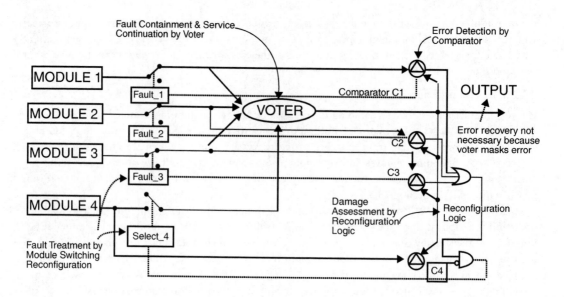

FIGURE 28.7   Hybrid TMR arrangement.

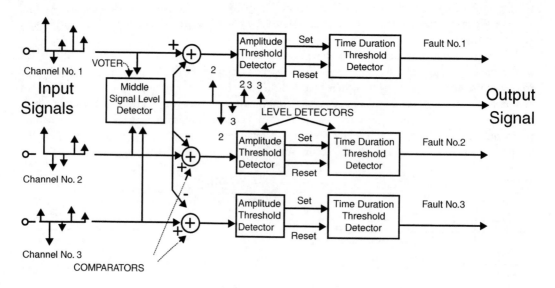

FIGURE 28.8   Triplex voter-comparator.

28.2.4   Integrated Mission Avionics

In military applications, redundant installations in some form will be made on opposite sides of the aircraft to avoid loss of functionality from battle damage to a single installation. Vulnerability to physical damage also exists in the integrated rack installations being used on commercial aircraft. Designers must take these vulnerabilities into account in the design of a fault-tolerant system.

28.2.5   System Self Tests

Avionics system reliability analyses are conditional on assumptions of system readiness at dispatch. For lower-criticality systems, certain reductions in redundancy may sometimes be tolerable at dispatch. For full-time flight-critical systems, however, a fully operable system with all redundancy intact is generally assumed in a system reliability prediction. This assumption places appreciable demands on system preflight self test in terms of coverage and confidence values. Such a test is typically an end-to-end test that exercises all elements in a phased manner that would not be possible during flight. The fault tolerance provisions demand particular emphasis. For example, such testing deliberately seeks to force seldom-used comparator trips to ensure the absence of latent faults, like passive hardware failures. Analysis of associated testing schemes and their scope of coverage is necessarily an ongoing design analysis task during development. These schemes must also include appropriate logic interlocks to ensure safe execution of the preflight test, e.g., a weight-on-wheels interlock to preclude testing except on the ground. Fortunately, the programming of system self-tests can be accomplished in a relatively complete and high-fidelity manner.

Because of the discrete-time nature of digital systems, all capacity is not used for application functions. Hence, periodic self tests are possible for digital components like processors during flight. Also, the processors can periodically examine the health status of other system components. Such tests provide a self-monitoring that can reveal the presence of a discrepancy before error states are introduced or exceed detection thresholds. The lead time afforded by self tests can be substantial because steady flight may not simulate comparator trips due to low-amplitude signals. Moreover, the longer a fault remains latent, the greater the possibility that a second fault can occur. Hence, periodic self-tests can significantly enhance system reliability and safety by reducing exposure to coincident multiple fault manifestations.

Self-monitoring may be employed at still lower levels, but there is a trade-off as to the granularity of fault detection. This trade-off keys on fault detection/recovery response and on the level of fault containment selected. In general, fault containment delineates the granularity of fault detection unless recovery response times dictate faster fault detection that is best achieved at lower levels.

28.3 Hardware-Implemented Fault Tolerance (Fault-Tolerant Hardware Design Principles)

28.3.1 Voter Comparators

Voter comparators are very widely used in fault-tolerant avionics systems, and they are generally vital to the integrity and safety of the associated systems. Because of the crucial role of voter comparators, special care must be exercised in their development. These dynamic system elements, which can be implemented in software as well as hardware, are not as simple as they might seem. In particular, device integrity and threshold parameter settings can be problematic.

Certain basic elements and concerns apply over the range of voter-comparator variants. A conceptual view of a triplex voter-comparator is depicted in Figure 28.7. The voter here is taken to be a middle signal selector, which means that the intermediate level of three inputs is selected as the output. The voter section precedes the comparators because the output of the voter is an input to each comparator. Basically, the voter output is considered the standard of correctness, and any input signal that persists in varying too much from the standard is adjudged to be erroneous.

In Figure 28.7, the respective inputs to each of the signal paths is an amplitude-modulated pulse train, as is normal in digital processing. Each iteration of the voter is a separate selection, so each voter output is apt to derive from any input path. This is seen in Figure 28.8, where the output pulse train components are numbered per the input path selected at each point in time. At each increment of time, the voter output is applied to each of the comparators, and the difference with each input signal is fed to a corresponding amplitude threshold detector. The amplitude threshold is set so that accumulated tolerances are not apt to trip the detector. As shown here, the amplitude detector issues a set output when an excessive difference is first observed. When the difference falls back within the threshold, a reset output is issued.

Because transient effects may produce short-term amplitude detector trips, a timing threshold is applied to the output of each amplitude detector. Typically, a given number of consecutive out-of-tolerance amplitude threshold trips are necessary to declare a faulty signal. Hence, a time duration threshold detector begins a count whenever a set signal is received, and in the absence of further inputs, increments the count for each sample interval thereafter. If a given cycle count is exceeded, a erroneous state is declared and a fault logic signal is set for the affected channels. Otherwise, the count is returned to zero when a reset signal is received.

The setting of both the timing and amplitude thresholds is of crucial importance because of the trade-off between nuisance fault logic trips and slow response to actual faults. Nuisance trips erode user confidence in a system, their unwarranted trips can potentially cause resource depletion. On the other hand, a belated fault response may permit an unsafe condition or catastrophic event to occur. The allowable time to recover from a given type of fault, which is application-dependent, is the key to setting the thresholds properly. The degree of channel synchronization and data skewing also affect the threshold settings, because they must accommodate any looseness. The trade-off can become quite problematic where fast fault recovery is required.

Because the integrity and functionality of the system is at stake, the detailed design of a voter comparator must be subject to careful assessment at all stages of development. In the case of a hardware-implemented device, its fault detection aspects must be thoroughly examined. Passive failures in circuitry that is not normally used are the main concern. Built-in test, self-monitoring, or fail-hard symptoms are customary approaches to device integrity. In the case of software-implemented voter comparators, their dependability can be reenforced through formal proof methods and in-service verified code.

28.3.2 Watchdog Timers

Watchdog timers can be used to catch both hardware and software wandering into undesirable states [Lala and Harper, 1994]. Timing checks are a form of assertion checking. This kind of check is useful because many software and hardware errors are manifested in excessive time taken for some operation. In synchronous data flow architectures, data are to arrive at a specific time. Data transmission errors of this type can be detected using a timer.

28.4 Software-Implemented Fault Tolerance—State Consistency

Software performs a critical role in digital systems. The term "software implemented fault tolerance" as used in this chapter is used in the broader sense indicating the role software plays in the implementation of fault tolerance, and not as a reference to the SRI International project performed for NASA in the late 1970s and referred to as SIFT.

28.4.1 Error Detection

Software plays a major role in error detection. Error detection at the system level should be based on the specification of system behavior. The outputs of the system should be checked to assure that the outputs conform to the specification. These checks should be independent of the system. Since they are implemented in software, the checks require access to the information to be checked, and therefore may have the potential of corrupting that information. Hence, the independence between a system and its check cannot be absolute. The provision of ideal checks for error detection is rarely practical, and most systems employ checks for acceptability [Anderson and Lee, 1981].

Deciding where to employ system error detection is not a straightforward matter. Early checks should not be regarded as substitute for last-moment checks. An early check will of necessity be based on a knowledge of the internal workings of the system and hence will lack independence from the system. An early check could detect an error at the earliest possible stages and hence minimize the spread of damage.

A last moment check ensures that none of the output of the system remains unchecked. Therefore, both last-moment and early checks should be provided in a system [Anderson and Lee, 1981].

In order to detect software faults, it is necessary that the redundant versions of the software be independent of each other, that is, of diverse design [Avizenis and Kelley, 1982] (See Section 28.5).

28.4.1.1 Replication Checks

If design faults are expected, replication must be provided using versions of the system with different designs. Replication checks compare the two sets of results produced as outputs of the replicated modules. The replication check raises an error flag and intiates the start of other processes to determine which component or channel is in error [Anderson and Lee, 1981].

28.4.1.2 Timing Checks

Timing checks are used to reveal the presence of faults in a system, but not their absence [Anderson and Lee, 1981] In synchronous hard real-time systems, messages containing data are transmitted over data buses at a specific schedule. Failure to receive a message at the scheduled time is an error. The error could be caused by faults in a sensor, data bus, etc. In this case, if the data were critical, a method of tolerating the fault may be to use a forward state extrapolation.

28.4.1.3 Reversal Check (Analytical Redundancy)

A reversal check takes the outputs from a system and calculates what the inputs should have been to produce that output. The calculated inputs can then be compared with the actual inputs to check whether there is an error. Systems providing mathematical functions often lend themselves to reversal checks [Anderson and Lee, 1981].

Analytic redundancy using either of two general error detection methods, multiple model (MM) or generalized likelihood ratio (GLR), is a form of reversal check. Both methods make use of a model of the system represented by Kalman filters. The MM attempts to calculate a measure of how well each of the Kalman filters is tracking by looking at the prediction errors. Real systems possess nonlinearity and the model assumes a linear system. The issue is whether the tracking error from the extended Kalman filter corresponds to the linearized model "closest to" the true, nonlinear system and is markedly smaller than the errors from the filters based on "more distant" models. Actuator and sensor failures can be modeled in different ways using this methodology [Willsky, 1980].

The Generalized Likelihood Ratio (GLR) uses a formulation similar to that for MM, but different enough that the structure of the solution is quite different. The starting point for GLR is a model describing normal operation of the observed signals or of the system from which they come. Since GLR is directly aimed at detecting abrupt changes, its applications are restricted to problems involving such changes, such as failure detection. GLR, in contrast to MM, requires a single Kalman filter. Any detectable failure will exhibit a systematic deviation between what is observed and what is predicted to be observed. If the effect of the parametric failure is "close enough" to that of the additive one, the system will work.

Underlying both the GLR and MM methods is the issue of using system redundancy to generate comparison signals that can be used for the detection of failures. The fundamental idea involved in finding comparison signals is to use system redundancy, i.e., known relationships among measured variables to generate signals that are small under normal operation and which display predictable patterns when particular anomalies develop. All failure detection is based on analytical relationships between sensed variables, including voting methods, which assume that sensors measure precisely the same variable. The trade-off using analytical relationships is that we can reduce hardware redundancy and maintain the same level of fail-operability. In addition, analytical redundancy allows extracting more information from the data, permitting detection of subtle changes in system component characteristics. On the other hand, the use of this information can cause problems if there are large uncertainties in the parameters specifying the analytical relationships [Willsky, 1980].

The second part of a failure detection algorithm is the decision rule that uses the available comparison signals to make decisions on the interruption of normal operation by the declaration of failures. One advantage of these methods is that the decision rule — maximize and compare to a threshold — are simple, while the main disadvantage is that the rule does not explicitly reflect the desired trade-offs. The Bayesian Sequential Decision approach, in which an algorithm for the calculation of the approximate Bayes decision, has exactly the opposite properties, i.e., it allows for a direct incorporation of performance trade-offs, but is extremely complex. The Bayes Sequential Decision Problem is to choose a stopping rule and terminal decision rule to minimize the total expected cost, and the expected cost that is accrued before stopping [Willsky, 1980].

28.4.1.4 Coding Checks

Coding checks are based on redundancy in the representation of an object in use in a system. Within an object, redundant data are maintained in some fixed relationship with the (nonredundant) data representing the value of the object. Parity checks are a well-known example of a coding check, as are error detection and correction codes such as the Hamming, cyclic redundancy check, and arithmetic codes [Anderson and Lee, 1981].

28.4.1.5 Reasonableness Checks

These checks are based on knowledge of the minimun and maximum values of input data, as well as the limits on rate of change of input data. These checks are based on knowledge of the physical operation of sensors, and employ models of this operation.

28.4.1.6 Structural Checks

Two forms of checks can be applied to the data structures in a computing system. Checks on the semantic integrity of the data will be concerned with the consistency of the information contained in a data structure. Checks on the structural integrity will be concerned with whether the structure itself is consistent. For example, external data from subsystems is transmitted from digital data buses such as MIL-STD-1553, ARINC 429 or ARINC 629. The contents of a message (number of words in the message, contents of each word) from a subsystem are stored and the incoming data checked for consistency.

28.4.1.7 Diagnostic Checks

Diagnostic checks create inputs to the hardware elements of a system, which should produce a known output. These checks are rarely used as the primary error detection measure. They are normally run at startup, and may be initiated by an operator as part of a built-in test. They may also run continuously in a background mode when the processor might be idle. Diagnostic checks are also run to isolate certain faults.

28.4.2 Damage Confinement and Assessment

When an error has been discovered, it is necessary to determine the extent of the damage done by the fault before error recovery can be accomplished. Assessing the extent of the damage is usually related to the structure of the system. Assuming timely detection of errors, the assessment of damage is usually determined to be limited to the current computation or process. The state is assumed consistent on entry. An error detection test is performed before exiting the current computation. Any errors detected are assumed to be caused by faults in the current computation.

28.4.3 Error Recovery

After the extent of the damage has been determined, it is important to restore the system to a consistent state. There are two primary approaches — backward and forward error recovery. In backward error recovery, the system is returned to a previous consistent state. The current computation can then be retried with existing components (retry)* with alternate components (reconfigure), or it can be ignored (skip frame).** The use of backward recovery implies the ability to save and restore the state. Backward error recovery is independent of damage assessment.

Forward error recovery attempts to continue the current computation by restoring the system to a consistent state, compensating for the inconsistencies found in the current state. Forward error recovery implies detailed knowledge of the extent of the damage done, and a strategy for repairing the inconsistencies. Forward error recovery is more difficult to implement than backward error recovery [Hitt et al., 1984].

28.4.4 Fault Treatment

Once the system has recovered from an error, it may be desirable to isolate and/or correct the component that caused the error. Fault treatment is not always necessary because of the transient nature of some faults or because the detection and recovery procedures are sufficient to cope with other recurring errors. For permanent faults, fault treatment becomes important because the masking of permanent faults reduces the ability of the system to deal with subsequent faults. Some fault-tolerant software techniques attempt to isolate faults to the current computation by timely error detection. Having isolated the fault, fault treatment can be done by reconfiguring the computation to use alternate forms of the computation to allow for continued service. (This can be done serially, as in recovery blocks, or in parallel, as in N-Version programming.) The assumption is that the damage due to faults is properly encapsulated to the current computation and that error detection itself is faultless (i.e., detects all errors and causes none of its own) [Hitt et al., 1984].

28.4.5 Distributed Fault Tolerance

Multiprocessing architectures consisting of computing resources interconnected by external data buses should be designed as a distributed fault-tolerant system. The computing resources may be installed in an enclosure using a parallel backplane bus to implement multiprocessing within the enclosure. Each enclosure can be considered a virtual node in the overall network. A network operating system, coupled with the data buses and their protocol, completes the fault-tolerant distributed system. The architecture can be asynchronous, loosely synchronous, or tightly synchronous. Maintaining consistency of data across redundant channels of asynchronous systems is difficult [Papadopoulos, 1985].

28.5 Software Fault Tolerance

Software faults, considered design faults, may be created during the requirements development, specification creation, software architecture design, code creation, and code integration. While many faults may be found and removed during system integration and testing, it is virtually impossible to eliminate all possible software design faults. Consequently, software fault tolerance is used. Table 28.5 lists the major fault-tolerant software techniques in use today.

The two main methods that have been used to provide software fault tolerance are N-version software and recovery blocks.

28.5.1 Multiversion Software

Multiversion software is any fault-tolerant software technique in which two or more alternate versions are implemented, executed, and the results compared using some form of a decision algorithm. The goal is to develop these alternate versions such that software faults that may exist in one version are not

---

*This is only useful for transient timing on hardware faults.

**For example, in a real-time system no processing for the current computation is accomplished in the current frame, sometimes called "skip frame."

TABLE 28.5   Categorization of Fault-Tolerant Software
Techniques

Multiversion Software
N-Version Program
Cranfield Algorithm for Fault-Tolerance (CRAFT) Food Taster
Distinct and Dissimilar Software
Recovery Blocks
Deadline Mechanism
Dissimilar Backup Software
Exception Handlers
Hardened Kernel
Robust Data Structures and Audit Routines
Run Time Assertions[a]
Hybrid Multiversion Software and Recovery Block Techniques
Tandem
Consensus Recovery Blocks

[a] Not a complete fault-tolerant software technique as it only detects errors.
Source: From Hitt, E. et al., Study of Fault-Tolerant Software Technology, NASA CR 172385.

contained in the other version(s) and the decision algorithm determines the correct value from among the alternate versions. Whatever means are used to produce the alternate versions, the common goal is to have distinct versions of software such that the probability of faults occuring simultaneously is small and that faults are distinguishable when the results of executing the multiversions are compared with each other.

The comparison function executes as a decision algorithm once it has received results from each version. The decision algorithm selects an answer or signals that it cannot determine an answer. This decision algorithm and the development of the alternate versions constitute the primary error detection method. Damage assessment assumes the damage is limited to the encapsulation of the individual software versions. Faulted software components are masked so that faults are confined within the module in which they occur. Fault recovery of the faulted component may or may not be attempted.

N-versions of a program are independently created by N-software engineering teams working from a (usually) common specification. Each version executes independently of the other versions. Each version must have access to an identical set of input values and the outputs are compared by an executive which selects the result used. The choice of an exact or inexact voting check algorithm is influenced by the criticality of the function and the timing associated with the voting.

28.5.2   Recovery Blocks

The second major technique shown in Table 28.5 is the recovery block and its subcategories — deadline mechanism and dissimilar backup software. The recovery block technique recognizes the probability that residual faults may exist in software. Rather than develop independent redundant modules, this technique relies on the use of a software module which executes an acceptance test on the output of a primary module. The acceptance test raises an exception if the state of the system is not acceptable. The next step is to assess the damage and recover from the fault. Given that a design fault in the primary module could have caused arbitrary damage to the system state, and that the exact time at which errors were generated cannot be identified, the most suitable prior state for restoration is the state that existed just before the primary module was entered [Anderson and Lee, 1981].

28.5.3   Trade-Offs

Coverage of a large number of faults has an associated overhead in terms of redundancy, and the processing associated with the error detection. The designer may use modeling and simulation to determine the amount of redundancy required to implement the fault tolerance vs. the probability and impact of the different types of faults. If a fault has minimal or no impact on safety, or mission completion, investing in redundancy to handle that fault may not be effective, even if the probability of the fault occuring is significant.

28.6 Summary

Fault-tolerant systems must be used whenever a failure can result in loss of life, or loss of a high-value asset. Physical failures of hardware are decreasing whereas design faults are virtually impossible to completely eliminate.

28.6.1 Design Analyses

In applying fault-tolerance to a complex system, there is a danger that the new mechanisms may introduce additional sources of failure due to design and implementation errors. It is important, therefore, that the new mechanisms be introduced in a way that preserves the integrity of a design, with minimum added complexity. The designer must use modeling and simulation tools to assure that the design accomplishes the needed fault tolerance.

Certain design principles have been developed to simplify the process of making design decisions. Encapsulation and hierarchy offer ways to achieve simplicity and generality in the realization of particular fault-tolerance functions. Encapsulation provides:

- Organization of data and programs as uniform objects, with rigorous control of object interaction.
- Organization of sets of alternate program versions into fault-tolerant program modules (e.g. recovery blocks and N-version program sets).
- Organization of consistent sets of recovery points for multiple processes.
- Organization of communications among distributed processes as atomic (indivisible) actions.
- Organization of operating system functions into recoverable modules.

Examples of the hierarchy principle used to enhance reliability of fault-tolerance functions include:

- Organization of all software, both application and system type, into layers, with unidirectional dependencies among layers.
- Integration of service functions and fault-tolerance functions at each level.
- Use of nested recovery blocks to provide hierarchical recovery capability.
- Organization of operating system functions so that only a minimal set at the lowest level (a "kernel") needs be exempted from fault tolerance.
- Integration of global and local data and control in distributed processors.

That portion of the operating system kernel that provides the basic mechanisms the rest of the system uses to achieve fault-tolerance should be "trusted." This kernel should be of limited complexity so that all possible paths can be tested to assure correct operation under all logic and data conditions. This kernel need not be fault tolerant if the foregoing can be assured.

28.6.2 Safety

Safety is defined in terms of hazards and risks. A hazard is a condition, or set of conditions that can produce an accident under the right circumstances. The level of risk associated with the hazard depends on the probability that the hazard will occur, the probability of an accident taking place if the hazard does occur, and the potential consequence of the accident [Williams, 1992].

28.6.3 Validation

Validation is the process by which systems are shown through operation to satisfy the specifications. The validation process begins when the specification is complete. The difficulty of developing a precise specification that will never change has been recognized. This reality has resulted in an iterative development and validation process. Validation requires developing test cases and executing these test cases on the hardware and software comprising the system to be delivered. The tests must cover 100% of the faults the system is designed to tolerate, and a very high percentage of possible design faults, whether hardware, software, or the interaction of the hardware and software during execution of all possible data and logical paths. Once the system has been validated, it can be put in operation. In order to minimize the need to validate a complete Operational Flight Program (OFP) every time it is modified, development methods attempt to decompose a large system into modules that are independently specified, implemented, and validated. Only those modules and their interfaces that are modified must be revalidated using this approach (see Chapter 29).

Rapid prototyping, simulation, and animation are all techniques that help validate the system. Formal methods are being used to develop and validate digital avionics systems. There are arguments both in favor of and against the use of formal methods [Rushby, 1993; Williams, 1992].

28.6.4 Conclusion

For safety-critical systems, fault tolerance must be used to tolerate design faults which are predominately software- and timing-related. It is not enough to eliminate almost all faults introduced in the later stages of a life cycle; assurance is needed that they have been eliminated, or are extremely improbable. Safety requirements for commercial aviation dictate that a failure causing loss of life must be extremely improbable, on the order of 10^{-9} per flight-hour. The designer of safety-critical fault-tolerant systems should keep current with new development in this field since both design and validation methods continue to advance in capability.

References

Anderson, T. and Lee, P. A., 1981. *Fault Tolerance, Principles and Practices*, Prentice-Hall, London.

Anderson, T., Ed., 1989. Safe and Secure Computing Systems, Blackwell Scientific, Oxford, U.K.

Avizienis, A., 1976. Fault-Tolerant Systems, *IEEE Trans. Comput.*, C-25(12):1304-1312.

Avizienis, A. and Kelly, J., 1982. Fault-Tolerant Multi-Version Software: Experimental Results of a Design Diversity Approach, UCLA Computer Science Department. Los Angeles, CA.

Avizienis, A., Kopetz, H., and Laprie, J.C., Eds., 1987. *The Evolution of Fault-Tolerant Computing*, Springer-Verlag, New York.

Best, D. W., McGahee, K. L., and Shultz, R. K.A., 1988. Fault Tolerant Avionics Multiprocessing System Architecture Supporting Concurrent Execution of Ada Tasks, Collins Government Avionics Division, AIAA 88-3908–CP.

Gargaro, A.B. et al., 1990. Adapting Ada for Distribution and Fault Tolerance, in *Proc.* 4th *Int. Workshop Real-Time Ada Issues*, ACM.

Gu, D., Rosenkrantz, D. J., and Ravi, S. S., 1994. Construction of Check Sets for Algorithm-Based Fault Tolerance, *IEEE Trans. Comput.*, 43(6): 641–650.

Hitt, E., Webb, J., Goldberg, J., Levitt, K., Slivinski, T., Broglio, C., and Wild, C., 1984. *Study of Fault-Tolerant Software Technology*, NASA CR 172385, Langley Research Center, VA.

Hudak, J., Suh, B.H., Siewiorek, D., and Segall, Z., 1993. Evaluation and Comparison of Fault Tolerant Software Techniques, *IEEE Trans. Reliability*.

Lala, J. H. and Harper, R. E., 1994. Architectural Principles for Safety-Critical Real-Time Applications, *Proc. IEEE*, 82(1): 25–40.

Lala, P. K., 1985. *Fault Tolerant & Fault Testable Hardware Design*, Prentice-Hall, London, ISBN 0-13-308248-2.

Papadopoulos, G. M., 1985. Redundancy Management of Synchonous and Asynchronous Systems, Fault Tolerant Hardware/Software Architecture for Flight Critical Functions, AGARD-LS-143.

Rushby, J., 1993. *Formal Methods and Digital Systems Validation for Airborne Systems*, NASA CR 4551, Langley Research Center, VA.

Shin, K. G. and Parmeswaran R., 1994. Real-Time Computing: A New Discipline of Computer Science and Engineering, *Proc. IEEE,* 82(1): 6–24.

Shin, K. G. and Hagbae, K., 1994. A Time Redundancy Approach to TMR Failures Using Fault-State Likelihoods, *IEEE Trans. Comput.,* 43(10): 1151–1162.

Sosnowski, J., 1994. Transient Fault Tolerance in Digital Systems, *IEEE Micro,* 14(1): 24–35.

Tomek, L., Mainkar, V., Geist, R. M., and Trivedi, K. S., 1994. Reliability Modeling of Life-Critical, Real-Time Systems, Proc. IEEE, 82(1): 108–121.

Vaidya, N. H. and Pradhan, D. K., 1993. Fault-Tolerant Design Strategies for High Reliability and Safety, *IEEE Trans. Comput.,* 42(10): 1195–1206

Williams, L. G., 1992. Formal Methods in the Development of Safety Critical Software Systems, UCRL-ID-109416, Lawrence Livermore National Laboratory.

Willsky, A. S., 1980. Failure Detection in Dynamic Systems, *Fault Tolerance Design and Redundancy Management Techniques,* AGARD-LS-109.

Further Information

A good introduction to fault-tolerant design is presented in *Fault Tolerance, Principles and Practices*, by Tom Anderson and P.A. Lee. Hardware-specific design techniques are described by P.K. Lala in *Fault Tolerant & Fault Testable Hardware Design*.

Other excellent journals are the IEEE publications, *Computers, Micro, Software,* and *Transactions on Computers,* and *Software Engineering.*

29

Boeing B-777

Michael J. Morgan

Honeywell

29.1 Introduction

The avionics industry has long recognized the substantial cost benefits which could be realized using a large-scale integrated computing architecture for airborne avionics. Technology achievements by airframe, avionics, and semiconductor manufacturers allow implementation of these integrated avionics architectures resulting in substantial life cycle cost benefits. The Boeing 777 Aircraft Information Management System (AIMS) represents the first application of an integrated computing architecture in a commercial air transport.

29.2 Background

Since 1988, the avionics industry has made a significant effort to develop the requirements and goals for a next-generation integrated avionics architecture. This work is documented in ARINC Project Paper 651. Top-level goals of the Integrated Modular Avionics (IMA) architecture are to reduce overall cost of ownership through reduced spares requirements (includes reduction in cost of spare Line Replaceable Modules [LRM]) and reduction in number of LRMs required), reduced equipment removal rate, and reduced weight and volume in both avionics and wiring. In addition, IMA addresses the airlines' demand for better MTBUR/MTBF (Mean Time Between Unscheduled Removals as a fraction of Mean Time Between Failures), improved system performance (response time), increased airborne functionality, better fault isolation and test, and maintenance-free dispatch for extended intervals.

Technology trends in microprocessor and memory technology demand that airborne computing architectures evolve if the avionics industry is to meet the goals of IMA. By exploiting these developments in the microprocessor and memory industries, very highly integrated architectures previously not technologically feasible or cost-effective may now be realized. These functionally integrated architectures minimize life cycle cost by minimizing the duplication of hardware and software elements (see Figure 29.1).

Hardware Resources	Software Resources
Power Supply	Operating System
Processor	
Memory System	Utility Software
Chassis	Hardware Bite
Shared I/O	
Unique I/O	Application Bite
	Application S/W

 Common Resource

In a typical LRU architecture, significant resource duplication exists between LRU's

FIGURE 29.1   Components of a typical LRU.

High levels of functional integration dictate availability and integrity requirements far exceeding the requirements for distributed implementations. Resource availability requirements must be sufficient to probabilistically preclude the simultaneous loss of multiple functions utilizing shared resources. These availability requirements imply application of fault-tolerant technology. Although fault tolerance is required to meet the integrity and availability goals of IMA, it is also directly compatible with the airline goal for deferred maintenance. Furthermore, since fault-tolerant technology requires high-integrity monitoring, it also is compatible with airline desires for improved fault isolation, better maintenance diagnostics, and reduced unconfirmed removal rate (MTBUR). Current IMA implementations are realizing a more than six times improvement in unconfirmed equipment removals over a typical federated LRU-based architecture.

High functional integration also implies the requirement to maintain functional independence for software utilizing any shared resource. Strict CPU separation is not sufficient to ensure that functions will not adversely affect each other. I/O resource sharing demands a backplane bus architecture which has extremely high integrity and enforces rigid partitioning between all users. Processor resource sharing requires a robust software partitioning system where all partition protection elements are monitored to ensure isolation integrity.

Robust partitioning protection must be performed as an integral part of the architecture, and isolation must not be dependent upon the integrity of the application software. In this environment, the robust partitioning architecture would be certified as a standalone element allowing functional software to be updated and certified independently of other functions sharing the same computational or I/O resources. Since it is anticipated that airborne functionality will continue to increase and that the majority of this increase will be accommodated via software changes alone, this partitioned environment will provide flexibility in responding to evolving system requirements (e.g., CNS/ATM).

29.3 Boeing 777 Airplane Information Management System (AIMS)

The Boeing 777 Airplane Information Management System implements the IMA concept in an architecture which supports a high degree of functional integration and reduces duplicated resources to a minimum. In this architecture the conventional Line Replaceable Units (LRUs) which typically contain a single function, are replaced with dual integrated cabinets which provide the processing and the I/O hardware and software required to perform the following functions (see Figure 29.2):

Flight Management
Display
Central Maintenance
Airplane Condition Monitoring
Communication Management (including flight deck communication)
Data Conversion Gateway (ARINC 429/629 Conversion)

The integrated cabinets are connected to the airplane interfaces via a combination of ARINC 429, ARINC 629, and discrete I/O channels (see Figure 29.3 Note that for clarity the 429 and discrete channels are not shown).

29.4 Cabinet Architecture Overview

The heart of the AIMS system consists of dual cabinets in the electronics bay that each contain four core processor modules (CPMs) and four input/output modules (IOMs), with space reserved in the cabinet to add one CPM and two IOMs to accommodate future growth (reference Figure 29.4). The shared platform resources provided by AIMS are

Common processor and mechanical housing,
Common input/output ports, power supply, and mechanical housing,
Common backplane bus (SAFEbus™) to move data between CPMs and between CPMs and IOMs,
Common operating system and built-in test (BIT) and utility software.

Instead of individual applications residing in a separate LRU, applications are integrated on common CPMs. The IOMs transmit data from the CPMs to other systems on the airplane, and receive data from these other systems for use by the CPM applications. A high-speed backplane bus, called SAFEbus™, provides a 60-Mbit/s data pipe between any of the CPMs and IOMs in a cabinet. Communication between AIMS cabinets is through four ARINC 629 serial buses.

The robust partitioning provided by the architecture allows applications to use common resources without any adverse interactions. This is achieved through a combination of memory management and

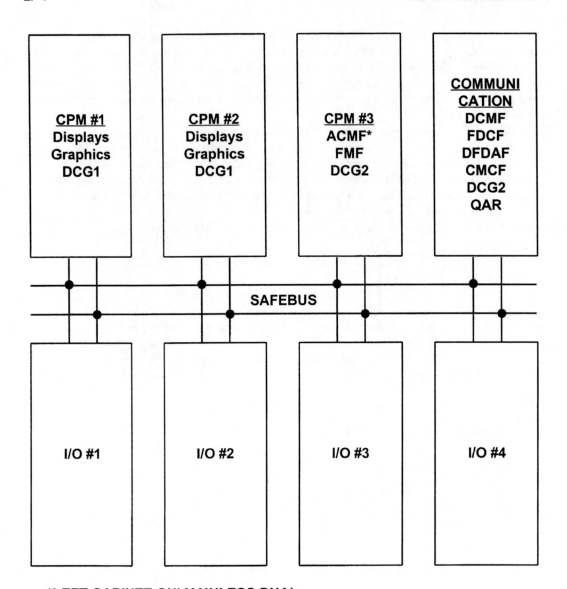

*LEFT CABINET ONLY UNLESS DUAL

FIGURE 29.2   AIMS baseline functional distribution.

deterministic scheduling of application software execution. Memory is allocated before run time, and only one application partition is given write-access to any given page of memory. Scheduling of processor resources for each application is also done before run time, and is controlled by a set of tables loaded onto each CPM and IOM in the cabinet. This set of tables operates synchronously, and controls application scheduling on the CPMs as well as data movement between modules across the SAFEbus™.

Hardware fault detection and isolation is achieved via a lock-step design of the CPMs, IOMs, and the SAFEbus™. Each machine cycle on the CPMs and IOMs is performed in lock-step by two separate processing channels, and comparison hardware ensures that each channel is performing identically. If a miscompare occurs, the system will attempt retries where possible before invoking the fault handling and logging software in the operation system. The SAFEbus™ has four redundant data channels that are compared in real time to detect and isolate bus faults.

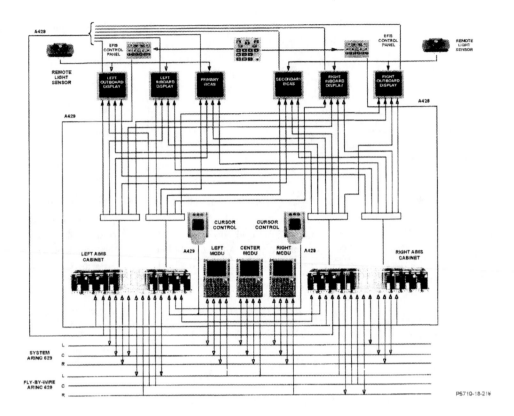

FIGURE 29.3 Airplane interface schematic.

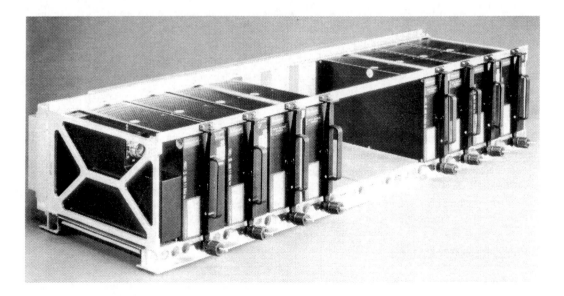

FIGURE 29.4 AIMS cabinet.

The applications hosted on AIMS are listed below, along with the number of redundant copies of each application per shipset in parentheses:

Displays (4)
Flight Management/Thrust Management (2)
Central Maintenance (2)
Data Communication Management (2)
Flight Deck Communication (2)
Airplane Condition Monitoring (1)
Digital Flight Data Acquisition (2)
Data Conversion Gateway (4)

All of the IOMs in the two AIMS cabinets are identical. The CPMs have common hardware for processor, memory, power, and SAFEbus™ interface, but have the capability to include a custom I/O card to provide specific hardware for an application "client." The client hardware in AIMS includes the displays graphics generator, the data communications management fiber optic interface, the digital flight data acquisition interface to the data recorder, ACARS modem interface, and the airplane condition monitoring memory.

The other flight deck hardware elements that make up the AIMS system are

Six flat panel display units
Three control and display units
Two EFIS display control panels
Display select panel
Cursor control devices
Display remote light sensors.

29.5 Backplane Bus

As stated previously, the cabinet LRMs are interconnected via dual high-speed serial buses called SAFE-bus™ (see Figure 29.5). These buses provide the only communication mechanism between the processing and I/O elements of the integrated functions. As such, extremely high availability and integrity requirements are necessary to preclude the simultaneous loss of multiple functions and to preserve robust partitioning of I/O resources. In addition, SAFEbus™ itself is required to provide and enforce the integrity of this key shared resource. Absolute data integrity must be ensured independent of hardware or software failures within any module. In this environment, SAFEbus™ behaves as a generic and virtual resource capable of supporting high levels of I/O integration.

The SAFEbus™ protocol is driven by a sequence of commands stored in each Bus Interface Unit's (BIU) internal table memory. Each command corresponds to a single message transmission. All BIUs are synchronized so that at any given point in time all BIU's "know" the state of the bus and are at equivalent points in their tables. Because buffer addresses are stored in tables they do not need to be transmitted over the bus, and since all transactions are scheduled deterministically, there is no need to arbitrate the bus. This allows for extremely high bus efficiency (>94%) with no bits required to be dedicated to address control and minimal bits required to control data. A more detailed description of SAFEbus™ operation can be found in ARINC Project Paper 659 and also in Reference 3.

29.6 Maintenance

The requirements for fault tolerance allow increased design flexibility and capability for deferred maintenance operation. By taking advantage of the high-integrity hardware monitoring which fault-tolerant design provides, the AIMS cabinets are capable of instantaneous fault detection and confinement. This increased fault visibility allows the cabinet to suppress most faults prior to producing a flight deck effect. This is an important step in reducing the mean time between removal (MTBR) of the equipment.

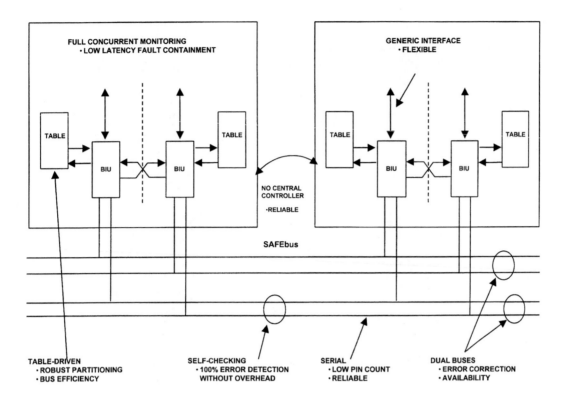

FIGURE 29.5 SAFE bus™ dual serial buses.

In addition, fault tolerance also provides the capability for deferring maintenance to regular (and thus schedulable) intervals. Depending upon the "fail-to-dispatch" probability that the airline is willing to endure, dispatch can continue for 10 to 30 days without maintenance following any first failure in the AIMS.

29.7 Growth

Functional growth is provided in the cabinets through two paths: spare computing and backplane resources provided as part of the baseline AIMS, and three spare LRM slots provided in each cabinet. Spare computing and backplane resources may be used by any function (new or existing) which requires additional throughput or I/O. Existing spare I/O hardware, for example 629 terminals, 429 terminals, and discrete I/O are also available for use by any function integrated into the cabinet. Spare LRM slots may be used for additional processing, I/O, or additional unique hardware which may be required for a specific function due to the generic backplane interface. Additional processing modules may be added as required without changes to existing cabinet hardware. Addition of I/O may require wiring changes if new airplane interfaces are needed.

References

1. Kelly, Michael R., Honeywell, Inc., "Airborne Computer Technology Initiatives," RTCA Paper, December 3, 1990.
2. ARINC Project Paper 651, Draft 6, Design Guidance for Integrated Modular Avionics.
3. Hoyme, Driscoll, Herrlin and Radke, Honeywell, Inc., "ARINC 629 and SAFEbus™: Data Buses for Commercial Aircraft," Scientific Honeyweller, Fall 1991.

Further Information

This chapter is substantially a reprint of material originally presented in:

Morgan, Michael J., Honeywell, Inc., "Integrated Modular Avionics for Next Generation Airplanes," IEEE AES Systems Magazine, August 1991.

Witwer, Robert, Honeywell, Inc., "Developing the 777 Airplane Information Management System (AIMS) A View from Program Start to One Year of Service," August 1996.

30

New Avionics Systems —Airbus A330/A340

J.P. Potocki de Montalk

Airbus Industrie

30.1 Overview

The A330/A340 project is a twin programme — the first time that an aircraft has been designed from the outset both with four engines, and also with two engines. Both aircraft types have essentially the same passenger and freight capacity. The four-engined A340 is optimized for long-range missions, but is also efficient at shorter ranges. With two engines, the A330 offers even better operating economics for the missions where an airline does not need the very long range of the A340.

The realization that on the two different aircraft very many features could in fact be engineered the same way without a penalty, was the key to obtaining substantial commonality between the two products. This approach has provided very substantial advantages for the operators, the airframe manufacturer, and for the equipment vendors. In effect, by designing for both sister aircraft from the outset, the requirements were engineered in common, and any added features for either of the two aircraft could be introduced at a point in the design where they cost very little extra in terms of price, weight, reliability/maintainability or fuel burn.

As a result, the two aircraft use the same parts (except the engine-related parts), can use the same aircrews, use the same airport and maintenance environment, and cost almost the same to develop as a single aircraft. And both are supremely efficient.

The A340 is offered in two configurations, allowing operators to tailor capacity and capability to demand. The larger A340-300 has the same fuselage length as the A330 and, while seating 300 to 350 passengers, has seat-mile costs close to those of the latest 747, making it an economical alternative on long-range routes with lower traffic densities.

The A340-200, seating 250 to 300 passengers, has the longest range capability of any commercial airliner available. Its low trip costs, coupled with the operating flexibility of four engines, make it an ideal aircraft for taking over when long-range twins become uneconomic.

While the A340 serves very long routles, the A330 is designed to serve high-growth, high-density regional routes. At the same time, it has the capability to operate economically on extended-range international routes. With typically 335 seats in a two-class arrangement, the A330 has a range of 4500 nmi with a full complement of passengers and baggage and 3200 nmi with maximum payload, making it ideal as a direct replacement for the costlier trijets and as a growth replacement for earlier twinjets.

30.2   Highlights

The A330 and A340 are built on the technological background established by two previous, complementary, product lines.

The A300/A310 series is the world's best-established twin-jet twin-aisle aircraft programme, with a very large number of technologically advanced features that transfer to larger, longer-range aircraft.

The A319/A320/A321 series is the world's most advanced single-aisle aircraft programme, again offering a large number of features that are found on much larger aircraft.

During the entire development process, there has been an insistence on securing the maximum commonality that could be achieved with the other programmes without loss of efficiency. Using selected features from each of these product lines, updated as needed, resulted in an all-new A330/A340 aircraft programme remarkably free of teething troubles, while at the same time providing a new benchmark for aircraft in this size category. As an added benefit, the technological features of the A330/A340 can, in many cases, be used to improve the established older product lines.

30.3   Systems

Before the entry into service of the A330/A340, the world's most technologically advanced airliner, in any category, was the A320. Its design formed the basis for the A330/A340 systems.

30.4   Cockpit

The A330/A340 cockpit is designed to be identical to that of the A320, from the point of view of the crew. The exceptions to this rule are associated with the size of the aircraft and to the added needs of the long-range mission, such as improved dispatchability, polar navigation capability, and of course, engine-related features.

The result is that the 130-seat-capacity short/medium haul A319 up to the 340-plus seat capacity very long-haul A340 have the most advanced flight deck of any airliner, enabling the same crews to fly any of these aircraft with minimum additional training needed.

30.5   User Involvement

The design of the A330/A340 cockpit has evolved from the same methods that were used successfully on the first Airbus Industrie A300.

The initial design of the cockpit (and the systems) was based on three features:

- The existing cockpit from the previous aircraft (the A320 in this case).
- The geometry of the A330/A340 nose section (which is based on the geometry of the A300, A310, and 300-600).
- Applicable new research and development work carried out since the A320 had been designed.

This initial design was reviewed by a task force consisting of pilots and engineers of each of the launch airlines in the light of their experience with the A320 or with other aircraft that were operating on the intended routes for the A330 and A340.

The task force met a number of times over a period of over a year. At each of these steps the design of the A330/A340 was refined, and certain features were mocked up for the next iteration in the review. The final design of the aircraft system and cockpit is essentially the one that the airline task force experienced and "flew" in the simulators during their final sessions.

30.6 Avionics

The avionics of the A330/A340 are highly integrated for optimal crew use and for optimal maintenance. As with all previous new and derivative aircraft since the A300FF of 1981, the primary data bus standard is ARINC 429 with ARINC 600 packaging. Other industry bus standards are used in specific applications where ARINC 429 is not suitable.

30.7 Instruments

The six CRTs on the main instrument panel display present flight and systems information to the pilots. This arrangement provides excellent visibility of all CRTs.

Flight information is provided by the Electronic Flight Instrumentation System (EFIS) consisting of a PFD and a ND in front of each pilot.

Systems information is provided by the Electronic Centralized Aircraft Monitor (ECAM) consisting of the engine/warning display on the upper screen and aircraft systems display on the lower screen.

Sensors throughout the aircraft continuously monitor the systems and if a parameter moves out of the normal range they automatically warn the pilot.

During normal flight the ECAM presents systems displays according to the phase of flight, showing the systems in which the pilot is interested, e.g., some secondary engine data, pressurization, and cabin temperature. The pilot can, by manual selection, interrogate any system at any time. Should another system require attention, the ECAM will automatically present it to the flight crew for action.

Should a system fault occur that results in a cascade of other system faults, ECAM identifies the originating fault, and presents the operational checklists without any need for added crew actions.

The information display formats currently in use enable the pilots to assimilate the operational situation of the aircraft much more easily than on the previous generation of aircraft.

There are substantial advantages on the maintenance side as well, in that the entire Electronic Instrument system consists of only three LRU types, enabling significant dispatchability and spare stocks availability. In fact, all the flight information (including standby) is presented on only 11 instruments of 6 types.

A new EIS, using liquid crystal displays, is being installed on the A330/A340 and A320 family of aircraft deliverable from 2001, offering improved capabilities and cost of ownership.

30.8 Navigation

Dual Flight Management Systems (integrated with the Flight Guidance and Flight Envelope computing functions; FMS) combine the data from the aircraft navigation sensors, including the optional GPS installation. Backup navigation facilities are included in each pilot's MCDU, allowing the aircraft to be dispatched with an inoperative FMS.

The FMS permits the crew to select an optimal flight plan for their route from a selection in the airline navigation data base, allowing the aircraft to fly automatically, through the autopilot or flight director, from just after take-off until the crew elects to carry out a precision approach and automatic landing. The "canned" flight plan captures the data needed for flight from the specifications entered by the crew prior to departure, as well as along the route as conditions change and more current information on weather and routing becomes available.

New FMSs, with improved cost of ownership and capability, are being installed on aircraft delivered from mid-2000. The same new FMSs are being installed on the A320 family.

30.9   Flight Controls

The flight control system of the A330/A340 is essentially the same as that of the A320, with five computers of two different types allowing the pilot to control the aircraft in pitch, roll, and yaw. The layout of the pilot controls is essentially the same as that of the A320 series, as are the handling qualities of the A330/A340. The technology features are also essentially the same, with extensive use of dissimilarity in the hardware and in the software, and extensive segregation in the hydraulic and electrical power supplies and signalling lanes. As with the A320 series, mechanical signalling is used for the rudder and for the horizontal stabilizer trim backup.

Detail changes have been introduced reflecting the longer mission times, specially of the A340, to provide better access to the system, and the opportunity has been taken to reduce the variety of backup submodes that the crew must use, making the aircraft even easier to fly.

Like the A320 series, the A330/A340 is a conventional, naturally stable airliner. The electronic flight controls offer a number of advantages to the pilot. There is a large reduction in manually operated mechanical parts, easier troubleshooting, and no need for rigging. Optimal use of the control surfaces is facilitated, as is the use of maneuver load alleviation.

The passengers and crew benefit as well, since the aircraft is more comfortable and easier to fly with precision in turbulence, while the flight envelope and structural protection features allow the crew to immediately use the whole capability of the aircraft should it be needed in an extreme situation.

30.10   Central Maintenance System

The A320, with its Central Fault Display System (CFDS), pioneered the industry standard for Central Maintenance Systems (CMS).

This industry background of experience has been built into the A330/A340 CMS. It enables trouble-shooting and return-to-service testing to be carried out rapidly and with a high degree of confidence, from the cockpit. Much of the CMS information may also be accessed remotely, via ACARS, giving the capability for the aircraft to be greeted upon arrival by a maintenance technician that already has a good idea of the exact nature of any defect, and who has likely been able to procure from stores the proper spare LRU required to resolve the fault.

Compared with the previous generation of CMSs, such as the A320 CFDS, the A330/A340 CMS has been improved in a number of respects, allowing trouble-shooting to take place on more than one system at a time, and with even clearer data available for the job. There has been a significant improvement in dependability as well, with great attention being paid to maintainability standards by the systems designers and the equipment vendors, and the incorporation of a maintenance message filter facility that enables known false message to be eliminated by the CMS, so that the mechanic does not apply the maxim *"Falsus in Uno, Falsus in Omnibus."**

30.11   Communications

There is a quiet revolution going on in the way that the crew communicates with the ground. This has been taking place in two ways. The A330/A340 uses the same full-capability standardized flight crew audio and frequency selections system as used on the A320 series, and also largely used on other recent derivative aircraft. This is a break from the traditional highly customized lower-capability systems.

---

*"False in one thing — false in all."

The other aspect of the revolution is more far-reaching, in that voice communication is giving way to data communications, with the advantages of lower error rates, more timely service, and lower costs. This started with highly customized ACARS systems for company communications, using VHF frequencies.

The A330/A340 is equipped with a standardized ACARS system that can be used by any customer, with allowance for each customer to easily introduce his own custom features to reflect his own needs.

These initial ACARS systems have been extended to offer worldwide coverage, even in mid-ocean and sparsely inhabited areas, using the Inmarsat facilities and HF data link, and to cover not only company communications but also ATC services, starting with predeparture and oceanic clearances.

On aircraft delivered since 1998, the ACARS unit has been replaced by the Air Traffic Services Unit (ATSU), which is designed to also accommodate safety-related ATC functions using the Aeronautical Telecommunications Network (ATN), offering the majority of ATC and other communication services now using voice, and more importantly, offering profitable migration to the ATN. The ATSU is the first unit to host software from a number of different vendors. The same ATSU is also used on the A320 family of aircraft.

The ATN upgrade is being implemented to be available when the corresponding communication and ATC services are in service.

30.12   Flexibility and In-Service Updates

The first generation of aircraft with widespread digital systems, such as the A300FF, A310, and B767, suffered from some of the same disadvantages as their analogue-system predecessors in that their avionics were not designed to accommodate unplanned change. Once a design change was made, equipment had to be removed from the aircraft, program memories had to be reloaded in the avionics workshops (sometimes by physically changing parts), and the equipment reinstalled. At some point the airframe manufacturer usually got involved to certify the change. There was an advantage in the avionics shop, because reloading a program and retesting is a faster and cheaper activity than installing a kit of new electronic parts, but the major cost of carrying out the change on the aircraft stayed the same.

The A330/A340 systems have to a large extent overcome this disadvantage, in that those digital LRUs that have been identified as requiring in-service change now have facilities for updates to be included *in situ* on the aircraft, at greatly reduced cost.

Two techniques are used, depending on the criticality of the LRU concerned, and on other practicalities.

1. On-Board Replaceable Memories (OBRMs) are memory modules that are located on the (accessible) front panel of an LRU. They come in industry-standard sizes, cost much less than the LRU itself, and can be "recycled" many times. The visible part of the OBRM contains the LRU's software part number section. OBRMs comply with the toughest criticality criteria, enable classic configuration control of the LRU and require no tools to change. They have been in use on the A320 since 1988.
2. The other technique in use is on-board data loading using 3.5 in. diskettes and other media. These are a little slower and are even less expensive than OBRMs, but do require a data loader to be carried to or installed on board the aircraft, and an adaptation to the aircraff's classic configuration control techniques. The same data loader is used for the FMS data base.

Both techniques enable software updates to be carried out overnight on the whole fleet.

Another aspect is flexibility in dealing with airlines' changing needs. The basic equipment for the aircraft is designed with a number of pin-programmable features that correspond to frequently requested airline changes, and other systems like the FMS where the airline loads a database which specifies its own preferences. These features allow airlines to pool databases and standard spares at outstations and still obtain the kind of operation that they need. Another feature is partitioned software, where heavily customized systems like ACARS can be certified just once for all users, with one set of "core" software. The airline may load its own additional operational software on top of this core to reflect its own needs.

Lastly, certain systems like the optional Aircraft Condition Monitoring System (ACMS) which used to be heavily customized, use a combination of these techniques to enable an airline to select the features that it needs out of a very powerful selection which forms a superset of the needs of all the customers.

30.13   Development Environment

The development of each Airbus Industrie aircraft has been supported by an Iron Bird whole-aircraft systems rig, and by supporting systems rigs that enable work to proceed simultaneously, without mutual interference. The A330/A340 model is no exception, and a number of facilities have been constructed specifically for this programme. These methods are now being used by other airframe manufacturers.

Proper software development is an essential part of systems development throughout the aircraft, and a number of software tools have been developed, notably in the areas of formal methods, rapid prototyping, automatic coding, and rapid data recovery and analysis. These are supplemented by large, fast data recording and telemetry facilities on the test aircraft fleet, associated with real-time and rapid-playback test data displays for the benefit of the flight test observers on board the test aircraft and for the test and systems engineers on the ground.

The result of this environment, the proper use of features from previous programmes, and the proper management of test data flow and the resulting decision process, has created an aircraft that has had a remarkable trouble-free period of introduction into service. This is true both in terms of customer satisfaction and in terms of measurable parameters such as delay rate, which have been up to an order of magnitude better for A340 than for the previous derivative long-range aircraft that entered service.

30.14   Support Environment

The A330/A340 airplanes have a number of unique support features, apart from those previously described.

As with other Airbus Industrie aircraft, an airframe-wide Automatic Test Equipment unit is available to customers, along with an airframe-wide test program suite. No other airframe manufacturer offers this facility for avionics.

The Aircraft Maintenance Manual and Trouble Shooting Manual have been carefully designed to integrate with the CMS for easier, faster fault rectification. For those airlines that wish to use it, a software package for a PC-compatible lap top is available to further speed fault-finding.

The documentation has also been designed to be compatible with open industry computer text and graphics standards, to facilitate the introduction of intelligent maintenance documentation systems.

31

McDonnell Douglas MD-11 Avionics System

Gordon R. A. Sandell
Boeing

31.1 Introduction

While the MD-11 is a derivative of the DC-10 airplane, the avionics system is an all-new system that represented the state of the art at the time of its introduction into service in December 1990. The MD-11 flight deck, shown in Figure 31.1 is designed to be operated by a two-pilot crew.

The crew is provided with six identical 8-in. color CRT displays, which are used to display flight instrument and aircraft systems information. A navigation system based on triple Inertial Reference Systems (IRS) and dual Flight Management Systems (FMS) is provided to automate lateral and vertical navigation, and reduce pilot workload. An Automatic Flight System (AFS) based on dual Flight Control Computers (FCC) is also installed to provide full flight regime autopilot and autothrottles, including fail-operational Category IIIb autoland capability.

The hydraulic, electrical, environmental, and fuel systems, that on previous aircraft were the responsibility of a flight engineer, were automated, with the system management now performed by Aircraft System Controllers.

The avionics equipment on a commercial airliner can be divided into two general categories, Seller-Furnished Equipment (SFE) and Buyer-Furnished Equipment (BFE). The BFE Avionics comprises the type of equipment that is largely standard from airplane to airplane, such as radios, sensors, and entertainment systems. Airlines generally buy this direct from competing suppliers, and may well use a common supplier for a particular piece of equipment for many different airplane types. Most of this BFE avionics is defined by industry standards, published by ARINC for the Airlines Electronic Engineering Committee (AEEC). The SFE avionics consists of the type of equipment that is specific to the airplane type, and is provided by the airframe manufacturer. It includes such systems as the Auto-Flight System (AFS), the Electronic Instrument System (EIS), Flight Management System (FMS), and the various system controllers. On the MD-11 most

FIGURE 31.1

of this SFE avionics is supplied by Honeywell under a partnership agreement which shared the systems integration function between McDonnell Douglas and Honeywell.

In commercial aviation, the various systems on an airplane are identified under chapter numbers that are defined by the Air Transport Association (ATA). The architectures of each of the systems (communication, navigation, displays, etc.) are discussed below under their respective ATA chapters. Simplified schematic diagrams are provided where appropriate. Note, though, that since the interfaces between the systems are largely ARINC 429 databuses (and in some cases discrete or video), they have been simplified for the purposes of illustration. Some of the data flows, for example, are shown as a single bi-directional arrow. Clearly this is not possible with the point-to-point ARINC 429 databuses, and the single line must therefore represent more than one databus.

31.2   Flight Controls (ATA 22-00 and 27-00)

A dual-dual (four-channel) Auto Flight System (AFS) is installed on the MD-11 to provide autopilot/autothrottle capabilities. The functions of the AFS include:

- Flight Director (FD)
- Automatic Throttle System (ATS)
- Automatic Pilot (AP)
- Autoland (to Cat IIIb minima)
- Yaw damper
- Automatic stabilizer trim control
- Stall warning
- Wind shear protection (detection and guidance)
- Elevator load feel
- Flap limiter
- Automatic ground spoilers
- Altitude alerting
- Longitudinal Stability Augmentation System (LSAS)

The system architecture of the AFS is shown in Figure 31.2. It is built around the dual-dual Flight Control Computers (FCC) and the Glareshield Control Panel (GCP). The GCP is used by the pilots to select AFS modes for pitch, roll, and thrust control and to select targets (e.g., speed or altitude) for those modes. The AFS Control Panel is used by the crew to reconfigure the system in the event of a failure.

The dual-dual FCC architecture is designed around the fail-operational Cat IIIb autoland requirement. Each FCC has two independent computational lanes to provide added integrity and redundancy. Each of these lanes consists of a power supply, two dissimilar microprocessors with dissimilar software (with one of these microprocessor types common to both lanes), and servo-electronics to drive the actuators that move the aircraft's control surfaces. This triple dissimilarity is intended to limit the possibility of software errors in one lane resulting in unsafe operation. This architecture is used for the functions that require high integrity (e.g., autoland and LSAS). Functions having a lower criticality only use a single version and are spread across the different processors. The system is designed to allow the airplane to be dispatched with only one FCC operational, though in this case it would not be able to perform a Cat IIIb autoland.

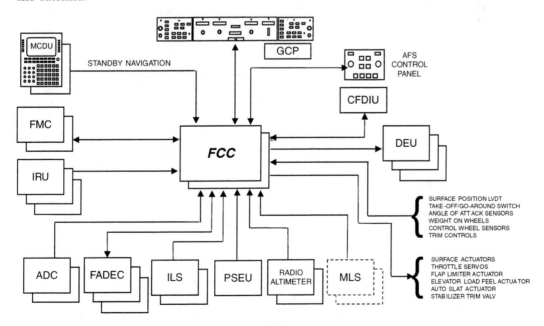

FIGURE 31.2 Auto Flight System (AFS) architecture.

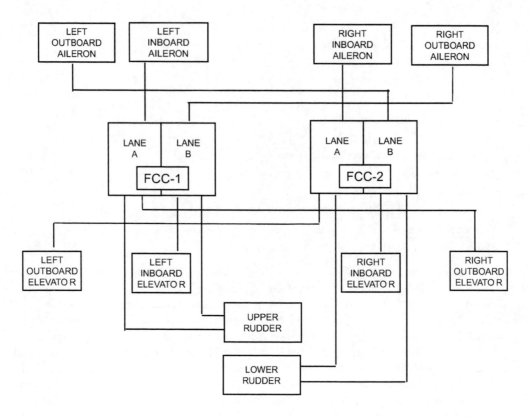

FIGURE 31.3   AFS actuator architecture.

A key element of flight control system design is the need to provide appropriate levels of redundancy in the interfaces to the actuators for the flight control surfaces. A number of related issues have to be considered:

- Dispatch with one Flight Control Computer (FCC) or one lane inoperative.
- Protection against both random and generic hardware and software failures/errors.
- Minimize the probability of a multi-axis hardover.

Figure 31.3 shows how the elevator, aileron and rudder actuators interface to the various channels of the FCC to satisfy these requirements. The control surfaces are also interconnected mechanically, so driving only one elevator, for example, will actually result in all elevator panels moving. Sufficient control authority is retained in the event of loss of a single channel or even of a complete FCC. The diagram does not show the stabilizer control (which is also driven by the FCC) or the spoilers, which are driven by the aileron/spoiler mechanical mixers and are thus driven indirectly by the aileron actuators.

31.3   Communications System (ATA 23-00)

The Communication System installed on the MD-11 is a highly integrated system, designed to reduce the workload of the two-man crew while providing the required levels of redundancy. It includes voice communication with the ground via VHF, HF, and SATCOM, as well as data link communications using an optional Aircraft Communications Addressing and Reporting System (ACARS) over the VHF radio, SATCOM, or HF data link (HFDL). The HF and VHF radios are controlled by the Communication Radio Panels located in the pedestal on the flight deck. Selective calling capability is provided by a SELCAL unit. The architecture is shown in Figure 31.4. The basic features of this architecture, in terms of the communication facilities provided, are dictated by Federal Aviation Regulations (FAR) Part 25, which mandate dual independent communication facilities be provided throughout the flight.

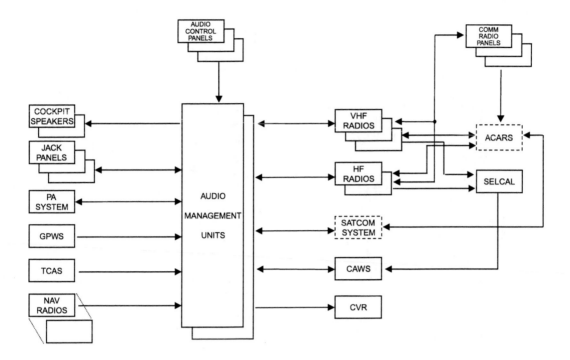

FIGURE 31.4   Communication system architecture.

The Audio Management Units (AMU) are the heart of the voice communication system for the pilots, and provide flight and service interphone capabilities, as well as supporting the aural alerts on the flight deck generated by the Central Aural Warning System (CAWS), Traffic Alert and Collision Avoidance System (TCAS), and Ground Proximity Warning System (GPWS). The Cockpit Voice Recorder (CVR) records all transmissions by the pilots. Audio Control Panels are provided for all crew on the flight deck (including the observer's station) to control volume, etc. Similarly, jack panels are provided for each crew member's headset.

One feature which is becoming more common on transport aircraft today is the SATCOM system, and the MD-11 has provisions to allow this to be installed. Either a single (6-channel) or a dual (12-channel) system may be installed to provide facilities for both passengers and crew transmissions. The data channel provided for crew use comes into its own for the new CNS/ATM environment (discussed below), while the other channels can be used to provide facsimile and telephone services for the passengers and the cabin crew. It is this passenger service use that typically pays for the system's installation on the airplane.

With all these communication systems, and the navigation systems described below, there is a need for a very large number of antennas on the airplane, and the total installation has to be designed to preclude interference between the different systems. The antenna layout on the MD-11 is shown in Figure 31.5.

Much of the communications equipment is defined by standard ARINC characteristics and is procured by the operators as BFE. Thus multiple suppliers are certified, and the operators may choose which they prefer. The same applies to the navigation radios described below.

31.4   Entertainment System (23-00)

On an airplane such as the MD-11, which can have as many as 410 seats, the entertainment system comprises the vast majority of the avionics Line Replaceable Units (LRUs). With one passenger control unit per seat, one seat electronics box per seat group, and one in-seat video monitor per seat (for those operators that provide in-seat video), it can amount to well over 1000 LRUs.

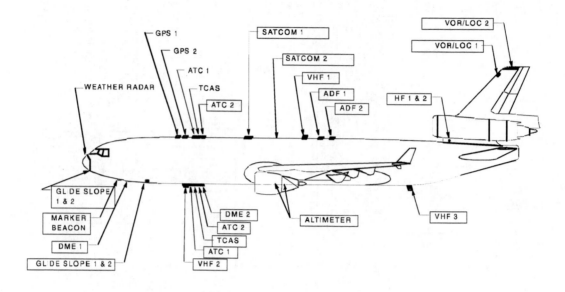

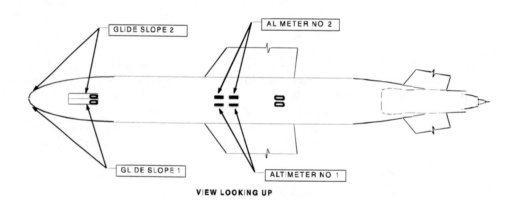

VIEW LOOKING UP

FIGURE 31.5   Antenna layout.

It is, however, typically a system that is almost totally outside the control of the airplane's manufacturer. It is usually BFE, and an airline may well upgrade it significantly (or replace it totally) several times during the life of the airplane. An airline will usually select a supplier for their entire fleet and then have the supplier adapt it to the particular airplane installation. It is therefore not really practicable to talk about a standard MD-11 entertainment system.

The only standard feature is the audio entertainment system/service system, which interfaces with the Passenger Address (PA) system to allow any safety-related announcements to override the entertainment audio.

31.5   Display System (ATA 31-00)

The most prominent feature of the MD-11 flight deck is the Electronic Instrument System (EIS). This consists of six 8-in. by 8-in. Cathode Ray Tube (CRT) Display Units (DU) arranged in two horizontal groups of three. The outer two DUs are Primary Flight Displays (PFD). Inboard of these are two Navigation Displays (ND), which can display any of five different formats. The center two DUs provide the Engine and Alert Display (EAD) and the System Display (SD). The SD has 10 selectable pages to

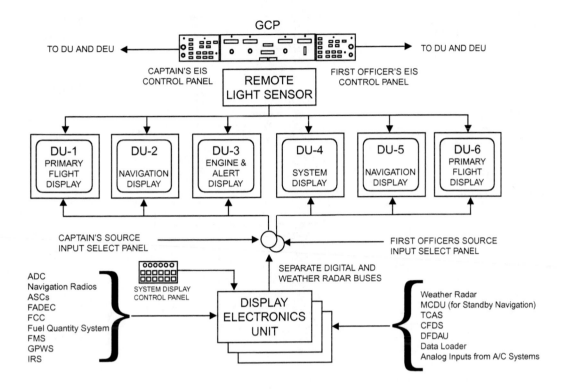

FIGURE 31.6   Display system architecture.

allow synoptic displays for any of the airplane systems to be presented. The SD pages are selected from the System Display Control Panel (SCP).

The architecture of the EIS is shown in Figure 31.6. This is a very simplified presentation. Any Display Electronics Unit (DEU) can support all six DUs, thus allowing the flight to continue in the event of a failure, and dispatch of the airplane with one inoperative. It is also possible to dispatch with one DU inoperative. In the event of loss of one or more DUs, the system will automatically reconfigure to provide the appropriate displays according to a fixed priority scheme. The lowest priority is accorded to the First Officer's Navigation Display (ND), and the highest priority to the Captain's Primary Flight Display (PFD).

In addition to these displays, standby displays of air data (airspeed and altitude) and a standby attitude indicator are provided on the main instrument panel. These are completely independent of the EIS, thus providing an additional level of backup. These standby displays are mandated by Federal Aviation Regulations (FAR).

On the MD-11 the Engine and Alert Display is part of the EIS. The DEUs thus contain all the alerting logic for the airplane and drive the Master Caution and Warning indicators. They also provide outputs to the Central Aural Warning System (CAWS) to generate voice alerts.

31.6   Recording Systems (ATA 31-00)

A number of in-flight recording capabilities are provided on the MD-11, for both voice and data storage. These include:

Cockpit Voice Recorder (CVR)
Flight Data Recorder (FDR)
Auxiliary Data Acquisition System (ADAS)
In-Service Data Acquisition System (ISDAS)

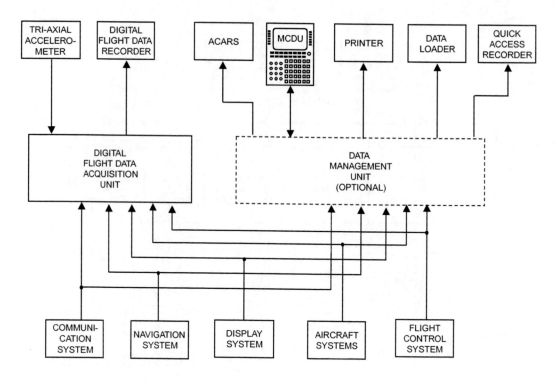

FIGURE 31.7   Recording system architecture.

The CVR was discussed along with the communications systems, to which group it properly belongs. The FDR is the recorder mandated by Federal Aviation Regulations to allow investigations of incidents that have occurred. Figure 31.7 shows how this is driven by a Digital Flight Data Acquisition Unit (DFDAU), which receives data from the other avionics systems in both digital and analog form. There is an FAA Notice of Proposed Rule-Making (NPRM) that mandates expanding the amount of data to be recorded. The MD-11 already records most of these parameters, but will be adapted as necessary to record the remainder.

The Auxiliary Data Acquisition System (ADAS) is also shown in the figure. It uses the Data Management Unit (DMU), and allows both the FAA mandatory data and additional data from the aircraft systems to be recorded for future access via the Quick Access Recorder (QAR). Typically an operator would have the DMU programmed to record specific data for specific events (e.g., recording data for analysis of an aircraft exceedance, or recording aircraft and engine performance parameters to allow trend analysis). The ADAS is an optional feature. The form of the ADAS shown in the figure is one version. There is also a version that uses a combined DFDAU/DMU.

The final recording system is the In-Service Data Acquisition System (ISDAS). Several of the major avionics LRUs have a databus that can be programmed to output an operator-defined set of parameters to allow in-service troubleshooting of the system. The FMS, EIS, and AFS are among those systems having this capability.

31.7   Navigation Systems (ATA 34-00)

The navigation system for the MD-11 is built around a triple Inertial Reference System (IRS) and a dual Flight Management System (FMS). The navigation system is shown schematically in Figure 31.8. The FMS is, of course, much more than just a lateral navigation system. It provides a number of functions that are central to operation of the airplane:

- Ability to create flight plans, including airways, Standard Instrument Departures (SIDs), and Standard Terminal Arrival Routings (STARs) by keyboard entry or data link.

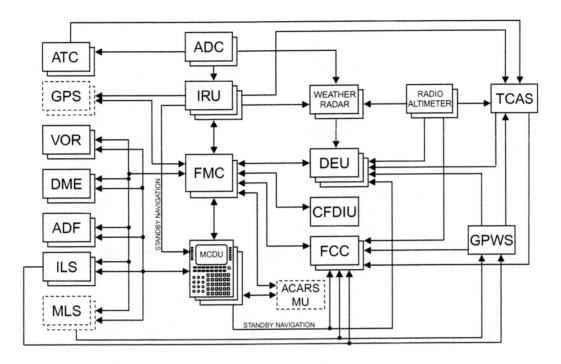

FIGURE 31.8 MD-11 navigation system architecture.

- Multisensor navigation using inertial reference data, together with inputs from GPS, DME, VOR, and ILS.
- Performance predictions for the complete flight plan, including altitude, speed, time of arrival, and fuel state.
- Guidance to the flight plan in three dimensions and controlling arrival time.
- Take-off and approach speed generation.
- Providing the VOR beam guidance mode.

On a long-range airplane, such as the MD-11, being able to dispatch the airplane when it is several thousand miles from the airline's maintenance facility and one navigation system has failed is very important to securing the bottom line for the operator. Such airplanes therefore usually have triple navigation systems. This capability to dispatch with a single failure is provided on the MD-11 by having triple IRS (thus allowing for a failure in this system) and having a standby navigation function provided in the Multipurpose Control/Display Units (MCDU), thus allowing for an FMS failure.

The Inertial Reference System provides a good independent position solution for short-term operation, or even for long-term operation within its capability of a drift of up to 2 nmi/h. However to provide the accuracy necessary for the area navigation required in today's airspace system or for terminal area operations, radio updating is necessary. This is provided on the MD-11 by having dual VHF Omni-Range Receivers (VOR) and dual Distance Measuring Equipment (DME) transceivers. Automatic Direction Finding (ADF) for flying nonprecision approaches and Instrument Landing System (ILS) for precision approach and landing are also provided. At the time that the MD-11 was designed, Microwave Landing System (MLS) was the up and coming system, and provisions are included to install this, although no airline has yet installed MLS. Global Navigation Satellite Systems for en-route operation and even in the future as a precision approach sensor are now the expected future means of navigation, and the option to install this on the MD-11 is now available. All of these navigation sensors are BFE equipment, and thus operators have a choice of suppliers to select from.

The antennas are not shown on the diagram, but one point that calls for a comment is that because of the geometry of the MD-11, the glideslope antennas for the ILS, which are installed in the radome, have to be replicated on the nose landing gear and the ILS must use the gear-mounted antennas on final approach. This is to meet the FAA requirement to have the antenna less than 19 ft above the wheels when crossing the runway threshold. The same rule, obviously, applies to the equivalent MLS antennas.

A dual air data system is also installed to provide airspeed, altitude, etc. for display to the crew and as inputs for the other systems (AFS, FMS, etc.) that need such data. Selection of baro reference is provided on the Glareshield Control Panel (GCP) which is part of the AFS (ATA 22-00) and is described there. Metric altitude displays and barosets are provided in addition to the English units normally used. There is an option to add a third air data system, in which case it is configured as a hot spare with a separate switching unit.

Additionally, dual weather radar systems (with a single flat plate antenna) are provided, together with radio altimeters, ATC transponders, and Traffic Alert and Collision Avoidance System (TCAS). The weather radar is now available with the capability to detect windshear ahead of the airplane. TCAS is a requirement for U.S. operators and foreign operators flying in U.S. airspace. Freighter aircraft (since the FAA regulation applies to aircraft with more than 30 seats) and government-operated aircraft do not require it, although most of these operators do, in fact, install it. These systems, again, are BFE, thus allowing the operators to select from competing suppliers.

All of this equipment is connected to the Centralized Fault Display System (CFDS) to provide fault reporting on each of the units, although for clarity only the FMCU is shown connected to the CFDIU in Figure 31.8. This system is discussed in more detail under ATA 45-00.

31.8 Maintenance Systems (ATA 45-00)

The maintenance system on the MD-11 consists of two main elements, the Centralized Fault Display System (CFDS) that is standard on the airplane, and the On-board Maintenance Terminal (OMT) which is available as a customer option.

The CFDS consists of a Centralized Fault Display Interface Unit (CFDIU) and any of the three MCDUs, with the capability to interface to all the major avionics subsystems on the aircraft, using ARINC 604 protocols, as shown in Figure 31.9. The functions provided by the CFDS are

- A summary of Line Replaceable Units (LRUs) that have reported faults on the last flight
- Capability to select individual "Reporting LRUs" for review of current faults and fault history
- Initiation of Return-to-Service testing of aircraft components (on-ground only)
- Capability to view sensor data
- Erasure of LRU maintenance memory (on-ground only)
- Ability to declare components inoperative for the Aircraft System Controllers (ASC)

The CFDS can also interface to an ACARS Management Unit to transmit fault data to the ground and to a printer on board the airplane.

The optional On-board Maintenance Terminal (OMT), shown in Figure 31.9, expands the capability of the CFDS and automates many of the required maintenance tasks. It contains a fault message database, which allows it to correlate each fault message to a specific flight deck effect. The displays on the OMT are customized by the individual airlines, but a typical display can show which LRU is involved, the fault message, the flight deck effect and alert, the Minimum Equipment List (MEL), documentation reference, and other useful information.

The OMT also incorporates a mass storage device, allowing it to store the aircraft maintenance documents, including the Aircraft Maintenance Manual (AMM), Fault Isolation Manual (FIM), MEL, Wiring Diagrams, etc. The built-in references to this documentation allow these data to be provided automatically as part of the fault-tracing process.

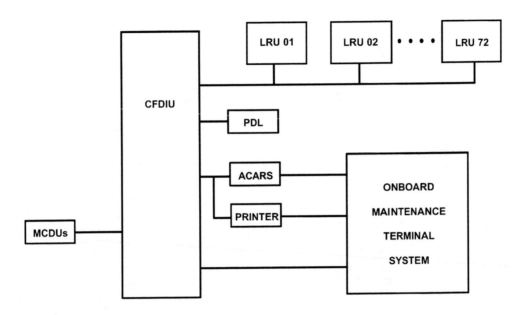

FIGURE 31.9   MD-11 maintenance system architecture.

31.9   Aircraft Systems

One of the major challenges in redesigning the DC-10 to become the MD-11 was to convert the three-crew flight deck of the DC-10 to a two-pilot flight deck, which is the current standard for the industry. To enable the MD-11 to be flown by a two-pilot crew, the normal, abnormal, and emergency system functions performed by the Flight Engineer on the DC-10 have been automated. This was made possible by advances in system control technology that enabled automatic execution of proven DC-10 procedures without extensive redesign or modification of the systems.

The status of the aircraft and its systems are provided to the crew on the main instrument panel Display Units (DUs) without any need to review the overhead panels, which provide backup annunciation and manual interface capability. Alert information is displayed on the Engine and Alert Display (EAD), while system status is displayed in schematic form as System Synoptics on the System Display (SD), allowing a rapid assessment of system failures and their consequences.

The aircraft systems are monitored for proper operation by the Automatic Systems Controllers (ASC). In most cases, system reconfiguration as a result of a malfunction is automatic, with manual input being required for irreversible actions, such as engine shutdown, fuel dumping, fire agent discharge, or generator disconnect. A "Dark Cockpit" philosophy has been adopted. During normal operations, all annunciators on the overhead panel will be extinguished, thus confirming to the crew that all systems are operating normally and are properly configured.

The MD-11 aircraft systems can be controlled manually from the overhead panel area of the cockpit. The center portion of the overhead panel is composed of the primary aircraft system panels, including Air, Fuel, Electric, and Hydraulic Systems, and are easily accessible by either pilot. Each panel is designed so that the left portion controls system #1, the center portion system #2, and the right hand portion system #3.

Each Aircraft System Controller (ASC) has two automatic channels and a manual mode. If the operating automatic channel fails or is shut off by its protection devices, the ASC will automatically select the alternate automatic channel and continue to operate normally. If both automatic channels fail, then the controller reverts to manual operation. The crew would then employ simplified manual procedures for the remainder of the flight. In order to allow operators to maximize the dispatch reliability of the airplane, the MD-11 is certified for dispatch with up to two systems operating in manual mode. The general architecture for Aircraft System Controllers is shown in Figure 31.10.

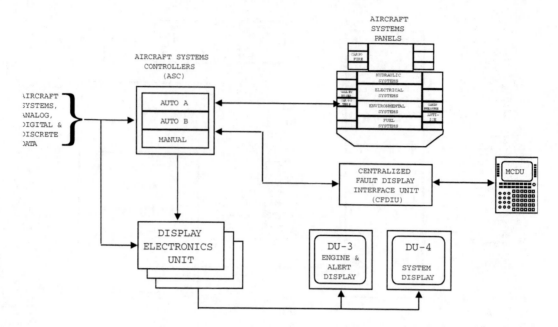

FIGURE 31.10   Generalized architecture for aircraft system controllers (ASC).

Automatic System Controllers (ASC) are provided for the primary systems as follows:

- Environmental System Controller (ESC)
- Hydraulic System Controller (HSC)
- Electrical Power Control Unit (EPCU)
- Fuel System Controller (FSC) and Ancillary Fuel System Controller (AFSC)

Pneumatic System Controller, Air Conditioning Controllers, and Cabin Pressure Controllers are also provided to control their respective subsystems.

31.10   Interchangeability

The MD-11 has been designed to allow interchangeability with the systems installed on other aircraft. The aircraft radios (HF, VHF, SATCOM, GNSSU, VOR, ILS, DME, ADF, ATC transponder, and radio altimeter) and other systems, such as ACARS, recorders, TCAS, and weather radar) are buyer-furnished equipment (BFE). They are all defined by industry standards (ARINC characteristics). Thus an operator is able to select a standard unit that is used on several airplane types and purchase in bulk for his entire fleet. With this standardization, he is also able to obtain spares from the vendor or other airlines at locations where he does not have a spares depot.

Another key element in providing interchangeability is with the MD-11's "single part number" philosophy. Under this philosophy, all the options required by various airlines for a particular system are incorporated in a single standard version of that system, and the individual airline then selects the appropriate options by means of a program pin or software option code. Additionally, as improved versions of the systems are developed, each MD-11 operator receives the latest version of that system (with the new features, if there are any, being selected again by program pin or option code). The advantages of this are considerable. At any time there is only one version of each system in the field to be supported, the airlines stay abreast of system improvements and avoid obsolescence, and there is a single part number available as a spare for airlines, thus improving their availability at remote locations.

Most of the changes to the various avionics systems that have to be kept updated under this single part-number philosophy are software changes. Most of the major avionics systems are software loadable, that is they can have their software loaded via the front or rear connector *in situ*. This simplifies the airlines' task in keeping their aircraft current.

31.11 CNS/ATM Architecture

One of the major changes affecting aircraft manufacturers and operators today is the need to operate in the new Communication, Navigation, Surveillance/Air Traffic Management (CNS/ATM) environment. This began with the ICAO Committee on Future Air Navigation Systems (FANS), and its first in-service application was on the Boeing 747-400 FANS-1 package in the South Pacific. This introduces a number of new CNS features in the airplane avionics systems:

- Controller/Pilot Data Link Communications (CPDLC) to communicate with ATC
- Global Navigation Satellite System (GNSS) navigation
- Required Navigation Performance (RNP) certification
- Required Time of Arrival (RTA) navigation to control arrival times at waypoints
- Automatic Dependent Surveillance (ADS) to provide surveillance data to ATC and the airline

The MD-11's original avionics architecture lends itself well to adaptation for these new functions. In the MD-11 CNS/ATM architecture, the FMC provides the computing resources for the new functions, with the ACARS MU (or CMU) used as a communications link to the ground via the SATCOM, VHF, and HF Data Link (HFDL) to the airline dispatch and ATC centers on the ground. The architecture is shown in Figure 31.11.

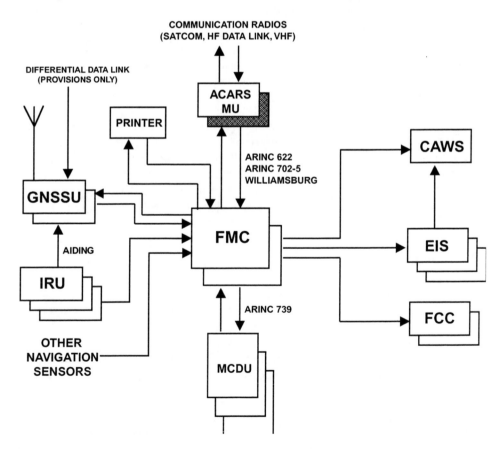

FIGURE 31.11 MD-11 CNS/ATM architecture.

Having the FMC host the applications for the new CNS/ATM functions, and in particular the communication/surveillance functions of CPDLC, AOC, and ADS, allows changes to these functions to be largely limited to the FMC. Similarly changes to the communications protocols can be kept to a single LRU, in this case the ACARS MU. With the migration to the Aeronautical Telecommunications Network (ATN), the ACARS MU will be replaced by an ARINC 758 Communications Management Unit (CMU), which will provide the necessary application gateway between the ATN Open System Interconnect (OSI) protocols and the FMS.

The GNSSU provides position, velocity, and time (PVT) to the Flight Management System (FMS), which determines the multisensor navigation solution based on best available data, i.e., the IRS combined with GNSS data or data from whichever other radio is most suitable. GNSS data are not provided to either the IRU or the MCDU, since standby GNSS navigation is not provided.

The GNSSU is provided today, and will be sufficient for the en route, terminal area, and non-precision approach accuracies required. In the next few years, however, as GNSS comes into use as a precision approach aid, it will be replaced by a unit such as the ARINC 755 Multi-Mode Receiver (MMR) to provide both precision approach capabilities and the outputs for the FMS.

The printer is interfaced directly to the FMC to allow the crew to print the flight plan clearance messages transmitted by ATC.

Obviously the architecture selected for the MD-11 was not the only possibility. AEEC has documented in ARINC 660 a number of possible architectures, some of which adopt a federated approach where the Air Traffic Services (ATS) functions are split between the FMC and the CMU.

McDonnell Douglas has elected to use the integrated approach described here for a number of reasons:

- Keeping the FMS/CMU interfaces simple
- Ability to re-use existing FANS-1 designs in the FMS
- Avoiding having to certify multiple different ATS applications with multiple ACARS/CMU suppliers
- Ability to provide a common ATS capability for all operators and provide commonality with other airplane types

The existing Flight Management Computer was designed to provide sufficient capabilities for the features required at the time of the MD-11's entry into service in 1990. It could probably provide sufficient computational resources (throughput and memory) to allow inclusion of the FANS-1 functionality, albeit with some degradation of the existing response times. However, growth in this configuration would be very limited. It therefore makes sense for such a major change to introduce the new hardware that will provide the necessary growth for the future. This is the Pegasus processor. The Pegasus processor is an Advanced Micro Devices (AMD) 29xxx family processor chip set, as used on the Airplane Information Management System (AIMS) designed by Honeywell for the Boeing 777.

31.12 Derivatives

The MD-11 flight deck was very much the standard to beat at the time of its introduction to service in 1990. In most respects, it still is, although technology developments in some areas, such as flat panel displays, have gone beyond the capabilities introduced by the MD-11. The flight deck layout is, however, still widely regarded as one of the best in the industry.

It was only to be expected, therefore, that the MD-11 flight deck would form the basis for future flight decks on McDonnell Douglas* commercial transport aircraft. For these aircraft, the Advanced Common

---

*In August 1997, The Boeing Company and McDonnell Douglas Corporation merged. Following that merger, the McDonnell Douglas name disappeared, and products of the former McDonnell Douglas Corporation now appear under the Boeing name. The term *McDonnell Douglas* has been used here where it would be historically inaccurate to use the term Boeing.

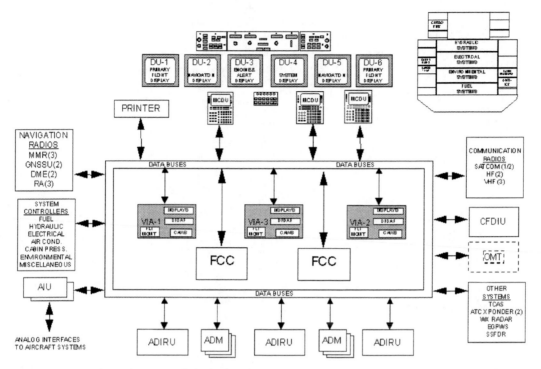

FIGURE 31.12   Advanced common flight deck (ACF) architecture.

Flight Deck (ACF) was developed, with the capability to be adapted to the specific needs of the airplane type. The ACF is based on MD-11 flight deck layout, and in many cases MD-11 components, and takes advantage of a number of technology advances that have occurred in the last few years:

- High-speed Reduced Instruction Set Computer (RISC) processors
- Dual lock-step processing to monitor for hardware errors
- Time and space partitioning to allow software of mixed criticality to exist in the same box
- High levels of functional integration, as on the AIMS cabinet
- Flat panel display technology for the main Display Units and for the standby unit
- Integrated Air Data and Inertial Reference System (ADIRS)
- CNS/ATM capabilities

The ACF uses these technologies not just for the sake of using technology, but because each offers real cost benefits to the operators of the airplane and to the manufacturer. The result is the architecture shown in Figure 31.12.

The heart of the ACF is the Versatile Integrated Avionics (VIA). This unit provides a similar level of integration as is provided by the 48" AIMS cabinets, but contained in an 8MCU Line Replaceable Unit (LRU). On a long-range aircraft, such as derivatives of the MD-11, three of these units will be installed to allow dispatch with one inoperative. On smaller, short-range aircraft, such as the MD-95, only two are installed. The VIAs will have the same hardware, and largely common software on all aircraft types. The key to this is the use of Aircraft Interface Units (AIU). These are data concentrator units that convert most of the analog data to digital form, and allow the VIA to process only digital data received via ARINC 429 databuses. MD-11-type control units (MCDU, Glareshield Control Panel, System Control Panel, etc.) are used to give the flight deck the look and feel of the MD-11.

This advanced flight deck is now in production for the Boeing 717-200 and a two-crew version of the DC-10 known as the MD-10. The diagram in Figure 31.12, in fact, shows the architecture for the MD-10 airplane. For application on a short-range airplane, such as the 717, the architecture is simplified by deleting one of the VIAs. Similarly, if Cat IIIb autoland capability is not required, some of the sensors (ADIRU, ILS, Radio Altimeter) can be reduced from triple to dual installations.

32

Lockheed F-22 Raptor

Ronald W. Brower
United States Air Force

32.1 F-22 Role and Mission

The F-22 will replace the F-15 as the U.S. Air Force's next generation air superiority fighter. With a first-look, first-shoot, first-kill capability it will maintain U.S. air supremacy in air-to-air and air-to-ground roles in the 21st century. It will deploy a wide mix of missiles and stand-off weapons which, under the guidance of the Integrated Avionics System (IAS), will provide the pilot with robust lethality and mission survivability.

32.2 IAS Hierarchical Functional Design

Behind this first-look, first-kill capability is the F-22's ability to establish superior situational awareness concerning target detection, location, identification, and lethality. The IAS provides the pilot situational awareness well Beyond Visual Range (BVR). Data fusion from multiple sensors is used to achieve long-range detection, high confidence BVR-Identification (BVRID) and highly accurate target tracking for BVR weapons employment and/or threat avoidance. The IAS directly contributes to increased survivability by providing threat warning and countermeasures against threat systems.

This first-look, first-kill requirement depends on the ability to collect data from multiple onboard sensors, to develop a highly accurate track file on enemy targets, and to do so before the F-22 is detected by enemy sensors. Each target track file is continually and automatically updated without pilot intervention. Targets receive increasingly tighter tracking accuracies as they penetrate a series of tactical engagement boundaries surrounding the F-22 as shown in Figure 32.1. From outermost inward, these "globes" are called (1) Situation Awareness Initial Track/ID, (2) Engage/Avoid Decision, (3) BVRID Initial AMRAAM Launch, (4) Initial Threat Missile Launch, and (5) Threat Missile Lethal Envelope. The globe boundary concept, inherent in the tactical software design, supports both (1) efficient sensor usage and (2) automated sensor tasking. It provides the pilot adequate time to make tactical decisions (such as engage, avoid, commit weapons, or expend countermeasures) instead of controlling sensors.

All multisensor information must be fused or correlated into a consistent, valid, integrated track file. This is done automatically by the sensor track fusion algorithms and the "smart" sensor-tasking

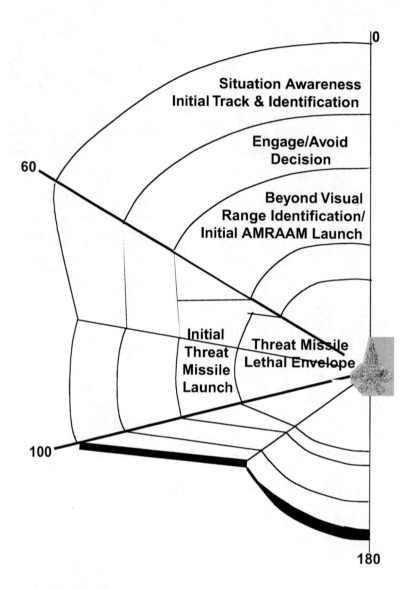

FIGURE 32.1   F-22 globe boundaries.

algorithms which are tailored to support each globe boundary's requirements. The integrated track file is then presented to the pilot on the integrated offensive, defensive, and area-wide situational awareness tactical displays.

Mission Software (MS/W) serves as the central controller of IAS operations, interfacing to all sensors, processors, pilot controls, and displays. It manages, coordinates, and supports the overall integrated capability to search, detect, track, identify, employ weapons, and expend countermeasures against airborne or ground threats. MS/W accomplishes this through a hierarchical functional tree consisting of three principle levels: the *integrating functions* level, the *decision-aiding functions* level, and the *mission functions* level (see Figure 32.2).

At the first level, the integrating functions manage and control the various onboard sensors. Sensor data are fused and correlated with navigation data to form integrated airborne and ground track files. This information is then sent to the second level of processing: the decision-aiding functions level.

At the second level, decision-aiding functions do critical assessments of the overall offensive and defensive tactical situation. Three key assessments are performed on the integrated track files received.

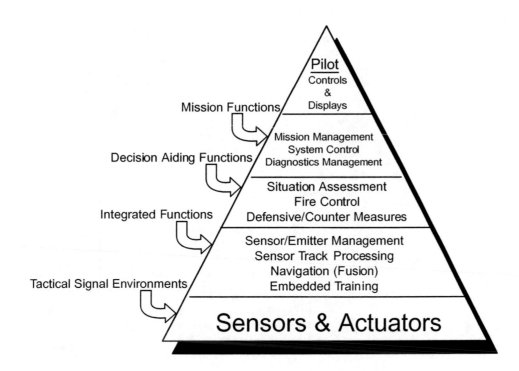

FIGURE 32.2 Avionics functional hierarchy.

First, *situation assessment* is done on the target track relative to the pilot's own ship threat and targeting environment. Consistent and timely information is continuously updated on the overall own ship situation, allowing the pilot to make key decisions to engage or avoid targets of interest. Second, *fire control assessment* calculates missile launch the envelopes against designated targets and controls launch and post-launch weapon support throughout the engagement. Finally, *continuous assessments* are made of the F-22's defensive tactical situation to assist the pilot in managing defensive countermeasures. These decision-aiding functions provide the pilot consistent and reliable tactical information to support war fighting decisions without the need to control individual sensors or correlate target track information in the heat of battle.

At the top level of the mission software hierarchy, mission functions control all avionics hardware and software and status the health of the IAS. These functions handle mission planning and system reconfiguration should hardware failures occur. Being at the top level of control for the IAS, the mission functions are the primary interface between the IAS and the pilot, who interacts with the IAS via *Controls and Displays*.

32.3   Integrated Avionics Architecture

The F-22 avionics architecture is characterized as a common, modular, highly integrated system. These characteristics result in increased performance, reliability, availability, and affordability. It is the first fully integrated avionics system in U.S. military aircraft, supplanting the *federated* architectures of the past. The F-22 does not employ traditional, single-function "black boxes" to perform basic avionics functions such as navigation, communications, threat warning, and fire control. Instead, these functions are implemented with common, programmable modules which are software-configured to process many different functions. This architecture not only allows increased mission effectiveness, but also allows significant flexibility in basic avionics design through: robust, fault-tolerant reconfiguration capabilities, higher reliability, easier supportability, higher availability, lower weight, extended growth capability, and lower acquisition and life cycle cost.

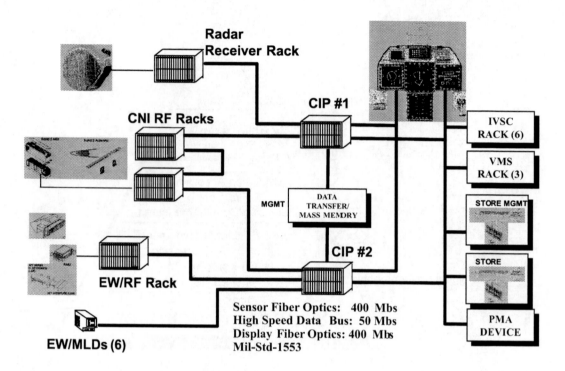

FIGURE 32.3   Avionics system architecture.

The IAS's system design features an interconnected set of high-speed, modular subsystems which use Standard Electronics Module Size E (SEM-E) modules. The core processing architecture is a distributed, parallel processing design that employs common, general purpose digital processor modules to perform all avionics functions. A common operating system is distributed on all core processing SEM-E modules to provide maximum flexibility and to support additional modules for new applications or for technology insertion.

IAS software, written mostly in Ada, is being built and integrated under a multiblock process to reduce development and integration risk. Each software block provides an increase in F-22 functional capability. Block 0 provides basic flight systems for initial flight qualification testing. Block 1 provides single-sensor threads of the IAS system. Block 2 provides multiple-sensor threads. Finally, Block 3 provides full F-22 functionality.

The IAS is partitioned into several interactive systems of antennas, sensors, processors, pilot interfaces, and high-speed interconnects (see Figure 32.3). The primary subsystems are Core Processing (consisting of two Common Integrated Processors, or CIPs), Electronic Warfare (EW), Radar, Communications/Navigation/Identification (CNI), Inertial Reference System (IRS), Stores Management System (SMS), and Controls and Displays (C&D).

By using low observable (LO) antennas and arrays, the sensors receive, measure, and extract both radio frequency (RF) and non-RF signals. Raw data are preprocessed, digitized, and routed to the CIPs via 400 Mbps fiber optic buses. Using digital and signal processor modules, the CIPs process raw data into sensor-level track reports which in turn are processed by sensor-track fusion algorithms residing on other digital and signal processor modules. Sensor-level reports are then combined into a single integrated track file and sent to the cockpit displays via fiber optic lines. The two CIPs are connected to one another via a 50 Mbps fiber optic High-Speed Data Bus (HSDB). Finally, the avionics architecture also features Mil-Std-1553 buses to interconnect to other aircraft systems.

32.3.1 Common Integrated Processor (CIP)

The Common Integrated Processor (CIP), developed by Raytheon Systems Company, provides the memory, I/O, data, and signal processing capability required for the IAS. It has an open, expandable architecture supporting radar, EW, CNI, mission software and Controls, and Displays processing requirements. The F-22 core processing system uses two installed CIPs (with growth space for a third). Each CIP contains 66 SEM-E slots in two rows. Due to the wide utilization of common modules, only 13 unique CIP module types are utilized. To provide for additional growth, each CIP is about two-thirds populated.

32.3.1.1 CIP LRM Types

The *Dual Data Processing Element* (DDPE) is the backbone of the CIP's digital processing capability. Each DDPE has two independent, 32-bit, 25-MHz Intel 80960 (i960®) microprocessors on each side of a SEM-E module. Each side of the DDPE operates as a general purpose computer executing Ada code. The DDPE module is Liquid Flow-Through (LFT) cooled, weighs 1.2 lb and is connected to the CIP backplane by a standard connector which uses 332 electrical pins and 4 fiber optic and two coolant connections. The IAS employs13 DDPEs to support radar, EW, CNI, and MS/W functions. Currently, product improvement programs plan to replace the Intel i960® with a state-of-art processor in 2005.

The *Dual Signal Processing Element* (DSPE) is a generic signal processor that executes mathematically intensive functions such as the state matrix multiplications used in Kalman filter propagation and Fast Fourier Transform (FFT) algorithms used in radar signal processing. Each DSPE uses two independent pipelines to perform high bandwidth signal processing. Each individual SPE can execute a fixed point instruction within one 25 MHz clock cycle and can operate at up to 18 operations per instruction. The DSPE consumes nearly 80 W of power, resides on a Liquid Flow-Through (LFT) cooled SEM-E module, and is connected to the CIP backplane by a standard connector (332 electrical pins and 4 fiber optic and two coolant connections). The IAS employs 9 DSPEs to support radar, EW, and MS/W functions.

The *DPE/Mil-Std-1553 I/O Port* (DPE/1553) features a Data Processing Element (DPE) on side A and a Mil-Std-1553 I/O interface port on side B.

The *Global Bulk Memory* (GBM) is a memory complex available to modules residing on the CIP backplane. Each GBM features 12 Mbytes of available bulk memory, consumes about 60 W of power, resides on a Liquid Flow-Through Cooled (LFT) SEM-E module, and is connected to the CIP backplane by a standard 360-pin connector.

The *Gateway* module (GWY) provides a bi-directional communications path between Parallel Interface (PI) bus segments within a CIP. The GWY module also provides communications between two CIPs via the fiber-optic HSDB.

The *Low Latency Signal Processor* (LLSP) uses a Texas Instruments SMJ320C31 (C-31) processor to provide the interface between the CNI front end and the CIP backplane via a fiber optic line. It performs low latency signal processing for the CNI system.

The *Graphics Processor/Video Interface* (GPVI) features a fiberoptic interface to the cockpit Multi-Function Displays (MFDs). One side of the GPVI module is a standard DPE, the other side performs graphics processing and I/O, generating up to 30 frames per second and supporting up to two MFD displays simultaneously.

The *Non-RF Signal Processor* (NRSP) is an Infra-Red (IR) signal processor that includes a pipeline processing structure optimized to perform IR impulse-response high-pass filtering, two-dimensional windowing for spatial filtering, data normalization, and thresholding for IR sensors. One NRSP can support up to three Missile Launch Detectors (MLDs).

The *Data Encryption/Decryption Device* (KOV-5) is an integrated Communications Security (COMSEC) unit housed in a SEM-E module. It can perform any 2 of 17 different COMSEC, data encryption, data decryption, cryptographic functions. The KOV-5 supports encryption/decryption functions of voice, text, data, and communications links. The encryption/decryption engine is National Security Agency (NSA) certified. The IAS employs 5 KOV-5 modules to support various crypto functions.

Voltage Regulator modules (VR) receive +270 VDC aircraft power and output +5 VDC and −5.2 VDC to the CIP backplane.

The *User Console Interface* (UCIF) is a two-sided LRM featuring a DPE on side A and UCIF hardware on side B. The UCIF is a nonproduction module which supports instrumentation and access to the CIP I/O backplanes during integration and test activities.

The *Fiber Optic Transmit/Receive Network Interface* module (FNIU) provides low latency, high bandwidth communications between a CIP processing cluster and the sensors. The FNIU supports bi-directional communications to the Parallel Interface bus or directly into the GBMs at the rate of 400 Mbps to the GBMs.

32.3.1.2 CIP Buses

The CIPs interconnecting communications paths consist of both internal and external buses. These communications paths are depicted in Figure 32.4.

The *Parallel Interconnect* (PI) Bus is a 32-bit, error-correcting, parallel, digital data bus that facilitates data and control exchange between modules within the CIP at a peak rate of 50 Mbytes per second. Each CIP contains three PI bus segments connected in a "triangle" by three Gateway modules. Each segment supports 22 modules. To optimize communications, subsystem processing (such as radar or EW) is usually clustered into modules within the same PI bus segment.

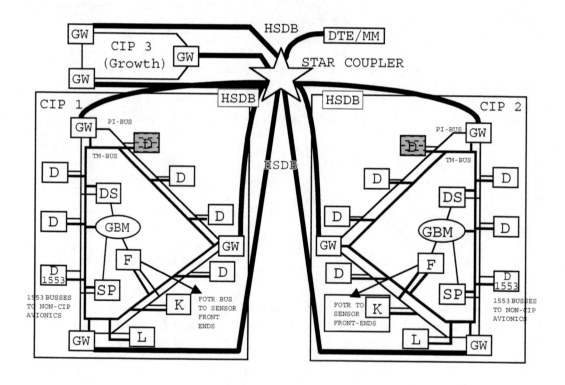

```
D  = DATA PROCESSOR ELEMENT
DS = DATA PROCESSOR/SERVER
K  = KOV-5
L  = LLSP
SP = SIGNAL PROCESSOR
F  = FNIU
GW = GATEWAY
```

FIGURE 32.4 Simplified CIP structure.

The *Test and Maintenance* (TM) Bus, like the PI bus, contains three segments connected by the Gateway modules. The TM bus is a 6.25 MHz bus primarily used for diagnostic monitoring of each module's health without interfering with either the PI bus or the internal processing within each module. The TM bus also supports fault reporting, isolation, and system reconfiguration. Via the TM bus, spare modules, typically DDPEs, can be commanded to reconfigure to maintain functionality lost due to a failed module.

The *High-Speed Data Bus* (HSDB) is a fiberoptic bus which provides 50 Mbps data transfer rate between the CIPs and the Data Transfer Cartridge or Mass Memory unit.

The *Fiber Optic Transmit-Receive* (FOTR) Bus supports low latency, high bandwidth (400 Mbps) data communications between the CIP and the sensors.

The *Mil-Std 1553* Bus provides I/O communications to standard interfaces such as weapons and aircraft flight control systems.

32.3.1.3 CIP Software

The CIP software has a layered architecture which provides a common set of utilities for the CIP application software and handles data transfer integrity and security. It consists of two principle software packages, the *Avionics Operating System* (AOS) and the *Avionics System Manager* (ASM). The layered architecture with AOS and ASM as an intermediary between the hardware and applications is shown in Figure 32.5.

The AOS provides operating system services to embedded avionics applications running on the CIP. The AOS resides and executes on each DPE-based module. It supports a multilevel secure execution environment in which multiple application programs may run and process concurrently at different security levels. This is enforced by the AOS Privilege Control Tables (PCTs) which require that data at a given security level not be processed by a program at a lower security level. Communications between application programs is restricted to those that are allowed by the PCT. The AOS provides four basic capabilities within the CIP: control of Ada application programs, control of the I/O interfaces to the DPE modules, debug capability, and PCT security access authorization.

The Avionics System Manager (ASM) is the central resource manager for the CIP, featuring three basic services: (1) system control, (2) module management, and (3) file services. It assigns global resources, such as memory and processing elements, to the application programs. ASM is also responsible for maintaining CIP health status, performing reconfiguration around failed modules, providing file management functions between applications and the DTC/MM, providing GBM file allocation services, and coordinating startup and shutdown of CIP functions.

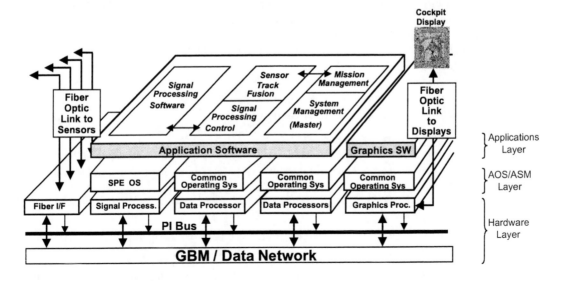

FIGURE 32.5 CIP processing architecture.

32.3.1.4 CIP Signal Flow

The sensor front-ends preprocess RF and non-RF data, converting raw sensor data into blocks of digitized data that are sent, via 400 Mbps FOTR lines, to the FNIUs. The FNIU modules route the raw digitized data to the GBM modules in real time for temporary buffering and storage. The data are then extracted over the Data Networks (DN) by either the DPE or the SPE modules for further processing and refinement. To reduce PI bus overloading, the GBM also supports intermediate buffering of data between the processors or between processing tasks on a processor. Processed data are then sent out on the PI bus to be further processed by other higher-order functions on other DPEs and SPEs. Once these DPEs and SPEs complete the higher-order functional tasks, the data are again sent on the PI bus to the GPVI modules, transferred to graphic format, and sent via FOTR lines to the cockpit displays.

This signal and data flow approach reduces bus loading and potential throughput problems. Sensor data are sent nearly real-time via FOTRs to GBMs and on through the "backdoors" via the DN bus to the processors. Only after the processors have completed their operational tasks on sensor data is the PI bus involved.

32.3.2 APG-77 Radar

The F-22's APG-77 radar is an advanced multimode, multitarget interleaved search/track, all-weather, fire control radar. Developed by Northrop Grumman, it incorporates the following design features: Active Electronically Scanned Array (AESA), low observability (LO), electronic counter-countermeasures (ECCM), and low probability of intercept (LPI). These features give the F-22 radar a major leap in combat capability. The main array, mounted in the nose radome, is composed of hundreds of Transmit/Receive (T/R) modules. Beam switching is performed by controlling each T/R module's phase characteristics, thus accomplishing a summed beam pattern of all T/R modules. These T/R modules are designed to operate for over 16,000 hours failure-free. The T/R module application features an extremely fault-tolerant design, where the system can lose numerous T/R modules before minimum required performance is affected. The system can continue to effectively operate with loss of even more T/R modules, however at reduced transmit power levels.

32.3.3 Communication, Navigation, Identification (CNI)

The F-22 CNI subsystem performs standard military communications, navigation, and identification functions. Developed by Lockheed-Martin Tactical Aircraft Systems (LMTAS) and TRW Military Electronics and Avionics Division, its primary functions consist of UHF/VHF secure and clear voice, Have Quick IIA, GPS, TACAN, ILS, MK XII Identification-Friend-or-Foe (IFF), JTIDS receive, and the Intra-Flight Data Link (IFDL). Like the CIP, the CNI architecture is also highly integrated and uses common SEM-E modules with diverse CNI functions sharing common hardware components. This integrated approach requires time-sharing and multiplexing of assets or system reconfiguration between mission phases (landing vs. engagement operations, for example).

The CNI subsystem is comprised of six major components:

1. Low observable apertures and arrays;
2. External Aperture Electronics (EAE) units (located near the arrays for low noise amplification),
3. RF filtering and switching; the Antenna Interface Unit (AIU) (to interface all RF lines to/from the RF/Preprocessor racks);
4. RF/Preprocessor Integrated Avionics Racks (IARs) which house SEM-E modules for RF transmission, reception, and processing;
5. Interphone/Intercom subsystem for voice communication and synthesis; and
6. CNI digital processing LLSP and KOV-5 LRMs resident on the CIP.

The RF/Preprocessor IARs consist of a pair of three-bay, SEM-E modular, liquid-cooled racks and the LRMs which perform CNI RF processing. The two IARs were originally fully redundant and identical, but to save weight each IAR is now more specialized. However, aircraft mission-critical functions such as UHF/VHF communications, ILS, TACAN navigation, or MK XII IFF transpond can be supported from either rack.

The CNI SEM-E modules which comprise the CNI IAR racks are as follows:

1. Eight L-Band tunable receivers with selectable IF bandwidths to support TACAN, MK-XII transponder, Mode S ATC, IFDL, and JTIDS receive.
2. Two 5-channel tunable L-Band receivers with common local oscillators used to support direction finding.
3. One L-Band transponder Carrier Generator/Power Amplifier (CG/PA) used to support low duty cycle pulse modulation of RF transmit power for MK XII IFF transponder and TACAN functions.
4. One Interrogator CG/PA used to provide low duty cycle RF pulse modulation for MK XII interrogate. This CG/PA is used as a backup for MK XII transpond and TACAN in the event of an L-Band Transpond CG/PA failure.
5. Four UHF/VHF single-channel tunable receivers which support U/VHF voice communications, ILS, and growth satellite communications.
6. Two UHF/VHF CG/PAs for supporting U/VHF transmit.
7. One GPS Receiver Processor to provide a complete decoded GPS navigation solution.
8. Two RF/FE controller LRMs which support all RF asset control, multiplexing, timing references, and reconfiguration.
9. Four Pulse Narrowband Processor (PNP) LRMs used to support L-Band programmable pulsed signal decoding such as TACAN, MK XII IFF transponder, and interrogate, and Mode SATC. The PNP LRM also supports A/D conversion of the U/VHF communications and ILS navigation.
10. One Pulse Environment AOA Processor (PEAP) LRM to convert measured pulse phase and magnitude data into a calculated Angle of Arrival (AOA) via algorithms which use real-time calibration data and prestored array characteristics.
11. Two CNI Bus Coupler LRMs to provide a FOTR bus interface to the FNIU module within the CIP.
12. One IFDL Mod/Synth LRM to provide the waveform generation, signal modulation and demodulation, and relative navigation processing for the F-22-to-F-22 Intra-Flight Data Link system.

Other SEM-E modules are the Air-Combat Maneuvering Instrumentation (ACMI) transceiver which provides ACMI signal modulation and demodulation; the Ovenized crystal oscillator LRM; a 5-volt backup battery LRM used to maintain system crypto keying and clocks with main power off; and various RF and digital power supply LRMs which provide up to seven different voltages required by the CNI system components.

32.3.4 Electronic Warfare (EW)

The EW subsystem provides Radar Warning (RW), Missile Launch Detection (MLD), and chaff and flare countermeasures. RW was developed jointly by Lockheed Martin Missiles and Fire Control, Lockheed Sanders, and LMTAS. It provides airborne and ground-based radar emitter detection, tracking, identification, and location to the mission software system for integrated target tracking. The Missile Launch Detector also provides a passive IR capability to detect, declare, track, and report missile launches to mission software. The defensive countermeasures function is responsible for timing and deploying chaff and flares. Deployment of countermeasures is programmable for fully automatic, semiautomatic, or manual.

The EW architecture, like the CIP and CNI, is an integrated architecture using common SEM-E modules. The EW subsystem employs resource sharing of common hardware components to perform the simultaneous search, detection, RF and non-RF measurement, signal analysis, direction finding, identification, and tracking of RF and non-RF signals. This integrated approach requires time and resource sharing of these common modules.

The EW subsystem is comprised of seven major components:

1. Low observable apertures and arrays;
2. Array Electronics (AEs) units near the arrays for low noise amplification, RF filtering, and switching;

3. Remote Antenna Interface Unit (RAIU) to interface all RF lines to/from the EW RF racks;
4. EW RF Integrated Avionics Racks (IARs) which house SEM-E modules for RF, reception, and processing;
5. Six Missile Launch Detector sensors;
6. The countermeasures controller and dispenser units to dispense MJU-7 and -10 standard flares, MJU-39 and -40 flares developed specifically for the F-22, and RR-170 and -180 chaff bundles, and
7. The CIP based DDPE, DSPE, NRSP, and GBM LRMs.

The EW RF IAR consists of a SEM-E modular liquid-cooled rack and the LRMs which perform the EW RF reception and processing. The EW SEM-E modules which comprise the EW RF IAR rack are as follows:

1. Six Narrow-Band Receiver (NBR) LRMs with selectable IF bandwidths to support signal analysis, emitter tracking, and emitter direction finding processing.
2. Six NBR Local Oscillator LRMs to tune the NBRs.
3. Six Pulse Measurement Units to extract RF characteristics from the NBR for signals analysis.
4. Four Wide-Band Receiver (WBR) LRMs to support wideband detection and acquisition of emitters in the environment.
5. Six signal frequency down converter LRMs to convert RF signals into a base frequency band.
6. Nine power supply LRMs to convert 270 V power to $+/-9$, $+/-15$ and $+/-5$ V.
7. One WBR asset controller LRM.
8. One NBR asset controller LRM.
9. Two RF Delay LRMs to support hand-off of signal analysis for direction finding processing.
10. One Reference Oscillator LRM for supplying a common local oscillator to the NBR and down converter LRMs.
11. One CIP Interface (CIPI) module to interface digitized EW information onto the CIP fiber optic FOTR bus.
12. One Data Converter LRM to reformat digitized RF data into pulse descriptor words for CIP processing.
13. Three Measurement Control Processor (MCP) LRMs to support timing and synchronization of receiver assets.
14. One Data Distribution Network (DDN) to provide DF triggers for supporting signal direction finding and angle binning to support pulse de-interleaving.
15. Two Compressive Receiver (CR) LRMs to support high probability of intercept against high priority signals.
16. One Array-RAIU Controller Interface (ARCI) to control RF line switching and filtering in the RAIU.

The EW RF IAR racks perform RF to digital conversion and then send the raw digitized information to the FNIU fiber optic interface module in the CIP for RF sorting, signal characteristic measurement, signal identification, and emitter tracking.

32.3.5 Stores Management System (SMS)

The SMS monitors, controls and statuses countermeasures, launchers, weapon bay doors, and the F-22 armament (AIM-9, AIM-120, gun). It also controls emergency jettison of stores. The SMS consists of two rows of SEM-E modules, two AIM-9 power supplies, a gun control unit, and the SMS Controller.

32.3.6 Inertial Reference System (IRS)

Developed by Litton Guidance and Control, the IRS is an advanced laser ring gyro with a common processor with flight controls. It provides position and rate information to the IAS mission software to support target location calculations.

32.3.7 Controls and Displays (C&DS)

Unlike today's generation of fighters, the F-22's tactical displays are not dedicated to providing sensor-only information such as radar-only displays. Instead, the F-22 has four active-matrix liquid-crystal Head-Down displays (HDD) and a Head-Up display (HUD) to provide highly integrated information concerning the overall tactical situation. The middle HDD, or Tactical Situation Display (TSD), is an 8 × 8 in. color display that provides the pilot with current situational awareness and enhanced navigational information, including location of airborne friendlies and threats, ownship heading, navigational waypoints, etc. The left 6 × 6 in. color HDD, or Attack Display, provides the pilot with the current offensive tactical situation and is tailored for weapons employment including target selection and offensive and defensive missile engagement ranges. The right 6 × 6 in. color HDD, or Defensive Display, provides the pilot with the current defensive tactical situation and is tailored for assessing threat capability against the F-22 and includes location and identification of airborne and ground-based engagement systems, missile engagement ranges, and countermeasures selection. The lower 6 × 6 in. color HDD provides status of aircraft expendables, stores, engine performance, and external doors status.

32.4   Fault Tolerance and Recovery

Another characteristic of F-22 avionics is its robust fault tolerance and fault recovery by means of reconfiguration — a mechanization that restores needed functionality after loss or failure of assets. Reconfiguration is achieved by reallocating and/or reprogramming modular resources. Recovery implies full operational capability or a degraded mode of operation, depending on the number of spare modular assets left to support reconfiguration. "Minor reconfiguration" occurs due to the loss of a module (or several) and "major reconfiguration" occurs due to the loss of an entire rack. Major reconfiguration can be caused by battle damage, loss of an engine and/or generator, or overheating due to loss of cooling. Major reconfiguration is the ability to reprogram and reconfigure a fully functional SEM-E rack to support emergency backup functions such as UHF voice communications, TACAN navigation, instrument landing system navigation, and MK XII identification for safe return to base or to complete a critical part of the mission. Minor reconfiguration is accomplished by reprogramming modules to perform functions lost by the failure of an identical common module. Minor reconfiguration is driven by a function prioritization table. Once all spare modules of a common type have been used, the lowest-priority function is dropped to support higher-priority functions. The common modular approach and reconfiguration flexibility results in outstanding mission availability and graceful degradation.

32.5   Summary

The F-22 program is completing the manufacture of its nine EMD aircraft and moving into production; 339 aircraft are scheduled to be produced. The F-22 IAS is developed in three major functional blocks. IAS flight testing will be complete in mid-2002. Flight testing on the first block is to begin on air vehicle number 4, which is the IAS-equipped aircraft, in mid-2000. F-22 Initial Operational Capability is scheduled for December 2005.

The IAS places robust, first-of-its-kind, fully integrated tactical war fighting capabilities into the hands of the pilot. The key avionics contribution to the F-22's unprecedented combat effectiveness is its ability to perform fusion of multisensor information to provide the pilot integrated target detection, identification, tracking, and threat warning information on his displays, significantly reducing the pilot workload during battle conditions. The common, modular, and open architecture allows the needed flexibility to handle growth in the face of ever-changing threat and advancements in avionics technology. The integrated IAS architecture and its superior functional capabilities in conjunction with the advanced F-22 capabilities in stealth, maneuverability, super cruise, and advanced armament will ensure F-22 air superiority and mission effectiveness well into the 21st century.

33

Advanced Distributed Architectures

Jim Moore
Smiths Industries Aerospace

In recent years a considerable momentum has built up on the assumption that future avionics will be based on centralized racks or integrated cabinets containing numbers of electronic modules of various types. Integrated Modular Avionics (IMA) in many sectors of the air transport industry has largely become the assumed way forward for the implementation of future avionics. Progress has already been demonstrated with first-generation civil IMA systems such as ELMS and AIMS on the B777. These are quite different implementations, having been optimized for their specific systems domains, and therefore appear to go only part of the way towards meeting the ultimate goals anticipated by the industry.

However, there are a number of reasons to suggest that relatively large groupings of tightly coupled and integrated computing resources may not provide the optimum architecture of future aircraft in the face of rapid technological change and ever-increasing business pressures.

This chapter will raise awareness of these trends and of alternative architectures in which the computing and interface resources are decentralized and operate autonomously within a networking communications environment.

33.1 Drivers and Trends

In addressing these issues it is important to identify and consider the changes that have taken place within the industry and the underlying trends to these changes and drivers

33.1.1 Technology Advance

Perhaps the greatest driver of change is the continuing rapid advance of electronics technology, affecting practically all areas of avionics systems — computers, sensors, displays, data buses, etc. Already there are examples of avionics computers introduced less than a few years ago that are now available at half the size, weight, and power consumption, but with considerably enhanced performance and function-ality. For instance, a Flight Management Computer that 10 years ago required a size 8 MCU size box (10 in. wide) can now be implemented as a single-card module (<1 in. wide). Likewise there have been comparable advances in display technology, data buses, control panels, bulk memory media, sensor technology, etc.

These technology advances are being driven primarily by the demands of the large commercial markets, e.g., information technology, communications, PCs, consumer products, automotive, etc. The compet-itiveness and size of these markets are driving down costs and more and more are setting the performance standards. Aerospace components and communication standards are now being increasingly replaced by commercial ones, e.g.

- Microprocessors, microcontrollers
- Data buses (e.g., Ethernet, CANbus)
- Flat panel AMLCD displays and interfaces (e.g., OpenGL graphics language)

These trends are now firmly established in both the civil and military aviation markets.

The principal drawback is the relatively short lifetime (availability) of commercial components, which can become obsolete within just a few years, in much less time than the production cycle of the average civil aircraft programme. Therefore appropriate strategies are required to deal with part obsolescence.

The rapid and continuing advance of electronic technology is constantly driving down the cost of hardware, as the number and cost of the components used falls and much more functionality is achieved with less and less hardware. Figure 33.1 illustrates the growth in microprocessor performance over time and adherence to Moore's law and the expectation that performance will continue to double every 18 to 24 months, i.e., grow by a factor of over 1000 times in the next 20 years.

This illustrates the need for care in choosing hardware architectures with interface boundaries between the principal elements, which are the least likely to be affected by changes in the hardware technology, i.e., to prevent obsolescence/upgrade repercussions from propagating from the affected element (e.g., module) to other elements. This also applies to the interface to the functional software to protect it from the impact of changes in the processing platform on which it resides.

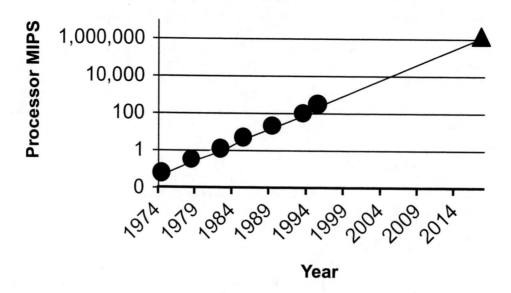

FIGURE 33.1   Increase in microprocessor performance.

33.1.2   Increasing Functional Complexity

Associated with the enhanced capability afforded by the technology, and as driven by the competitive pressures of the civil transport aircraft market, the functionality of avionics systems has continued to escalate. It is evident that avionics have long since become central to the ability of a manufacturer to manufacture and sell competitive aircraft and therefore are key to ongoing business viability. For example;

- Fly-by-wire flight controls
- FMS, full flight regime 4D flight management
- Full glass cockpits, large multifunction displays
- FANS capability to operate in the new air traffic management environment
- Passenger entertainment systems and commercial/business services
- On-board central maintenance computers and electronic documentation

Most of this added functionality is, of course, effected in software, and so the quantity of code embedded in avionics continues to rise. For the suppliers, more software generally leads to increased development and certification cost. High-level languages such as Ada and C are widely used to simplify the programming effort, and where possible, automatic coding is used to further simplify the process and reduce cost.

Much effort is being applied these days to ensure that application software can be reused on different hardware, i.e., be independent of the hardware. The primary driver for this is to avoid the high cost of new software for the application being hosted on different processors, for example, in the event that during the life cycle of the product the processor becomes obsolete or has insufficient capability to support growth in required functionality.

Hardware-independent application software requires an embedded software Operating System (OS) or executive which provides a generic interface to the application code and which translates it for the particular processor and hardware architecture being used. By the use of such an executive and a standardized application interface definition it is possible to achieve compatibility of applications developed by different suppliers capable of being used on a variety of different hardware platforms. Hence the concept of APEX (Application-Executive interface) as documentation ARINC 653.

33.1.3   Hardware/Software Cost Ratio Continually Falling

The combination of the above two trends suggests that, for the foreseeable future at least, the computing hardware cost is going to be less and less significant as a component of the overall cost of systems. The main cost drivers will be for software and integration. Already, today, software typically consumes 70 to 80% of the development budget and represents by far the highest schedule and cost risk to any new programme.

It is also important to recognize that there are recurring elements of software cost. Historically suppliers have included amortized software development costs in the equipment recurring price. But the actual value or market worth of the embedded software has been invisible to the purchaser. However, as the material (hardware) cost falls and as suppliers use or adapt outsourced COTS software elements (e.g., embedded operating systems, data-bus protocol software, networking software) the recurring cost and value of such software will become more significant. In the limit where the product is only software, for example an FMS to be integrated into the customers host system, the price will be driven by the market value per copy of that function, assuming the development costs have already been recovered. After all, it is the software that embodies the supplier's know-how, systems expertise, and would carry copyright.

The lesson this suggests as we consider the next generation of avionics systems architectures is to beware of overcomplicating the software and level of integration just in order to minimize the quantity of hardware employed.

33.1.4   Integration

As technology has advanced, so there has been a continuing trend of avionics integration. In the past most integration has been concerned with processes or functions which were already interdependent, and by so doing savings could be made in interface hardware, aircraft wiring, power supplies, etc. with little risk to the certification of the system. This is usually referred to as "vertical" integration. The integration of inertial sensors with computers to produce Inertial Reference Systems (IRS) is an example of vertical integration.

In other cases separate processors have been colocated into a single box because they can share the same I/O hardware on which they both depend, thereby eliminating one set of I/O hardware and simplifying the aircraft wiring.

More latterly there is a trend to go one step further by the integration of largely unrelated avionics functions onto shared processors. This may be referred to as "horizontal" integration. There are significantly higher risks because, while additional hardware resource may be saved, there are added complexities to provide "equivalent independence" or partitioning within the processing platform. Partitioning is used to ensure that malfunctions within one application (function) cannot affect the others, or that modifications made to one may be certified without the need to revalidate the others. A further complication may be that additional levels of hardware redundancy and associated monitoring and configuration management facilities may be needed to offset the risk of failures within the shared processor, causing simultaneous loss of all the otherwise unrelated application functions.

The trade between savings of hardware (e.g., processor modules) on the one hand, and the escalation in cost to achieve acceptable levels of segregation and integrity on the other, needs to be carefully weighed. This is a trade that would appear to be more difficult to support as the hardware proportion of overall cost continues to fall with time. For example a "cluster" of separate low-cost processor chips, each dedicated to a system function and each including it's own program memory, communicating via shared data memory areas and external bus interfaces, may become a cheaper and better alternative.

33.1.5   Modularity

The concept of modular design for both hardware and software has been with us for a long time. Most avionics Line Replaceable Units (LRUs) employ a modular approach to some extent or other. The object is to build up the design by repetitively using building blocks of a minimum number of different types. A supplier can substantially reduce costs if new designs reuse existing modules. There are benefits of scale in purchased parts and manufacturing. The modules used in greater numbers more rapidly reach maturity and modifications and repairs can be more readily implemented and tested. The number and cost of spare modules needed to support maintenance shops is much reduced.

However modularity also generally adds some complexity, e.g., for connectors and interface circuits that otherwise would not be needed. Generally, there will also be elements added to allow a generic design to meet specific requirements. Thus there is an overhead in modularity which tends to reduce the benefits of integration.

A primary goal of IMA is to establish the application of a standard set of hardware modules, directly line-replaceable, encompassing as much of the total avionics suite as possible. However, the adherence to prescribed module boundaries, e.g., packaging processors separately from I/O modules and from power converter modules, etc. but all within a centralized cabinet, sets up real barriers to the benefits that can be gained from the rapid advances in electronics technology.

This concept stems from previous generations of equipment in which these distinct electronic functions each required relatively large amounts of real estate and could not be economically integrated into the same circuit board or module. Also, it was believed that the industry standards would lead to a market for competitive sources of such modules and therefore yield freedom of choice. However the reality is that this approach delivers only a "closed" architecture, no effective industry standards, and no market for choice of modules.

33.1.6   Business Pressures

The aircraft manufacturers are constantly driving to reduce manufacturing lead times and cost at the same time as the product complexity increases and technology continues to change.

Manufacturers place contracts for large packages of systems to single suppliers or consortia, i.e., the "one-stop shop." This not only reduces the overhead and diversity but also reduces the OEMs integration task and engineering burden. For the smaller manufacturers who may be unable to support large engineering resources this is perhaps the only way to cope with the problem of increasing functional complexity and technological change.

For the larger manufacturers there is the added question of the extent to which they want to also be able to manage the internal "package" integration and modification task independently of the supplier. Coupled with this may be the need to ensure long-term freedom of choice of suppliers and ability to exercise control. This is a key issue bearing upon the degree to which the package of systems or functions has open internal interfaces and is supported with reliable tools.

Because of the high value and high degree of dependence upon such suppliers the manufacturers are increasingly requiring risk-sharing partnerships. This is also evident in the trend towards partner design and manufacture of major airframe parts, or modular aircraft manufacture. Costs, build time, and risk can be reduced when the interfaces between the parts support rapid assembly and the parts are delivered fully dressed and pretested for their systems content.

From a systems viewpoint this is made considerably easier when the distributed elements located in the various major airframe parts are interfaced via serial data buses, perhaps through Remote Data Concentrators (RDCs) or local controllers.

33.2 Integrated Modular Avionics (IMA)

33.2.1 The Concept

The concept of integrated modular avionics has come about in response to the above trends and pressures.

Various implementation architectures are described in ARINC 651, the industry's overall design guidance document for integrated modular avionics. They share the common theme of a number of "cabinets" that are connected together and with other peripheral equipment by means of a number of multiple-access serial data buses (refer to Figure 32.2)

Each cabinet is seen as a high-power computing center which, in essence, replaces a number of today's application-specific avionics computers (line replaceable units, LRU). Each cabinet contains a selected mix of line replaceable modules (LRM) e.g., core processor, input-output (I/O), power supply, and other

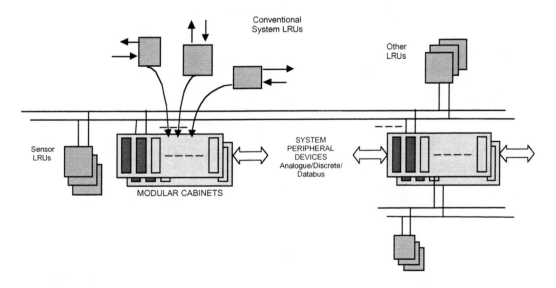

FIGURE 33.2 IMA overall architecture.

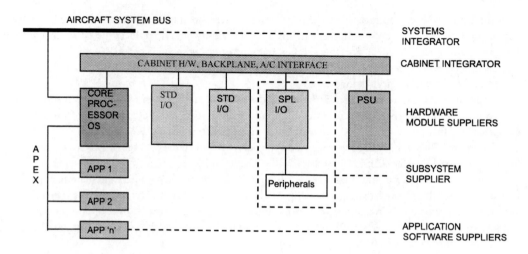

FIGURE 33.3   IMA concept.

special LRMs, all interconnected via an internal backplane bus. The cabinet enclosure is designed to provide a standard environment for the LRMs and effectively replaces conventional avionics equipment racks. Each cabinet is intended to have adequate processing and interface capacity to support the required integration of avionics functionality with spare capacity to meet life cycle growth requirements. The modules and backplane need to be fault tolerant to protect the group of integrated functions against single failures, or to provide high dispatch reliability, particularly for long-range or large aircraft. The system-specific application software is designed to meet the standard application-executive interface specification (APEX) to allow reusability on core processors of different hardware design.

33.2.2   Modular Architecture and Supplier Roles

The goal of standard, reusable, and interchangeable modules is central to the concept of IMA (refer to Figure 33.3). By rigorous definition and control of each module, both hardware and software, and of its interfaces, the aim is to permit a minimum number of different module types to support the broad range of current and future avionics system functions. The concept is expected to allow the systems integrator to procure each type of module from the most competitive source and to allow the user to support the aircraft with a greatly reduced number of spares. The concept, therefore, promises significant reductions in overall life-cycle cost.

It also will have a major impact on the way avionics business is conducted, since it opens up the possibility for LRMs to be supplied separately from the cabinet, software supplied separately from the hardware, and the whole cabinet to be integrated by an outside party.

To date, civil aircraft solutions like the B777 ELMS and AIMS have gone some way towards the envisaged IMA goals. Both approaches were optimized for the application domain and have brought benefits to the manufacturer and airline alike. It is important to recognize that neither AIMS nor ELMS are "open" architectures and all the parts and functions are manufactured, qualified, and integrated by the system supplier.

A key question now is whether and how a truly open common architecture and module suite can be developed and applied across the wider spectrum of avionics system domains. The following issues are considered relevant to this discussion.

33.2.3   Industry Standard Modules

Despite the original intentions, it is now fairly obvious that the industry will not be able to produce standard modules independently of the cabinet supplier/integrators. The specification process is too complex and the means are not available to independently validate and qualify modules for compatibility with the platform. The best that industry committees and agencies can do is to define form, fit, and interface requirements as a means of guidance and recommendations. It is also unlikely that the industry will be able to mandate a particular cabinet architecture. Apart from the complexities of these processes the time involved to achieve mature specifications is generally too long and will be overtaken by technology advance and real program-driven solutions.

33.2.4 Commercial Modules

The parallel is often drawn with the PC industry. The assumption being that because commercial interface standards have become widely established and the large market provides a wide choice of PC-compatible cards for the PC supplier and user, that the same "plug and play" approach will apply to the civil aircraft industry. In fact, for some avionics LRUs PC interfaces and modules are being used. However, the key difference is that the aircraft system integrator will generally have to prove and validate the use of any component used in the platform to a far higher level than used in the nonaviation world. This is likely to include a significant level of testing with the applications for a particular platform. The integrator carries the certification responsibility and will have to approve any alternative module sources.

33.2.5 Achieving the Wider Goals for IMA

The only realistic possibilities for achieving a wider application of common modules and integrated platforms appear to be the following:

- One platform supplier establishes his solution and module suite as the *de facto* standard, or
- A collaborative basis (probably aimed at a new aircraft program) in which the manufacturer and suppliers work together to define the standards and qualify a sufficient choice of interoperable parts.

Clearly the former is unlikely to generate freedom of choice and competitive component supply, and may simply establish a monopolistic market. However, some manufacturers may opt for this approach if they are not concerned about dependence on one supplier.

It is difficult to see how the larger vision of common modules across the wider systems domain along with freedom of choice and competition can be totally realized. Even the collaborative approach is likely to yield only a limited choice of supply. During the formative years the hardware and avionics functions will still be linked, if only to establish the certification principles. Then the technology will progress and the modules are likely to be obsolete for the next aircraft generation, requiring the cycle to be repeated. However, it is possible to see for a new aircraft programme that some module commonality could be achieved across platforms of different application domain, by collaboration between the platform suppliers.

33.2.6 Control of the Interfaces—Open Systems

A related issue is the question of the degree to which the manufacturer intends to be able to exercise control over the interfaces within the integrated system, e.g., within the platform between the hardware modules and between the application software and the operating system. For example he may wish to be in control of the integration of third-party software and variants arising from customer changes during the aircraft life cycle.

This is understood to be a clear objective for at least the manufacturers of large civil transport aircraft. In which case the same principles and features are needed as for integration management at the intersystems/aircraft level. It is this requirement, probably above all others, that will determine whether open architectures really do become established, and also the extent to which the platform internal architecture integration is maximized.

The key interfaces requiring control are the intermodule databuses (backplane) and the application software interface (APEX). It is vital for the aircraft buses to be truly open the communication over these interfaces (at the interoperability level) must be independent of the technology used in the system. Also, the interfaces must place minimum possible constraints on the functioning of the modules and equipment operation.

33.2.6.1 Software Interface

For the application software interface the manufacturer will require the same programming and integration tools as used by the platform supplier integrator. These tools would have been qualified by the platform supplier and would need to be maintained through the life cycle of the aircraft. Appropriate contractual arrangements will be necessary to define responsibility and liability associated with modifications or added functionality implemented by the manufacturer independently of the platform supplier.

33.2.6.2 Hardware Interface

For the module interfaces, it will be much more difficult for the manufacturer to add modules or introduce an upgraded performance module (for example) and then revalidate the platform if the platform architecture is tightly coupled, i.e., one in which the modules are controlled in lock-step fashion by the applications as an extension of the processor internal bus. In effect this is a wholly integrated platform, with the modules having only physical independence from each other. It may be open in the sense that the backplane bus meets a standard such as ARINC 659 or VME, but it does not readily permit the user to change or add hardware functionality independently of the integrator.

33.3 Aircraft and Systems Architecture Issues

33.3.1 "Smart" Peripherals

The technology advances that have enabled rack-mounted LRUs to be reduced in size or be integrated together over the years have also, in combination with serial data bus interfaces and other advances, enabled electronics to be used increasingly in various locations around the aircraft. Some examples of "smart peripherals" or "distributed systems" are

Full Authority Engine Control (FADEC) — Engine mounted
Air Data Modules (ADM) — Fuselage skin
Cabin entertainment systems — Cabin distributed
Smart autothrottle actuators — under floor
Smart display heads — Flight deck
Smart systems panels — Flight deck/cabin

With today's advanced microcircuits it is already possible to provide considerable functionality within many system peripheral devices without significantly affecting the space they occupy or significantly reducing their inherent reliability. As costs fall we are likely to see data-bus-interfaced smart peripherals being used increasingly, e.g., primary and secondary displays, flight control actuators, landing gear sensors/actuators, tank-mounted fuel quantity processing, remote data concentrators, radio sensors/transceivers, etc.

The main drivers are

- Reduction in installation complexity — wires, connectors.
- Robustness of peripheral interfaces to noise, improved HIRF/EME immunity.
- Reduced uncertainties in identifying fault location, i.e., whether fault is in the peripheral, the computer, or the wiring.
- Simplification in data interfaces, by incorporating special peripheral processing at the "point of action" and reducing communication data flow to higher-order parameters.

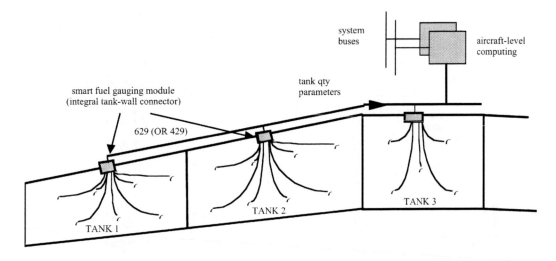

FIGURE 33.4   Distributed fuel gauging system.

- Architectural flexibility; the I/O interface is available on aircraft-level data buses for direct use elsewhere.
- Enabling the interface to be independent of the technology of the peripheral.
- Simplified procurement boundary promoting freedom of choice and control by the integrator.

Electronics is used at the peripherals and serial data buses provide the interface, so the systems become digital from end to end.

Figure 33.4 illustrates an example of a distributed Fuel Gauging System using the principal of smart peripherals. In a conventional high accuracy system the individual tank probes would be wired back to a gauging computer in the avionics bay. In a large aircraft there could be over a 100 probes and therefore at least 200 wires running over a considerable distance through the airframe, carrying sensitive analogue signals of very low strength. By locating highly reliable electronics close to the tanks this wiring is greatly simplified. All the probe signals for each tank are now combined onto a serial data bus. The signals are preprocessed into a form which is now independent of the probe technology and, being digital, are much less sensitive to wiring abnormalities and electrical interference. The processing task for the central computer is also simplified so that the gauging computation can now more easily be accomplished on a standard computing resource such as an Avionics Computer Resource (ACR).

The greater use of smart peripherals and remote data concentrators will also significantly diminish the need for analogue and discrete I/O modules within IMA cabinets and computer LRUs (see Figure 33.5).

33.3.2   High Speed Serial Data Buses

Serial digital data buses have been used on both civil and military aircraft of all types for many years. ARINC 429, which is a single-transmitter multiple-receiver topology bus operating at either 12 Kbps or 100 Kbps, is used on all modern civil transport aircraft. The disadvantage of this topology is that a single bus can carry data only from one source, and so each item of equipment needs as many inputs as the number of buses (sources) that it expects to receive data from. Accordingly, there are numerous bus spur connections and wires in the aircraft associated with most of the avionics boxes.

The Boeing 777 was the first civil transport aircraft to use a multiple access data bus in which all connected units can both transmit and receive via a single port connection (there are, of course, multiple redundant buses). This bus operates at 2 Mbps and uses active couplers to connect the unit port via a stub cable to the linear bi-directional bus itself. This is ARINC 629. This topology simplifies equipment

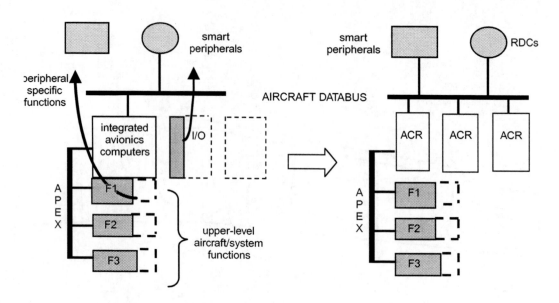

FIGURE 33.5   RDCs reduce need for analogue and discrete I/O modules.

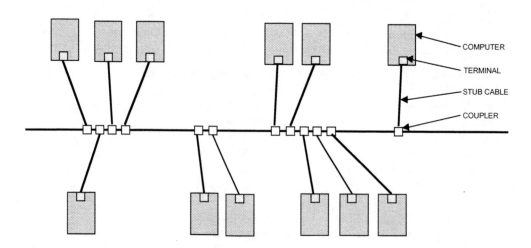

FIGURE 33.6   ARINC 629 topology.

and the aircraft wiring installation because one port and one bus stub connection only are needed for each unit to communicate to all the other units on the same bus. The bus is very deterministic in its operation and each terminal has high integrity control over the associated unit's access to the bus, the correct use of allocated bandwidth, and the selection of transmitted and received data. Its principal limitation, however, is the 2 Mbps rate. Also, the components are relatively expensive and there can be electrical performance factors which limit the physical spacing of coupler connections along the bus. Figure 33.6 illustrates the ARINC 629 topology.

More recently, fast networking data buses based on commercial standards are starting to appear in civil avionics systems. The industry has started standardization work on Ethernet and the first example of FDX (full duplex Ethernet) applied on a civil transport aircraft is for the interfaces between the various display units and computers in the cockpit primary display system on the B767-400 aircraft.

Airbus is also planning to use a variant of FDX (AFDX) as the main avionics and systems buses on the A3XX. These buses typically have a STAR topology in which the equipment ports are connected to

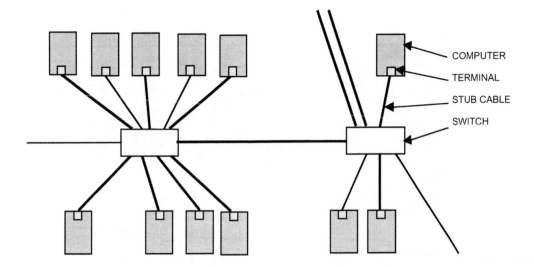

FIGURE 33.7 FDX (AFDX) topology.

each other via stub cables and the star point (Switch Unit). Again each item of equipment can communicate with all the others on the same bus through one port. In this case the data rate may be typically 100 Mbps, much faster than ARINC 629, although some of this bandwidth is used by higher formatting and protocol overheads. The concept is potentially much cheaper, since it can use the commercial standard directly (at least for the physical layer components), and it is likely to become the next civil aircraft industry bus standard. For most aircraft applications, however, modifications to the services and functions of standard Ethernet are needed to ensure a statically configured network, additional integrity, and controls on message routing and channel bandwidth usage. These changes impact upon the design of both the switch and the user terminal. The topology is illustrated in Figure 33.7.

An important advantage of buses like AFDX is that they can form a continuous network covering both the aircraft level (e.g., communication between LRUs, IMA centers, and other smart peripheral equipment) and the systems domain level (e.g., communication between modules associated with a particular group of systems). In an IMA context the switch provides the backplane connectivity for a domain and provides a direct interface to the aircraft-level part of the network, without the need for conversion gateways. In addition, only one set of bus management tools is needed to integrate the communication at all levels and for all domains across the aircraft.

33.3.3 Procurement Boundaries

An important architectural driver for the aircraft manufacturer is the procurement boundary, i.e., the interfaces through which a system or equipment to be procured operates in relation to its neighbours. Simplifying these interfaces is one of the goals inherent in determining where the boundaries should be. This is a necessary process to minimze the specification and integration task, reduce aircraft wiring and data flow, and achieve correct interoperability with minimum risk. Figure 33.8 illustrates these principles with a hypothetical example.

System functions A, B, and C (FA, FB, and FC) are highly related to each other and have complex interfaces between them, but have relatively simple or standard interfaces with sensors S1, S2, and S3. In addition FB and FC communicate with FD via a small number of readily specified signals. It is assumed that suppliers exist who are competent in all these system functions. The grouping together of functions A, B, and C makes sense because it significantly simplifies the manufacturer's engineering and procurement workload compared, say, to specifying and procuring FA combined with its associated sensors S1 and S2, separately from FB and FC. The (procurement) boundaries around the FA, FB, and FC group

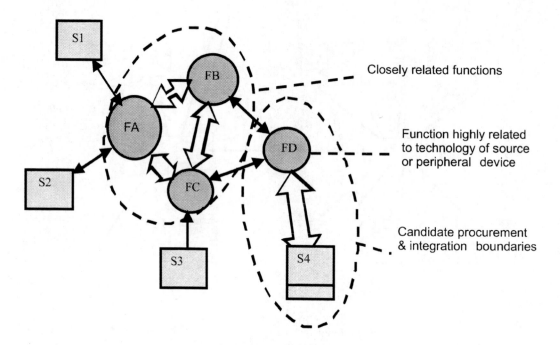

FIGURE 33.8   Integration and procurement boundaries.

are all relatively simple, easily controlled, and not dependent on the technology of the peripheral sensors. Therefore these functions are also likely to be good candidates for integration within a common computer, with least risk of requirements capture errors or overall integration problems within the aircraft.

On the other hand, FD is very closely related to sensor 4. The signal interfaces are complex or the function is highly dependent on the technology or specific design of the sensor. The simplest procurement boundary, therefore, encloses, both, and once again there is the potential for further simplification through integration of the function with the sensor.

Analyses of functional relationships and interdependencies across all the aircraft systems are required to determine the best options for functional grouping and integration. A top-down hierarchical approach is needed to establish the optimum groupings at the global systems level and the systems domains level. This approach assists the manufacturer to ensure that the functional groupings simultaneously meet technical and commercial requirements.

33.4   Conclusions

The foregoing acknowledges that benefits can be achieved by the careful application of cabinet-based integrated modular avionics for any given aircraft programme. The degree of integration, as distinct from modularity within an IMA platform or cabinet, and the degree of openness of the internal platform interfaces for any particular aircraft programme, will depend largely upon the purchaser's (e.g., aircraft manufacturer's) desire for control over these interfaces and the integration process.

If the aircraft manufacturer is to readily fulfil the role of platform integrator, a less tightly integrated backplane and module communication concept is needed. Here the modules would have autonomy of operation and communicate with identified data sets or messages. Each would be specified for its functionality and communication requirements. Each could be validated to a greater extent independently of the application environment. This concept is not dissimilar to the way LRUs interoperate. Indeed, it is to some considerable advantage to the manufacturer and the end user if the same data bus standard can be used for the backplane as is used for aircraft-level communications. Then exactly the same integration tools and maintenance tools can be used at both the aircraft and system domain level.

In the longer term, as the aircraft systems become more and more digital end-to-end through the use of smart peripherals and remote data concentrators, the need for analogue and discrete IO interfaces within centralized computing platforms will diminish. In addition, the processing within these platforms need not be burdened with functions associated with peripheral devices and can be simplified to that needed to perform upper-level system functions only.

It is also evident that the conventional segregation of processors, IO, and power converter hardware into distinct and separate modules is too restrictive and uneconomic compared with single modules containing all of this capability using the latest advances in technology. Hence, low-cost compact LRMs or LRUs providing a computing platform capability with substantial capacity and without internal architectural constraints are now achievable and are likely to become increasingly more competitive as time passes. Though these may differ internally from different suppliers, they can be designed to meet performance and interface standards (e.g., as for the ACR) including the APEX interface for abstracted application software and hardware interfaces for standard and special IO, both digital and analogue.

It is possible, therefore, to envisage the advantages hoped for from IMA to be obtained by the use of multiple ACRs, each hosting a number of system applications, RDCs, and smart peripherals, networked together using high-speed data buses such as AFDX.

Index

A

Absolute evaluations, comparative vs., 9-11
Absolute navigation systems, 13-2
Abstract
   Data Type (ADT), 26-16, 26-27
   specification, 21-2
Abstraction, 26-16
ACARS, *see* Aircraft Communications Addressing and
    Reporting System
Accelerated testing, 22-14, 22-15
Access types, 26-14
ACMS, *see* Aircraft Condition Monitoring System
Acquisition and supply processes, 24-3
ACR, *see* Avionics Computer Resource
Active Matrix Electroluminescent (AMEL), 5-5
Actuator force-flight elimination, Boeing 777, 11-11
Ada, 26-1 to 38
   abstraction, 26-16 to 30
     classes and polymorphism, 26-21 to 25
     freedom and, 26-2 to 4
     genericity, 26-25 to 26
     object oriented terminology, 26-26 to 27
     objects and inheritance, 26-18 to 21
     tasking, 26-27 to 30
   from Ada 83 to Ada 95, 26-4
   key concepts, 26-4 to 16
     access types, 26-14
     arrays and records, 26-11 to 13
     error detection, 26-14 to 16
     errors and exceptions, 26-9 to 10
     overall structure, 26-5 to 9
     scalar type model, 26-10 to 11
   programs and libraries, 26-30 to 37
     input-output, 26-31 to 33
     numeric library, 26-33 to 34
     running of program, 26-34 to 37
   software engineering, 26-2
ADI, *see* Attitude Direction Indicator
ADIRU, *see* Air Data and Inertial Reference Unit
ADS, *see* Automatic Dependence Surveillance
ADT, *see* Abstract Data Type
Advanced distributed architectures, 33-1 to 13
   aircraft and systems architecture issues, 33-8 to 12
     high speed serial data base, 33-9 to 11
     procurement boundaries, 33-11 to 12
     smart peripherals, 33-8 to 9
   drivers and trends, 33-1 to 5
     business pressures, 33-4 to 5
     hardware/software cost ratio continually falling, 33-3

     increasing functional complexity, 33-3
     integration, 33-4
     modularity, 33-4
     technology advance, 33-1 to 2
   integrated modular avionics, 33-5 to 8
     achieving wider goals for IMA, 33-7
     commercial modules, 33-7
     concept, 33-5 to 6
     control of interfaces, 33-7 to 8
     industry standard modules, 33-6 to 7
     modular architecture and supplier roles, 33-6
AEEC, *see* Airlines Electronic Engineering Committee
Aeronautical Telecommunications Network (ATN), 30-5,
    31-14
Aerospace Recommended Practice (ARP), 19-2
Aerospatiale, 12-1
AFDC, *see* Autopilot Flight Director Computers
AFS, *see* Auto-Flight System
AGC, *see* Automatic gain control
AIMC, *see* Boeing 777, Airplane Information Management
    System
AIMS, *see* Airplane Information Management System
Airbus A330/A340, 30-16
   avionics, 30-3
   central maintenance system, 30-4
   cockpit, 30-2
   communications, 30-4 to 5
   development environment, 30-6
   flexibility and in-service updates, 30-5
   flight controls, 30-4
   highlights, 30-2
   instruments, 30-3
   navigation, 30-3 to 4
   support environment, 30-6
   systems, 30-2
   user involvement, 30-2 to 3
Aircraft
   batteries
     commercial, 10-16 to 22
     military, 10-24 to 27
   Communications Addressing and Reporting System
     (ACARS), 31-4
   Condition Monitoring System (ACMS), 30-5
   flight test, 4-16
   lighting, evaluating, 7-18
   military, 10-23
   mission, 19-1
   movement simulation, 12-13
   programme, twin-jet twin-aisle, 30-2
   Retained Unit (ARU), 6-4

W

Y